Strategic Adoption of 5G Technology:

New Applications and Services

Vincent Sabourin
School of Management, University of Quebec at Montreal, Canada

IGI Global
Scientific Publishing
Publishing Tomorrow's Research Today

Vice President of Editorial Melissa Wagner
Director of Acquisitions Mikaela Felty
Director of Book Development Jocelynn Hessler
Production Manager Mike Brehm
Cover Design Jose Rosado

Published in the United States of America by
IGI Global Scientific Publishing
701 East Chocolate Avenue
Hershey, PA, 17033, USA
Tel: 717-533-8845
Fax: 717-533-7115
Website: https://www.igi-global.com E-mail: cust@igi-global.com

Library of Congress Cataloging-in-Publication Data

LCCN: 2025007417 (CIP Data Pending)
ISBN13: 979-8-3373-0497-7
Isbn13Softcover: 979-8-3373-0498-4
EISBN13: 979-8-3373-0499-1
British Cataloguing in Publication Data
A Cataloguing in Publication record for this book is available from the British Library.

Table of Contents

Preface xi

Section 1
Overview of 5G

Chapter 1
Overview of 5G Technology 1

Chapter 2
Development of 5G and Smartphones 37

Chapter 3
Deployment of 5G 63

Chapter 4
Availability of 5G by Country 85

Section 2
The Technology of 5G

Chapter 5
The Technological Perspective of 5G 121

Chapter 6
Evolution of Mobile Telecommunications With 5G 161

Chapter 7
Spectrum, Requirements, and Trends of 5G 195

Section 3
Applications

Chapter 8
Promising Applications and Services for the 5G Market 231

Chapter 9
Risk and Challenges in Application Development 257

Chapter 10
5G Applications for Logistics, Energy, and Financial Services 287

Section 4
Research

Chapter 11
Emerging Research for 5G 325

Chapter 12
5G Technology: Critical Research for Applications and Product Development 347

Section 5
Competition

Chapter 13
5G Ecosystem and Supply Chain 385

Chapter 14
Strategies of Leading Companies in 5G 419

Chapter 15
5G Future 457

Conclusion 493

Compilation of References 503

About the Author 531

Index 533

Detailed Table of Contents

Preface.. xi

Section 1
Overview of 5G

Chapter 1
Overview of 5G Technology 1

The advent of 5G marks a significant shift in mobile telecommunications, offering unmatched speed, low latency, and scalability. As the fifth generation of wireless networks, 5G introduces network slicing, enhanced mobile broadband (eMBB), and ultra-reliable low-latency communication (URLLC). These innovations enable customized connectivity for sectors such as healthcare, manufacturing, and public safety. 5G supports applications that require fast data transfer, low latency, and large-scale device connectivity. It drives digital transformation across smart cities, autonomous vehicles, and advanced IoT systems. Industries benefit from real-time data exchange, remote control, and immersive tech.

Chapter 2
Development of 5G and Smartphones 37

The global rollout of 5G in 2019 marked a turning point in mobile connectivity, offering enhanced bandwidth and ultra-low latency. This fueled rapid adoption of 5G-enabled devices, with 56 manufacturers introducing 129 models by year's end. These included smartphones, hotspots, and other equipment, showcasing 5G's wide-ranging impact. The launch of 5G-compatible chipsets played a key role, enabling high-speed, high-performance devices and paving the way for models like the Samsung Galaxy S20. 5G has since redefined mobile technology and expanded its role in consumer and industrial applications.

Chapter 3
Deployment of 5G 63

The deployment of 5G marks a major shift in telecommunications, turning years of planning into real-world innovation. More than just an upgrade, 5G introduces a new network architecture delivering ultra-fast speed, low latency, and improved service quality. It supports IoT, smart cities, autonomous vehicles, and AR, creating vast opportunities across sectors. Powered by small cells, edge computing, and network slicing, 5G meets varied use case demands. Chen and Zhao note that network slicing enables dynamic, tailored connectivity, optimizing resources for services like video surveillance, real-time data, and AI-powered applications.

Chapter 4
Availability of 5G by Country 85

The deployment of 5G marks a global shift in connectivity, offering high speed, ultra-low latency, and the ability to connect billions of devices. It enables transformative uses in smartphones, autonomous vehicles, and IoT. However, access varies widely. Countries like the U.S., South Korea, and China lead with extensive 5G networks, while many developing nations remain in planning or pilot phases. Infrastructure, regulatory, and financial barriers slow progress. By 2025, over 10% of mobile connections are expected to use 5G, reflecting its key role in innovation and digital growth—despite persistent deployment challenges.

Section 2
The Technology of 5G

Chapter 5
The Technological Perspective of 5G 121

5G is more than a mobile broadband upgrade, it's a transformative force enabling innovation across industries. With features such as ultra-reliable low-latency communication (URLLC), massive machine-type communication (mMTC), and enhanced mobile broadband (eMBB), 5G addresses the unique needs of sectors including automotive, healthcare, utilities, and finance. eMBB boosts speed, latency, and user density for seamless connectivity. mMTC supports vast networks of low-cost, battery-powered devices, enabling smart meters, IoT sensors, logistics systems, and wearables. 5G empowers both individuals and organizations with next-level connectivity and efficiency.

Chapter 6
Evolution of Mobile Telecommunications With 5G 161

Spectrum allocation is vital to 5G deployment, impacting coverage, capacity, and performance. It requires balancing low (sub-1 GHz), mid (1–6 GHz), and high-frequency (above 6 GHz) bands. Low bands offer wide rural coverage, while high-frequency millimeter waves deliver ultra-fast speeds for urban and industrial applications like AR and autonomous vehicles. Global regulators, including the ITU-R, shape spectrum frameworks—especially in the 24–86 GHz range—to support varied use cases. Their focus includes fair access, efficient allocation, and enabling both mobile operators and new industry-specific players to deploy 5G effectively.

Chapter 7
Spectrum, Requirements, and Trends of 5G 195

The spectrum allocation for 5G is a critical factor in the successful deployment of next-generation wireless networks, influencing the efficiency, coverage, and capacity of 5G services. It involves balancing sub-1 GHz, mid-range (1-6 GHz), and high-frequency millimeter-wave bands (above 6 GHz) to meet diverse application needs. While lower frequency bands provide broader coverage for rural areas, high-frequency millimeter waves enable ultra-fast data rates essential for urban and industrial use cases like augmented reality and autonomous vehicles. Global regulatory bodies, such as the ITU-R, are pivotal in defining the spectrum frameworks for 5G, focusing on the 24-86 GHz range to support various applications. These efforts ensure timely deployment and address challenges like spectrum allocation, fair access, and non-discriminatory authorization models. Regulatory bodies face the challenge of establishing equitable spectrum authorization models to ensure fair access for stakeholders" (ITU-R, 2019)

Section 3
Applications

Chapter 8
Promising Applications and Services for the 5G Market 231

5G technology has ushered in a transformative era in mobile communication, redefining connectivity through unprecedented speed, ultra-low latency, and exceptional reliability. Designed to meet the stringent demands of diverse Internet of Things (IoT) use cases, 5G enables gigabit-level data rates, massive connection densities, and real-time operational capabilities. By seamlessly connecting devices, machines, and systems, 5G facilitates innovation and efficiency across industries, paving the way for advanced applications that need to be more attainable with previous network generations. Recent analyses, including those conducted by IoT Analytics, predict IoT-driven use cases will contribute to 76 million 5G connections by 2025. These use cases span critical sectors like smart cities, industrial automation, healthcare, automotive systems, and public safety, showcasing the versatility and scalability of 5G networks.

Chapter 9
Risk and Challenges in Application Development 257

5G is ushering in a new era of mobile communication, offering unmatched speed, ultra-low latency, and high reliability. Designed for the complex demands of the Internet of Things (IoT), 5G enables gigabit speeds, massive device connectivity, and real-time operations. It connects machines, devices, and systems, driving innovation across industries. IoT Analytics projects 76 million 5G IoT connections by 2025.

These use cases span smart cities, industrial automation, healthcare, automotive, and public safety—demonstrating 5G's versatility and scalability for mission-critical applications previously out of reach with earlier networks.

Chapter 10
5G Applications for Logistics, Energy, and Financial Services 287

5G is transforming logistics and shipping by boosting efficiency, connectivity, and innovation. Its high-speed data transfer, low latency, and massive connectivity enable real-time communication across transportation systems. Applications like vehicle-to-vehicle (V2V) communication support autonomous vehicle testing, fleet coordination, and route optimization. 5G enables seamless data exchange between vehicles, infrastructure, and central systems, resulting in safer and smarter logistics. When combined with IoT, 5G enhances supply chain visibility, allowing real-time tracking of location, temperature, and condition, while predictive analytics help prevent delays and disruptions.

Section 4
Research

Chapter 11
Emerging Research for 5G 325

This chapter examines cutting-edge research that is shaping the evolution of 5G networks, with a focus on advanced technologies that address their complex demands. Key innovations include the deployment of small cells and peer-to-peer communications (D2D, M2M) in heterogeneous networks. Enabling technologies such as full-duplex communication, massive MIMO, and millimeter-wave systems are highlighted for their role in enhancing network performance. Energy efficiency emerges as a core priority, with energy-aware communication and energy-harvesting strategies being crucial to the user experience. Additionally, cloud-based radio access networks (C-RAN) and spectrum virtualization are presented as essential for managing resources and ensuring network scalability.

Chapter 12
5G Technology: Critical Research for Applications and Product Development 347

The rapid evolution of 5G technology has led to a surge in innovative applications, underscoring the need for targeted research to address critical challenges and unlock its full potential. This chapter provides a comprehensive framework for identifying technological areas that can drive innovation in 5G applications and product development. It aims to guide nations, organizations, academic institutions, and research entities in contributing to developing a robust and standardized 5G ecosystem. The chapter begins by examining key foundational technologies of 5G, including system ubiquity, interoperability, and IP-based network architectures. It highlights cognitive radio technology, integrated with artificial intelligence, as a pivotal component for enhancing terminal performance and optimizing system efficiency across diverse access systems. The discussion also addresses technological challenges, such as limited frequency spectrum resources and security vulnerabilities requiring focused research to ensure seamless implementation.

Section 5
Competition

Chapter 13
5G Ecosystem and Supply Chain 385

The emergence of 5G technology is transforming global supply chain management through ultra-fast connectivity, low latency, and real-time automation. It enables enhanced visibility, predictive analytics, and seamless integration with IoT, AR, and cloud platforms. Businesses benefit from smart tracking, efficient logistics, and autonomous systems such as drones and AGVs. 5G also supports network slicing and edge computing, allowing tailored, high-performance services across complex operations. While deployment involves regulatory and cost challenges, the long-term advantages include improved resilience, cost efficiency, and collaborative agility. As 5G networks mature, their role in driving supply chain innovation becomes central to digital transformation.

Chapter 14
Strategies of Leading Companies in 5G 419

The strategies of leading companies in the 5G ecosystem reflect diverse and innovative approaches to capturing opportunities in this transformative technology. Spanning telecommunications giants, hardware manufacturers, and technology providers, these organizations invest in infrastructure, research, and partnerships to maintain a competitive edge. U.S.-based companies such as AT&T and Verizon focus on robust network deployment, while Chinese leaders like Huawei and ZTE spearhead advancements in 5G equipment. European players, including Nokia, Ericsson, and Orange, emphasize network solutions and service innovation. Meanwhile, Asian powerhouses like Samsung Electronics and NEC Corporation integrate 5G with cutting-edge technologies like IoT and AI, driving new capabilities and market applications.

Chapter 15
5G Future 457

The competitive landscape of 5G technology is defined by its convergence with transformative innovations, including IoT, AI, and edge computing. These technologies rely on 5G's ultra-fast, low-latency, and high-capacity infrastructure to deliver scalable, real-time solutions. Industry leaders, including Oracle, Cisco, and Intel, drive this evolution through strategic R&D and partnerships, targeting key verticals such as healthcare, automotive, and manufacturing. Meanwhile, regulatory compliance and national policy frameworks significantly influence deployment speed and scope. Together, these dynamics underscore the importance of integrated strategies and cross-sector collaboration to maximize the economic and technological potential of 5G worldwide.

Conclusion 493

Compilation of References 503

About the Author 531

Index 533

Preface

The fifth generation of mobile networks—5G—is not just another technological leap; it is a transformative force that is reshaping how societies connect, communicate, and innovate. This book, uniquely designed for researchers, practitioners, and policymakers, offers a comprehensive and multidisciplinary understanding of the technological, industrial, and strategic dimensions of 5 G. As the digital economy gains momentum, 5G emerges as the pivotal point where telecommunications, artificial intelligence, cybersecurity, logistics, finance, healthcare, and public infrastructure converge.

The genesis of this volume lies in years of academic research, collaboration with global industry experts, and engagement with technology policy initiatives. While the technical literature on 5G is extensive, few resources integrate the broader ecosystem perspective—linking emerging technologies with organizational strategies, deployment models, and socio-economic impact. This book, a product of collective efforts, aims to bridge that gap and invite you into the global conversation about 5 G.

Each chapter of this book serves as a practical guide, reflecting the contributions of specialists from multiple disciplines and geographies. It provides a clear understanding of how 5G is being deployed, tested, and challenged in real-world contexts. The book pays particular attention to global disparities in access, spectrum management, emerging applications, security challenges, and the competitive dynamics among leading companies. The insights shared throughout are supported by practical frameworks, comparative case studies, and original data.

We aim to provide a resource that informs, challenges, and inspires. Whether the reader is an academic, an industry executive, a student, or a public-sector decision-maker, this book offers valuable guidance for navigating the opportunities and risks of the 5G era.

I would like to sincerely thank Mustafa Malik and Aarti Vijay Chaudhari for their invaluable contributions as research assistants during the initial phase of this book. Their insights and dedication were instrumental in shaping the foundational content. I am deeply grateful for their support and expertise, and their contributions are a testament to the collaborative nature of this project. I want to thank Wessam Ajib, Professor in the Department of Computer Science at UQAM, and Soumaya Ben Letaifa, Professor in the Department of Strategy, Social and Environmental Responsibility at ESG UQAM, for their valuable feedback.

We extend our heartfelt thanks to IGI Global for their support in this project and for providing a platform that enables interdisciplinary and forward-looking scholarship to reach a global audience. The insights presented here are just the beginning of the broader conversation about the future of connectivity, a conversation that we hope you, our readers, will be an integral part of.

ORGANIZATION OF THE BOOK

As the author of this volume, I present an in-depth exploration of the revolutionary shift brought about by 5G technology, a pivotal advancement reshaping the fabric of modern communication and digital infrastructure. This book is structured to offer a holistic perspective on the technical, developmental, regulatory, and applicational dimensions of 5G, aiming to bridge the knowledge gap between theory, industry practice, and global implementation.

Chapter 1: Overview of 5G Technology

We begin with a foundational overview of 5G, highlighting how this fifth-generation wireless technology transcends its predecessors through unparalleled speed, scalability, and ultra-low latency. Emphasizing core innovations such as network slicing, enhanced mobile broadband (eMBB), and ultra-reliable low-latency communication (URLLC), the chapter showcases 5G's ability to tailor connectivity solutions for diverse sectors, from smart cities and public safety to autonomous vehicles and industrial IoT ecosystems. This technological leap underpins the digital transformation of society, enabling real-time data exchange, immersive technologies, and remote capabilities like never before.

Chapter 2: Development of 5G and Smartphones

The global deployment of 5G in 2019 catalyzed a wave of device innovation, particularly in smartphones. This chapter explores the rapid proliferation of 5G-enabled devices, made possible by the concurrent development of advanced chipsets and mobile hardware. From flagship releases like the Samsung Galaxy S20 to a broad range of connected gadgets, we examine how this ecosystem growth has propelled

5G into the consumer mainstream and expanded its influence across business and industrial applications.

Chapter 3: Deployment of 5G

Turning plans into reality, the deployment of 5G presents a reimagined network architecture built for performance, adaptability, and resilience. We delve into the infrastructure behind 5G's rollout—including small cells, edge computing, and network slicing—and examine how these components synergize to support mission-critical services. The chapter highlights real-world implementations, showing how 5G fosters breakthroughs in autonomous systems, real-time analytics, and immersive AR/VR experiences.

Chapter 4: Availability of 5G by Country

Though global in scope, 5G's availability is uneven, with stark contrasts between leading adopters and countries still navigating initial phases. This chapter provides a comparative analysis of 5G availability across nations, identifying key infrastructural, financial, and regulatory barriers. While nations like the United States, China, and South Korea lead the way, others face considerable hurdles. Despite this, projections indicate a growing global footprint, with over 10% of mobile connections expected to leverage 5G by 2025.

Chapter 5: The Technological Perspective of 5G

We further dissect the technological backbone of 5G, focusing on its three pillars: eMBB, mMTC, and URLLC. Each pillar addresses specific use-case demands—from delivering high data rates to supporting vast IoT deployments and enabling low-latency critical operations. This chapter illustrates how 5G unlocks new levels of productivity and connectivity in industries such as automotive, healthcare, and energy, offering detailed insights into its transformative capacity.

Chapter 6: Evolution of Mobile Telecommunications with 5G

Understanding 5G's evolution requires a look at the radio spectrum that fuels it. This chapter outlines the allocation of low, mid, and high-frequency bands and their implications for rural and urban coverage. Special emphasis is placed on millimeter-wave technologies and the roles of international regulatory frameworks in optimizing spectrum distribution. These technical and policy considerations are vital for realizing 5G's full potential in both densely populated and remote areas.

Chapter 7: Spectrum, Requirements, and Trends of 5G

Expanding on spectrum-related issues, this chapter examines the strategic importance of frequency management. Balancing sub-1 GHz bands for widespread coverage with higher frequencies for speed-intensive tasks is a recurring theme. We also explore the role of regulators like the ITU-R in defining equitable access and spectrum-sharing models to ensure fair participation among telecom operators and emerging service providers.

Chapter 8: Promising Applications and Services for the 5G Market

This chapter captures the vibrancy of the 5G-enabled services landscape. From smart grids and remote healthcare to industrial automation and intelligent transportation systems, 5G enables a new generation of services characterized by high reliability and instantaneous communication. Citing forecasts from IoT Analytics, we anticipate explosive growth in use cases that exploit 5G's unique features, further integrating digital systems with everyday life.

Chapter 9: Risk and Challenges in Application Development

While the promise of 5G is immense, it is not without challenges. This final chapter addresses the risks inherent in application development for 5G environments, including concerns around interoperability, data security, and service complexity. Developers and stakeholders must navigate technical, regulatory, and economic constraints to build robust, scalable applications. We conclude by reflecting on the critical importance of resilience and foresight in ensuring that 5G not only connects but empowers.

Chapter 10: 5G Applications for Logistics, Energy, and Financial Services

In this chapter, we examine how 5G is transforming the backbone of modern industry by reshaping logistics, energy, and financial services. The logistics sector, in particular, is leveraging 5G's high-speed data transmission and ultra-low latency to support real-time vehicle-to-vehicle (V2V) communication, enabling autonomous driving, fleet coordination, and route optimization. When combined with IoT, 5G allows unparalleled visibility into supply chains—tracking the condition, location, and movement of goods with precision. Meanwhile, in the energy sector, 5G enables smart grid operations and predictive maintenance. Financial institutions are also adopting 5G for secure, real-time transactions and improved customer engagement. This chapter demonstrates how seamless connectivity and intelligent automation are creating a new era of operational efficiency and service delivery across key sectors.

Chapter 11: Emerging Research for 5G

This chapter explores the forefront of 5G research, where innovation continues to push boundaries to meet the complex and dynamic demands of next-generation connectivity. We highlight how technologies such as device-to-device (D2D) and machine-to-machine (M2M) communication are gaining prominence in heterogeneous network environments. Full-duplex systems, massive MIMO, and millimeter-wave spectrum use are transforming performance metrics, while energy efficiency remains a top concern, addressed through energy-aware communication protocols and harvesting strategies. Cloud-based radio access networks (C-RAN) and spectrum virtualization emerge as critical enablers of scalability and adaptability. As we transition into a hyper-connected world, this chapter underscores the importance of continuous R&D to drive sustainable, high-performance 5G infrastructure.

Chapter 12: 5G Technology: Critical Research for Applications and Product Development

In this chapter, we shift focus to the foundational research required to shape practical applications and guide product development in the 5G era. I present a comprehensive framework for identifying core technologies—such as IP-based architectures, system interoperability, and ubiquitous network access—that underpin successful 5G deployment. The chapter emphasizes cognitive radio technologies enhanced by artificial intelligence, which enable smart terminal behavior and dynamic optimization across network systems. Recognizing the limitations of current frequency spectrum and existing security frameworks, I call for intensified research on these challenges to support seamless application development and adoption. Ultimately, this chapter provides direction for researchers, developers, and policymakers aiming to build a resilient and future-ready 5G ecosystem.

Chapter 13: 5G Ecosystem and Supply Chain

This chapter explores the profound influence of 5G on the global supply chain, where its low latency, high reliability, and real-time processing capabilities redefine how goods and services are managed and delivered. Through smart tracking, automation, and predictive analytics, 5G empowers businesses to respond dynamically to disruptions and optimize operational flow. The integration of 5G with

IoT, augmented reality (AR), and cloud platforms fosters collaborative agility and decision-making. Furthermore, innovations such as edge computing and network slicing tailor services to meet the demands of specific industries. While challenges related to regulation and infrastructure costs persist, the long-term benefits—ranging from improved visibility to enhanced customer responsiveness—demonstrate the pivotal role of 5G in supply chain digital transformation.

Chapter 14: Strategies of Leading Companies in 5G

In this chapter, we take a close look at the strategic initiatives pursued by the global frontrunners in the 5G space. Telecommunications leaders like AT&T, Verizon, Huawei, and ZTE are investing in nationwide rollouts and infrastructure scaling, while technology giants such as Samsung, Nokia, and Ericsson are combining hardware innovation with software development to unlock new capabilities. I highlight how diverse geographic and regulatory environments shape each company's approach—from AI-driven service models in Asia to cloud-native infrastructure solutions in Europe and North America. This comparative analysis offers readers a deeper understanding of how companies position themselves in an increasingly competitive 5G marketplace and what lessons can be drawn from their evolving strategies.

Chapter 15: 5G Future

We conclude the volume with a forward-looking chapter that maps the trajectory of 5G as it converges with other transformative technologies—most notably IoT, AI, and edge computing. These technologies will depend on 5G's capacity, speed, and responsiveness to deliver next-generation services in sectors such as healthcare, transportation, and advanced manufacturing. I discuss the pivotal role of companies like Cisco, Oracle, and Intel, whose innovations and partnerships are driving this convergence. At the same time, national regulatory frameworks and global policy considerations are recognized as major influencers of rollout speed and scalability. This chapter underscores the necessity of integrated thinking, cross-industry collaboration, and adaptive regulation to harness the full societal and economic promise of 5G in the coming decade.

Section 1

Overview of 5G

Chapter 1
Overview of 5G Technology

ABSTRACT

The advent of 5G marks a significant shift in mobile telecommunications, offering unmatched speed, low latency, and scalability. As the fifth generation of wireless networks, 5G introduces network slicing, enhanced mobile broadband (eMBB), and ultra-reliable low-latency communication (URLLC). These innovations enable customized connectivity for sectors such as healthcare, manufacturing, and public safety. 5G supports applications that require fast data transfer, low latency, and large-scale device connectivity. It drives digital transformation across smart cities, autonomous vehicles, and advanced IoT systems. Industries benefit from real-time data exchange, remote control, and immersive tech.

INTRODUCTION

The advent of fifth-generation (5G) technology marks a transformative era in global telecommunications. It builds on the foundations of its predecessors—1G, 2G, 3 G, and 4 G. As the first wireless standard tailored to the digital age, Engineers designed 5G to revolutionize connectivity by integrating advanced intelligence, mobility, and diverse technologies. With capabilities far surpassing those of fourth-generation (4G) systems, 5G offers unprecedented enhancements in bandwidth, ultra-low latency, and machine-to-machine communication. It enables seamless connectivity for countless devices, creating a unified system supporting emerging innovations such as the Internet of Things (IoT), smart cities, and autonomous vehicles (Egea-Roca et al, 2022)

DOI: 10.4018/979-8-3373-0497-7.ch001

OVERVIEW AND DEFINITION

To fully grasp the significance of 5G, it is essential to contextualize its evolution within the broader history of wireless communication. From the analog 1G networks of the 1980s, which facilitated basic phone calls, to the digital breakthroughs of 2G and the internet-enabled capabilities of 3G and 4G, each generation has laid the groundwork for the unparalleled advancements of 5 G. As the cornerstone of future telecommunications, 5G redefines connectivity, fueling innovation across industries and reshaping global digital ecosystems.

Definition

5G is a new global wireless standard that follows 1G, 2G, 3 G, and 4G networks. It enables a brand-new system to link practically everything and everyone, including devices, objects, and machines.

In telecommunications, 5G is the 5th generation technology standard for cellular networks. Cellular telephone companies began deploying globally in 2019, the planned successor to the 4G networks that provide connectivity to most existing cell phones.

5G networks are cellular networks structured into small geographical areas, known as cells. Radio waves from a local antenna link most 5G wireless devices in a cell to the Internet and the telephone system (Henley Business School, 2021).

The brand-new networks' main benefit is their larger bandwidth, which provides high download speeds and, ultimately, as many as ten gigabits per second (Gbit/s) (ITU-R, 2017)

Context and Stakeholders

Fifth-generation (5G) technology is the first in the digital era, where different networks converge to form a single program (Osseiran et al., 2014). *The 5G networks significantly enhance performance compared to the current fourth-generation (4G) system, supporting increased bandwidth, low latency, and machine-type communications*. They also provide seamless connectivity to several devices, integrating various technologies, intelligence, and mobility.

The 5G ecosystem comprises various actors, each playing a critical role in deploying, operating, and adopting this next-generation technology. These include operators who build and manage networks, service providers that enable functionalities, equipment suppliers who develop infrastructure, and end-users who consume the services. To better understand this complex ecosystem, a summary table is provided below for quick reference.

The table below presents a summary of the key stakeholders involved in the 5G ecosystem. Each category plays a distinct and essential role in enabling the deployment, operation, and application of 5G technology. From operators managing networks to service providers offering advanced digital platforms, this overview highlights the interconnected nature of the ecosystem. Equipment suppliers ensure the infrastructure backbone, while end-users drive demand through innovative applications across various sectors. Understanding these roles is crucial to grasping the full scope and impact of 5 G.

Table 1. Summary of key stakeholders in the 5G ecosystem

Stakeholder Category	Examples	Roles and Responsibilities
Operators	AT&T, Orange, MTN, Airtel	Deploy, maintain, and manage 5G networks.
Service Providers	AWS, Microsoft Azure, Google Cloud	Offer services like cloud, AI, and edge computing.
Equipment Suppliers	Huawei, Nokia, Ericsson, Qualcomm	Provide 5G infrastructure, modems, and antennas.
End Users	Industries, Healthcare, Smart Cities, Consumers	Leverage 5G for automation, connectivity, and innovation.

Based on the alliance of the mobile community NGMN (next-generation mobile network), the 5G networks of a cellular border operator of a minimum of fifty megabits per second (Mbit /s) is probably the lowest problem with gigabit per second (Gbit /s) peaks; the ultra-low end to end latency (possible under ten ms) (NGMN, 2015).

BRIEF HISTORY OF THE 5G

The development of 5G technology marks a significant milestone in the evolution of wireless communication. From the initial rollout of 1G in the early 1980s to the lightning-fast and ultra-reliable networks of 5G today, each generation has brought transformative improvements in speed, latency, and connectivity. 5G stands apart for its capacity to support massive IoT deployments, smart city infrastructure, autonomous vehicles, and real-time data transmission in critical sectors such as healthcare and manufacturing.

Singh and Debra (2017) provide a historical overview of the evolution of cellular networks, tracing advancements from the early analog-based 1G systems to the current IP-based 4G networks. The authors emphasize the rapid technological progression that has allowed mobile devices to maintain call connectivity and support high-speed data activities such as streaming. Initially, 1G networks offered basic voice services with limited speed, while 2G introduced digital communication and text messaging, leading to a surge in mobile demand. The deployment of 3 G in 1998 enabled data transfer rates starting from 200 kbit/s and introduced incremental upgrades like 3.5G and 3.75 G. These developments paved the way for 4G, which utilizes packet-switched technology for faster data transmission and integrates standards such as LTE, UMB, and WiMAX to enhance mobile broadband capabilities (Fowdur, Indoonundon, & Milovanovic, 2024).

5G Promises End-to-End Latency

According to the Next Generation Mobile Network (NGMN) Alliance, 5G promises end-to-end latency under ten milliseconds and gigabit-per-second (Gbit/s) download speeds, a quantum leap compared to the current networks' 50 megabits per second (Mbit/s) thresholds. A network architecture based on small, densely distributed cells connected by radio waves to local antennas achieves these breakthroughs.. It ensures access to robust, high-speed Internet and telecommunication systems.

Cellular operators globally introduced 5G in 2019, paving the way for the proliferation of new devices and applications. Unlike traditional networks that primarily cater to mobile phones, 5G extends its reach to general internet service providers, enabling high-speed internet for laptops, desktops, and a wide range of Internet of Things (IoT) devices. This larger bandwidth and enhanced connectivity unlock new possibilities for machine learning, real-time analytics, and industrial automation.

The worldwide push to advance and implement 5G started in the early 2010s with initial trials and spectrum auctions. It gained momentum through collaborations between governments and private companies, key technological innovations, and progress in standardization. By 2019, early rollouts had begun, with nations like South Korea, the U.S., and China emerging as front-runners in paving the way for broad commercial use.

Today, 5G is evolving through innovations like network slicing, edge computing, and Open RAN, laying the groundwork for 6G and future hyperconnected ecosystems. This timeline highlights key milestones from concept to deployment, providing a visual synthesis of how 5G has become the cornerstone of modern digital infrastructure.

The following figure provides a concise timeline of key milestones and developments that have defined the rise of 5 G.

Figure 1. Brief history of the 5G

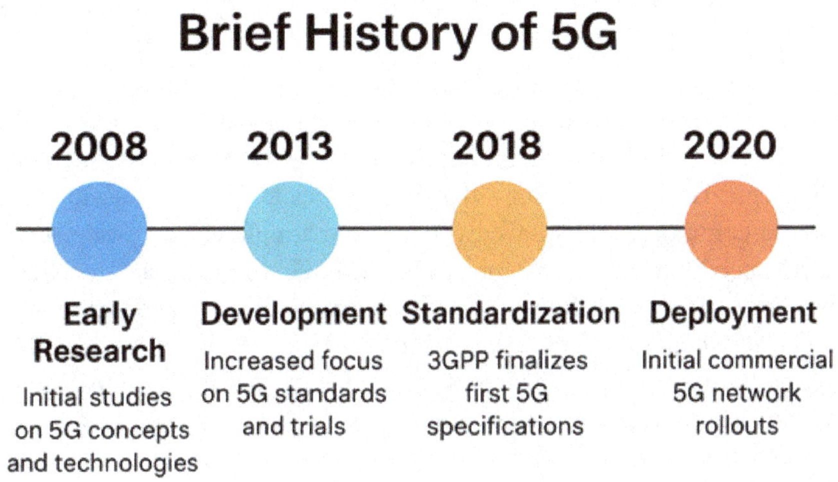

Key Advancement Illustrates the Alignment

This advancement illustrates the alignment of scientific research, industry funding, regulatory backing, and growing consumer expectations. It also highlights the collective global commitment necessary to realize such a groundbreaking technology worldwide.

As a reminder, TeliaSonera introduced LTE in December 2009. We have reached 2.1 billion LTE products globally in approximately seven years, with almost 600 networks deployed. Global Mobile Suppliers Association (2015) states that 22 operators will introduce 5G networks between today and 2021.

To know 5G, its relevance, and its benefits, you must first know its predecessors: 1G, 3G, 2G, and 4 G. So, we need to go through them immediately, one by one:

1G (Analog): Introduced in the first to mid-1980s, 1G was the first generation of wireless communications, enabling individuals to make straightforward phone calls between mobile phones (Awwwards, 2019).

Information Transfer Speed

The information transfer speed in 1G was approximately 1.9 kbit/second 2G (GPRS): Introduced in 1991, 2 G is the first electronic cellular technology to supply owners with voice calls, text-only SMS messages, picture messages, and internet connection facilities, with a maximum information transfer speed of ten kbit/second.

Table 2 presents the historical timeline of mobile network evolution, highlighting the technological milestones achieved over each decade. This progression illustrates how mobile communications have evolved from simple analog voice services to today's advanced, ultra-fast, and interconnected 5G systems. Each generation not only introduced new capabilities but also fundamentally reshaped how individuals, businesses, and societies interact, access information, and deliver services. The table underscores the exponential growth in speed, capacity, and application diversity that has defined each new mobile generation.

Table 2. Timeline of mobile network evolution

Decade	Generation & Key Features
1980s	1G - Analog Voice Communication
1990s	2G - Digital Voice and SMS
2000s	3G - Mobile Data and Internet Access
2010s	4G - Broadband Mobile Internet, HD Video Streaming
2020s	5G - Ultra-fast Speed, Low Latency, IoT Connectivity

Evolution of Mobile Networks From 1G to 5G

The evolution of mobile networks from 1G to 5G marks significant technological advancements (Admin, 2022). This timeline provides a graphical representation of the key milestones in the development of mobile communication.

Today, we intend to enter a brand-new golden era of 5 G. The 5th generation of wireless mobile telecommunications builds upon the existing options of its current predecessors. It provides a significantly faster data transfer speed, ranging from a minimum of 1 Gbit/s to a maximum of 50 Gbit/s.The global rollout of 5G networks involves extensive collaborations, emphasizing spectrum allocation, infrastructure investment, and regulatory alignment" (GSMA, 2020).

5G has reached approximately one terabit per second (1Tbps), which is about 65,000 times faster than a standard 4G connection. Nevertheless, 5G can provide a regular information transfer speed of 2.8 Gbps in a real-world environment. It is a game-changing technology!

The launch of 5G Mobile communication networks has achieved massive recognition over the last thirty years and has been shaped by successive generations. For instance, the first model (1G, AMPS) was an analog structure created in the 1980s.

For instance, the next generation (2G, GSM) made brief messages (SMS) readily available to the broad public in the 1990s. It was initially built to facilitate telephone calls but was later enhanced to support the transmission of information. The third generation (3G, UMTS) considered telephony and information transmission equally in its design and became commonly deployed in the 2000s.

The Previous Generations (4G, For Instance, LTE-A)

The 4th generation (4G LTE-A), dating back to the 2010s, was designed for information transmission and supports telephony through data packets, including Voice over IP (VoIP) and Voice over LTE. Each brand-new model introduced more significant data fees and features to the public. As a result, all generations for comparatively few devices have reasonable data rates. Still, in the future, machine-to-machine (M2M) interaction from numerous small devices will be an essential component of the network. The succeeding generation (5G) is dealing with this problem (among others).

The 5th Generation Mobile Communication System (5G)

The design of the 5th-generation mobile communication structure (5G) led to its deployment in 2020. Telecommunication hardware companies and telecommunication system operators have massively supported research on 5G, for instance, with 700 million euros from the EU (through the 5G PPP).

The term "5G" is added to several wireless communication concepts as a buzzword and explains numerous concepts, but many suggestions share similar ideas at their core. Although the core technologies will likely drive the market, fringe features of 5G still need to be developed. These need to present us with extensive specifications later, though they allow us to foresee the core features.

The Speed of the 5G

5G technology delivers a transformative improvement in network speed, reaching up to ten gigabits per second (Gbps). *This exceptional performance represents a significant upgrade, making 5G networks 10 to 100 times faster than 4 G.* Such speeds enable seamless streaming of ultra-high-definition content, instant downloads, and real-time communication, even in data-intensive scenarios. This leap in speed is driven by the need for convenience and the ability to support innovative applications that require massive bandwidth, such as augmented reality, virtual reality, and smart city infrastructure.

The unparalleled speed of 5G ensures that users experience minimal delays, fostering smoother connectivity for emerging technologies like autonomous vehicles and industrial IoT. 5G networks empower businesses to enhance productivity and efficiency by drastically reducing the time it takes to transfer data. This performance unlocks the potential for advanced remote applications, such as telemedicine and remote surgery, where precision and speed are paramount.

Overall, 5 G's unprecedented speed redefines the possibilities of wireless technology, creating opportunities for innovation across industries and enhancing everyday digital experiences.

Significant Improvements in Network Speeds

The evolution of wireless communication technology has showcased significant improvements in network speeds across successive generations. Advancements from earlier technologies with limited data rates have led to exponential increases, culminating in speeds of up to 10,000 Mbps with the latest generation. This progression highlights the leap in performance capabilities, offering users faster downloads, smoother streaming, and real-time connectivity. Such improvements have enabled innovative applications like IoT, augmented reality, and smart infrastructure. These milestones reflect the increasing demand for high-speed, reliable networks to meet the evolving needs of modern technology and users.

SPECIFICATIONS OF THE 5G

Beyond speed improvement, 5G is expected to unleash a massive IoT (Internet of Things) ecosystem, where networks can serve the communication needs of billions of connected devices, with the right trade-offs between speed, latency, and cost (Kantenna Technology Limited, 2025).

The specifications of 5G technology highlight its groundbreaking performance enhancements. 5G technology's integration with edge computing enhances low-latency applications, enabling real-time analytics and decision-making processes. *It delivers data rates of up to 10 Gbps, representing a 10 to 100-fold improvement over 4G networks, with an ultra-low latency of 1 millisecond.* 5G supports 1000 times more bandwidth per unit area and enables 100 times more connected devices than 4G LTE. It offers 99.999% availability and a 90% reduction in network energy usage, ensuring reliability and efficiency. 5G extends battery life for low-power IoT devices to up to 10 years, making it ideal for supporting next-generation connectivity and innovative solutions.

Table 3. Core specifications of 5G technology

Parameter	Value
Peak Data Rate	Up to 20 Gbps
User Experienced Speed	100 Mbps to 1 Gbps
Latency	As low as 1 ms
Mobility	Up to 500 km/h
Connection Density	Up to 1 million devices/km^2
Spectrum	FR1 (Sub-6 GHz), FR2 (mmWave)

Specification Requirements Drive 5G Technology

Among the specifications of 5G, we have a data range of up to 10 Gbps. It represents a hundred times better than the 4G with the following characteristics:

- The latency of one millisecond
- Broadband – a thousand times faster per device area
- Up to a hundred connected products in a unit area (compared to 4G LTE networks)
- Availability of 99.999%
- 100%Coverage

- of a 90% reduction in the power consumption of the network
- About ten years of battery life on low-power IoT (Internet of Things) devices

There is no unique definition of 5G technology. However, 5G merely integrates several techniques, scenarios, and use cases rather than introducing a new single radio access technology (Osseiran et al. 2014).

This creates the foundation for a truly connected planet, where smart cities, intelligent infrastructure, and responsive healthcare systems can thrive. With its capabilities, 5G is unlocking entirely new services and business models, driven by data, automation, and distributed intelligence. It paves the way for innovations like augmented reality, remote surgery, and seamless industrial automation. By combining speed, reliability, and density, 5G is transforming industries and redefining how we live, work, and interact.

Main Categories of Use Situations for 5G

5G technology enables a wide range of transformative applications, which can be broadly grouped into three main categories based on their technical requirements and use contexts. These categories — massive machine-type communications (mMTC), ultra-reliable low-latency communications (URLLC), and enhanced mobile broadband (eMBB) — define the foundational pillars of 5G's value and illustrate its potential to reshape industries, infrastructure, and everyday life. You find three main categories of use situations for 5G:

- Massive Machine-Type Communications (MMTC): Massive machine-to-machine communications, also known as the Internet of Things (IoT), involve linking vast numbers of devices with minimal human intervention on a scale unprecedented in previous years. This technology can revolutionize contemporary industrial applications and processes, including agriculture, manufacturing, and business communications.
- Ultra-Reliable, Lower-Latency Communications: Ultra-reliable, lower-latency communications are mission-critical, including real-time control of products, industrial robotics, car-to-vehicle communications, safety systems, autonomous transportation, and safer transportation networks. Conversely, adequate latency in communications prepares a new world in which remote medical care, methods, and therapies are possible.
- Enhanced Mobile Broadband: Enhanced mobile broadband offers substantially faster data speeds, higher capacity, and a more immediate connection, keeping the world connected. Innovative uses include wireless internet ac-

cess for homes, backyard casting without the need for broadcast vans, and enhanced connectivity for individuals.

For communities, 5G enables the fastest production of products for our smart cities, homes, and schools, safer and more intelligent cars, improved education and healthcare, and more secure supply chains. It also provides a safer, more effective planet to live on.

Figure 2. High-level 5G use-case classification

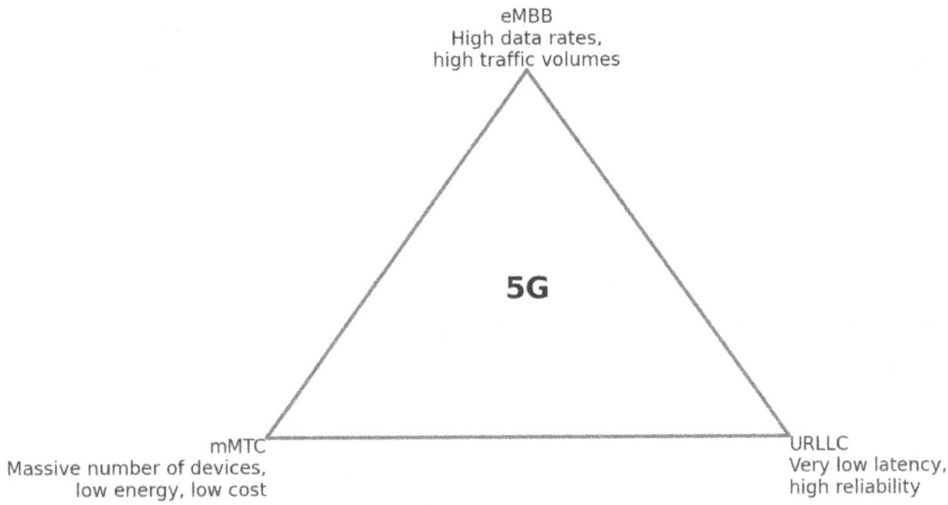

Availability of the 5G

According to a report by PCMag (Fischler, 2020), AT&T currently offers a low-band 5G network available in sixteen cities, as well as a high-band 5G network deployed in twenty-one cities. The low-band network is accessible to most consumers and is compatible with devices like the Samsung Galaxy Note 10 5 G. In contrast, the high-band network is more limited and primarily targets enterprise users, supporting devices such as the Samsung Galaxy S10 5G and the Netgear Nighthawk 5G hotspot.

Sprint has launched a mid-band 5G network that covers approximately sixteen million people across nine metropolitan areas. However, further expansion has been delayed due to the company's ongoing merger discussions with T-Mobile. Sprint's

5 G-compatible devices include the HTC Hub hotspot, LG V50, OnePlus 7 Pro 5G, and the Samsung Galaxy S10 5 G.

Benefits for Users

Working with higher-frequency stereo waves, compared to current cellular networks, partially achieves the improved speed. Nevertheless, higher-frequency radio waves have a shorter range than the lower-frequency waves used by previous mobile phone towers, so they require small cells. Therefore, 5G networks operate across three frequency bands: low, moderate, and high, which confirms their great potential.

A 5G community would consist of three types of cells, each requiring multiple antennas, and each style provides a unique trade-off between download speed, service, and distance. 5G cellphones, as well as wireless products, hook up to the system with the largest antenna within range at their location:

5G Provides Five Key Benefits for Users

One of the most important is its ability to deliver a high-quality broadband experience, anytime and anywhere. Unlike previous generations, 5G networks are designed to operate alongside 4G infrastructure using a combination of macrocells, small cells, and in-building systems (Osseiran et al., 2014). This hybrid architecture supports greater flexibility and performance, particularly in densely populated areas.

Small cells are a crucial component of this system. These compact base stations provide localized coverage, typically ranging from about ten meters to a few hundred meters. They complement the wider macro network by covering areas where high-frequency millimeter-wave (mmWave) signals—used in 5G—cannot reach far due to their short range. Small cells are essential for effective spectrum reuse, enabling high-capacity, high-speed connections in urban environments.

Additionally, enhanced mobile broadband (eMBB) represents a significant advancement from 4G LTE mobile services to 5 G. It delivers faster connection speeds, increased throughput, and improved capacity, supporting data-intensive applications such as high-definition video streaming, real-time cloud computing, and immersive digital environments.

Enhanced Capacity for Speed

Speed is thrilling, but analysts' questions and business executives' minds are this: How does 5G help companies scale their technology initiatives? The exponential increase in network speed with 5G, reaching up to 10,000 Mbps, fosters advancements in augmented reality and high-definition streaming applications.

5G offers up to 1,000 times more capabilities than 4G, creating fertile ground for the development of the Internet of Things (IoT). *As a result, 5G and IoT formed an ideal fit, redefining how people use wireless networks and the web.* With the capacity for thousands or hundreds of pieces of equipment to communicate seamlessly, new use cases for cities, schools, farms, factories, and homes are flourishing.

Picture 5G pick cases involving a considerable number of receptors on a massive selection of various models, automating supply chain management processes, making sure of just-in-time delivery of substances while utilizing predictive maintenance to reduce breakdowns

Innovative Applications

With the advent of 5G, innovative applications are becoming increasingly feasible across various sectors due to its unprecedented speed, ultra-low latency, and high connection density. One prominent example is the enablement of autonomous, or self-driving, vehicles. 5G networks provide the necessary capabilities for real-time interaction and rapid data processing among vehicles, infrastructure, systems, and even pedestrians. This connectivity is essential for creating safer and more intelligent transport ecosystems.

For instance, Lagorio, Sianesi, and Miragliotta (2023) explore how 5G technology can significantly enhance logistics operations. Their study highlights several areas where 5G can drive substantial efficiency improvements, including real-time communication, greater automation, and improved system-wide connectivity. Specific logistical functions—such as inventory management, tracking, and dynamic route optimization—stand to benefit from the responsiveness and scalability provided by 5G networks.

However, the same study emphasizes the importance of addressing critical implementation challenges. These include the need for widespread infrastructure development, rigorous data security protocols, and significant investment. Lagorio, Sianesi, and Miragliotta (2023) conclude that while 5G presents a powerful enabler of innovation, its success will depend heavily on strategic planning, cross-sector collaboration, and long-term commitment.

Robustness and Security Challenges

Ahmad et al. (2019) analyze the expanding security challenges associated with 5G wireless networks and their future extensions, highlighting the critical need for robust security measures in light of the technology's pervasive role in modern life. The paper discusses how the complexity and diversity of 5G services and connected devices have significantly broadened the threat landscape. The authors systematically review security vulnerabilities in key technological innovations, such as Software-Defined Networking (SDN), Network Function Virtualization (NFV), and cloud-based architectures integral to 5 G. They offer current solutions and propose future research directions to address these threats, emphasizing the importance of proactive security frameworks. Additionally, the paper addresses the emerging security concerns in post-5G networks, referred to as "XG," and outlines their potential vulnerabilities, highlighting the need for forward-looking security strategies.

For instance, significant developments in autonomous automobile engineering are achievable with 5G, increasing individuals ' possibilities for new professional and personal flexibility levels.

Interaction Among Everything

Networks facilitate interaction among everything on the highway—from lamp-posts to gasoline stations—to ensure security and traffic control.

5G offers a striking edge over earlier technologies, with limitless capacity and brief lag times. Moreover, with significantly improved quality and faster streaming, 5G promises groundbreaking immersive experiences, including multisensory electronic information, thanks to enhanced capabilities that support technologies such as virtual reality and augmented 3D.

In addition to some small cell communications service providers, most US wireless carriers plan to deploy 5G mobile networks that provide broadband-like services, such as HD streaming, without fear of buffering. With a vastly improved network capacity, people expect 5G to lower slowdowns during use spikes—for instance, sports fans can continue to stream throughout the big game.

Critical Control of Remote Device

Significant changes are on the horizon for smartphones -5G products will quickly become the majority. The primary benefit of 5G over 4G for consumers is improved coverage, i.e., signals reach previously hard-to-reach locations with link assurance service programs. Moreover, online subscribers often finally have the remarkable quality they expect on their devices, as downloads are predicted to lack perceptible delay.

Enhancing Safety, Health, and Security

The introduction of 5G technology significantly enhances mission-critical services that directly impact public safety, health systems, and urban security. With its ultra-low latency and high reliability, 5G enables real-time communication, which is essential for applications that require immediate responsiveness.

Key opportunities include the development of smart cities, where 5G connectivity supports real-time surveillance systems, emergency response, and public services. In healthcare, 5G opens the door to possibilities such as **remote surgery**, where surgeons can operate across long distances with near-zero latency. Enhanced traffic management is another promising area where connected infrastructure and autonomous vehicles interact dynamically to prevent accidents and reduce congestion.

According to Kshetri (2014), 5 G-powered innovative environments—from intelligent cities to adaptive homes—represent a significant leap forward. These systems rely not only on 5G's network speed and density but also on edge computing and artificial intelligence (AI), allowing real-time decision-making. For example, smart homes can adjust energy consumption based on usage patterns and environmental data, while traffic lights can autonomously adapt to real-time traffic conditions. The result is a more responsive, sustainable, and secure urban ecosystem.

Figure 3. The five major benefits of 5G for users

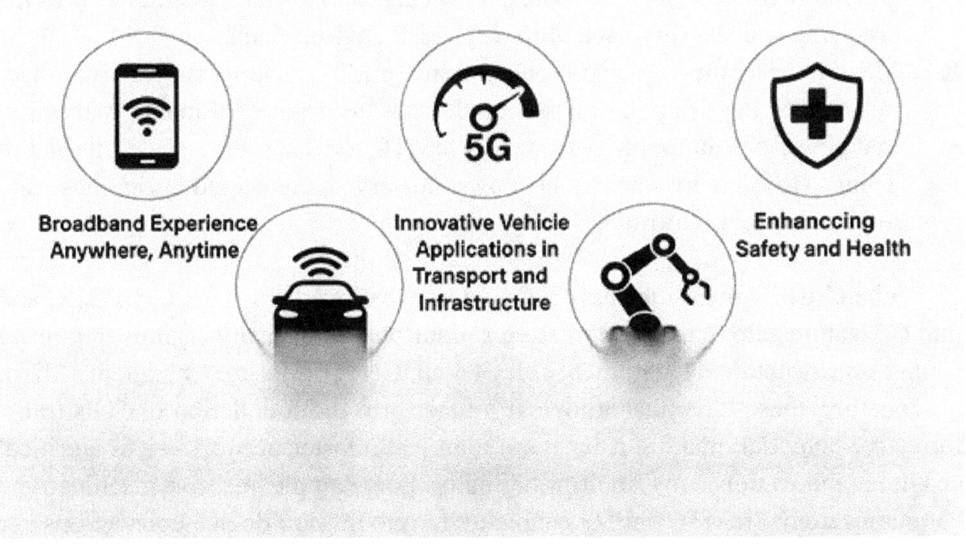

TECHNICAL CHARACTERISTICS

The exceptional performance of 5G—its ultra-fast speed, extremely low latency, and unmatched connectivity—relies on the integration of three foundational technologies: enhanced Mobile Broadband (eMBB), Ultra-Reliable and Low-Latency Communications (URLLC), and massive Machine-Type Communications (mMTC). These three components define the operational capabilities of 5G and are often referred to as the "5G triangle" (Osseiran et al., 2014).

The 5G Triangle

The International Telecommunication Union (ITU-R) has formally identified these three pillars as central to the 5G ecosystem (McClellan, Cervelló-Pastor, & Sallent, 2020):

- Enhanced Mobile Broadband (eMBB): Delivers ultra-fast data speeds—up to 20 Gbps—and sustained performance at the network edge (around 100 Mbps). *It supports high-bandwidth applications, such as seamless video streaming, 8K video, cloud gaming, and immersive virtual or augmented reality experiences.*
- Ultra-Reliable Low-Latency Communications (URLLC): Designed for mission-critical applications that require real-time responsiveness and near-perfect reliability. With latency as low as 1 millisecond, URLLC enables safe deployment of autonomous vehicles, remote surgery, industrial automation, and intelligent energy systems. As ITU (2020) notes, this feature ensures the precision and stability needed for high-risk environments.
- Massive Machine-Type Communications (mMTC): Supports the simultaneous connection of up to one million devices per square kilometer with minimal power consumption. This makes mMTC ideal for large-scale Internet of Things (IoT) environments, such as smart cities, connected agriculture, and infrastructure monitoring.

While eMBB capabilities have been widely deployed since 2020, URLLC and mMTC features are still in progressive rollout phases in many regions, requiring further infrastructure development (McClellan, Cervelló-Pastor, & Sallent, 2020).

Together, these three technological pillars form the foundation of 5G's transformative potential, making it far more than just a faster network—it is a critical enabler of innovation across multiple domains. For example, massive machine-type communications (mMTC) in 5G enable up to one million device connections per square kilometer, driving the proliferation of the Internet of Things (IoT) across various industries.

Enhanced Mobile Broadband (eMBB)

Enhanced Mobile Broadband (eMBB) is designed to service much more heavily populated metropolitan areas with downlink speeds of up to 1 Gbps (gigabits per second) indoors, as well as 300 Mbps (megabits per second) outdoors. Enhanced Mobile Broadband (eMBB) utilizes high-frequency millimeter-wave (mmWave) antennas on lampposts, the sides of buildings, the limbs of trees, pre-existing electric towers, and in a novel situation suggested by AT&T, on the tops of town buses.

Since every one of the antennas inside the metro uses position, dealing with space most likely no more significant than a baseball diamond, hundreds, maybe thousands, are required to service any heavily populated city region entirely. Moreover, because most would not be omnidirectional, the maximum beam width of approximately four mmWave antennas would cause them to bounce indicators off each other's reflectors until they arrived at their intended customer locations. For many rural and remote areas, eMBB aims to replace 4G's current LTE process with a new community of lower-power, omnidirectional antennas, providing a 50 Mbps downlink service.

Technical Characteristics: Enhanced Mobile Broadband (eMBB) – Performance

Enhanced Mobile Broadband (eMBB) is one of the core pillars of 5G, designed to deliver vastly improved speed, capacity, and user experience compared to 4G LTE. In terms of performance, 5G networks have demonstrated speeds ranging from approximately 50 Mbit/s to well over 1 Gbit/s, depending on the frequency band and deployment environment.

The highest speeds are achieved using millimeter wave (mmWave) technology, which offers significant bandwidth but has a limited coverage area. For example, AT&T's mmWave 5G network reached peak speeds of 1.8 Gbit/s as early as July 2019. This makes mmWave ideal for dense urban areas or specific high-demand applications such as ultra-HD video streaming or AR/VR environments.

More commonly, sub-6 GHz (mid-band) 5G is used in mainstream deployments, offering a balance between speed and coverage. Typical speeds range around 400 Mbit/s, and this band allows for much broader geographic coverage than mmWave. Meanwhile, low-band 5G provides the widest range of signal penetration and reach, though at lower speeds compared to other bands (GSMA, 2020).

It is worth noting that in some cases, advanced 4G technologies, such as LTE combined with License Assisted Access (LAA), still deliver high speeds. T-Mobile's LTE/LAA network has achieved speeds of up to 500 Mbit/s in major U.S. cities, occasionally outperforming early-stage 5G deployments, (Rysavy, 2018).

Technical Performance: Focus on Enhanced Mobile Broadband (eMBB)

Enhanced Mobile Broadband (eMBB) represents one of the core pillars of 5G technology, designed to deliver significantly faster data speeds, increased capacity, and a more stable mobile internet experience compared to previous generations. This use case focuses on high-speed, high-capacity services, such as 4K/8K video streaming, virtual and augmented reality, and seamless mobile connectivity in densely populated urban environments.

Actual 5G speeds range from approximately 50 Mbit/s to over 1 Gbit/s depending on the frequency band. eMBB performance is particularly optimized in millimeter wave (mmWave) deployments, where ultra-high data rates are achievable, reaching up to 1.8 Gbit/s on AT&T's 5G network as of July 2019. This spectrum enables eMBB to unlock immersive media experiences with virtually no buffering.

However, mmWave's limited range and poor penetration through obstacles restrict its deployment primarily to dense urban zones. To expand eMBB coverage, operators increasingly rely on sub-6 GHz (mid-band) frequencies, which offer a more balanced performance profile. These bands generally deliver speeds around 400 Mbit/s and perform better than mmWave in terms of range, particularly in outdoor environments.

At the lower end of the spectrum, low-band 5G provides the most extensive geographic coverage, although with reduced speeds. As noted by the GSMA (2020), low-band eMBB supports basic broadband access in rural or remote areas, but does not fully showcase the potential of 5G speed and capacity.

While 5G's eMBB can exceed 4G LTE in most cases, some advanced 4G networks—such as T-Mobile's LTE/LAA system—still reach impressive speeds of up to 500 Mbit/s in cities like Manhattan and Chicago. Although 5G standards include License Assisted Access (LAA) capabilities, this feature is not yet fully implemented in most deployments, (Rysavy, 2018).

The performance similarity between 4G and 5G in specific scenarios is partly due to 4G nearing the Shannon limit for data transmission. Nonetheless, eMBB's use of broader bandwidths and advanced modulation in 5G—especially in mmWave—positions it as the driver of a radically improved user experience, offering high-resolution video, real-time gaming, and enhanced mobile internet even during network congestion.

Latency: Impact on Enhanced Mobile Broadband (eMBB)

Latency—the time it takes for devices to communicate within a network—is a critical performance metric that significantly influences the responsiveness of applications, particularly those that rely on real-time data transmission. In the context of Enhanced Mobile Broadband (eMBB), low latency is essential for delivering seamless user experiences in applications such as 8K video streaming, cloud-based gaming, augmented reality (AR), and virtual reality (VR).

Historically, latency has improved significantly with each generation of wireless technology. Third-generation (3G) networks exhibited average response times of approximately 100 milliseconds, while 4G networks reduced this to around 30 milliseconds. With 5G, latency can drop as low as 1 millisecond, making it nearly instantaneous and unlocking new categories of applications that were previously infeasible.

In real-world conditions, the "air latency" observed in early 5G equipment deployments in 2019 was approximately 8 to 12 milliseconds, (Rysavy, 2018). *However, this figure does not account for server-side latency, which can significantly impact end-to-end communication performance*. For example, Verizon reported an overall latency of approximately 30 milliseconds during its initial 5G deployments (Verizon, 2020), which includes both network and server latency.

To further minimize latency and enhance eMBB experiences, edge computing has emerged as a vital solution for enhancing mobile broadband (eMBB) experiences. By deploying edge servers closer to the user, often near base stations or cell towers, latency can be reduced to 10-20 milliseconds in typical operating environments. However, ultra-low latency values of 1 to 4 milliseconds remain rare outside of controlled laboratory settings (Sauter, 2021).

These developments in latency performance enable eMBB to support advanced mobile broadband services that require not only high throughput but also real-time responsiveness. As such, latency optimization remains a key area of ongoing innovation within the 5G ecosystem, enabling the full realization of next-generation mobile experiences.

MASSIVE MACHINE-TYPE COMMUNICATIONS (MMTC)

Massive Machine-Type Communications (MMTC) represents one of the three core pillars of 5G technology, enabling the large-scale connection of devices without human intervention. It is specifically designed to support the explosive growth of the Internet of Things (IoT) by offering efficient, low-latency, and low-power communication. MMTC opens new possibilities for innovation across various industries, including smart cities, agriculture, and logistics.

Description

Massive Machine-Type Communications (MMTC) is a foundational pillar of 5G technology, designed to support vast networks of interconnected devices with minimal human intervention. MMTC facilitates large-scale machine-to-machine (M2M) communication, underpinning the growing Internet of Things (IoT) ecosystem. These systems involve billions of low-power, low-data-rate devices operating simultaneously within the same network infrastructure, all without overwhelming the system's capacity.

Earlier cellular generations, such as 2G, were initially adequate for narrowband M2M applications; however, as device density increased, network signaling congestion and added latency became significant constraints. MMTC addresses these limitations by offering a dedicated communication layer optimized for devices that typically require downlink speeds around 100 Kbps, while keeping latency below 10 milliseconds. This enables critical yet low-data devices, such as sensors in agriculture, logistics, environmental monitoring, and intelligent infrastructure, to operate effectively without compromising overall network performance.

Efficiency is central to the monetization strategy of 5G networks. Unlike previous generations, 5G's architectural design allows service providers to segment the network and customize performance across different use cases. This segmentation, often referred to as network slicing, enables optimal resource allocation for massive IoT deployments, balancing reliability, latency, and bandwidth needs according to specific application requirements.

Moreover, the utilization of millimeter-wave frequency bands in 5G enhances potential data throughput, although these higher frequencies require dense network architectures to overcome signal attenuation and propagation challenges (3GPP, n.d.). This is particularly relevant in urban IoT environments where high device density and interference risk require robust and localized infrastructure.

In this way, MMTC supports not just technological innovation but also economic transformation. It shifts telecom providers from merely offering connectivity to actively shaping and deriving value from the digital economy, leveraging the vast data generated by IoT devices to drive analytics, automation, and new services.

Technical Characteristics

Massive Machine-Type Communications (MMTC) enables connectivity for a vast number of low-power devices with minimal human intervention. It supports the proliferation of IoT by providing scalable, energy-efficient communication over enormous areas. According to Jovović, Forenbacher, and Periša (n.d.), 5G is expected to connect a significant portion of the estimated 50 billion IoT devices, many of which use cost-effective Wi-Fi solutions.

In disaster recovery operations, drones equipped with 4G or 5G can transmit real-time data to emergency responders (FlytBase, 2025). *Most modern vehicles now feature 4G or 5G cellular connectivity, enabling advanced features such as navigation, diagnostics, and entertainment.* However, fully autonomous vehicles must also function in areas without network coverage, meaning 5G is beneficial but not always essential, (RantCell, 2019).

MMTC also supports critical applications such as remote surgeries. While fiber-optic connections remain faster and more reliable for such procedures, 5G's ultra-low latency can serve as a complementary solution in some scenarios. Overall, MMTC provides the infrastructure for the large-scale integration of smart sensors, meters, and machines across industries, supporting the IoT revolution.

ULTRA-RELIABLE AND LOW LATENCY COMMUNICATIONS (URLLC)

Ultra-Reliable and Low Latency Communications (URLLC) focuses on mission-critical applications where data transmission must be both highly dependable and nearly instantaneous. Unlike applications where high bandwidth is the priority, URLLC emphasizes reducing latency to less than one millisecond, making it essential for scenarios like autonomous driving, where decisions must occur faster than human reaction time.

The ultra-low latency of URLLC also positions 5G as a competitor to satellite technology for functions such as geolocation, potentially expanding 5G's role in navigation systems. In this context, URLLC enables real-time control for critical infrastructure and processes, including telemedicine, robotics, and safety systems.

Latency measures the time it takes for a signal to travel from source to destination and back. 5G achieves significantly lower latency than previous generations, where 3 G and 4G LTE average 100 ms and 30 ms, respectively. 5G can reduce this to under 5 ms, and even as low as 1 ms in controlled environments. According to Verizon (2020), early deployments have demonstrated latency around 30 ms, while edge computing servers located closer to users can reduce latency to 10–20 ms. An actual latency of 1–4 ms is rare outside of laboratory conditions (Sauter, 2021; Rysavy, 2018).

The combination of low latency and high speed allows real-time machine-to-machine (M2M) communications without human delay. For example, logistics, manufacturing, and agriculture sectors benefit from rapid decision-making and automation. Gamers and users of immersive technology also anticipate improvements, as virtual and augmented reality (VR/AR) demand fast and seamless data transmission for fully immersive experiences.

MASSIVE MIMO

Massive MIMO (Multiple-Input, Multiple-Output) is a foundational technology for the ultra-fast, high-capacity 5G networks that began rolling out at the end of 2019. It plays a crucial role in enhancing the capacity, efficiency, and reliability of wireless networks, particularly as demand for data continues to increase.

Definition

MIMO technology enables the simultaneous transmission and reception of multiple data signals over the same radio channel. While traditional MIMO systems typically use two or four antennas, Massive MIMO systems employ tens or even hundreds of antennas to support much larger-scale deployments. There is no strict threshold, but systems with 64, 96, or 128 antennas are typically considered "massive."

According to Manivasagam and Shukla (2023), Massive MIMO technology supports 5G networks by significantly boosting spectral efficiency and enabling robust data throughput. Major industry players such as Huawei, ZTE, and Facebook have demonstrated systems with up to 128 antennas. For example, Ericsson's AIR 6468, promoted as the first 5G NR radio, utilizes 64 transmit and 64 receive antennas.

Advantages of Massive MIMO

Massive MIMO enhances wireless network performance without requiring additional spectrum by increasing the capacity of a connection. Reports suggest that capacity improvements could be up to 50 times greater than those of conventional systems. These gains result from the ability to send more signals simultaneously and to target them more precisely.

The increased number of antennas enhances signal strength, improves data rates, and bolsters connection reliability. It also supports better indoor signal reception—crucial given the penetration challenges associated with 5G's high-frequency bands.

Additionally, the extensive antenna arrays make Massive MIMO more resistant to interference and deliberate jamming than legacy systems.

Another key component is beamforming—a technology that allows focused signal delivery to specific users rather than broadcasting in all directions. This optimizes the use of spectrum, especially in densely populated urban areas, where current systems face congestion. As Larsson et al. (2014) note, Massive MIMO and beamforming technologies ensure more uniform data rates and reduced latency throughout the network.

Cross-Layer MIMO

Cross-layer MIMO is an advanced concept that enhances performance by addressing interactions across different protocol layers. This approach aims to refine MIMO performance by optimizing various technical aspects simultaneously. Notable cross-layer techniques include:

- Joint source-channel coding
- Adaptive Modulation and Coding (AMC)
- Hybrid Automatic Repeat Request (HARQ)
- User scheduling

These strategies enable seamless adaptation to fluctuating channel conditions, thereby maximizing throughput and reliability.

In highly interconnected wireless ad hoc networks, multi-user-to-multi-user interactions become increasingly complex. According to Meza (2007) these environments introduce additional challenges such as self-interference and overlapping transmissions. Rather than avoiding these issues, advanced cooperative systems may decode strong interfering signals before decoding their own, thus mitigating low Carrier-over-Interference (CoI) ratios across physical (PHY), medium access control (MAC), and application layers.

Figure 4. Massive MIMO

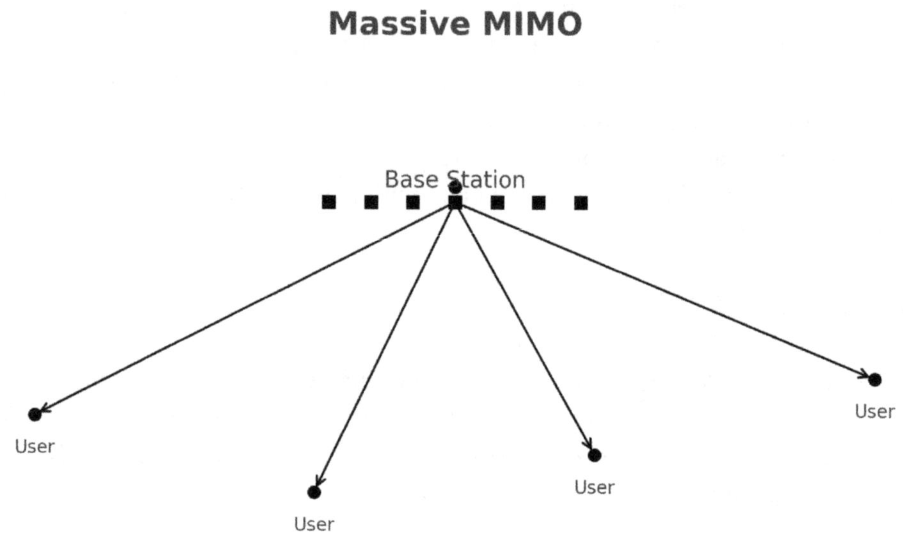

COOPERATIVE MULTIPLE ANTENNA RESEARCH

Cooperative multiple antenna research explores how multiple antenna technologies can be extended beyond a single device, leveraging antennas from nearby wireless terminals to enhance communication performance. These cooperative strategies are particularly valuable in dense wireless environments or in situations where individual devices may lack the physical space or power to support multiple antennas. By collaborating, wireless nodes can share resources, increasing reliability, range, and data throughput through diversity and multiplexing gains.

Cooperative Techniques

Several core cooperative antenna techniques underpin this research:

- Cooperative Multiple Antenna Systems: These systems apply MIMO principles to scenarios in which antennas are distributed across neighboring wireless terminals, enabling enhanced performance. The goal is to emulate the performance of a centralized MIMO system through collaboration among nodes.

- Cooperative Diversity: This technique achieves antenna diversity gain by enabling distributed antennas—each belonging to independent nodes—to share signals. The cooperation increases link reliability, particularly in fading environments.
- Cooperative MIMO: A more advanced technique where distributed transmitters or receivers work together to mimic a large-scale MIMO system. This cooperation enables spatial multiplexing, which improves the overall throughput by transmitting multiple data streams simultaneously across different spatial paths.

Together, these methods increase the robustness and spectral efficiency of 5G networks and are considered crucial to future 6G and beyond network architectures.

Cooperative Relay

Cooperative relaying builds on the concept of using intermediate nodes to support communication between a source and a destination. The aim is to enhance the trade-off between performance (e.g., throughput, error rates) and delay by applying collaborative strategies.

- Cooperative Relay Systems combine ideas from cooperative diversity and Multiple-Input Multiple-Output (MIMO) technology, where relay nodes assist in the transmission process. These systems can enhance coverage and system performance with only minor losses due to the need to coordinate transmission among nodes (Wikipedia, n.d.).

Key relaying techniques include:

- Store-and-Forward (S&F): The relay stores the received data and forwards it later, improving reliability at the cost of added delay.
- Amplify-and-Forward (A&F): The relay amplifies the incoming signal and forwards it without decoding, maintaining simplicity.
- Decode-and-Forward (D&F): The relay decodes the message, re-encodes it, and then forwards it, thereby improving performance at the expense of increased processing time.
- Coded Cooperation: Involves splitting a message between the source and relay to provide coding gains.
- Compress-and-Forward (C&F): The relay compresses its observation of the received signal before forwarding it.

- Non-Orthogonal Methods: These methods enable simultaneous transmission by the source and relay, thereby increasing spectral efficiency at the expense of more complex interference management.

These cooperative communication techniques are foundational to the development of resilient, high-performance wireless systems and will play a vital role in the evolution of post-5G architectures.

MULTI-ACCESS EDGE COMPUTING (MEC)

Multi-Access Edge Computing (MEC)—originally known as Mobile Edge Computing—is an architecture defined by the European Telecommunications Standards Institute (ETSI) that brings cloud computing capabilities closer to end-users by integrating computing and storage resources at the edge of the mobile network. The primary purpose of MEC is to reduce latency, enhance application performance, and mitigate network congestion by processing data closer to its point of generation.

Concept and Objectives

MEC is deployed at the edge of the Radio Access Network (RAN), typically co-located with cellular base stations or aggregation points. By bringing computing power closer to users, MEC enables real-time data processing, essential for applications that require ultra-low latency and high reliability. This is especially critical in emerging technologies such as:

- Autonomous vehicles
- Augmented and Virtual Reality (AR/VR)
- Smart manufacturing (Industry 4.0)
- Remote healthcare (e.g., telesurgery)
- IoT-based monitoring systems

MEC reduces the time it takes for data to travel between users and the data center, thereby improving response times and reducing reliance on centralized cloud infrastructures.

Technical Capabilities

Key features and advantages of MEC include:

- Ultra-low latency: Since computation happens closer to the user, data does not need to travel to a distant cloud server, allowing for near real-time processing.
- High bandwidth efficiency: MEC optimizes the usage of available bandwidth by minimizing unnecessary data backhaul to the core network.
- Context-aware services: By operating at the edge, MEC systems can leverage local context (e.g., location, user behavior, network conditions) to offer tailored and intelligent services.
- Scalability and flexibility: MEC supports rapid deployment of new services without requiring extensive infrastructure changes at the core.

Integration With 5G

MEC is considered a foundational element of the 5G ecosystem. The convergence of MEC and 5G enhances network capabilities by:

- Supporting network slicing, where specific MEC instances serve particular applications or industries.
- Enabling Quality of Service (QoS) guarantees for critical applications ensures optimal performance.
- Allowing real-time analytics and AI at the edge for dynamic decision-making in connected environments.

MEC and 5G together facilitate a distributed computing paradigm that meets the requirements of URLLC (Ultra-Reliable Low Latency Communications) and mMTC (Massive Machine-Type Communications) use cases.

Business and Innovation Opportunities

One of MEC's most transformative aspects is its potential to open the RAN (Radio Access Network) to third-party developers, such as application vendors, content providers, and enterprises. This openness enables:

- New business models for mobile network operators (MNOs) that can monetize network edge services.
- Customized services for enterprise clients in sectors like healthcare, logistics, and entertainment.
- Enhanced user experiences, especially in innovative city applications and mobile content delivery (e.g., streaming media, gaming).

By blending information technology (IT) with telecommunications infrastructure, MEC acts as a bridge between the telecom network and cloud computing environments.

The European Telecommunications Standards Institute created technical norms for MEC and submitted a white paper.

Distributed Computing in the RAN

MEC gives a distributed computing domain for application and administration hosting. It can store and deliver content near cell users for quicker reaction time. Real-time radio access network (RAN) data enables the development of innovative applications.

The critical component is the MEC application server, which integrates with the RAN component. The MEC application server provides computing assets, storage, and access to RAN data, ensuring availability. The MEC application supports multitenancy run-time and facilitates application conditions. These applications are transferred as packaged operating system virtual machine (VM) images or containers that include working frameworks and applications. The stage also provides numerous middleware applications and foundational services. Gear merchants, specialist organizations, and outsider players in the telecom industry provide applications.

CONCLUSION

The deployment of 5G technology represents more than a technical milestone—it marks a pivotal transformation in how society connects, communicates, and operates. Throughout this chapter, we have explored the defining features that distinguish 5G from its predecessors, including its ultra-low latency, exceptional speed, and the capacity to connect billions of devices simultaneously. These capabilities are not merely enhancements but enablers of unprecedented innovation across various industries, including healthcare, manufacturing, logistics, and transportation.

The Table 4 summarizes 5G's performance and benefits relative to 4G across several key dimensions, providing a clear comparative understanding of its advantages.

Table 4. Comparative summary of 5G performance and benefits over previous generations

Aspect	4G (LTE)	5G	Benefits of 5G
Speed	100 Mbps – 1 Gbps	1–10 Gbps (up to 20 Gbps)	Faster downloads, seamless 4K/8K streaming
Latency	30–50 ms	<1 ms	Real-time applications, including remote surgery
Device Capacity	10^4 devices/km^2	10^6 devices/km^2	Massive IoT connectivity
Use Cases	Streaming, web browsing	Autonomous driving, smart cities, industrial IoT	Broad transformation across sectors
Energy Efficiency	Moderate	Up to 90% improvement	Sustainable for long-term and large-scale use

According to Mujahid (2024), the fifth generation of mobile communication, 5G, represents a monumental leap in the evolution of wireless networks, seamlessly bridging the digital and physical worlds. Its unprecedented capabilities, including enhanced bandwidth, ultra-low latency, and seamless machine-to-machine communication, have redefined the possibilities for connectivity across various industries. Unlike its predecessors, 5G is not merely an upgrade but a revolutionary platform that supports a wide range of applications—from IoT and smart cities to autonomous vehicles and advanced industrial automation. Mujahid (2024) argues that this transformative technology has laid the foundation for a new era of innovation, enabling global economies to embrace digital transformation on an unprecedented scale.

The architecture of 5G networks, built on small cells and robust infrastructure, ensures unparalleled connectivity for billions of devices. Its ability to deliver gigabit-per-second speeds and near-instantaneous responsiveness has introduced a performance previously unimaginable in telecommunications. This advancement paves the way for breakthroughs in telemedicine, real-time analytics, and immersive virtual and augmented reality experiences, fostering innovation across sectors and enhancing the quality of life for individuals worldwide.

However, the successful deployment of 5G also presents challenges. Significant investment in infrastructure, compliance with evolving regulatory frameworks, and the development of 5 G-compatible devices require a collaborative effort among governments, technology providers, and industry stakeholders. Addressing these challenges is crucial to unlocking 5G's full potential and ensuring equitable adoption across regions and demographics.

In conclusion, 5G represents the next step in mobile communication, catalyzing societal and economic transformation. It builds on the legacy of previous generations, driving us toward a future of unparalleled connectivity and innovation (Colin, 2022). As we move forward, the sustained evolution of 5G will continue to reshape industries, bridge digital divides, and empower humanity to explore new frontiers of technology and communication.

REFERENCES

Admin. (2020). *The Evolution of Mobile Networks: From 1G to 5G – Centech 5G Hub*. Centech5ghub.com. https://centech5ghub.com/the-evolution-of-mobile-networks-from-1g-to-5g/

Ahmad, I., Shahabuddin, S., Kumar, T., Okwuibe, J., Gurtov, A., & Ylianttila, M. (2019). Security for 5G and Beyond. *IEEE Communications Surveys and Tutorials, 21*(4), 1–1. DOI: 10.1109/comst.2019.2916180

Awwwards. (2019). *How 5G Will Change the Mobile App Development Landscape*. Awwwards.com. https://www.awwwards.com/5g-web-design-mobile.html

Colin. (2022, April 12). *No Moment's too big for all-new Lexus LX 600 SUV*. Reel 360 News. https://reel360.com/article/no-moments-too-big-for-all-new-lexus-lx-600-suv/

Egea-Roca, D., Arizabaleta-Diez, M., Pany, T., Antreich, F., Lopez-Salcedo, J. A., Paonni, M., & Seco-Granados, G. (2022). GNSS User Technology: State-of-the-Art and Future Trends. *IEEE Access : Practical Innovations, Open Solutions, 10*, 39939–39968. DOI: 10.1109/access.2022.3165594

El-Shorbagy, A. (2021). 5G Technology and the Future of Architecture. *Procedia Computer Science, 182*, 121–131. DOI: 10.1016/j.procs.2021.02.017

Fischler, J. (2020, January 6). *Where is 5G available in the US? The biggest providers, including Verizon and T-Mobile, all offer 5G*. (Updated for 2025). Lifewire. https://www.lifewire.com/5g-availability-us-4155914

FlytBase. (2025). *Drone for Disaster Management – How Drones are Used for Emergency Response*. Www.flytbase.com. https://www.flytbase.com/blog/drone-disaster-relief

Fowdur, T. P., Indoonundon, M., & Milovanovic, D. A. (2024). *5G NR modelling in MATLAB: Network architecture, protocols, and physical layer.* MathWorks. https://www.mathworks.com/academia/books/5g-nr-modelling-in-matlab-fowdur.html

Global Mobile Suppliers Association (GSA). *(2015). The Road to 5G: Drivers, Applications, Requirements and Technical Development*. https://gsacom.com/content/uploads/2015/11/151118-GSA-the-Road-to-5G.pdf

GPP – The Mobile Broadband Standard Partnership Project. (n.d.). 3GPP. https://www.3gpp.org

GSMA. (2020). *5G Spectrum Requirements and Deployment Strategies*. GSMA Intelligence. https://www.gsma.com

Henley Business School. (2021). *Henley research to explore factors impacting Huawei 5G rollout*. Henley Business School. https://www.henley.ac.uk/news/2021/henley-research-to-explore-factors-impacting-huawei-5g-rollout

International Telecommunication Union Radiocommunication Sector (ITU-R). *(2017). Minimum requirements related to technical performance for IMT-2020 radio interface(s)* (Report ITU-R M.2410-0). Itu.net. https://www.itu.int/pub/R-REP-M.2410-2017

Jovović, I., Forenbacher, I., & Periša, M. (n.d.). *Massive Machine-Type Communications: An Overview and Perspectives Towards 5G*. https://www.fpz.unizg.hr/ikp/upload/RCITD_2015_Massive%20Machine%20Type%20Communications%20An%20Overview%20and%20Perspectives%20Towards%205G.pdf

Kantenna Technology Limited. (2025). *HOME.* Kantennatech.com. https://www.kantennatech.com/

Kshetri, N. (2014). The emerging role of Big Data in key development issues: Opportunities, challenges, and concerns. *Big Data & Society*, *1*(2). DOI: 10.1177/2053951714564227

Lagorio, A., Cimini, C., Pinto, R., & Cavalieri, S. (2023). 5G in Logistics 4.0: Potential applications and challenges. *Procedia Computer Science*, *217*, 650–659. DOI: 10.1016/j.procs.2022.12.261

Larsson, E. G., Edfors, O., Tufvesson, F., & Marzetta, T. L. (2014). Massive MIMO for next-generation wireless systems. *IEEE Communications Magazine*, *52*(2), 186–195. https://ieeexplore.ieee.org/document/6736761. DOI: 10.1109/MCOM.2014.6736761

Manivasagam, G., & Shukla, M. (2023). Massive MIMO in 5G Networks: Trends and Challenges. *2023 International Conference on Power Energy, Environment & Intelligent Control (PEEIC)*, 1456–1460. DOI: 10.1109/peeic59336.2023.10450543

McClellan, M., Cervelló-Pastor, C., & Sallent, S. (2020). Deep Learning at the Mobile Edge: Opportunities for 5G Networks. *Applied Sciences (Basel, Switzerland)*, *10*(14), 4735. DOI: 10.3390/app10144735

Meza, R. R. P. (2007). *Interference and power control in wireless ad hoc networks*. University of New Mexico UNM Digital Repository. https://digitalrepository.unm.edu/cgi/viewcontent.cgi?article=1202&context=ece_etds

Mujahid. (2024, August 19). *The Future of 5G and Beyond: Transforming Connectivity*. Zoeteck.com. https://zoeteck.com/the-future-of-5g-and-beyond-transforming-connectivity/

NGMN. (2015). *NGMN 5G White Paper*. NGMN. https://www.ngmn.org/work-programme/5g-white-paper.html

Osseiran, A., Boccardi, F., Braun, V., Kusume, K., Marsch, P., Maternia, M., Queseth, O., Schellmann, M., Schotten, H., Taoka, H., Tullberg, H., Uusitalo, M. A., Timus, B., & Fallgren, M. (2014). Scenarios for 5G mobile and wireless communications: The vision of the METIS project. *IEEE Communications Magazine*, *52*(5), 26–35. DOI: 10.1109/mcom.2014.6815890

RantCell. (2019). Rantcell.com. https://rantcell.com/do-autonomous-vehicles-need-5g.html

Rysavy, P. (2018). *Industry Voices—Rysavy: Midband spectrum for 5G is needed now*. Fierce Network. https://www.fierce-network.com/5g/industry-voices-rysavy-mid-band-spectrum-for-5g-needed-now

Sauter, M. (2021). *From GSM to LTE-Advanced Pro and 5G*. DOI: 10.1002/9781119714712

Singh, K., & Shivinder Devra. (2017, October 12). *Evolution of Cellular Network: From 1G to 5G*. Academia.edu. https://www.academia.edu/34839361/Evolution_of_Cellular_Network_From_1G_to_5G

Verizon. (2020). *What is the Latency of 5G?* Verizon. https://www.verizon.com/about/our-company/5g/5g-latency?msockid=3c2aa922788360e430f3bcaa79416152

Wikipedia. (n.d.). *Multi-user MIMO*. Wikipedia. https://en.wikipedia.org/wiki/Multi-user_MIMO

ADDITIONAL READING

archit_tandon. (2024, July 23). *5G Testing: Safeguarding the future of smart cities and IoT | Communications Today*. Communications Today. https://www.communicationstoday.co.in/5g-testing-safeguarding-the-future-of-smart-cities-and-iot/

Brake, D. *(2020)*. A U.S. National Strategy for 5G and Future Wireless Innovation. Information Technology & Innovation Foundation. https://itif.org/publications/2020/04/27/us-national-strategy-5g-and-future-wireless-innovation/

Center for a New American Security. (2019). *Securing Our 5G Future*. Cnas.org. https://www.cnas.org/publications/reports/securing-our-5g-future

Cisco. (2023, March 9). *Cisco Annual Internet Report (2018–2023) White Paper*. Cisco. https://www.cisco.com/c/en/us/solutions/collateral/executive-perspectives/annual-internet-report/white-paper-c11-741490.html

Congressional Research Service (CRS). (2019). *Fifth-Generation (5G) Telecommunications Technologies: Issues for Congress*. Congress.gov. https://www.congress.gov/crs-product/R45485

Cybersecurity and Infrastructure Security Agency (CISA). *(2019)*. Overview of Risks Introduced by 5G Adoption in the United States. https://www.cisa.gov/sites/default/files/publications/19_0731_cisa_5th-generation-mobile-networks-overview_0.pdf

Ericsson. (2019, October 2). *5G - The 5G switch made easy*. Ericsson.com; Ericsson. https://www.ericsson.com/en/5g

Federal Communications Commission (FCC). *(2018). Accelerating Wireless and Wireline Broadband Deployment by Removing Barriers to Infrastructure Investment*. Federal Register. https://www.federalregister.gov/documents/2018/10/15/2018-22234/accelerating-wireless-and-wireline-broadband-deployment-by-removing-barriers-to-infrastructure

Intelligence, G. S. M. A. (2014). *Understanding 5G: Perspectives on Future Technological Advancements in Mobile*. https://www.gsma.com/futurenetworks/wp-content/uploads/2015/01/Understanding-5G-Perspectives-on-future-technological-advancements-in-mobile.pdf

Khan, W. Z., Ahmed, E., Hakak, S., Yaqoob, I., & Ahmed, A. (2019). Edge computing: A survey. *Future Generation Computer Systems*, *97*, 219–235. DOI: 10.1016/j.future.2019.02.050

Meesha Sundarum. (2025, April 16). *The State of 5G: Growth, Challenges, and Opportunities in 2025*. 5G Americas. https://www.5gamericas.org/the-state-of-5g-growth-challenges-and-opportunities-in-2025/

Newswire, P. R. (2015). *The 5G Wireless Ecosystem: 2015 - 2025 - Technologies, Applications, Verticals, Strategies & Forecasts*. Cision PR Newswire. https://www.prnewswire.com/news-releases/the-5g-wireless-ecosystem-2015---2025---technologies-applications-verticals-strategies--forecasts-300027177.html

NGMN – We make better connections. (2025, July 9). NGMN. https://www.ngmn.org

Office, U. S. G. A. (2020, November 24). *5G Wireless: Capabilities and Challenges for an Evolving Network*. Www.gao.gov. https://www.gao.gov/products/gao-21-26sp

Segan, S. (2016, June 21). *What Is 5G?* PCMAG; PCMag. https://www.pcmag.com/news/what-is-5g

Sudarsanam, P., Dwarakanatha, G. V., Anand, R., Shah, H., & Jayashree, C. S. (2021). Learning of Advanced Telecommunication Computing Architecture (ATCA)-Based Femto Gateway Framework. *Lecture Notes in Networks and Systems*, 375–392. DOI: 10.1007/978-981-16-2126-0_32

Weisman-Pitts, J. (2025, January 27). *5G Technology: A Comprehensive Report on its Impact by 2025*. Global Banking and Finance Review. https://www.globalbankingandfinance.com/5g-technology-a-comprehensive-report-on-its-impact-by-2025

Winmate, U. S. A. (2021, January 15). *New Horizons for Industrial Automation | Winmate USA Blog*. Winmate-Rugged.com. https://www.winmate-rugged.com/en-US/blog/blog-5g-for-industrial-automation

Xenakis, C. (2008). *Security in 2.5G Mobile Systems*. In Y. Zhang, J. Zheng, & M. Ma (Eds.), Handbook of Research on Wireless Security (pp. 351–363). IGI Global. DOI: 10.4018/978-1-59904-899-4.ch022

Yukawa, T., & Sanada, Y. (2017). *Offloading effect of heterogeneous network with non-orthogonal multiple access*. 1–5. DOI: 10.23919/apcc.2017.8303972

Chapter 2
Development of 5G and Smartphones

ABSTRACT

The global rollout of 5G in 2019 marked a turning point in mobile connectivity, offering enhanced bandwidth and ultra-low latency. This fueled rapid adoption of 5G-enabled devices, with 56 manufacturers introducing 129 models by year's end. These included smartphones, hotspots, and other equipment, showcasing 5G's wide-ranging impact. The launch of 5G-compatible chipsets played a key role, enabling high-speed, high-performance devices and paving the way for models like the Samsung Galaxy S20. 5G has since redefined mobile technology and expanded its role in consumer and industrial applications.

INTRODUCTION

The development of 5G technology has revolutionized telecommunications and redefined the role of smartphones in modern connectivity. Enhanced Mobile Broadband (eMBB) capabilities in 5G networks enable 10 times faster download speeds compared to 4G technologies.

As 5G networks began rolling out globally in 2019, the Global Mobile Suppliers Association (GSA) tracked the launch of 5 G-enabled devices, marking a critical milestone in the industry's transition. By the end of 2019, the number of announced 5G devices rose to 129 across 56 vendors, showcasing the rapid adoption and diversification of 5G applications. These devices spanned a variety of form factors, including smartphones, hotspots, customer-premises equipment, and dongles, underscoring the extensive possibilities offered by 5G technology.

The introduction of 5 G-compatible chipsets also played a vital role in accelerating this evolution. At the start of 2019, four certified 5G modem chipsets were available, with more anticipated shortly. It enabled manufacturers to develop devices

DOI: 10.4018/979-8-3373-0497-7.ch002

with unprecedented speed and capability. It laid the groundwork for flagship 5G smartphones, such as the Samsung Galaxy S20, which debuted in March 2020. Despite its premium price compared to its 4G predecessors, the device demonstrated the transformative potential of 5G for enhanced data speeds and low latency.

The development of 5 G-enabled smartphones has necessitated a redesign of hardware and software architectures to meet the performance and connectivity demands of next-generation networks. These devices integrate advanced chipsets, multi-band antennas, optimized battery systems, and seamless cloud interaction to enable real-time applications. Figure 1 provides an overview of the key functional components within a 5G smartphone ecosystem.

Figure 1. Architecture of a 5G smartphone and its supporting components

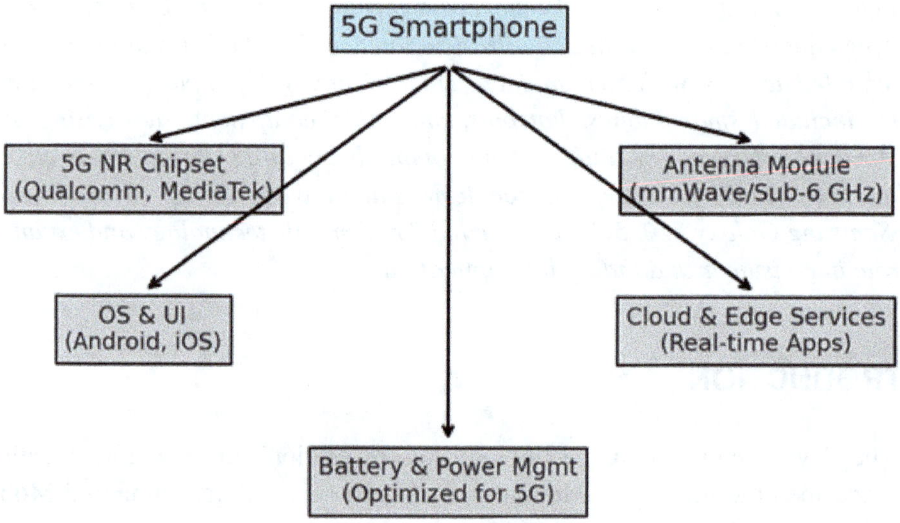

HMD Global further expanded the accessibility of 5G with the Nokia 8.3, which boasted broad compatibility with 5G bands from 600 MHz to 3.8 GHz. This midrange model highlighted how 5G technology became increasingly accessible to consumers, catering to diverse price points and geographic markets. As smartphones evolve alongside 5G networks, their role in driving innovation and connectivity across industries becomes ever more critical, paving the way for new applications and unparalleled user experiences. The authors argue that 5G technology can significantly enhance logistics operations by improving real-time data exchange, automating processes, and increasing overall system connectivity —essential components of modern supply chain efficiency (Lagorio et al., 2023).

TRACKING OF INTERNATIONAL 5G UNIT LAUNCHES

In 2019, the Global Mobile Suppliers Association presented its first data source tracking 5G unit launches internationally. The GSA identified twenty-three vendors with the accessibility of impending 5G devices, encompassing thirty-three different devices, including regional variants. There were seven announced 5G devices from elements: telephones (×12 products), hotspots (×4), outdoor and indoor customer premises equipment (×8), modules (×5), Snap-on dongles, adapters (×2), and with USB terminals (×1). By the end of 2019, the number of announced 5G deployments had risen to 129 from 56 vendors.

At the beginning of 2019, four certified 5G modem chipsets and one commercial processor were available in the 5G IoT chipset arena, with additional launches expected shortly.

On March 6, 2020, the first 5G smartphone, Samsung Galaxy S20, was released. According to Business Insider, the 5G characteristic is more expensive than the 4G; the lineup begins at US$1,000, compared with the Samsung Galaxy S10e, which starts at US$750.On March 19, HMD Global, the current developer of Nokia-branded cell phones, announced the Nokia 8.3, which is said to offer a broader range of 5G compatibility than any other phone released (HandWiki, 2025). *The Nokia 8.3 was among the first 5G smartphones to offer compatibility with various frequency bands, enhancing accessibility (Nokia Corporation, 2025).* The mid-range design supports 5G bands from 600 MHz to 3.8 GHz, with a significant expense in the Eurozone.

Mobile Connectivity: Primarily by Devices Like Smartphones

One of the primary applications of 5G is mobile connectivity, with smartphones serving as the most common access point. In addition to smartphones, 5G connects "less mobile" devices, such as 5G modems and mobile hotspots. These compact wireless access systems connect to 5G networks and then share connectivity with other devices via Wi-Fi.

A third and increasingly important category is 5G Fixed Wireless Access (FWA). *These systems utilize antennas mounted on windows or exterior building walls to provide broadband service to homes or small businesses, offering performance comparable to that of wired connections.* FWA is particularly promising in underserved or rural areas, where traditional broadband infrastructure is limited. According to the Ericsson Mobility Report (Ericsson, 2020), 5 G-powered FWA delivers a cost-effective broadband solution for such regions.

These devices operate across the entire spectrum of cellular radio frequencies:

- Low-band (sub-1 GHz, e.g., 700 MHz),

- Mid-band (1.6–3.8 GHz), and
- Millimeter wave (mmWave) bands (e.g., 28 GHz).

While smartphones, modems, and hotspots primarily utilize low- and mid-band frequencies, 5G FWA devices often depend on mmWave technology, which offers significantly higher bandwidth. However, mmWave signals have poor penetration through walls and glass, necessitating the installation of external antennas for optimal signal reception.

The 5G Smartphone Context

The transition to 5G has introduced a new level of complexity and innovation in smartphone design. Compared to 4G devices, 5G smartphones require specialized hardware, including advanced modems and high-performance antennas that support both sub-6 GHz and mmWave frequency bands.

These technical requirements bring significant engineering and cost challenges. Antenna design, in particular, is more complicated due to the higher frequencies used, and the cost of 5G components can be $40 to $50 higher than those used in 4G models. However, advances in chip design and battery efficiency have helped mitigate concerns about power consumption, thereby improving the overall user experience.

Designing a 5 G-ready handset involves major adaptations in two key areas:

- The radio modem is responsible for establishing a connection to the network.
- The antenna system must accommodate the directional and frequency-specific demands of 5 G.

This section highlights the technical intricacies and market implications of 5G smartphones, which are at the heart of consumer adoption and the broader success of the 5G ecosystem.

Mobile Network Architecture

The architecture of 5G networks represents a significant departure from previous generations, delivering capabilities such as ultra-low latency, massive device connectivity, and high-speed data throughput. The network is broadly divided into:

- The Radio Access Network (RAN) typically consists of gNodeBs that connect user devices to the core.
- The 5G Core Network (5GC).

- Integration with Multi-Access Edge Computing (MEC) for decentralized processing.

5G is built upon a Service-Based Architecture (SBA), which decouples control plane and user plane functions. This modular approach supports advanced features, including network slicing, virtualization, and edge computing, enabling the delivery of tailored services for various industries and applications.

According to Gupta and Jha (2015), this flexible architecture enables support for the three core 5G use cases:

- Enhanced Mobile Broadband (eMBB)
- Ultra-Reliable Low Latency Communications (URLLC)
- Massive Machine-Type Communications (mMTC)

In this context, security and resilience are also emphasized. Advanced techniques such as multi-layer attack graph analysis are used to dynamically detect and mitigate vulnerabilities in the 5G edge computing environment (Kholidy, 2021).

Table 1 provides an overview of the main components of the 5G mobile network architecture, highlighting their functions and the key innovations introduced with 5G technology.

Table 1. Key components of 5 G mobile network architecture

Main Component	Function	Key Innovations
User Equipment (UE)	Devices connected to the network (e.g., smartphones, IoT sensors)	Supports 5G NR, low latency, energy optimization
gNodeB (RAN)	Acts as the interface between user devices and the core network	Massive MIMO, beamforming, mmWave compatibility
5G Core (5GC)	Manages the control and routing of network communications	Service-based architecture (SBA), network slicing, virtualization
MEC (Multi-access Edge Computing)	Processes data near the end-user to reduce latency	Ideal for URLLC applications, edge-based service execution
Network Slicing	Creates virtual network segments for specific use cases	Service isolation, resource optimization, and tailored SLAs
Virtualization & Cloud-Native Infrastructure	Deploys network functions on generic hardware	Dynamic orchestration, scalability, and agility

THE BUILDING OF 5G SMARTPHONES

The development of 5G smartphones represents a complex convergence of advanced hardware, network standards, and power efficiency strategies. At the core of these devices, the modem is typically integrated onto the same chip as the processor. In comparison to 4G smartphones, which already require tightly engineered components, 5 G-ready devices introduce higher costs and design challenges.

A bundled 4G chip for a premium smartphone model in earlier generations already carried a notable price tag. However, 5 G-enabled components—especially those released in early 2020—further increased the hardware cost. A leading chipset manufacturer announced that its 5G system-on-chip (SoC) was ready for deployment in 2019. However, supply constraints indicated that full availability might not be realized until the second half of the year.

Challenges in 5G Smartphone Design

One of the most technically demanding aspects of 5G smartphone development lies in antenna design. Unlike 4G, 5G relies on dual-band communication, operating at sub-6 GHz frequencies as well as high-frequency millimeter-wave (mmWave) bands, such as 28 GHz. The use of mmWave necessitates narrow beams and high-gain antenna arrays, making antenna engineering significantly more complex than in 4G systems.

In 2018, the antenna and front-end module in a premium 4G smartphone cost around US$20. The comparable 5G modules released in 2019 carried higher costs, contributing to a total component cost increase of US$40–$50 per device compared to equivalent 4G handsets. These cost increases are partly offset by improved power management. Unlike during the 4G transition, battery life is no longer a major limiting factor for early 5G phones. Chip manufacturers have indicated that the battery performance in the initial 5G devices will be comparable to that of 4G equivalents.

5G Modems and Hotspots

The evolution of modems and mobile hotspots provides an essential bridge between 4G and 5G deployments. The first 4G networks were launched in December 2009. By 2019, global 5G deployments began to accelerate, particularly in urban centers, as device manufacturers introduced the first wave of 5 G-compatible smartphones, (Stewart, 2018).

Initially, 4G hotspots—small, puck-sized devices—were sold for US$200–300, but their prices quickly declined to under US$100. These lightweight devices allowed users to share internet access with phones, tablets, and laptops. However, as more smartphones gained built-in 4G hotspot capabilities, the demand for standalone modems diminished.

Looking ahead, 5G equivalents of these hotspots are expected to play a transitional role. While full-featured 5G smartphones remain expensive and are not yet widely accessible globally, 5G modems can offer a cost-effective interim solution. Leading U.S. carriers have emphasized plans to launch modems ahead of smartphones, anticipating growing demand from both enterprises and consumers.

Modems are fundamentally simpler than smartphones—they consist of a radio, antennas, and a battery. They do not include displays, cameras, or premium casings, which significantly reduces their cost. Even as smartphones eventually dominate the market, modems will remain integral to the early adoption of 5G, especially in enterprise and fixed wireless contexts.

One important distinction is that while a 5G modem connects to the network via 5G, the connected device (e.g., a laptop or phone) communicates via Wi-Fi or other local wireless protocols, which can reduce the effective data rate due to local bandwidth limitations.

In recent research, Tao and Chen (2022) demonstrated that 5G New Radio (NR) signals, when processed using fingerprinting techniques such as K-Nearest Neighbor (KNN) and Convolutional Neural Networks (CNN), can support accurate indoor localization. CNN models were particularly effective in both static and dynamic environments, reinforcing the multi-functional potential of 5G smartphones beyond connectivity alone.

THE 5G FIXED WIRELESS ACCESS (FWA) PRODUCTS

Fixed Wireless Access (FWA) represents a transformative application of 5G technology, delivering high-speed broadband to fixed locations through mobile networks. This approach bypasses traditional wired infrastructure, offering an efficient solution for underserved areas. The following section explores its definition, growth drivers, and production outlook.

Definition of FWA

Fixed Wireless Access (FWA) provides broadband connectivity using mobile network-enabled customer premises equipment (CPE). *This includes various device types such as indoor (desktop or window-mounted) and outdoor (rooftop or wall-mounted) units*. It excludes portable, battery-powered devices, such as Wi-Fi routers or USB dongles.

FWA enables wireless broadband data transmission between two fixed locations, serving as an alternative to traditional copper or fiber infrastructure. It connects subscribers to communication services, such as the internet or telephony, through wireless links established by Customer Premises Equipment (CPE) devices (Ericsson, 2025). The improved design of termination points at the user premises has enhanced performance and service reliability.

Primary Factors Driving the FWA Market

According to Ericsson (2020), the FWA market is expanding rapidly due to three primary factors:

- The increasing demand for digital solutions among consumers and enterprises drives the need for reliable broadband connectivity.
- Technological advances and the broader availability of spectrum in 4G and 5G networks make FWA a cost-efficient alternative to DSL, fiber, or cable in areas lacking adequate infrastructure. Enhanced network performance enables operators to deliver more data at lower per-gigabit costs.
- Government initiatives and funding programs that prioritize broadband deployment to support national digital agendas and economic development.

According to a recent global study, 185 out of 309 mobile operators have introduced FWA services, nearly doubling since late 2018. However, differences in how FWA is defined and reported across regions lead to variations in global connection estimates.

Production Estimates and Market Impact

Production of FWA devices is expected to triple by 2025, reaching nearly 160 million units. By 2019, FWA already accounted for around 50% of global mobile data traffic. Forecasts suggest that by 2025, FWA data volumes could reach 53 exabytes annually, representing 25% of total mobile network traffic.

FWA is now widely recognized as a scalable, economical solution to bridge the digital divide. In areas where wired broadband is either unavailable or impractical, FWA powered by 5G delivers high-speed internet for both residential and enterprise use. It also supports a growing range of applications, including IoT deployments and innovative services.

As reported in the *Ericsson Mobility Report* (Ericsson, 2020), FWA continues to play a pivotal role in enabling connectivity across underserved regions, making digital access more equitable and supporting global digital transformation.

Figure 2. Fixed Wireless Access (FWA) products and their role in expanding 5G broadband access

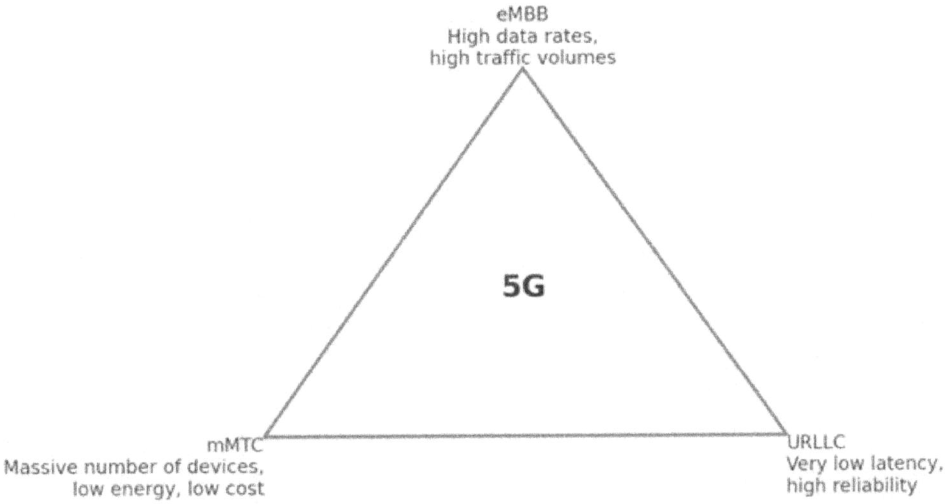

OVERVIEW OF 5G APPLICATIONS

5G is transforming the digital landscape by enabling smart cities, connected schools, and intelligent homes. As a next-generation cellular network, 5G introduces a paradigm shift in mobile communications, allowing applications and services that were once unimaginable. At the core of this transformation is a dense network of small cells that replaces the broader cell structure of earlier generations, improving both speed and capacity.

Al-Dujaili and Al-dulaimi (2022) provide a detailed analysis of 5G's technical foundation, tracing its evolution and identifying the architectural features, such as smart antennas, small cells, and virtualization, that enable enhanced data throughput and greater spectrum efficiency. Their work also highlights the challenges still facing 5G, including infrastructure rollout, service optimization, and system integration, while offering strategies for overcoming these barriers. Despite significant progress, the authors emphasize that continued research is crucial to leverage 5G's capabilities fully.

From a functional standpoint, 5G relies on radio waves to transmit digitized voice and data from wireless devices to local antennas, which are connected to the broader network via high-capacity fiber or wireless backhaul links. As devices move between cells, the network ensures seamless handoff, maintaining connectivity without interruption.

The capacity of 5G networks far exceeds that of 4G, with support for up to one million devices per square kilometer, ten times the density of 4 G. New 5G devices often retain 4G LTE compatibility, falling back on it in areas where 5G access is limited.

Telecom providers, such as Verizon, are deploying millimeter wave (mmWave) technologies in urban areas. Operating at high frequencies, mmWave enables ultra-fast data rates but has a short transmission range and limited wall penetration. These limitations necessitate the use of smaller antennas and a more densely packed network layout. Antennas just a few centimeters in size are strategically placed to support mmWave transmission.

Massive MIMO (Multiple Input, Multiple Output), first implemented in 4G systems, plays a crucial role in 5G performance. Using arrays of 32 to 128 antennas per cell, it enables simultaneous transmission of multiple data streams. Beamforming enhances this capability by directing radio signals through the most efficient paths, using phased arrays to ensure targeted, high-speed connectivity to individual devices.

Together, these technologies create a robust and flexible foundation for 5G applications, ensuring reliable performance across a wide range of environments and use cases.

Application Areas

The International Telecommunication Union Radiocommunication Sector (ITU-R) outlines three core application domains for 5G's advanced capabilities: Enhanced Mobile Broadband (eMBB), Ultra-Reliable Low-Latency Communications (URLLC), and Massive Machine-Type Communications (mMTC). While commercial rollout began with eMBB in 2020, full-scale deployment of URLLC and mMTC remains limited in most regions.

- Enhanced Mobile Broadband (eMBB) builds upon 4G LTE by offering faster data speeds, increased capacity, and enhanced user experiences in high-traffic environments, such as urban centers and stadiums. With faster data rates and more efficient spectrum utilization, eMBB supports bandwidth-intensive applications, including high-definition streaming, cloud gaming, and immersive virtual or augmented reality experiences. Basu and Das (2021) emphasize that 5G's low latency is crucial for real-time processing in VR applications, while Choudhury and Banerjee (2020) highlight 5G's role in enabling immersive AR/VR environments. For instance, A cloud-assisted key agreement protocol has been proposed to enhance security and privacy within E-healthcare systems, leveraging cloud infrastructure to facilitate efficient key exchanges, (Keshta, 2025).

- Massive Machine-Type Communications (mMTC) refers to 5G's ability to connect vast numbers of low-power devices, supporting the rapidly expanding Internet of Things (IoT). *With a capacity for up to one million connected devices per square kilometer (Basu & Das, 2021), mMTC enables smart city infrastructure, agriculture sensors, industrial monitoring, and more.* While many IoT devices still use low-cost Wi-Fi, 5G is increasingly applied in critical scenarios such as drones assisting in disaster response by delivering real-time data to emergency personnel, (FlytBase, 2025).

Modern vehicles also demonstrate a practical application of mMTC and eMBB, with most models equipped with 4G or 5G connectivity for services such as navigation, diagnostics, and infotainment. However, autonomous vehicles are not entirely dependent on 5G due to the necessity of functioning in low-coverage areas, (RantCell, n.d.).

- Ultra-Reliable Low-Latency Communications (URLLC) is designed for mission-critical applications that require extremely low latency and high reliability, such as autonomous vehicle control, industrial automation, and remote surgery (Zaol et al., 2024). While 5G promises wireless support for remote medical procedures, current implementations often still rely on fiber-optic links for the speed and reliability required in surgical environments (AT&T Business Editorial Team, 2018).
- As 5G evolves, its application areas will continue to diversify across industries. However, security remains a growing concern. Ahmad et al. (2019) note that the expanded use of technologies like Software-Defined Networking (SDN), Network Functions Virtualization (NFV), and cloud platforms increases exposure to cyber threats. Their work outlines vulnerabilities and emerging strategies to fortify 5G systems, underscoring the need for resilient infrastructure in the face of rising complexity.

Manufacturing

The convergence of 5G, AI, and IoT can completely transform factory floors. Beyond predictive maintenance, which helps control costs and reduce downtime, factories also utilize 5G to maintain and evaluate manufacturing tasks accurately. Choudhury et al. (2020) explain that the integration of 5 G with IoT devices transforms healthcare, particularly in remote patient monitoring.

With the connectivity boost offered by 5G, companies can also streamline conventional quality assurance procedures using sensor technology and AI.

In logistics and shipping, monitoring inventory is expensive, time-consuming, and complex. 5G presents the possibility for more significant interaction amongst cars, involving infrastructure and vehicles. According to Lagorio et al. (2023), successfully deploying 5G in logistics necessitates overcoming critical challenges, including robust infrastructure requirements, data protection concerns, and high implementation costs.

Fleet monitoring and course plotting become considerably easier at scale with 5 G. Additionally, driver navigation is integrated with an augmented reality system that identifies and flags potential dangers without diverting the driver's attention from the path. Gupta and Jha (2015) emphasize the scalability of 5G networks, which is crucial for addressing the increasing demand for device connectivity.

Healthcare

5G healthcare utilizes cases that enable patients and doctors to stay more connected. Wearable devices could alert healthcare providers every time an individual experiences symptoms, much like an inner defibrillator, which instantly alerts a group of ER cardiologists prepared for an incoming individual with an extensive history of information collected by the unit.

Retail

For 5G list applications, the buyer experience will be everything. Retailers of tomorrow might not look like present-day aisles of stocked shelves. Instead, imagine a showroom that allows you to add items to a virtual cart, rather than shopping with an actual one.

Stores might also utilize 5G to manage inventory stocking on time, which is accurate. Consumers may even see modifications like cashier-less stores that track everything placed in your cart, rather than the standard checkout line.

Agriculture

Farms of the future will use far more data and fewer chemicals. With information from receptors in fields, growers can pinpoint which areas require clean water, have an ailment, and need pest control.

As wearables become more affordable and 5G facilitates the scaling of networks that contain large numbers of IoT systems, monitoring livestock's wellness is also a viable option. With more accurate health information, growers can reduce antibiotic usage without compromising the security of the food source.

Virtual Reality and Gaming

For gamers, 5G promises a far more immersive long-term experience. High-definition live streaming experiences are expected to see a massive increase with 5G speeds; moreover, 5G gaming is closely tied to products with adequate computing power. Instead, the cloud enables retrieval, storage, and processing (even though a mobile unit displays and manages the game).

THE 5G INDUSTRY STANDARDS, DEPLOYMENT, AND SPECTRUM ANALYSIS

The progress of 5G technology is driven by three critical pillars: the establishment of industry standards, efficient spectrum allocation, and strategic deployment methodologies. These strategic elements play a crucial role in shaping the performance, scalability, and global adoption of 5G networks. Industry standards are essential for ensuring interoperability and integration across a wide range of devices and infrastructures, enabling a cohesive and functional ecosystem.

This section provides an in-depth review of these core components, emphasizing that their alignment is crucial for the global deployment of 5 G.

Spectrum Allocation

Equally important, spectrum allocation determines the effectiveness of coverage and the capacity of 5G services. Allocating the right frequency bands—low, mid, and high—is vital for supporting diverse use cases, from widespread rural coverage to ultra-high-speed urban connectivity. Meanwhile, strategic deployment approaches, including phased rollouts and infrastructure planning, are key to unlocking 5G's full transformative potential. These strategies support the revolution of industries, enhance global connectivity, and serve as catalysts for technological innovation.

Coordinated Efforts in Standardization

Coordinated efforts in standardization, spectrum management, and strategic rollout are crucial to unlocking 5G's full potential and realizing its long-term economic and societal benefits. These critical factors play a vital role in shaping the performance, efficiency, and broad scope adoption of 5G networks. Industry standards ensure interoperability and seamless integration across a diverse range of devices and infrastructures, fostering a unified and scalable ecosystem. Effective spectrum allocation is essential for achieving optimal coverage and network capacity, supporting various use cases across urban, suburban, and rural areas. Additionally, strategic deployment approaches unlock the full transformative potential of 5G, driving industrial innovation, enhancing global connectivity, and accelerating digital advancements. This section explores the foundational elements that support the successful implementation of 5G on a global scale.

5G Industry Standard, Deployment, and Spectrum Analysis: The advancement of 5G technology is propelled by the development of industry standards, spectrum allocation, and strategic deployment processes.

Industry Standard

The business consortium setting standards for 5G is the 3rd Generation Partnership Project (3GPP) (3GPP, 2020). *It describes some methods utilizing 5G NR (5G New Radio) applications as "5G", a standard definition in late 2018.*

The International Telecommunications Union (ITU) establishes the minimum standards. Previously, several reserved the phrase "5G" for devices that provide download speeds of 20 Gbit/s, as specified in the ITU's IMT-2020 document. The introduction of 5G chipsets, like the Qualcomm Snapdragon X55, has enabled devices to achieve unprecedented speed and efficiency (Qualcomm, 2017).

At first, the word involved the International Telecommunication Union's IMT 2020 standard format, which required a theoretically good download speed of twenty gigabits per second and a ten-gigabits-per-second upload speed We have among other requirements.5G technology, the fifth generation of cellular standards, delivers significantly higher data speeds, and lower latency, and the ability to connect a vast number of devices simultaneously (Everipedia, 2018).

Several telecom operators completed the first stage of 3GPP 5G specifications (Release 15) in 2019, and Release 16 is planned for 2020.

Nevertheless, the speed and latency in initial FR1 deployments using a 5G NR application on 4G hardware (non-standalone) are somewhat better than those of brand-new 4G methods, estimated at 15 to 50% better (Sauter, 2021). Omar and

Ali (2022) argue that deploying small cells is crucial for achieving the promised speeds of 5 G in urban settings.

Standards for Numerous Parts of 5G

IEEE addresses numerous aspects of 5G, with a primary focus on the wireline areas between the Remote Radio Head (RRH) and Baseband Unit (BBU). *The standard 1914.1 target divides the network structure into two critical areas: the Baseband Unit (BBU) and the Remote Radio Unit (RRU).*

The radio Unit (RU) to the distributor Unit (DU) becomes the NGFI I (Next Generation Fronthaul Interface), and the DU on the central unit (CU) serves as the NGFI II user interface, enabling a cost-effective and diverse community. NGFI-II and NGFI-I have defined performance values developed and outlined by the ITU. 1914.3 standard is developing an innovative

Based on the purposeful split, the Ethernet frame format can have IQ information in a more practical option—this 3GPP Ethernet frame format characterizes functional breaks. Additionally, multiple network synchronization requirements within the IEEE groups update norms and standards to ensure community timing precision in the RU and the necessary quality. Yang, Chen, Liu, and Zhu (2024) propose a dynamic feature-fused localization (DFF-Loc) framework that integrates 5G NR synchronization signal blocks (SSB) with Wi-Fi signals, significantly enhancing indoor positioning accuracy through multivariate fingerprinting and dynamic signal processing. The authors report that DFF-Loc achieved average positioning accuracies of 1.87 m, 3.97 m, and 1.23 m in various indoor scenarios, outperforming traditional 5G- or Wi-Fi-only methods by 28.23% and 20.38% respectively, thus demonstrating its robustness for real-world applications (Yang et al., 2024).

A Progressive Journey

The development of 5G standards has been a progressive journey shaped by key milestones established by the 3rd Generation Partnership Project (3GPP). *This evolution began with early study items and gradually advanced to the formulation of 5G New Radio (NR) specifications, leading to commercial deployments.* Notable innovations, including the implementation of standalone (SA) and non-standalone (NSA) modes, have been instrumental in enabling 5G to deliver enhanced performance, flexibility, and a broader range of applications. Parallel efforts to enhance LTE technology ensure seamless integration with the wider 5G platform. Qualcomm highlighted that these advancements reflect a coordinated effort to establish a robust global framework for 5G connectivity.

SPECTRUM AND THE 5G

Spectrum is a foundational component of wireless communication networks, encompassing the radio frequencies that power technologies like mobile phones, Wi-Fi, and broadcast television. For 5G networks to fulfill their promise of ultra-fast, reliable, and low-latency connectivity, they rely on a strategic mix of spectrum bands, each with distinct characteristics suited to specific applications.

5G spectrum is typically divided into three primary bands:

- Low-band spectrum (sub-1 GHz), such as 600 MHz or 700 MHz, offers broad coverage and intense penetration through obstacles like walls. It is ideal for rural and suburban areas where range is more important than speed.
- The mid-band spectrum (1–6 GHz), including the 3.5 GHz range, offers a balance between coverage and performance. It supports higher speeds than low-band and can still cover considerable distances, making it a key layer in 5G deployments.
- The high-band spectrum, also known as the millimeter wave (mmWave) spectrum (above 24 GHz), delivers extremely high speeds and massive capacity, but with limited range and poor penetration. It is well-suited for dense urban areas and specialized applications such as stadiums, airports, and industrial automation zones.

The U.S. Federal Communications Commission (FCC) and other regulatory bodies globally have emphasized the importance of a diverse spectrum portfolio to support 5G's broad range of use cases (FCC, 2018). This layered approach ensures that both coverage and capacity challenges are addressed across various geographic and industrial settings, enabling 5G to serve a wide range of applications, from remote farms to smart factories and metropolitan centers.

Spectrum at Lower and High Frequencies

Spectrum at lower frequencies, particularly under one GHz, allows 5G coverage. Spectrum at higher frequencies with relatively large bandwidths below 6 GHz provides the necessary capacity to support many devices.

The spectrum of high frequencies, above 24 GHz, with huge bandwidths, provides ultra-high speed and low latency. Cells at these frequencies have a smaller coverage radius (between 50 and 200 m), and 5G networks in millimeter-wave bands focus on areas of high traffic demand or specific locations or premises requiring services with very high capacity and peak data rates (Gbps) (Sauter, 2021).

Among the various issues are the vast number of new stereo spectrum (5G NR frequency bands). *For instance, in July 2016, the U.S.* Federal Communications Commission (FCC) freed up vast quantities of bandwidth in underused high-band spectrum for 5 G.

The Spectrum Frontiers Proposal (SFP) doubled the millimeter-wave unlicensed spectrum to 14 GHz, four times the supply and mobile use spectrum the FCC had licensed to date [xv]. In 2018, European Union regulators agreed to start the 3.6 and twenty-six GHz bands by 2020.

The Three Primary Frequency Ranges

The radio spectrum for 5G is divided into three primary frequency ranges—low, mid, and high bands—each serving distinct applications. Low-frequency bands, such as 700 MHz, offer extensive coverage and deep signal penetration, making them ideal for large-scale events and densely populated areas where reliable and widespread connectivity is crucial. Mid-frequency bands, generally ranging from 3.4 to 3.8 GHz, offer a balanced compromise between coverage and capacity, making them particularly valuable for applications like vehicle-to-everything (V2X) communication, environmental monitoring, and the deployment of smart city infrastructure.

High-frequency millimeter-wave (mmWave) bands, such as 26 GHz, support ultra-high-speed data transmission and are ideal for enhanced residential broadband, advanced energy management systems, and high-capacity networks in densely populated urban areas. This layered spectrum strategy ensures that 5G caters to diverse needs, enhancing efficiency and connectivity across various sectors.

The Spectrum Frontiers Proposal (SFP) of the USA

The Spectrum Frontiers Proposal (SFP), introduced by the U.S. Federal Communications Commission (FCC), marked a significant advancement in unlocking millimeter-wave (mmWave) spectrum for 5G applications. By expanding the unlicensed mmWave spectrum to 14 GHz, the SFP provided four times the total mobile-use spectrum previously licensed by the FCC. This significant increase was designed to accommodate the growing demand for ultra-high-speed wireless communication and enable 5G networks to deliver fiber-optic-level performance over wireless links.

The proposal's impact extended across various sectors and use cases. Entrepreneurs and established carriers were granted access to this newly available bandwidth, helping to fulfill the core promise of 5G: ultra-fast connectivity with data rates 10 to 100 times faster than 4G LTE. Some of the key beneficiaries and applications include:

- Public initiatives, such as New York City's LinkNYC, deploy high-speed Wi-Fi kiosks throughout urban areas.
- Wireless carriers are providing gigabit-per-second download speeds to mobile users in densely populated metropolitan areas.
- Affordable delivery of high-bandwidth services—both mobile and fixed—for underserved and rural communities.
- Vehicle-to-vehicle communication systems supporting collision avoidance technologies that require near-zero latency.
- Health applications enabling remote diagnosis and real-time transmission of high-resolution medical imaging (e.g., CT scans).
- Support for the exponential growth of connected devices, with projections of over 50 billion IoT systems and wearables globally by 2020.
- High-density venues, such as stadiums, where early 5G adoption began as early as 2017, require managing vast volumes of simultaneous device connections.

The SFP was a pivotal move in securing the spectral foundation necessary for scaling 5G innovation and meeting future digital infrastructure needs in both urban and remote areas.

The FCC's Spectrum Frontiers Report and Order (R&O) outlines key frequency bands allocated for 5G deployment in the United States. These bands, spanning from 28 GHz to 71 GHz, support a diverse range of 5G use cases, including licensed high-capacity backhaul links and unlicensed short-range communications. Each band addresses different technical needs and regulatory structures, enabling more flexible and scalable wireless services. Table 2 summarizes the principal characteristics and applications of the spectrum bands included in the FCC's initiative.

Table 2. Bands Addressed by the FCC Spectrum Frontiers R&O

28 GHz Band (27.5-28.35 GHz)	Currently licensed for Local Multipoint Distribution Service (LMDS)operations. Existing licensees receive two 425 MHz authorizations on a county-by-county basis, indexed to their current 850 MHz licenses, issued on a Basic Trading Area (BTA) basis.	Used primarily for LMDS services, this technology enables broadband wireless access.
39 GHz Band (38.6-40 GHz)	They are currently licensed for fixed microwaves (point-to-point and point-to-multipoint operations). Existing licensees can repack the band to create 200 MHz-wide channels (rather than the current 50 MHz-wide channels). Licenses are reissued on a Partial Economic Area (PEA) basis.	It supports high-capacity fixed wireless links that are commonly used in backhaul networks.
37 GHz Band (37-38.6 GHz)	There are no current non-federal terrestrial operations in the T-band. The lower segment (37-37.6 GHz) will be made available on a shared basis to both federal and non-federal users. The government auctioned the upper part (37.6-38.6 GHz) in 2000 MHz blocks on a PEA basis.	Future shared-use spectrum for both federal and non-federal users.
64-71 GHz Band	The band is available for unlicensed use, and the same rules (Part 15) apply to the unlicensed 57- 64GHz band.	Supports short-range, high-speed wireless communication, including WiGig and other unlicensed technologies.

Spectrum In EU Countries

EU nations and lawmakers have struck a deal to regain radio frequencies that have been valid for twenty years for next-generation 5G solutions. The bloc is trying to catch up with the U.S. and China in developing internet-connected products (Vinocur, 2018).

Telecoms lobbying group ETNO, whose members include Telecom Italia, Telefonica, and Orange, criticized the offer, thinking it lacked ambition. The requirements have a timeframe of at least twenty-five years.

The temporary contract was signed late on Thursday after weeks of squabbling between EU governments, who sought to protect a key revenue source for cash-strapped telecom providers, and revenue seekers seeking new ways to increase earnings.

Next-generation 5G services promise quicker video, faster phones, and improved connectivity for automobiles, as well as internet-connected manufacturing sensors, within the next ten years. Nair and Prasad (2021) explain that 5G's high bandwidth enables seamless video streaming even in high-density areas.

Under the agreement, the bloc will make 5G available by 2020 and guarantee the spectrum for at least twenty years.

In Europe, the 37–43.5 GHz frequency range accommodates a diverse mix of terrestrial and satellite services, reflecting a carefully structured balance between mobile broadband needs and scientific, broadcasting, and satellite operations. The bands serve fixed, mobile, and satellite communications, while also supporting space research and Earth exploration services. Some allocations are primary, while others operate on a secondary or shared basis, ensuring spectrum efficiency and coexistence. Table 3 details the specific services assigned across this mmWave range.

Table 3. Services allocated to 37-43.5 GHz in Europe

37.0 – 38.0 GHz	Space Research Service
37.0 – 39.5 GHz	Fixed and Mobile Service
37.5 – 39.5 GHz	Fixed Satellite Service
37.5 – 39.5 GHz	Earth Exploration Satellite Service, on a secondary basis
37.0 – 40.5 GHz	Earth Exploration Satellite Service, downlink operations, on a secondary basis
37.5 – 42.5	Fixed Satellite Service, downlink operations
39.5 – 40.5 GHz	Mobile Satellite Service, downlink operations
40.0 – 40.5 GHz	Earth Exploration Satellite Service, uplink operations Space Research Service, uplink operations
40.5 – 42.5 GHz	Broadcasting and Broadcasting Satellite Services
42.5 – 43.5 GHz	Fixed Satellite Service, uplink operations, Radio Astronomy Service

Source: 5G America

Spectrum and Scenario of IMT 2020 and Beyond

According to Recommendation ITU-R M.2083, titled *IMT Vision – Framework and Overall Objectives of the Future Development of IMT* for 2020 and Beyond, three primary usage scenarios have been defined for 5G: Enhanced Mobile Broadband (eMBB), Ultra-Reliable and Low-Latency Communications (URLLC), and Massive Machine-Type Communications (mMTC) (ITU-R, 2015).

As 5G becomes available, buyers must consider its impact on efficiency, safety, and revenue. Increased visibility for fleet and cargo owners immediately translates to decreased losses and delays, ultimately saving money and time, and protecting the workers involved in cargo transport. In addition, superior solutions to complement 5G networks lead to more effective Internet business practices throughout logistics and supply chain companies. Fang and Li (2020) argue that 5G enables edge computing, which is crucial for reducing network congestion and latency.

ORGANIC EVOLUTION OF 4G/LTE TECHNOLOGIES

The enhanced mobile broadband internet platform may be the latest in the organic evolution of 4G/LTE technologies, which facilitate mobile broadband, except that the newest development is quick.

San Diego-based Qualcomm tests simulations of 5 G's real-world functionality in a few cities. It works with the current spectrum and cell site allocation for mobile networks.

Based on Qualcomm's simulation for San Francisco, subscribers may appreciate a significant increase in speeds (from 71 megabits per second (Mbps) for 4G to 1.4 gigabits per second for 5G) and download speeds (from 10 Mbps to 186 Mbps).

However, there is more: video streaming reaches an average quality of 8K resolution, 120 frames per second, 10-bit, and two-tone. This video streaming does not portend a simple enhancement in the press's streaming but unlocks real-world uses on a large scale. Virtual and augmented reality even gain from the growth of mobile broadband enhancements.

The future use of 5G spans a wide range of transformative applications, driven by improvements in performance over LTE and tailored to meet diverse technological demands. Enhanced Mobile Broadband (eMBB) supports high-bandwidth activities such as 3D video streaming, UHD screen usage, and cloud-based work and entertainment. Ultra-Reliable Low Latency Communication (URLLC) is designed for mission-critical applications.

This is particularly true for use cases such as self-driving cars, industrial automation, and augmented reality, where 5G ensures seamless, real-time interactions with high reliability. Massive Machine-Type Communication (mMTC) supports the proliferation of smart cities, intelligent buildings, and IoT ecosystems by enabling high device density and energy-efficient connectivity. Together, these capabilities underscore 5 G's pivotal role in driving technological advancements across industries and enhancing everyday life.

CONCLUSION

The advancement of 5G has significantly reshaped the smartphone sector, introducing powerful new dimensions of performance, speed, and functionality. The launch of 5 G-capable devices as early as 2019 marked a significant turning point in the evolution of mobile technology, signaling the beginning of a new era in connectivity and performance. This chapter has illustrated how flagship smartphones, such as the Samsung Galaxy S20, and mid-range models, like the Nokia 8.3, helped bring 5G connectivity to a wide range of consumers. With enhanced data speeds, reduced latency, and expanded spectrum support, these devices significantly elevated the mobile user experience. Beyond consumer applications, 5G has also unlocked new opportunities across various sectors, including logistics, manufacturing, and innovative infrastructure, driving innovation and operational efficiency on a broader scale.

However, this evolution has been accompanied by manufacturing and design challenges, including increased component costs and the complexity of antenna integration. Despite these obstacles, technological innovation in chip design and battery efficiency has helped manufacturers respond swiftly. Mobile devices have evolved beyond simple communication tools to become essential nodes within the broader 5G ecosystem, supporting cloud computing, edge processing, and real-time service delivery. As the chapter demonstrates, mobile architecture must evolve in tandem with network capabilities to realize the full benefits of 5G.

To summarize the key contributions of 5G in the context of smartphones, Table 4 presents a synthesis of the main topics discussed in this chapter. It outlines how 5G has impacted smartphone architecture, use cases, and industrial applications.

Table 4. Summary of 5G impacts on smartphone development and application

Aspect	Impact of 5G	Implications for Smartphones
Network Capability	High-speed, low-latency communication	Enhanced streaming, VR/AR, and real-time interaction
Architecture	Advanced chipsets, antenna design, sub-6 GHz, and mmWave bands	Complex design requirements and increased production costs
Use Cases	Cloud gaming, remote healthcare, and IoT integration	Expansion beyond traditional communication to multifunctional platforms
Accessibility	Mid-range models like the Nokia 8.3 support broadband bands	Wider market reach and global deployment
Device Role in Ecosystem	Integration with innovative infrastructure and edge computing	Smartphones as mobile control and data processing hubs

The evolution of 5G technology has marked a pivotal shift in the global smartphone industry, setting new benchmarks for connectivity and technological advancement. Since the rollout of 5G networks in 2019, manufacturers have responded rapidly by introducing a diverse portfolio of 5 G-enabled devices. Premium models, such as the Samsung Galaxy S20, alongside more accessible alternatives like the Nokia 8.3, reflect the industry's agility in addressing the dynamic demands of next-generation wireless communication. Spanning smartphones, mobile hotspots, and IoT modules, these devices highlight the adaptability and far-reaching impact of 5G in transforming the broader consumer electronics ecosystem.

The emergence of 5G chipsets has been instrumental in driving a transformative shift in the mobile industry, enabling manufacturers to incorporate advanced functionalities far beyond those of previous mobile technologies. With increased bandwidth, ultra-low latency, and broader application compatibility,5 G-enabled smartphones have become central to global digital transformation. Remote rendering in virtual reality can be effectively achieved using synchronous methods, which enhance the immersive experience by minimizing latency issues (Kelkkanen, Fiedler, Lindero, & Katchabaw, 2021).

Despite the initial challenges posed by higher production costs and premium pricing, consumer adoption of 5G devices has skyrocketed, reflecting the perceived value and long-term potential of the technology. As network infrastructure continues to expand and mature, market experts anticipate more competitive pricing, which will further accelerate the adoption of these technologies. The wide-ranging capabilities of 5G smartphones are already enabling smarter cities, more efficient industrial operations, and enhanced user experiences, highlighting their pivotal role in the evolving landscape of digital connectivity.

In conclusion, as emphasized by Legacy (NYC, 2025), the integration of 5G into smartphones represents more than a simple technological upgrade—it marks a fundamental transformation in how devices connect, communicate, and function within a digitally interconnected world. This chapter has illustrated the smartphone industry's remarkable agility in embracing 5G, showcasing its ability to innovate and adapt to the demands of an ever-evolving technological ecosystem. As 5G technology continues to advance, it will serve as a cornerstone of modern telecommunications, reshaping both consumer technology and the broader digital economy.

REFERENCES

Al-Dujaili, M. J., & Al-dulaimi, M. A. (2022). Fifth-Generation Telecommunications Technologies: Features, Architecture, Challenges and Solutions. *Wireless Personal Communications*. DOI: 10.1007/s11277-022-09962-x

AT&T Business Editorial Team. (2018). *5 Ways 5G will Transform Healthcare | Improving patient experience with personalized, preventative care*. AT&T Business. https://www.business.att.com/learn/updates/how-5g-will-transform-the-healthcare-industry.html

Ericsson. (2020). *3GPP releases 16 & 17 overview – 5G NR evolution.* Eriicsson.com. Retrieved from https://www.ericsson.com/en/reports-and-papers/ericsson-technology-review/articles/5g-nr-evolution

Ericsson. (2025). *Ericsson Fixed Wireless Access Handbook 2025. GSA*. https://gsacom.com/paper/ericsson-fixed-wireless-access-handbook-2025/

Everipedia. (2018). *5G*. Everipedia. https://everipedia.io/wiki/5g/

GPP. (2020). *About 3GPP*. 3gpp.org. https://www.3gpp.org/about-3gpp

Gupta, A., & Jha, R. K. (2015). A Survey of 5G Network: Architecture and Emerging Technologies. *IEEE Access : Practical Innovations, Open Solutions*, *3*(1), 1206–1232. DOI: 10.1109/access.2015.2461602

HandWiki. (2025). 5G. Handwiki.org. https://handwiki.org/wiki/5G

Keshta, I. (2025). A cloud-assisted key agreement protocol for the E-healthcare system. *PLoS One*, *20*(6), e0322313. DOI: 10.1371/journal.pone.0322313

Kholidy, H. A. (2021). Multi-Layer Attack Graph Analysis in the 5G Edge Network Using a Dynamic Hexagonal Fuzzy Method. *Sensors (Basel)*, *22*(1), 9. DOI: 10.3390/s22010009

Lagorio, A., Cimini, C., Pinto, R., & Cavalieri, S. (2023). 5G in Logistics 4.0: Potential applications and challenges. *Procedia Computer Science*, *217*, 650–659. DOI: 10.1016/j.procs.2022.12.261

Nokia Corporation. (2025). *Nokia 8.3 5G specifications. HMD - Human Mobile Devices.* https://www.hmd.com/en_us/nokia-8-3-5g/specs

Qualcomm. (2017). *Wireless Technology & Innovation | Mobile Technology | Qualcomm*. Qualcomm. https://www.qualcomm.com/

Sauter, M., & Wiley, J. (2021). *From GSM to LTE-Advanced Pro and 5G : an introduction to mobile networks and mobile broadband*. Wiley.

Tao, X., & Chen, L. (2022). *5G NR indoor localization with smartphones*. 25–25. https://doi.org/DOI: 10.1117/12.2640925

Yang, H., Chen, L., Liu, H., & Zhu, G. (2024). Dynamic Feature-fused Localization with Smartphones Exploiting 5G NR SSB and Wi-Fi for Indoor Environments. *IEEE Transactions on Instrumentation and Measurement*, *73*, 1–14. DOI: 10.1109/tim.2024.3352695

ADDITIONAL READING

Adebusola, J. A., Ariyo, A. A., Elisha, O. A., Olubunmi, A. M., & Julius, O. O. (2020, March 1). *An Overview of 5G Technology*. IEEE Xplore. https://doi.org/DOI: 10.1109/ICMCECS47690.2020.240853

Bhandari, N., Devra, S., & Singh, K. (2017). Evolution of Cellular Network: From 1G to 5G. *International Journal of Engineering and Techniques, 3*. https://oaji.net/articles/2017/1992-1515158039.pdf

Cortés-Polo, D., & Ignacio, L., José-Luis González-Sánchez, & Jesús Calle-Cancho. (2020). A Methodology for Network Analysis to Improve the Cyber-Physicals Communications in Next-Generation Networks. *Sensors (Basel)*, *20*(8), 2247–2247. DOI: 10.3390/s20082247

Deloitte. (2019). 5G: the new network arrives. The Deloitte Consumer Review. https://www.deloitte.co.uk/consumer-review-digital-predictions/assets/img/download/the-deloitte-consumer-review-digital-predictions-2019-5g-prediction-article.pdf

Ericsson (2025). *5G NR: The Next Generation Wireless Access Technology*. Ericsson.com. https://www.ericsson.com/en/reports-and-papers/books/5g-nr-the-next-generation-wireless-access-technology

Federal Communications Commission (FCC). *(2017). FCC Takes Next Steps on Facilitating Spectrum Frontiers Spectrum*. Fcc.gov. https://www.fcc.gov/document/fcc-takes-next-steps-facilitating-spectrum-frontiers-spectrum

GSMA. (2022). *5G mmWave Deployment Best Practices*. GSMA. https://www.gsma.com/futurenetworks/wp-content/uploads/2022/10/GSMA-5G-mmWave-Deployment-Best-Practices-White-Paper-Nov-2022.pdf

Intelligence, G. S. M. A. (2019). *The Mobile Economy 2019*. GSMA Intelligence. https://www.gsmaintelligence.com/research/the-mobile-economy-2019

Manivasagam, G., & Shukla, M. (2023). Massive MIMO in 5G Networks: Trends and Challenges. *2023 International Conference on Power Energy, Environment & Intelligent Control (PEEIC)*, 1456–1460. https://doi.org/DOI: 10.1109/peeic59336.2023.10450543

Sultan, A. (2022, August 8). *5G System Overview*. Www.3gpp.org. https://www.3gpp.org/technologies/5g-system-overview

Zhao, B., Yang, W., Du, W., Ren, Y., Sun, J., Wu, Q., & Zhou, X. (2024). A multipath scheduler based on cross-layer information for low-delay applications in 5G edge networks. *Computer Networks*, *244*, 110333. DOI: 10.1016/j.comnet.2024.110333

Chapter 3
Deployment of 5G

ABSTRACT

The deployment of 5G marks a major shift in telecommunications, turning years of planning into real-world innovation. More than just an upgrade, 5G introduces a new network architecture delivering ultra-fast speed, low latency, and improved service quality. It supports IoT, smart cities, autonomous vehicles, and AR, creating vast opportunities across sectors. Powered by small cells, edge computing, and network slicing, 5G meets varied use case demands. Chen and Zhao note that network slicing enables dynamic, tailored connectivity, optimizing resources for services like video surveillance, real-time data, and AI-powered applications.

INTRODUCTION

The deployment of 5G represents a groundbreaking transformation in telecommunications, turning long-term plans and research into a tangible reality. Shallow machine learning techniques are nearly as effective as deep learning models for critical 5G tasks. They offer much faster processing times, making them more suitable for real-time network applications (Teixeira et al., 2024). *After more than a decade of meticulous planning and development, 5G networks have moved from theoretical blueprints to limited service deployments worldwide.* This next-generation wireless technology introduces unparalleled advancements in connectivity, offering significantly faster speeds, reduced latency, and enhanced service quality compared to its predecessors. These benefits are poised to revolutionize industries, improve consumer experiences, and unlock new opportunities in the digital economy.

DOI: 10.4018/979-8-3373-0497-7.ch003

5G DEPLOYMENT ARCHITECTURE

The architecture of 5G deployment consists of a multi-layered network composed of various cell types, each optimized for different environments. From expansive macrocells that provide long-range connectivity to compact femtocells designed for indoor applications, each layer is crucial in delivering consistent and scalable service. This tiered approach enhances network capacity, reduces latency, and supports a wide range of diverse use cases. Figure 1 illustrates these cell types and their respective deployment characteristics in the 5G infrastructure.

Figure 1. Overview of multi-tier cell types in 5G deployment architecture

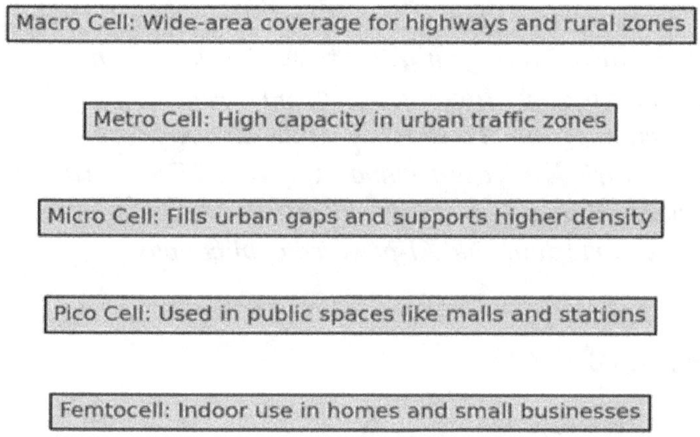

Testing and Measurement Setup for 5G Systems

Testing and measurement in 5G systems are crucial for validating performance, ensuring compliance with standards, and optimizing network efficiency. Due to the complexity of 5G technologies, including higher frequency ranges, massive MIMO, beamforming, and network slicing, a comprehensive testing setup is required. This involves components for signal generation, channel emulation, RF measurements, and end-to-end system validation. Proper testing enables vendors and operators to verify key performance indicators (KPIs) such as throughput, latency, coverage, and reliability across various deployment scenarios, ranging from lab simulations to live field trials.

Table 1. Key components of a 5G testing and measurement setup

Component	Function
Signal Generator	Generates 5G NR waveforms across FR1 and FR2 bands for RF testing.
Spectrum Analyzer	Measures signal strength, spectrum emissions, and interference.
Channel Emulator	Simulates real-world propagation environments including fading and mobility.
Network Emulator	Mimics 5G core and RAN functionalities to test user equipment.
Test UE (User Equipment)	Used to evaluate downlink/uplink performance under various conditions.
Over-the-Air (OTA) Chamber	Enables radiated testing of antennas, beamforming, and MIMO.
Data Analytics Tools	Analyze key performance indicators (KPIs), including throughput, latency, jitter, and packet loss.

The rollout of 5G goes beyond a simple upgrade; it represents a significant evolution in network architecture, incorporating a multi-layered and intricate framework. This advancement integrates cutting-edge technologies, including small cells, network slicing, and edge computing. For instance, combining artificial intelligence with 5G facilitates the development of more innovative IoT systems, enhancing operational efficiency. *These elements work harmoniously to enable seamless services across various domains and applications, from connected devices and intelligent urban infrastructure to self-driving vehicles and augmented reality.* Moreover, 5G networks introduce ultra-reliable low latency communication (URLLC) and massive machine-type communication (MTC), addressing the unique demands of diverse use cases. Integrating artificial intelligence with 5G facilitates the development of more innovative Internet of Things (IoT) systems. The integration of artificial intelligence, 5G, and IoT is driving the evolution of intelligent systems by enhancing their efficiency, intelligence, and responsiveness, enabling them to process massive data flows and make complex decisions in real-time (Adewale & Paul, 2024).

Global Rollout of 5G

The global rollout of 5G involves extensive collaboration between the public and private sectors, as well as rigorous coordination among spectrum regulators, network operators, and technology providers. Adopting a millimeter-wave spectrum is pivotal in achieving the high-speed capabilities of 5G networks. These initiatives aim to address challenges such as spectrum allocation, infrastructure investment, and regulatory compliance, which are crucial for achieving the widespread availability and scalability of 5G services.

As 5G deployment progresses, it will extend its impact beyond connectivity, shaping the future of industries, economies, and societies. Patwary et al. (2020) emphasize that while the infrastructure costs of 5G deployment pose a significant barrier—particularly in countries like the UK—the use of shared public infrastructure such as streetlights and telephone poles could reduce costs by up to 60%, potentially transforming both business models and stakeholder roles in the telecom industry.

Network slicing in 5G provides dynamic resource allocation, tailoring connectivity to specific use cases and applications. This chapter examines the complexities of 5G deployment, analyzing the technological innovations, industry collaborations, and regulatory frameworks that are driving this transformative era in telecommunications.

The promise of a new 5G deployment has transitioned from the drawing board to reality. Researchers and engineers planned and developed the succeeding generation of wireless technology a decade ago, which has begun limited service. 5G signal eventually brings quicker speeds per second, decreased latency, and service upgrades.

The fundamental architectural transformation that makes this deployment possible is a multi-layered and complex endeavor. The worldwide private and public networks, as well as the spectrum sectors, are working on an architectural change. The authors also suggest that simplifying infrastructure and enabling spectrum sharing could encourage the rise of localized networks, such as private, community, or micro-operator networks, thereby challenging traditional telecom business models and introducing new actors into the 5G ecosystem (Patwary et al., 2020).

Technical Deployment

The movable operator network structures individual networks by integrating communications, enterprise networking, and industrial IoT solutions. Shoaib et al. (2024) highlight that implementing 5G technologies—particularly those involving millimeter-wave communication and massive MIMO—faces significant measurement and modeling challenges, requiring advanced techniques such as recurrent neural networks and spectrogram-based learning to optimize performance in real-world scenarios.

Original 5G NR launches trusted current LTE (4G) infrastructure in a non-standalone (NSA) function (5G NR application on LTE stereo hardware) before maturing the standalone (SA) mode (5G NR software on 5G NR radio hardware) together with the 5G core system.

In 2019, the Global Mobile Suppliers Association identified 224 operators in eighty-eight countries that are trialing or testing, or are qualified to conduct field trials of 5G technologies, are deploying 5G networks, or have announced service launches. The equivalent amounts in November 2018 were 192 operators in 81 nations.

South Korea was the first nation to follow. The widespread deployment of 5G began in April 2019. Swedish telecom giant Ericsson predicted that the 5G web would cover as much as 65% of the world's population by 2025. Additionally, it plans to commit one billion reais (approximately $ 238.30 million) in Brazil to install a new assembly line dedicated to fifth-generation technology (5G) for its Latin American operations.

South Korea as a Pioneer Nation

When South Korea introduced its 5G network, most carriers utilized base facilities and equipment from Samsung, Ericsson, and Nokia, except for LG U+ Plus, which also used Huawei gear. As a result, Samsung was the most prominent provider of 5G base facilities in South Korea at launch, delivering 53,000 out of the 86,000 base stations installed nationwide.

The first moderately significant deployments occurred in April 2019. In South Korea, SK Telecom claimed 38,000 base stations, KT Corporation 30,000, and LG U+ 18,000, with 85% of these to be located in six major cities. They have a 3.5 GHz (sub-6) spectrum in non-standalone (NSA) mode, and evaluated speeds ranged from 193 to 430 Mbit/s. In the first month, 2,100, 60,000, and 1,000 devices were registered, and by the end of 2019, the total reached 4.7 million.

5G NR Equipment Manufacturers and Standardization

As of today, nine major companies are leading the manufacturing and supply of 5G radio hardware and network equipment for telecom operators: Altiostar, Samsung, Qualcomm, Nokia, Huawei, Ericsson, Datang Telecom/Fiberhome, Cisco Systems, and ZTE.

At the heart of the 5G wireless ecosystem lies 5G New Radio (NR), a radio access technology developed by the 3rd Generation Partnership Project (3GPP) to serve as the global standard for 5G air interfaces (3GPP, 2021). The technical framework for this new RAT is outlined in the 3GPP 38 series of specifications, which go beyond the LTE standard.

Work on 5G NR began within 3GPP in 2015, and initial specifications were released by the end of 2017. Even before final standardization, industry players began implementing preliminary agreements based on draft versions. The first large-scale commercial deployment of 5G New Radio (5 G NR) was successfully targeted and achieved in 2019.

5G NR Frequency Bands

The dual-band architecture of 5G NR allows network operators to tailor their deployments to specific geographic and use-case requirements. Frequency Range 1 (FR1), which encompasses sub-6 GHz bands, is particularly well-suited for wide-area coverage and indoor penetration. These bands are instrumental for applications that require stable connectivity across large areas, including mobile broadband in suburban zones and seamless coverage along transportation corridors.

In contrast, Frequency Range 2 (FR2), which includes millimeter-wave (mmWave) frequencies, offers significantly higher bandwidth but with a more limited range. This makes FR2 ideal for dense urban environments and high-demand areas, such as stadiums, airports, and industrial campuses, where data throughput requirements are exceptionally high.

The ability of 5G NR to operate across these distinct spectrum ranges ensures it can meet diverse performance needs, from massive machine-type communications (mMTC) to enhanced mobile broadband (eMBB) and ultra-reliable low-latency communication (URLLC), thereby enabling the full spectrum of 5G use cases envisioned by the International Telecommunication Union (ITU).

Scenario: Smart Hospital Connectivity With Band-Specific Optimization

Different 5G NR frequency bands are strategically used in a smart hospital to serve critical functions. Low-band ensures reliable coverage for IoT devices, such as wearable monitors, throughout the building. Mid-band supports high-speed transmission of imaging data (e.g., MRI scans) between departments. Meanwhile, high-band (mmWave) is reserved for latency-sensitive applications, such as remote robotic surgery. This tiered frequency approach enables hospitals to strike a balance between coverage, capacity, and precision.

Network Deployments

Ooredoo was the first telecommunications provider to launch a commercial 5G New Radio (NR) service in Qatar in May 2018. Since then, other providers worldwide have adopted similar deployment strategies.

Development Timeline

In 2018, the 3rd Generation Partnership Project (3GPP) released Release 15, marking the first standardized phase of 5G New Radio (5 G NR) ("Stage 1"). *This was followed by Release 16, representing "Stage 2" of 5G development, which reached its functional freeze in March 2020 and was finalized by June of the same year*. The process continued with Release 17, which was scheduled for publication in September 2021.

Deployment Modes

Initial 5G NR rollouts have relied on existing 4G LTE infrastructure in non-standalone (NSA) mode. This transitional phase precedes the full implementation of standalone (SA) mode, which is supported by a dedicated 5G core network. Additionally, frequency bands will be gradually shared between 4G LTE and 5G New Radio (5 G NR) technologies.

Non-Standalone (NSA) Mode

The Non-Standalone (NSA) mode of 5G New Radio (NR) represents an initial deployment strategy that leverages existing 4G LTE infrastructure. In this configuration, the 4G LTE network handles the control plane (signaling), while 5G NR manages the user plane (data transmission). This hybrid approach accelerates the adoption of 5G by enabling mobile operators to deploy NR without initially building a complete standalone core network.

While NSA facilitates a faster rollout, it has been criticized by some vendors and operators who argue that it delays the transition to the fully independent 5G architecture. Nevertheless, NSA has played a vital role in the early expansion of 5G services by maximizing current infrastructure investments.

Dynamic Spectrum Sharing (DSS)

An essential feature of NSA deployments is Dynamic Spectrum Sharing (DSS). *DSS enables mobile operators to dynamically share spectrum resources between 4G LTE and 5G NR, based on real-time user demand.* This is made possible through software upgrades, which enable the simultaneous use of the same frequency band. Only the 5G NR devices need to be DSS-compatible, making it a cost-effective way to bridge the transition between network generations.

CELL TYPES AND DEPLOYMENT ENVIRONMENT

New Radio Frequencies

The 5G air interface, known as New Radio (NR) and defined by 3GPP, operates across two main frequency ranges, each offering unique performance characteristics:

- Frequency Range 1 (FR1): Sub-6 GHz frequencies, typically ranging from 3.3 to 4.2 GHz, are used in this range. The maximum channel bandwidth in FR1 is 100 MHz, reflecting the scarcity of available spectrum in this band. For example, Korean operators currently utilize the 3.5 GHz band but have yet to allocate lower millimeter-wave (mmWave) spectrum for 5 G.
- Frequency Range 2 (FR2): This range includes millimeter-wave (mmWave) bands above 24 GHz, with bandwidths ranging from 50 MHz to a maximum of 400 MHz. Notably, U.S. carriers such as Verizon and AT&T utilize the 28 GHz and 39 GHz bands, respectively. These higher frequencies enable ultra-fast data transfer but require denser infrastructure due to their limited range and penetration.

mmWave deployment is essential for unlocking 5G's full high-speed potential, though it introduces new logistical and engineering challenges.

Cell Types and Deployment Environment in 5G Networks

The deployment of 5G networks leverages a variety of cell types, each optimized for specific environments and user densities. This hierarchical architecture ensures that both coverage and capacity are tailored to the diverse demands of various settings, ranging from residential zones to large metropolitan hubs.

Femtocells are designed for highly localized indoor coverage, making them ideal for homes and small businesses. Typically supporting between 4 and 8 users in residential settings and up to 32 in commercial environments, femtocells operate with transmission power levels ranging from 10 to 100 mW indoors and up to 1000 mW outdoors. Their effective range spans tens of meters, making them suitable for enclosed spaces with limited user activity.

Picocells serve high-traffic indoor public venues, including shopping centers, airports, and train stations. These cells support approximately 64 to 128 users, with power outputs ranging from 100 to 250 mW indoors and up to 5,000 mW outdoors. Their coverage spans tens of meters, offering stable and high-capacity connections in densely populated areas.

Microcells are deployed to bridge coverage gaps within urban landscapes, offering connectivity to 128 to 256 users. With transmission power ranging from 5,000 to 10,000 mW, microcells can maintain reliable connections over a few hundred meters, particularly in areas where macrocell signals are inconsistent or obstructed.

Metrocells, often positioned in high-density urban areas, enhance network capacity in locations with intense traffic demand. They are capable of supporting more than 250 users and typically operate at power levels between 10,000 and 20,000 mW. Their effective range is approximately one hundred meters, offering a strategic solution for maintaining performance in busy downtown districts and city centers.

Together, these cell types form a heterogeneous 5G network, ensuring seamless connectivity and efficient use of spectrum resources across both indoor and outdoor environments.

Scenario: Enhancing Urban Connectivity With Metro Cells

In urban settings, Metro cells enhance capacity, particularly in high-traffic zones. These cells accommodate more than 250 users and have an output power ranging from 10,000 to 20,000 mW, delivering efficient service over a 100-meter range. They are typically deployed on lamp posts, rooftops, or urban fixtures to fill coverage gaps left by macro cells. Metro cells are crucial during peak hours in crowded areas, such as transit stations, business districts, and stadium entrances. Offloading traffic from the macro layer ensures low latency and sustained data rates in dense user environments.

For comparison, Wi-Fi networks primarily serve homes and businesses with fewer than 50 users. They operate with an output power of 20–100 mW indoors and 200–1000 mW outdoors, providing coverage within a few tens of meters.

This classification underscores the strategic deployment of 5G cells to ensure efficient and scalable connectivity across varied environments.

Table 2 provides a comparative overview of different 5G cell types and their typical deployment environments. It highlights the maximum number of users supported, the output power range, and the effective distance from the base station for each device type. These characteristics demonstrate how 5G networks adapt to a variety of contexts, from private indoor settings to dense urban landscapes. The table also includes Wi-Fi specifications for comparison, highlighting the relative scale and capabilities of 5G cells. Such a layered approach ensures both capacity and coverage in next-generation wireless networks.

Table 2. Cell types and deployment environment

Cell types		**Deployment environment**	**Max number of users**	**Output power (mW)**	**Max. distance from the base station**
5G NR FR2	Femtocell	Homes, businesses	Home: 4–8 Businesses: 16–32	indoors: 10–100 Outdoors: 200–1000	10s of meters
	Pico cell	Public areas like shopping malls, airports, train stations, skyscrapers	64 to 128	indoors: 100–250 Outdoors: 1000–5000	10s of meters
	Microcell	Urban areas to fill coverage gaps	128 to 256	Outdoors: 5000–10000	A few hundred meters
	Metro cell	Urban areas are to provide additional capacity	more than 250	Outdoors: 10000–20000	hundred meters
Wi-Fi (for comparison)		Homes, businesses	less than 50	indoors: 20–100 Outdoors: 200–1000	A few 10s of meters

Source: adapted from rf wireless-world.com

MOTIVES FOR 5G DEPLOYMENT

The deployment of 5G networks is motivated by a range of strategic goals, with network leadership emerging as the primary driver for many operators. According to a McKinsey analysis, telecommunications operators view 5G as a pivotal opportunity to reinforce their market position, secure competitive advantage, or reclaim leadership in a rapidly evolving industry landscape, (Bandyopadhya et al., 2023).

Around half of the surveyed operators emphasize this aspect, underlining the importance of competitive positioning as the foremost priority for 5G implementation. This focus reflects the intense competition in the global market, where operators aim to distinguish themselves through superior infrastructure and innovative service offerings.

One of the most significant drivers of 5G deployment is enhancing the consumer experience, which includes providing faster speeds, ultra-low latency, and seamless connectivity. In an increasingly data-driven world, user expectations for reliability and performance are at an all-time high. Operators view 5G as a means to meet and exceed these expectations, thereby creating a transformative user experience that fosters customer loyalty and drives revenue growth.

Network capacity ranks third among the objectives, with approximately one-third of operators prioritizing this aspect. As data consumption continues to rise exponentially, particularly in urban and industrial environments, operators recognize the need to increase capacity to support the growing traffic. It ensures the network remains efficient and scalable in handling the surge in demand driven by emerging technologies such as augmented reality, autonomous vehicles, and cloud gaming. Interestingly, though many discuss the Internet of Things (IoT) as a key application for 5G, most operators do not view it as a primary driver.

Current IoT capabilities provided by 4G and other technologies are sufficient for many existing use cases. Similarly, Fixed Wireless Access (FWA), which offers a broadband alternative to underserved areas, is ranked lower in priority. Only 22% of operators consider it a top concern.

In summary, while 5G's transformative potential spans multiple domains, its deployment is primarily shaped by the need to maintain competitive advantage, enhance consumer experience, and expand network capacity. These priorities reflect operators' strategic focus on positioning 5G as a cornerstone for additional connectivity and innovation.

RISKS AND CHALLENGES

The deployment of 5G technology promises revolutionary advancements in telecommunications, offering unprecedented speed, low latency, and increased connectivity. However, realizing this transformative technology comes with significant challenges and risks that span technical, regulatory, and societal domains, (Holtrup et al., 2021). It is essential to tackle these concerns to guarantee the successful adoption and optimization of 5G networks globally.

The Challenge of the Spectrum Allocation Management

One of the main difficulties involves spectrum allocation management, where efficient allocation of low, mid-, and high-frequency bands is essential to meeting 5G's diverse speed, capacity, and coverage requirements. Despite making strides, countries like Canada still lag behind global leaders such as the United States and Australia in mid-band spectrum allocation. Regulatory bodies, such as Innovation, Science, and Economic Development Canada, balance competing interests while ensuring equitable spectrum allocation to support the growth of 5 G. Additionally, the increasing need for small cell antennas, often deployed on public infrastructure, has raised concerns about outdated and ill-suited regulatory frameworks that are not well-suited for 5G's complex infrastructure requirements.

The rise of 5G also introduces security risks, as the expansion of IoT-enabled devices increases the potential attack surface for cyber threats. Integrating 5G technology with the Internet of Things (IoT) enables high-speed data transfer, ultra-low latency, massive device connectivity, and enhanced security, facilitating advanced applications across various industries.

Critical Privacy Concerns

According to Ahmed et al. (2024), the integration of 5G technologies with the Internet of Things (IoT) introduces substantial opportunities for advanced connectivity and responsiveness but also raises critical privacy concerns, The rise of 5G has prompted telecom operators to address security vulnerabilities within authentication protocols. *Recent studies have highlighted vulnerabilities in 5G authentication protocols, which could expose networks to new security challenges, including massive Distributed Denial-of-Service (DDoS) attacks and data breaches.* The increasing number of IoT devices powered by 5G raises concerns about cybersecurity risks, including DDoS attacks and data breaches (IoT Analytics, 2018). Furthermore, geopolitical tensions surrounding hardware suppliers, particularly those from China, have led to restrictions and heightened scrutiny of 5G infrastructure, aimed at mitigating concerns about surveillance.

Public health is another critical area of concern. Misinformation about the potential health effects of 5 G has fueled resistance to its rollout in some regions. Although scientific studies overwhelmingly support the safety of 5G technology, concerns about radiation exposure have led to regulatory delays and public opposition in countries such as Belgium and Switzerland. This misinformation underscores the importance of transparent communication and rigorous scientific evaluation in addressing public concerns.

Lastly, the environmental impact of 5G infrastructure development has become a significant concern. Courts have mandated ecological impact assessments for 5G tower installations in some regions to ensure sustainable development. This reflects a growing need for green technologies and practices to align with 5G's ambitious deployment solutions and informed public discourse. Effectively resolving these challenges will be crucial for maximizing the potential of professionals.

As the deployment of 5G continues to accelerate, these risks and challenges underscore the need for robust regulatory frameworks and innovative technological solutions to realize 5 G's full potential while mitigating its associated risks.

CHALLENGES IN SPECTRUM MANAGEMENT

A mix of low (less than a single gigahertz [GHz]), moderate (between one GHz and ten GHz), and excessive (above ten GHz) wavelengths must satisfy the coverage capacity, along with the speed needs of 5G technology. In Canada, spectrum management is the duty of the Department of Innovation, Science, and Economic Development (ISED). *Various telecommunications companies have requested that ISED issue numerous additional licenses for 5G technology*. According to a worldwide report on 5G technology, Canada lags behind several countries in the mid-band spectrum, including Australia, Japan, the United Kingdom, and the United States.

Plans are in place to reserve a portion of the 600-megahertz (MHz) band for 5G technology in Canada. Nevertheless, the 3,500 MHz and 3,800 MHz bands postponed their project, as they are already in use, especially for fixed wireless services in remote areas. However, ISED continues to consult to help these companies meet the requirements of all parties. Additionally, it is preparing to hold a public auction in 2020 to allocate some 5G engineering resources. ISED has also suggested previewing many bands above 20 GHz to designate a portion of them for 5G technology within five years.

Regulation of Wireless Infrastructure

Banafaa et al. (2024) emphasize that integrating uncrewed aerial vehicles (UAVs) into 5G and beyond wireless networks offers vast potential for applications ranging from emergency response to real-time surveillance. However, it also raises complex challenges regarding security, regulatory compliance, and the reliability of high-frequency communication.

In 2018, there were around 13,000 wireless antenna locations in Canada. This number is likely to increase significantly as the Internet of Things (IoT) evolves with the launch of 5G technology. The number of related items might jump by a few billion in the following decade, leading to a tremendous rise in connectivity must-haves. Small cell antennas (the shoebox dimensions) grow the network and increase the must-have connectivity. Additionally, numerous infrastructures, such as street lighting, bus shelters, and public buildings, rely on small antennas.

Many telecommunications companies have raised concerns regarding ISED's management of tiny cell antenna sites. From their perspective, it can be challenging to utilize these locations, which belong to numerous stakeholders, which could impede the large-scale deployment of this kind of antenna. Additionally, stakeholders believe the antenna regulations are inadequate for the development of 5G technology. These laws were created for earlier networks that used fewer, but larger, antennas. Therefore, some businesses have advised ISED to develop a national framework outlining the best methods for facilitating the deployment of small-cell antennas.

The architecture of 5G wireless networks integrates advanced features, including beamforming directional antennas, hybrid and standalone mm-wave networks, and user-centric network designs, to enhance connectivity and performance. The network ensures efficient resource allocation and seamless communication by separating user and control planes. Additionally, cloud radio access network (C-RAN) architectures leverage virtual base station clusters and optical distribution networks to optimize load balancing and switching. These advancements contribute to the development of scalable, efficient, and high-speed wireless communication systems, (Agarwal, Roy, & Saxena, 2016).

Risks

Due to the risks of likely surveillance of Chinese hardware merchants' clients, several nations (starting in mid-2019) have restricted or eliminated the use of Chinese gear in their 5G systems, including the USA, the UK, Canada, and Australia. However, Chinese sellers and the Chinese government have denied these cases.

A report distributed by the European Commission and European Agency for Cybersecurity details the security issues surrounding 5G while attempting to avoid referencing Huawei. The report cautions against using a single provider, primarily based outside the European Union, for a transport company's 5G foundation. (Nokia and Ericsson are the leading European producers of 5G gear.)

According to some reports, intelligence agencies such as the FBI in the United States and GCHQ in the United Kingdom allegedly influenced the development of 5G standards through 3GPP, aiming to ensure access to extensive metadata for surveillance purposes.

Environmental Impact

The deployment of 5G networks will bring transformative advancements in telecommunications, but also raise significant concerns regarding their environmental impact. Environmental assessments have become mandatory for 5G infrastructure projects, highlighting the importance of sustainable development practices.

As 5G infrastructure expands globally, there is a surge in new development ventures, including the installation of 5G towers, small cell antennas, and other network components. In August 2019, a U.S. court mandated that these ventures assess their ecological implications, signaling the importance of sustainable practices in the deployment of 5 G. This decision highlights the importance of environmental safeguards to ensure that the development and deployment of 5G infrastructure do not compromise long-term ecological stability.

A significant environmental concern associated with 5G is its substantial energy consumption. The increased data rates, enhanced capacity, and broader connectivity of 5G networks demand substantial energy resources. This consumption raises concerns about the carbon footprint of telecommunications networks and their contribution to global climate change. Moreover, the deployment of small cell networks, which require a dense network of antennas, may exacerbate these issues by increasing the energy demands for powering and maintaining the infrastructure.

Additionally, the physical development of 5G infrastructure, including the construction of towers and the installation of equipment, can disrupt local ecosystems. The use of rare earth materials in 5G devices and antennas also has environmental consequences, as their extraction and processing often result in habitat destruction and pollution. These activities necessitate careful planning and regulatory oversight to minimize ecological harm.

To address these concerns, stakeholders in the telecommunications industry must prioritize adopting green technologies and sustainable practices. Innovations such as energy-efficient network equipment, renewable energy sources, and electronic waste recycling programs can help mitigate the environmental footprint of 5 G. Furthermore, as mandated by the U.S. court, comprehensive environmental impact assessments should become a standard practice to evaluate and address the ecological implications of 5G infrastructure projects.

The rollout of 5G offers immense benefits in connectivity and innovation. However, its environmental challenges underscore the importance of adopting a comprehensive approach that balances technological advancements with ecological responsibility. By adopting sustainable strategies, the industry can ensure that the development of 5G networks aligns with broader global efforts to combat environmental degradation.

Security Concerns in 5G Deployment

The advent of 5G technology has opened new opportunities for connectivity, but it also presents significant security challenges that require urgent attention. A primary concern is the anticipated increase in distributed denial-of-service (DDoS) attacks. Security organizations such as Fortinet, Arbor Networks, A10 Networks, and Voxility have warned about the need for tailored security solutions to counter these threats. The high-speed, low-latency nature of 5G networks enables faster and more coordinated attacks, thereby raising the stakes for operators and users.

Another pressing issue is the exponential growth in Internet of Things (IoT) devices powered by 5G, which IoT Analytics estimates will increase. The number is expected to grow from 7 billion in 2018 to approximately 21.5 billion by 2025. While this proliferation enhances connectivity and innovation, it also amplifies the potential for cyberattacks. IoT devices, often vulnerable to exploitation due to inadequate security measures, can become vectors for large-scale breaches, including cryptojacking and other malicious activities.

The broader implications of these security concerns underscore the urgency for robust countermeasures. Enhanced encryption protocols, real-time threat detection, and AI-driven security systems are essential for mitigating the risks associated with

5G deployment. Moreover, collaboration among telecom operators, technology providers, and regulators is critical to developing a comprehensive security framework.

By proactively addressing these vulnerabilities, stakeholders can better safeguard 5G networks against evolving threats and ensure the secure and reliable deployment of this transformative technology.

Health

The logical consensus is that 5G innovation is safe, and contentions to the contrary depend on a conspiratorial distraction that uses the innovation's novelty as an explanation for not trusting it. A misunderstanding of 5G design has led to the rise of fear-inspired notions, allowing its adversaries to impact human well-being.

In 2019, 180 researchers from 36 countries maintained contact with the European Union, expressing concerns about potential health risks associated with the rollout of 5 G. In April 2019, Brussels, Belgium, halted a 5G pilot due to radiation laws. For the same reason, officials in Geneva, Switzerland, arranged a move to 5 G.

The Swiss Telecommunications Association (ASUT) has stated that reviews have been unable to demonstrate a negative impact of 5G frequencies on well-being. As indicated by CNET, Members of Parliament in the Netherlands are also approaching the government to investigate 5 G. A few heads in Congress have written to the Federal Communications Commission, expressing their concern about potential health risks. In Mill Valley, California, the city committee hindered the organization of new 5G small cells. To support 5G deployment, small cell antennas need revised regulatory frameworks to manage the complexities of their installation.

For instance, in Vermont and New Hampshire, we have similar concerns. Following campaigns by extremist groups, a series of small cell areas in the UK, including Totnes, Brighton and Hove, Glastonbury, and Frome, passed resolutions against implementing a 5G network.

There have been various concerns over the spread of disinformation in the media and on the web, regarding the potential health impacts of 5G technology. IAM Broad announced that RT America started airing programming connecting 5G to unsafe well-being impacts, which "need logical help, for example, mind malignancy, barrenness, mental imbalance, heart tumors, and Alzheimer's disease." Expansive coverage of the topic significantly increased, as demonstrated by RT America airing seven dedicated programs on the subject by mid-April 2019, compared to only one in 2018, reflecting how the issue spread rapidly across numerous online platforms.

CONCLUSION

The advancement of 5G has significantly reshaped the smartphone industry, introducing powerful new dimensions of performance, speed, and functionality. As early as 2019, the introduction of 5 G-capable devices signaled a turning point in mobile innovation. This chapter highlighted how flagship smartphones, such as the Samsung Galaxy S20, and mid-range models, like the Nokia 8.3, brought 5G capabilities to a broad consumer base. Enhanced data rates, lower latency, and broader spectrum support contributed to a vastly improved mobile experience. Beyond consumer use, 5G has opened up new frontiers in logistics, manufacturing, and smart infrastructure.

However, this evolution has been accompanied by manufacturing and design challenges, including increased component costs and the complexity of antenna integration. Despite these obstacles, technological innovation in chip design and battery efficiency has helped manufacturers respond swiftly. Smartphones have evolved beyond simple communication devices to become essential components within expansive 5G ecosystems, supporting cloud computing, edge processing, and real-time applications. As this chapter has demonstrated, mobile device architecture must evolve in tandem with network innovations to fully leverage the potential of 5G technology. For instance, the Huawei Themes App offers users a full range of customization options, including themes, for a personalized smartphone experience.

To summarize the key contributions of 5G in the context of smartphones, Table 3 presents a summary of the main themes discussed in this chapter. It outlines how 5G has impacted smartphone architecture, use cases, and industrial applications.

Table 3. Summary of 5G impacts on smartphone development and application

Aspect	Impact of 5G	Implications for Smartphones
Network Capability	High-speed, low-latency communication	Enhanced streaming, VR/AR, and real-time interaction
Architecture	Advanced chipsets, antenna design, sub-6 GHz, and mmWave bands	Complex design requirements and increased production costs
Use Cases	Cloud gaming, remote healthcare, and IoT integration	Expansion beyond traditional communication to multifunctional platforms
Accessibility	Mid-range models like the Nokia 8.3 support broadband bands	Wider market reach and global deployment
Device Role in Ecosystem	Integration with innovative infrastructure and edge computing	Smartphones as mobile control and data processing hubs

The deployment of 5G marks a monumental leap in the evolution of wireless networks, transforming the digital landscape with unprecedented capabilities. This technological advancement is not merely an extension of previous generations but a comprehensive overhaul of network architecture, enabling faster speeds, reduced latency, and enhanced reliability. The transition from theoretical designs to real-world implementation underscores the extraordinary efforts of global stakeholders, including network operators, technology providers, and spectrum regulators, in overcoming the complexities of deploying 5G at scale.

At the heart of 5G deployment is its multi-layered infrastructure, which integrates cutting-edge technologies such as small cells, network slicing, and edge computing. These innovations provide the foundation for seamless connectivity across diverse applications, from autonomous vehicles and smart cities to industrial automation and immersive augmented reality experiences. The introduction of ultra-reliable low-latency communication (URLLC) and massive machine-type communication (MMTC) further highlights 5 G's ability to adjust and cater to the specific needs of different industries.

Despite its transformative potential, the deployment of 5G has faced significant challenges, including the allocation of spectrum, high infrastructure development costs, and the need for global regulatory alignment. However, these hurdles have been met with unprecedented collaboration between public and private entities, ensuring steady progress in bringing 5G services to consumers and industries worldwide. This spirit of partnership has been instrumental in driving the success of early deployments and laying the groundwork for broader rollouts in the years to come.

In conclusion, the deployment of 5G marks a critical milestone in the digital revolution, transforming the way people, devices, and systems interact. As this technology advances and grows, it will serve as the backbone for a hyper-connected future, driving innovation, economic growth, and societal progress. The story of 5G deployment is not just one of technological achievements, but also a testament to the power of collaboration in overcoming complexity and realizing a shared vision of global connectivity.

REFERENCES

Adewale, T., & Paul, J. (2024). *AI, 5G, and IoT: How These Technologies Are Creating the Perfect Storm for Smart Systems*. ResearchGate. https://www.researchgate.net/publication/385855348_AI_5G_and_IoT_How_These_Technologies_Are_Creating_the_Perfect_Storm_for_Smart_Systems

Agiwal, M., Roy, A., & Saxena, N. (2016). Next Generation 5G Wireless Networks: A Comprehensive Survey. *IEEE Communications Surveys and Tutorials, 18*, 1617–1655. DOI: 10.1109/COMST.2016.2532458

Banafaa, M., Ömer, P., Shayea, I., Alhammadi, A., Shamsan, Z., Razaz, M. A., Majid, A., & Al-Sowayan, S. (2024). A comprehensive survey on 5 G-and-beyond networks with UAVs: Applications, emerging technologies, regulatory aspects, research trends, and challenges. *IEEE Access : Practical Innovations, Open Solutions, 12*, 7786–7826. DOI: 10.1109/ACCESS.2023.3349208

Bandyopadhyay, S., Jain, P., Leing, J., & Prisacaru, S. (2023) *Navigating the three horizons of 5G business building*. McKinsey & Company. https://www.mckinsey.com/industries/technology-media-and-telecommunications/our-insights/navigating-the-three-horizons-of-5g-business-building

Holtrup, G., Lacube, W., David, D. P., Mermoud, A., Bovet, G., & Lenders, V. (2021, August 20). *5G System Security Analysis*. ArXiv.org. https://doi.org//arXiv.2108.08700DOI: 10.48550

IoT Analytics. (2018). *The Rise of Connected Devices and the Role of 5G Networks*. Retrieved from https://iot-analytics.com

Patwary, M. N., Nawaz, S. J., Rahman, M. A., Sharma, S. K., Rashid, M. M., & Barnes, S. J. (2020). The potential short- and long-term disruptions and transformative impacts of 5G and beyond wireless networks: Lessons learnt from developing a 5G testbed environment. *IEEE Access : Practical Innovations, Open Solutions, 8*, 11352–11379. DOI: 10.1109/ACCESS.2020.2964673

Ahmed, S. F., Alam, M. S. B., Afrin, S., Rafa, S. J., Taher, S. B., Kabir, M., ... & Gandomi, A. H. (2024). Toward a secure 5G-enabled internet of things: A survey on requirements, privacy, security, challenges, and opportunities. IEEE Access, 12, 13125-13145.

Teixeira, L., Almeida, P., Rodrigues, M., Antunes, D., Gomes, D., & Aguiar, R. L. (2024). *Shallow vs.* Deep Learning: Prioritizing Efficiency in Next Generation Networks. 2024 11th International Conference on Future Internet of Things and Cloud (FiCloud), Vienna, Austria, 308–315. https://doi.org/DOI: 10.1109/FiCloud62933.2024.00055

ADDITIONAL READING

ADTRAN. INC. (2020). *Comments to NTIA on 5G Implementation Plan.* https://www.ntia.gov/sites/default/files/publications/adtran_comments_to_ntia_on_5g_plan_6-25-20_0.pdf

5G. Americas (2015). *5G Technology Evolution Recommendations.* 5G Americas. https://www.5gamericas.org/5g-technology-evolution-recommendations/

Capizzi, K. *(2015) Mobilizing America: Making 5G an Opportunity for Broad-Based Progress.* Interdigital.com. https://www.interdigital.com/post/mobilizing-america-making-5g-an-opportunity-for-broadbased-progress-

Congressional Research Service (CRS). *(2019). Fifth-Generation (5G) Telecommunications Technologies: Issues for Congress.* Congress.gov. https://www.congress.gov/crs-product/R45485

Consultancy.in. (2019, October 21). *Nearly 90 million Indians could be connected on 5G by 2025.* Consultancy.in. https://www.consultancy.in/news/2567/nearly-90-million-indians-could-be-connected-on-5g-by-2025

Cybersecurity and Infrastructure Security Agency (CISA). *(2019). Overview of Risks Introduced by 5G Adoption in the United States.* Cybersecurity and Infrastructure Security Agency (CISA). https://www.cisa.gov/sites/default/files/publications/19_0731_cisa_5th-generation-mobile-networks-overview_0.pdf

Deloitte. (2019). *Our services | Deloitte. (2025).* Deloitte. https://www.deloitte.com/us/en/services.html?stream=top-stories

developercgit. (2024, January 29). *How Does 5G Technology Enhance the Internet of Things IoT.* Convert Generation Information Technology. https://cgit.pk/5g-technology/

Ericsson. (2020). *Enhanced 5G User Experience - Mobility Report.* https://www.ericsson.com/en/reports-and-papers/mobility-report/articles/differentiating-5g-experience

Federal Communications Commission. (2020). *Federal Communications Commission FCC 20-50*. Federal Communications Commission. https://docs.fcc.gov/public/attachments/FCC-20-50A1.pdf

Fisher, T. (2023). *5G Challenges: Why It Isn't Rolling Out Faster*. LifeWire. https://www.lifewire.com/5g-challenges-4580354

General Services Administration (GSA). *(2023). GSA Supports National Strategy to Secure 5G with New Acquisition Guidance*. GSA. https://gsablogs.gsa.gov/technology/2023/03/28/gsa-supports-national-strategy-to-secure-5g-with-new-acquisition-guidance/

Intelligence, G. S. M. A. (2025). *The Mobile Economy 2025*. GSMA. https://www.gsma.com/solutions-and-impact/connectivity-for-good/mobile-economy/wp-content/uploads/2025/02/030325-The-Mobile-Economy-2025.pdf

Juniper Research. (2015). *5G Service Revenues Forecast to Exceed $ 65 Bn Globally by 2025*. Juniper Research. https://www.juniperresearch.com/press/5g-service-revenues-forecast-to-exceed-65bn-globally-by-2025-contributing-7-of-operator-revenues/

MarketsandMarkets. (2021). *5G Core Market Size & Share - Growth Analysis Report - 2032*. Marketsandmarkets.com. https://www.marketsandmarkets.com/Market-Reports/5g-core-market-136573849.html

Milo Medin & Louie. Gilman (2019). *The 5G Ecosystem: Risks & Opportunities for DoD*. Defense Innovation Board https://media.defense.gov/2019/Apr/04/2002109654/-1/-1/0/DIB_5G_STUDY_04.04.19.PDF

National Telecommunications and Information Administration (NTIA). (2020). *A U.S. National Strategy for 5G and Future Wireless Innovation*. Information technology & Innovation Foundation. https://www.ntia.doc.gov/files/ntia/publications/itif-06252020.pdf

NGMN Alliance. (2015). *5G White Paper*. NGMN. https://www.ngmn.org/work-programme/5g-white-paper.html

Nokia. (2015). *Ten Key Rules of 5G Deployment*. GSA. https://gsacom.com/paper/nokia-ten-key-rules-of-5g-deployment/

Rafique, W., Barai, J. R., Fapojuwo, A. O., & Krishnamurthy, D. (2025). A Survey on Beyond 5G Network Slicing for Smart City Applications. *IEEE Communications Surveys and Tutorials*, *27*(1), 595–628. DOI: 10.1109/COMST.2024.3410295 shafiee.htc. (2025). *Our 5G world in 2025*. Scribd. https://pt.scribd.com/document/828176750/Our-5G-world-in-2025

Shoaib, M., Ghassan Husnain, Sayed, N., & Lim, S. (2024). Unveiling the 5G frontier: Navigating challenges, applications, and measurements in channel models and implementations. *IEEE Access : Practical Innovations, Open Solutions*, *12*, 59533–59560. DOI: 10.1109/ACCESS.2024.3392761

Teixeira, L., Almeida, P., Rodrigues, M., Antunes, D., Gomes, D., & Aguiar, R. L. (2024). *Shallow vs.* deep learning: Prioritizing efficiency in next-generation networks. 2024 11th International Conference on Future Internet of Things and Cloud (FiCloud), 308–315. https://doi.org/DOI: 10.1109/FiCloud62933.2024.00055

U.S. Government Accountability Office (GAO). (2020). *5G Wireless: Capabilities and Challenges for an Evolving Network*. Gao.gov. GAO-21-26SP. https://www.gao.gov/assets/gao-21-26sp.pdf

Chapter 4
Availability of 5G by Country

ABSTRACT

The deployment of 5G marks a global shift in connectivity, offering high speed, ultra-low latency, and the ability to connect billions of devices. It enables transformative uses in smartphones, autonomous vehicles, and IoT. However, access varies widely. Countries like the U.S., South Korea, and China lead with extensive 5G networks, while many developing nations remain in planning or pilot phases. Infrastructure, regulatory, and financial barriers slow progress. By 2025, over 10% of mobile connections are expected to use 5G, reflecting its key role in innovation and digital growth—despite persistent deployment challenges.

INTRODUCTION

The rollout of 5G, the latest advancement in wireless networking technology, marks a significant milestone in global connectivity. The development of the first international 5G modem illustrates the hardware advancements driving next-gen connectivity (Evans, 2017). *With the potential to revolutionize industries and everyday life, 5G technologies offer unparalleled speed and ultra-low latency, enabling the seamless connection of billions of devices.* 5G promises to enable many transformative applications, from smartphones and smartwatches to connected vehicles and IoT devices. A report details the growing ecosystem of 5G devices, including smartphones and industrial modules. For example, various 5G use cases, including industrial automation and telemedicine, illustrate its transformative potential across different sectors, (DIGI, 2025). Monetization strategies for 5G use cases hinge on convergence with AI, edge computing, and vertical-specific applications such as smart grids and connected vehicles (Chhabra & Rogers Communications Inc., 2023). Monetizing 5G requires alignment of use cases with AI, IoT, and edge computing,

DOI: 10.4018/979-8-3373-0497-7.ch004

enabling revenue generation in smart cities and industrial automation, (Chhabra & Rogers Communications Inc., 2023).

Global 5G research efforts focus on innovations in radio access, low-latency communication, and device connectivity across various industries, including automotive and healthcare, (Pirinen, 2014).

However, the availability of this cutting-edge technology varies widely across countries, reflecting disparities in infrastructure readiness, economic resources, and regulatory frameworks. Transport Control Protocol (TCP) and its multipath variant (MP-TCP) must be re-evaluated for compatibility with the high-speed, low-latency nature of 5G mmWave networks (Polese, Jana, & Zorzi, 2017). *For instance, Edge computing enabled by 5G unlocks real-time analytics and low-latency applications, redefining operational models* (Deloitte, 2025).

Furthermore, Field tests comparing 5G speed across global markets reveal significant disparities based on infrastructure and spectrum availability (Dolcourt, 2019). *Real-world trials and global measurements of 5G network speeds reveal performance variations driven by infrastructure, spectrum, and urban deployment models* (Dolcourt, 2019).

The deployment of 5G is expected to significantly contribute to global GDP, while also driving innovation, job creation, and economic transformation in emerging markets (GSMA, 2020). *The exponential rise in 5G subscriptions and infrastructure investments signals the technology's central role in future digital economies and service ecosystems (GSMA, 2023).* For instance, the adoption of massive MIMO in 5G networks dramatically increases spectral efficiency and user capacity by employing large-scale antenna arrays that enable spatial multiplexing and improved beamforming (Bishen, 2019).

With 5G adoption gaining momentum, projections show that by 2023, more than 10% of global mobile connections will rely on this technology. This rising adoption rate highlights 5G's pivotal role in advancing economic development, innovation, and digital transformation. 5G is identified as one of the top 10 strategic technology trends influencing enterprise innovation in 2020, (Panetta, 2019). *However, the pace and breadth of its rollout are shaped by various factors, including governmental policy, spectrum availability, and the level of investment by telecom providers.* 5G is emphasized as a strategic asset that can drive national competitiveness and technological leadership over the coming decade (Deloitte, 2025). Leading nations, such as the United States, South Korea, and China, have already established robust 5G infrastructures. The 2020 Global Telecommunications Outlook identifies 5G as a catalyst for industry transformation and new consumer experiences (Deloitte, 2025). In this context, many developing countries are still in the preliminary phases of planning and experimentation.

Global 5G research activities focus on the development of enabling technologies, including millimeter-wave communication, network densification, and device-to-device connectivity, to enhance performance and support emerging applications across various sectors, such as healthcare, automotive, and smart cities. For instance, a technical primer explains the fundamentals of 5G, including frequencies, latency, and public health considerations, (EMF Explained, 2019).

Simultaneously, partnerships among governments, tech companies, and global institutions are actively addressing these obstacles, with a focus on expanding 5G access to more regions and diverse populations.

This chapter examines the availability of 5G in various countries, exploring the factors that influence its rollout and adoption. Analyzing global trends and regional variations sheds light on the progress, challenges, and opportunities associated with the worldwide deployment of 5G technology. Real-world measurements of 5G latency reveal discrepancies between theoretical and observed performance, underscoring deployment challenges (Greene, 2018).

A comprehensive overview of 5G includes its speed, deployment challenges, and network structure, structure (De Looper, 2019). *The deployment of 5G systems introduces technical challenges related to high-frequency propagation, device heterogeneity, and backward compatibility* (Thompson et al., 2014). For instance, the COVID-19 pandemic was found to accelerate 5G deployment as businesses sought resilient digital infrastructure, (Fieldfisher, 2020).

5G represents the latest advancement in wireless communication technology, expected to power smartphones, wearables, vehicles, and various connected devices in the coming years. However, its rollout will not occur uniformly across all countries, (Satariano & Bayer,, 2019). *Most countries had initiated limited 5G rollouts by early 2020, despite disparities in coverage and access,* (Fieldfisher, 2020). A 5G readiness report evaluates how businesses are adopting 5G to drive innovation and competitiveness (Nokia, 2020). Small and medium-sized enterprises must leverage digital tools, including 5G, to remain competitive in increasingly connected markets (OECD, 2021). The digital transformation of SMEs represents a critical vector for inclusive economic growth and integration into global supply chains (OECD, 2021).

The digital economy's rapid evolution demands policy responses that foster innovation, secure infrastructure, and broaden access (innovationoecd, 2017). *Publications analyze digital development indicators, economic projections, and global readiness for 5G adoption* (ITU, 2017, 2018, 2020, 2021). A few evaluations estimate that by 2025, 5G will support more than 10 percent of the world's mobile connections.

NORTH AMERICA

The deployment of 5G technology in North America has progressed, but the expansion has faced natural challenges related to network infrastructure and system complexity. By 2023, experts estimate that 32% of mobile connections in North America will be operating on 5G networks, indicating a steady rate of adoption. Significant developments across the United States, Canada, Mexico, and Puerto Rico showcase varying levels among countries.

United States

Leading telecom providers, including Verizon, AT&T, Sprint, and T-Mobile, have substantially advanced 5G deployment. Verizon, C Spire, and Starry have launched fixed wireless broadband services, while mobile 5G networks are available in various cities. Foundational reports outline the transformative potential of 5G in enabling innovative services, autonomous systems, and high-speed communication, while highlighting policy and infrastructure challenges (ITU, 2017).

Accenture Strategy (2021) argues that the advent of 5G will reshape business models and operations, enabling new value streams and digital capabilities. *Cloud-native principles in 5G architecture are enabling network agility, elastic scalability, and the implementation of network slicing to support diverse service demands* (Bishen, 2019). LTE and Wi-Fi technologies remain essential in the transition to 5G, with hybrid architectures offering continuity and expanded access (Selinis et al., 2018). The global trajectory of 5G subscriptions from 2020 to 2027 confirms its pivotal role in driving digital connectivity and service expansion, (Petroc, 2025).

The Cisco Annual Internet Report outlines the traffic growth, IoT expansion, and 5G adoption trends shaping global digital infrastructure from 2018 to 2023 (Cisco, 2020). *Mobile industry growth and the expanding economic contribution of mobile technologies, including 5G, are projected to significantly influence global GDP and digital inclusion,* (GSMA, 2020). Recent trends indicate accelerated 5G adoption and subscriber growth, with strategic implications for digital infrastructure and industry verticals, (GSMA, 2023).

Global 5G subscription growth and economic impact projections reflect its expanding role in digital economies (GSMA, 2020, 2023). *Reports project exponential 5G user growth and spotlight use cases in business digitization and fixed wireless access* (Ericsson, 2025, 2019, 2023). Foundational frameworks for 5G emphasize its role in enabling next-generation services, highlighting both opportunities and implementation challenges for global telecom systems (ITU, 2017). However, monitoring digital development indicators remains essential for assessing the global readiness for 5G and bridging the digital divide, (ITU, 2020).

A clear explanation of 5G performance metrics helps demystify consumer expectations for speed and reliability (Hoffman, 2019). *For instance, it has been argued that cloud integration in 5G networks enables dynamic scalability and operational efficiency, serving as a foundational element for flexible network slicing (Bishen, 2019).* Real-world testing of 5G networks across major cities reveals significant variability in speed and latency, influenced by spectrum availability and deployment maturity, (Dolcourt, 2019).

As of 2023, the rollout continues to expand, with U.S. Cellular and other operators working to bring at-home and portable 5G to more regions. Cloud integration in 5G architectures is pivotal for enabling network scalability, automation, and service agility, (Bishen, 2019). *5G technologies are reshaping network architectures by integrating cloud-native principles, thereby supporting network slicing and scalable services,* (Bishen, 2019).

Enhanced productivity and innovation across verticals are driven by 5G's capabilities to support automation, real-time analytics, and ubiquitous connectivity, (Accenture Strategy, 2021). Benchmarking digital connectivity and broadband infrastructure is essential to understanding the readiness of global regions to implement scalable 5G technologies, (ITU, 2020).

Cloud-native 5G architecture redefines network flexibility, service provisioning, and dynamic orchestration, (Bishen, 2019). Business models are being redefined through 5G-enabled transformation, which facilitates agility, new value chains, and connected enterprise operations, (Accenture Strategy, 2021).

The economic impact of 5G deployment in the United States is substantial, with significant gains in productivity, job creation, and industry transformation, (Accenture Strategy, 2021). *This rapid deployment underscores the country's commitment to becoming a global leader in 5G technology*, (Accenture Strategy, 2021). The deployment of 5G is expected to generate substantial economic growth in the United States by enhancing productivity and creating new job opportunities, (Accenture Strategy, 2021). 5G technology accelerates innovation and productivity across industries, serving as a foundation for economic transformation, (Accenture Strategy, 2021). Several reports contextualize 5G within broader trends in digital economy growth and SME transformation (innovationoecd, 2017; OECD, 2021).

Canada

Canadian telecom providers have invested heavily in 5G infrastructure. Rogers Communications launched its 5G network in mid-2020, following an investment of over $4 billion in network development and deployment. They also established a 5G testing site at the University of British Columbia. Telus Mobility joined the effort, offering 5G accessibility to its customers nationwide. These initiatives reflect Canada's dedication to advancing 5G connectivity across its vast geographic landscape.

Mexico

América Móvil, a leading telecommunications provider in Mexico, announced the launch of 4.5G networks in 2017 as a precursor to 5 G. The company's CEO projected that, depending on technological advancements, 5G would become accessible by 2020. As of 2023, efforts to expand 5G coverage continue, with gradual adoption across urban and industrial centers.

Puerto Rico

Claro, a primary telecommunications provider, began testing 5G technology in 2019. Since then, the company has made strides toward deploying a functional 5G network. Puerto Rico's adoption of 5G reflects its ambition to stay aligned with global technological advancements.

EUROPE

European countries are at various stages of 5G deployment. Leading nations, such as Germany, the United Kingdom, and Sweden, have established functional 5G networks in urban areas, while others are gradually expanding their coverage. Most European nations expect 5G to be operational by 2023, driven by collaboration between telecom operators and regulatory bodies.

Norway

Norway has been a proactive player in deploying 5G networks, with Telenor spearheading efforts. After extensive testing since 2017, Telenor officially launched 5G in March 2020 across several areas, including Kongsberg, Elverum, Bodø, and Trondheim. These deployments demonstrate Norway's commitment to rolling out advanced connectivity infrastructure. Meanwhile, Telia, another key operator, conducted its first 5G test in 2018 and initiated 5G services for customers in Lillestrøm and parts of Oslo in 2020, with further expansion planned.

Germany

The German survey established robust estimates for time and reliability valuation, supporting cost–benefit analysis in national transport planning.

Deutsche Telekom launched its 5G network in major cities such as Berlin, Munich, and Bonn in 2019, supporting devices like the Samsung Galaxy S10 5 G. Vodafone Germany followed suit with 25 operational base stations by mid-2019, aiming for rapid network expansion. Additional players, such as Telefónica Germany and United Internet AG, have contributed to the competitive landscape through innovation clusters and spectrum auctions. 5G spectrum bands (low, mid, high) each offer trade-offs in speed and coverage, influencing deployment strategies, (Horwitz, 2019).

Czech Republic

Vodafone introduced 5G services in Karlovy Vary in mid-2020, building on prior tests that showcased innovative applications, including 5G holographic calls. These developments signal the Czech Republic's readiness to embrace 5G's transformative capabilities.

United Kingdom

The UK is among Europe's most advanced 5G markets. EE, the largest network operator, became the first to launch 5G in May 2019, covering cities like London and Manchester. Vodafone UK followed, focusing on automotive applications and conducting holographic call trials. Three, BT Mobile, and O2 have also deployed 5G, providing extensive coverage across urban centers and bringing new possibilities for both industries and consumers. 5G.co.uk (2025): Massive MIMO is a cornerstone of 5G technology that enhances spectral efficiency and network capacity by leveraging a high number of antennas at base stations, (Bishen, 2019.).

Italy

Italy's 5G rollout commenced in June 2019, with Vodafone launching services in cities such as Milan and Rome. The country aims to expand to 100 cities by 2021. Telecom Italia (TIM) also commenced 5G deployment in 2019, targeting broader coverage in 120 cities by 2021. In partnership with Nokia, the Iliad further reinforces Italy's 5G expansion.

Switzerland

Swisscom launched 5G in April 2019, achieving coverage in 54 towns, including Zurich and Geneva. The company plans to cover 90% of the population, offering devices such as the OPPO Reno 5 G. Sunrise has deployed 5G in over 200 locations. At the same time, Salt collaborates with Nokia to upgrade its systems for flexible 5G services.

Spain

Vodafone Spain led the 5G deployment in June 2019, launching services in 15 cities, including Madrid and Barcelona. Orange began rolling out 5G in 2020, following trials in Valencia and Seville, reflecting Spain's strategic focus on advanced connectivity.

Austria

Austria's 5G journey started with A1 establishing its first 5G connection in January 2019. T-Mobile Austria, now known as Magenta Telekom, and Hutchison Three Austria have made significant contributions, deploying 5G base stations across urban centers, including Linz. It demonstrates Austria's ambition to become a leader in advanced connectivity.

Finland

In Finland, Elisa Oyj, a broadcast communications company, opened a 5G business array in Tampere in June 2018, claiming to be the "first in the world to deploy business 5G." According to their 5G inclusion map, Oulu, Kuopio, Helsinki, Jyväskylä, and different zones have deployed the 5G system.

Elisa offers 5G Unlimited designs for 44.90 EUR/month at 1 Gbps and 34.90 EUR/month at 600 Mbps. Elisa's 5G network also supports the use of the ZTE Axon 10 Pro 5G telephone, (Chhabra & Rogers Communications Inc., 2023).

Near the beginning of 2018, Telia announced that Helsinki Airport had become the world's leading 5G air platform and its first client to utilize its pre-commercial 5G arrangement. In mid-2019, the organization started its 5G network in three urban areas and has since expanded to include a few more. Discover the Telia 5G devices you can purchase now for versatile and FWA 5 G.

DNA is another Finnish media communications organization bringing 5G to Finland. In December 2019, they started offering fixed remote access administration. DNA mentioned that even with the presentation of 5G, 4G remains the dominant technology in Finland.

Russia

In 2018, Russia's most prominent and versatile administrator, Mobile TeleSystems (MTS), and Samsung collaborated to conduct 5G tests, including video calls, ultra-low latency computer games, and 4K video streaming.

These 5G tests demonstrate that 5G remains available in the Russian Federation, and Samsung's 5G switches, tablets, and other devices are well-suited for operation on a 5G network.

As suggested by the GSMA, 5G methods are expected to cover more than 80% of the Russian population by 2025. Therefore, we anticipate a significant portion of the nation arriving much sooner.

However, another indicator that 5G in the Russian Federation is arriving sooner rather than later is the 5G analysis concentration available in Innopolis, a leading-edge town in Tatarstan.

Tele2 Russian Federation is another telecom business that brings 5G to the country. In February 2019, the company announced it would mail 50,000 base stations to the Country in collaboration with Ericsson. However, Tele2's CEO states, "Before moving to 5G methods, the Russian Federation must first deal with a couple of framework issues." Therefore, customers might have to wait a while to acquire 5G services.

Luxembourg

Orange remains a key European broadcast communications provider and has announced comprehensive 5G plans for Luxembourg. The rollout is part of its broader strategy to enhance connectivity across Europe.

Slovakia

While the timeline for 5G deployment in Slovakia remains unclear, progress is underway. In early 2019, SWAN Mobile signed a commercial agreement with ZTE to initiate its rollout plans. In July 2019, the two conducted the country's first 5G video call, (Fieldfisher, 2020).

Orange Slovensko seleaims to be completed by 2022 cted Nokia to deploy its 5G Radio Access Network (RAN) infrastructure in early 2020, signaling significant advancements in the nation's 5G development.

Ireland

Vodafone's telecom company went live with a 5G arrangement in Cork, Dublin, Galway, Limerick, and Waterford on August 13, 2019.

Dublin, Limerick, Kilkenny, Waterford, Cork, and Galway have all dispatched their 5G cooperation. A 5G inclusion map with all the areas is available.

Vodafone is the same, chipping away at fixed remote access. By December 2018, the company had begun preliminary work for the provincial 5G broadband rollout in Clonmel, Dungarvan, Gorey, and Roscommon, covering approximately 20,000 premises.

Envision is yet another telecom company bringing 5G to Ireland. In February 2019, it claimed to have dispatched its 5 G-prepared fixed broadband internet service and planned to serve over 300 destinations, covering more than 1 million premises by autumn 2020.

Chief Sean Bolger Had This to Express Regarding Their 5G Plan

As an Irish organization, we are delighted to announce this significant investment and a new methodology that, at last, addresses this issue and delivers genuinely necessary, fast, and dependable high-speed broadband to homes, businesses, and communities across rural and provincial Ireland today and into the future.

Three Ireland is rolling out a 5G network nationwide, but it needs to be more reliable.

Faroe Islands

Føroya Tele, the Faroe Islands' biggest telecom supplier, launched 5G on June 6, 2019.

Channel Islands

Jersey Telecom (ZTE and JT) has announced the introduction of 5G to the Channel Islands. In mid-2019, they started a pre-business deployment to test 5 G.

Belarus

Minsk's capital is testing the 5G technology initiatives through the cloud.

Romania

The Romanian government endorsed the 5G Strategy in June 2019.

The 5G Strategy for Romania extends an activity plan with reliable undertakings and implementation deadlines. It focuses on the dispatch of administrations in 2020 and 5G services, including all urban centers and major land transport routes, by 2025.

Vodafone Romania launched the nation's first 5G memberships on June 26, 2019. Their Supernet 5G page provides details and lists the available 5G phones for purchase.

RCS and RDS are the most prominent Romanian link and satellite TV organizations. They also offer 5G access through the Digi Mobil 5G Smart system.

Orange launched 5G in Romania in November 2019. Inclusion began in Bucharest, Cluj-Napoca, and Iasi and will expand to more regions through 2020. For details, see the Orange 5G inclusion map.

Hungary

Magyar Telekom, the largest Hungarian telecom company, launched 5G in Hungary in April 2020. The administration started in parts of Budapest and Zalaegerszeg.

Vodafone is another organization that is driving the adoption of 5G in Hungary. In May 2019, Vodafone installed Hungary's first live 5G station in Zalaegerszeg.

Croatia

Croatia is expected to launch commercial 5G services before the end of 2020. The administration begins in Osijek, inside Slavonia.

Sweden

Tele2 launched a 5G network in Sweden on May 24, 2020. Stockholm's downtown area was the first to see it; however, other regions, including Gothenburg and Malmö, will have access through 2020.

Telia also launched business 5G in Sweden on May 25, 2020. The system is accessible in Stockholm City Center, including Norrmalm, Östermalm, and Vasastan. Additionally, they are expected to have 5G in an additional 12 urban areas by 2021, including Gothenburg and Malmö.

Another Swedish 5G network, Three, can easily arrange services. On its 5G pages, you can see the diverse 5G deployment stages it has implemented, as well as where you can access 5 G.

Poland

On December 7, 2018, T-Mobile Polska reported the deployment of the nation's first 5G networks. Initially, the network was accessible only in Warsaw's focal point using five base stations; however, as of June 9, 2020, clients can use the system across its 1,600 base stations.

Polkomtel's Plus cell phone organization administrator started offering 5G in Poland on May 11, 2020. The initial 100 5G base stations, deployed in these seven urban communities, have secured around 900,000 individuals: Warsaw, Gdańsk, Katowice, Łódź, Poznań, Szczecin, and Wrocław. As a result, officials expect to add six hundred extra pinnacles throughout 2020.

5G is available in Warsaw, Wrocław, and other areas through Play. Check their inclusion map for subtleties.

Orange is another versatile European system administrator that will deploy 5G in Poland at some point before 2021.

Portugal

Vodafone welcomed Portugal into its first 5G association on December 12, 2018, using Qualcomm's 5G cell phone model. The organization additionally tried a computer-generated experience game and video approach to Ericsson's 5G arrangement.

ANACOM, Portugal's telecommunications regulator, has approved 5G tests and expects business rollouts to commence before the end of 2021.

France

Orange has launched 5G pilots in France in Lille, Douai, Marseille, and Châtillon. . French regulator Arcep expected the business 5G rollout in France to begin in 2020.

In an organization with Nokia, Iliad sends 5G across France (Simnovus, 2024).

Estonia

Estonia is expected to have an industrially available 5G network by 2020.

Telia Company, TalTech University, and Ericsson organized the nation's first 5 G test, conducted live at the end of 2018.

According to Telia Company, the 5G testbed is organized for development and exploration by industry partners and the academic world. TalTech University, like other organizations and new businesses, can create and test arrangements that require rapid and significant data integration.

Latvia

One organization bringing 5G to Latvia is Tele2. They deployed their first 5G base station in Riga in July 2019, made their first 5G video call a couple of months later, released 5G telephones, and started offering their 5G service in Daugavpils and Jelgava in mid-2020.

Denmark

Telenor Group conducted 5G tests in Denmark in 2019 through Nokia's AirScale Base Stations.

As per Telenor, the preliminaries are:

For example, bolstering 5G use cases, mechanical autonomy control, modern robotization, dual availability of 5G/LTE, and Fixed Wireless Access for a superior last-mile network.

Telia Company is another portable system administrator that started conducting 5G tests in Denmark in June 2019.

Azerbaijan

Azercell is a portable telecom organization deploying 5G in Azerbaijan. In 2019, it selected Ericsson for a two-year 5G agreement to furnish it with radio hardware and related services.

Iceland

Síminn is the leading supplier of remote correspondence services in Iceland. They reported in mid-2019 that they would collaborate with Ericsson to deliver 5 G-ready hardware to the nation.

The Netherlands

T-Mobile promised to include nationwide 5G coverage in the Netherlands in 2020. In late spring 2019, it reported three 5G research areas: Scheveningen, The Hague Tech, and the T-Mobile administrative center in The Hague.

With Ericsson Spectrum Sharing, Vodafone Ziggo launched 5 G in the most significant part of the Netherlands in April 2020. The company expects it to be available nationwide by the end of July 2020.

Monaco

Monaco Telecom clients accessed 5G coverage throughout the Principality of Monaco using the Huawei 20 X and the Xiaomi Mi MIX. The country completed the full inclusion of all its territories in July 2019.

Belgium

Proximus is the most prominent provider of portable media communications in Belgium. Investigate their 5G and arrange an inclusion map for all the details on where you can get administration. You will require a 5G versatile membership and a 5G telephone that supports the Proximus arrangement, such as the Oppo Find X2 Pro 5 G.

In 2019, Orange and Proximus agreed to establish a universal mutual access arrangement, which could accelerate the rollout of 5G in Belgium.

Malta

Vodafone Malta has yet to announce a robust 5G rollout date, but it is working on bringing 5G to Malta. On October 4, 2019, Vodafone Malta began allowing clients to evaluate the 5G coverage in the areas of Birkirkara, St. Julian's, and Sliema.

CENTRAL AMERICA AND SOUTH AMERICA

The rollout of 5G in Central American nations is progressing steadily, albeit slowly. Several South American countries with larger populations began seeing sporadic 5G deployments in late 2019, with notable progress continuing into 2020 and beyond.

Honduras

In December 2018, Ericsson announced that Tigo had selected them to modernize its radio access infrastructure, including a 5 G-ready multi-standard mobile system. Although 5G has not officially launched in Honduras, this modernization is a crucial first step toward its adoption.

Chile

Entel is Chile's leading telecommunications provider, offering 5G services to its customers in collaboration with Ericsson. This partnership has significantly advanced 5G adoption in the country.

Argentina

Movistar and Ericsson deployed 5G infrastructure in 2017. However, these services will become fully available to customers once Argentina finalizes its broader 5G rollout plan.

Brazil

- Vivo: 5G services were available in eight cities by the end of July 2020.
- Tim: Expected to have fully operational 5G networks in three cities by September 2020.
- Claro launched its 5G network on July 14, 2020, initially focusing on São Paulo and Rio de Janeiro.

Colombia

Telefónica Telecom, Colombia's largest telecommunications provider, is expected to offer 5G services to its customers. In 2018, Ericsson and Telefónica Telecom demonstrated a groundbreaking 5G data transfer speed of 27 Gbps in Bogotá.

Paraguay

Tigo partnered with Ericsson to prepare its network for 5 G. In December 2018, Ericsson announced plans to expand and modernize Tigo's existing 2G, 3 G, and 4G sites, paving the way for 5G and IoT services. This development is a strong foundation for future 5G deployment in Paraguay.

Aruba

Aruba's leading communications provider, SETAR, partnered with Nokia to achieve full 5G coverage on the island. This initiative aims to be completed by 2022 (Nokia, 2019). Successful full-island 5G implementation in Aruba demonstrates the potential of strategic partnerships for complete national coverage, (Nokia, 2019).

ASIA

5G is already a reality in several countries, with broad adoption anticipated for the current year.

South Korea

On December 1, 2018, three South Korean companies — SK Telecom, LG Uplus, and KT — collaborated to introduce flexible 5G to the country. They began by administering 5 G for select businesses. However, on April 5, they started offering 5G to others, including the Samsung Galaxy S10 5 G.

The SK Telecom specialist organization began offering 5G administration to the public on April 5, using its four 5G plans. This offer came after the organization started a restricted 5G administration with the Myunghwa industry-producing organization. SKT's 5G administration plans follow the trial of 5G in 2017 and its deployment at the self-driving test site, K-City.

LG Uplus' 5G organizes living in Seoul and surrounding areas, with LS Mtron as the first client. With over 4,000 5G base stations in Gyeonggi, the company planned to transport an additional 7,000 to Seoul and Incheon by 2018.

KT Corporation launched pre-business 5G services at Lotte World Tower in Seoul, as well as at six other locations, including Ulleungdo, Jeju, and Dokdo. On April 5, the company introduced boundless 5G services, known as KT 5G Super Plans, and expanded its coverage in Korea to a total of eighty-five cities by 2019. In addition, KT teamed up with Intel to showcase 5G at the 2018 Olympic Winter Games and has deployed over $20 billion through 2023 in 5G and other imaginative advancements.

As the ICT and Broadcasting Technology Policy executive at the Ministry of Science and ICT, Mr. Heo Won-Seok mentioned that five percent of the nation's versatile clients will be on a 5G network by 2020 and 90 percent by 2026, (Chhabra & Rogers Communications Inc., 2023).

Japan

NTT DOCOMO is Japan's leading remote transporter. It has been considering and exploring various 5G options since 2010 and launched pre-business 5G services in September 2019, preceding the launch of complete 5G services on March 25, 2020.

The administration utilizes the most substantial info speed of 3.4 Gbps; however, it has since expanded to 4.1 Gbps. Check the NTT DOCOMO 5G cellular phone page for gadget options.

In September 2018, NTT DOCOMO achieved download speeds of 25-27 Gbps in a 5G preliminary test with Mitsubishi Electric. The test might establish a quick 5G network that operates with cars.

KDDI and Rakuten are also offering 5G administrations in 2020. KDDI launched its 5G organization on March 26.

SoftBank began its 5G services on March 27 for 1,000 yen per month ($9).

China

On October 31, 2019, three major telecom companies—China Mobile, China Telecom, and China Unicom—launched 5G in China.

These organizations do not currently offer broad 5G inclusion. However, Beijing, Shanghai, and Shenzhen are currently China's most prominent 5G territories.

Hong Kong

launched 5G on April 1, 2020. According to a public statement, 5G was deployed in both indoor and open-air zones across all areas of Hong Kong in 2020, (Fieldfisher, 2020).

SmarTone is another organization that propelled 5G in Hong Kong in May 2020. Domain-wide inclusion is expected to be regular by mid-2021.

Qatar

According to Ooredoo, a Qatari telecom organization that has been pursuing 5G since 2016, it was the first organization on the planet to offer business 5 G.

5G is currently only accessible in Qatar; however, since Ooredoo has markets in several countries, such as Iraq, Oman, Palestine, the Maldives, Singapore, Algeria, and others, it is anything but a stretch to believe that we will see 5G arrive in those territories in 2020.

Ooredoo, in association with Ericsson, likewise utilizes 5G for home broadband help.

Vodafone is another organization providing 5G in Qatar. In late 2018, the organization started a 5G service in Katara Cultural Village and Souq Waqif. Before that, it was Abu Hamour, Salwa Road, Al Rayyan, Al Maamoura, Azizya, and Umm Salal Mohammed. Vodafone Qatar offers free 5G plans and a range of 5G smartphones.

Kuwait

Two media transmission organizations in Kuwait have launched a 5G network.

Zain was the first to report the 5G deployment in June 2018. You can join here for a 500 GB, 1 TB, 2 TB, or 4 TB plan, (Fieldfisher, 2020). *The main application offered by Zain's 5G network in Kuwait is the 5G Bolt switch for home broadband.*

The 5G plans available from Ooredoo include a 500 GB plan at 45 KWD per month and a 1 TB plan at 65 KWD per month.

STC (previously known as VIVA) is another telecommunications organization in Kuwait that has launched 5G services. For details, refer to the 5G inclusion map on their website.

STC launched a 5G Innovation Center in 2019 to explore and develop 5G use cases in Kuwait. As of February 2019, they had over 1,000 5G NR destinations worldwide and offered 5G administrations in partnership with Huawei.

Arab Emirates

Etisalat UAE launched its 5G services in the UAE on May 30, 2019. Several 5G phones are available for procurement.

In mid-2019, Etisalat UAE concluded an arrangement with Huawei to "offer its most recent best-in-class organization arrangements, including 5G remote, 5G administration situated center, and 5G prepared vehicle system to encourage smooth 5G innovation adoption." Etisalat UAE has also chosen Ericsson to provide 5G, enabling portable broadband and fixed remote access in the United Arab Emirates.

5G is also being introduced to the United Arab Emirates by Du. Formerly known as EITC, or Emirates Integrated Telecommunications Company, it announced in mid-2019 the rollout of 700 5G sites. Its accomplices include Nokia and Huawei.

Virgin Mobile is another 5 G-prepared organization deploying this cutting-edge technology across the UAE. Discover the 5G phones available from Virgin Mobile.

India

The Telecom Regulatory Authority of India tracked the development of the 5G standard draft and identified when 5G became widely used.

Manoj Sinha, the Minister of Telecommunications, India, is expected to receive 5G this year. When the world turns 5G in 2020, India will be at the same standard as them.

On top of that, in 2018, among India's major telecom suppliers, Vodafone Idea Limited converged with Vodafone (the world's second-biggest telephone business before the merger). Vodafone was preparing for 5G, getting future development in 2017 by updating its entire stereo phone system to support 5 G.

Jio, a system administrator in India, intends to offer 5G services in 2020, alongside their 5G handsets.

Bharat Sanchar Nigam Limited (BSNL) is another Indian telecom business experimenting with 5 G. In mid-2019, BSNL upgraded its partnership with Ciena, a systems integration framework company, to establish its 5G system.

Cavil Wireless completed a 5G test arrangement in 2020.

Indonesia

A Telkomsel SIM card was required to associate with the 5G system.

It is unclear whether Indonesia will see a business 5G rollout in 2020 or later; yet, a pilot of this size was a significant marker that they are on track. Additionally, the organization has collaborated with Ericsson to update its system in preparation for 5 G.

Turkey

Turkey's Cellular is the country's most visible mobile phone operator. In mid-2017, the company completed a 24.7 Gb/s 5G preliminary agreement with Ericsson, and in September 2018, it reported a collaboration with Nokia to develop 5G technologies.

In November, the company trialed 5G fixed remote access arrangements with Samsung in Istanbul. Turkcell's CEO remarked that today, with 5G, we have shown that the most recent era of fast remote access is now practical for our clients. Our objective is clear: to make Turkey among the leading nations in the world with 5G innovation.

In mid-2019, Turkey's Communication and Information Technologies Authority (BTK) affirmed 5G preliminaries in Ankara, Izmir, and Istanbul. The businesses included Turkcell, TT Mobil, and Vodafone Turkey.

Turk Telekom is yet another company exploring the introduction of 5G to Turkey. In September 2019, the CEO stated the company is the "most well-prepared administrator for 5G as much as fiber foundation commonness. «Turkcell is on the right track to furnishing Turkey with 5G, yet it is hazy when clients can precisely anticipate a live system.

Vietnam

As indicated by Viettel, the nation's state-owned and most prominent telecom organization, 5G trials were conducted in 2019 and were intended to have a system operational by mid-2020.

Mobifone is another versatile system administrator that intends to deploy 5G in Vietnam in 2020.

Iran

Several portable system administrators, including those from the Mobile Telecommunication Company of Iran (MCI), are currently in Iran. MCI offers a "4.5G" web, indicating that they are on the path to providing 5G in Iran. In 2017, they also agreed to a deal with Nokia to develop 5G technology in Iran.

Iran's second-largest supplier, Irancell, offers a range of versatile and fixed remote Internet services. In late 2017, Ericsson, along with two of Iran's most prominent players, deployed its first 5G test in Tehran and announced that 5G would be accessible in Iran by 2020.

Thailand

Propelled Info Service (AIS), the nation's largest mobile operator, launched 5G in Thailand in 2020.

Taiwan

Chunghwa Telecom, Taiwan's most prominent company, launched 5G on July 1, 2020.

FarEasTone intends to deploy 2,000 5G base stations by the end of 2020.

Taiwan Star offers its clients a 5G redesign experience, which they can select to take advantage of discounted costs once the new system is released in Taiwan, expected to occur in August.

Singapore

StarHub reported in November 2018 that Nokia had completed its first 5G pilot on the 3.5 GHz frequency band. There is no information available on when StarHub plans to prepare a 5G network for Singaporean clients.

According to the Singaporean government's structure, the IMDA (Info-communications Media Development Authority) expects a 5G rollout to occur in 2020. Two systems have been introduced to Singapore since IMDA plans to allocate millimeter bands for 5 G, which «will be adequate for at least two nationwide 5G systems."

The four telcos that may bring 5G to Singapore include Singtel, M1, and TPG Telecom.

Philippines

The small interchanges organization Smart has been trying 5G since 2016 and declared 2018 the deployment year for 5G at the TehnoLab. Their 5G development lab, Savvy, aims to have a 5 G-prepared system live for clients.

In November 2018, Smart announced its first 5G cell sites in the Philippines. The Makati Central Business District (Makati CBD) is located in the Clark Freeport Zone, Pampanga.

Their 5G deployment plan aims to achieve rates exceeding 14 Gbps, and they have successfully completed a 5G-powered video call.

Globe Telecom offers a 5G fixed remote broadband service called Globe at Home Air Fiber 5 G. It provides speeds of up to 100 Mbps and information bundles as extensive as 2 TB.

Telecom is another organization that wants to roll out 5G in the Philippines.

PLDT aims to have 5G fully operational by the end of this year.

Bangladesh

Even though Bangladesh is one of the most crowded nations on the planet, the delay was due to 4G, and it probably took longer than other countries to deploy 5 G.

In mid-2018, the nation's telecom controller, BTR, C said, "The world will grasp 5G in 2020. In this way, we should acknowledge innovation and proceed to 55 G. There is no choice for tarrying."

BTRC relies on the selling range for 5G services before 2020 to achieve widespread 5G coverage by 2026.

BTCL and Bangla Link are two organizations looking for 5G in Bangladesh.

Malaysia

5G in Malaysia is expected to become accessible in select areas, (Chhabra & Rogers Communications Inc., 2023).

In 2019, Huawei, U Mobile, ZTE, and Maxis worked on Malaysia's 5G arrangements. U Mobile intends to go live before the end of 2020.

In late 2019, TM declared its interest in the 5G Demonstration Project, which would test new 5G features and determine the best way to implement them in Malaysia. For example, it has tried utilizing 5G for intelligent traffic lights, bright well-being and security, and keen stopping.

The Malaysian Communications and Multimedia Commission (MCMC) anticipates that businesses will be offering 5G services in Malaysia by the next quarter of 2020.

Pakistan

5G may go live for buyers this year, given that the Pakistan Telecommunication Authority arranged for open 5G trials in 2019.

Pakistan is set to introduce 5G because its 4G pioneer, Zong, also became the nation's leading organization to test 5G services in August 2019.

Telenor Pakistan is also investigating the introduction of 5G to the nation, but has not announced a specific date.

Saudi Arabia

5G has been accessible in Saudi Arabia through the Saudi Telecom Company (STC) since June 2019. However, it is only in explicit zones of significant urban communities and only through home switches (a stable choice, yet).

As indicated by the organization's CEO:

5G was considered a significant advance toward digitalization and the interconnection of everything, supporting Saudi Arabia's Vision 2030 in its pursuit of economic expansion.

The 5G rollout in Saudi Arabia occurred long after Ericsson's arrangement to deploy 5G in the country was finalized. STC has likewise joined forces with Nokia to deploy 5G in Saudi Arabia; they reported in mid-2019 that the initiation stage had just begun.

Another 5G supplier in Saudi Arabia is Zain. An aggregate of 30 areas has 5G aaccess This would include Riyadh, Al Khobar, Dhahran, Tabuk, Hail, Sabya, Ahad Rafidah, Taif, Bisha, Najran, and Jaizan; The Samsung Galaxy Note10+ 5G and ZTE Axon 10 Pro 5G are two of Zain's 5G items.

Mobily 5G is another alternative. Use their inclusion checker to determine if you can access 5G where you are. Notwithstanding, their versatile 5G plan is a 5G home web plan.

Bahrain

5G in Bahrain was introduced in July 2019 when STC (now known as VIVA) launched its 5G data plans. You can get versatile 5G from STC with one of their 5G telephones or 5G at home with the 5G switch. The administration is accessible in areas such as Reef Island, Amwaj Islands, and Riffa Views.

Batelco is another company deploying 5G in Bahrain, picking Ericsson to provide the 5G hardware. However, only a few territories, such as the Amwaj Islands and Reef Island, have access to this service.

Zain launched its system on June 29, 2020.

Kazakhstan

Kazakhstan and 5G are the ninth-biggest nations on the planet. Nonetheless, as Prime Minister Askar Mamin indicated, 5G is unquestionably insightful and requires an "Activity Plan for the execution of 5G in Kazakhstan."

In 2019, 5G deployed its technology to cover all of Kazakhstan's settlements, which had a population of over 50,000 individuals.

In October 2019, VEON, a telecom organization, launched a 5G pilot on a live system.

Sri Lanka

The nation's second-biggest versatile system administrator, Mobitel, invested USD 50 million in 2019 to produce 5G in Sri Lanka.

Mobitel has worked to bring 5G to Sri Lanka for a long time. In mid-2017, they joined forces with Ericsson to mark the 5G with a Memorandum of Understanding (MoU) to bring 5G to Sri Lanka.

Cambodia

The specialist organization Smart Axiata is likely the leading organization to deploy 5G in Cambodia. In 2014, they were the main specialist co-op to make 4G accessible nationwide, and in 2019, they followed up by exhibiting their first 5G live demonstration. The organization plans to focus the initial limited phases of its 5G rollout in Cambodia on hotspots in major urban areas.

Viettel Cambodia (Metfone) is another organization associated with Telecom Cambodia that is driving the rollout of 5G in Cambodia.

AFRICA

5G is accessible in a few African nations.

South Africa

Vodacom Group was the first to introduce 4G, 3 G, and 2G in South Africa and is now expanding with the arrival of 5 G. In May 2020, Vodacom launched 5G services in Johannesburg, Pretoria, and Cape Town. More extensive inclusion is consistent.

Downpour is another South African telecom offering 5 G. As of November 11, 2019, Rain clients can access the 5G network from home in Johannesburg and Tshwane, with more regions set to open throughout 2020. The Rain inclusion map shows the 5G administration in South Africa.

In November 2018, MTN Group Limited collaborated with Ericsson to deploy a fixed remote access 5G site in Midrand, South Africa. Their organization operates 5G networks in several zones and will expand further throughout 2020.

Nigeria

The Nigerian Communications Commission (NCC) plans to roll out 5 G in 2020.

Senegal

Tigo Senegal Telecom and Ericsson are redesigning their systems to deliver LTE across 1,000 sites.

Egypt

Although Egypt's 5G release date remains unknown, Telecom Egypt and Nokia agreed in mid-2019 to develop and present a 5G plan.

Morocco

The telecom organization provides more than 90 percent of the nation with mobile web access and is bringing 5G to Morocco.

Two authorized telecom organizations in Morocco are Orange Morocco and Maroc Telecom. However, both have been quiet on a 5G arrangement.

Republic of the Congo

The discharge date for 5G in the Congo is unclear; however, as indicated by Léon Juste Ibombo, the nation's Minister of Posts, it involves Applus and Congo Telecom. Interest in Congo's media communications division is immense, and the state needs trustworthy organizations to assist it in implementing an advanced economic environment. Therefore, this organization supports our officeholder administrator, Congo Telecom, in implementing 5 G.

Tunisia

Ooredoo Tunisia has partnered with Nokia, using its AirGile cloud-local center, to advance to the point where it can provide Tunisia clients with 5 GG.

Kenya

Safaricom probably deployed 5G in Nairobi in 2020.

Madagascar

5G was introduced in Antananarivo and Toamasina on June 26, 2020, using the Telma network.

Uganda

ZTE and MTN Uganda launched the first 5G SA arrangement in Nyonyi Garden, Kampala, Uganda, on January 21, 2020.

Lesotho

Vodafone Lesotho has a guidelines-based 5 G business organized and accessible to two banking and mining venture customers. A system extension will also benefit other sectors.

OCEANIA

Most wealthy nations in Oceania experienced a limited rollout of 5G in 2019, with more widespread availability in 2020.

Australia

Australia's second-largest media communications organization, Optus, has launched 5G for mobile and home use. It was used on November 4, 2019. The dispatch required more than 290 5G sites in several locations, including Sydney, Canberra, Adelaide, Brisbane, Melbourne, Perth, and other areas in New South Wales, Victoria, and Queensland. See the Optus 5G inclusion map for details, (Fieldfisher, 2020). Telstra reported in August 2018 that it had enabled 5G innovation in areas of the Gold Coast, making Queensland the first state in the nation to launch a 5G pilot initiative. Additionally, it deployed 5G base stations in Adelaide, Canberra, Perth, Melbourne, and Sydney in October 2018, and Melbourne and Sydney in December.

In November 2018, Telstra completed Australia's first 5G association with a. The Network Engineering Executive said they would "keep testing over the coming years to improve information rates."

In December 2018, Telstra announced that its first client to utilize a 5G device—the HTC 5G Hub—was FKG Group in Toowoomba.

Vodafone has given a 2020 rollout date for 5G in Australia. It is a reasonable date, considering Vodafone is not the nation's largest mobile supplier exclusively, as they partnered with Nokia in 2019 to agree on a 2020 rollout. So, Vodafone clients can expect 5G in these territories.

New Zealand

The New Zealand Minister of Broadcasting, Communication, and Digital Media, Kris Faafoi, completed the first designation of the 5Grange as the 3.5 GHz band. The government planned to unload national rights in 2020.

Having laid the groundwork for 5G and initiated discussions on 5G in 2018, New Zealand launched its plans to establish a 5G network in 2020. In September 2019, they turned on 5G without precedent for a predetermined number of clients and launched it in five different cities in December 2019.

Vodafone is a significant player in New Zealand's 5 G market. On December 10, 2019, it turned on 5G in different parts of Auckland, Wellington, Christchurch, and Queenstown. Vodafone New Zealand's 5 G-ready phones are listed below.

Guam

5G fixed remote access was introduced in Guam during the second half of 2019 through SK Telecom and IT&E.

According to SK Telecom, the rollout of 5G has begun in Guam, focusing on highly concentrated urban areas that serve local business clients and areas requiring fixed-line infrastructure. Coverage will then extend to more extensive regions.

THE 5G ADOPTION BY COUNTRY: KEY FINDINGS AND COMPARATIVE INSIGHTS

The global adoption of 5G is accelerating, with countries at varying stages of deployment and integration. This section provides a comparative overview of 5G adoption by selected countries, including key findings, benefits realized, and existing shortcomings compared to previous mobile network technologies, such as 4 G. Despite the global expansion of 5G networks, most countries had only limited, early-stage access as of 2020, highlighting disparities in technological rollout, (Fieldfisher, 2020).

Table 1 summarizes the state of 5G deployment and adoption across different regions.

The comparative analysis of 5G adoption across key global markets reveals how different countries are leveraging the technology in line with their strategic priorities, infrastructure readiness, and regulatory frameworks. According to Thompson et al. (2014), the outcomes of 5G deployment vary significantly depending on national policies, investment levels, and spectrum availability. The following table summarizes significant findings from leading nations, identifying key benefits and challenges in comparison to 4G networks.

Table 1. 5G adoption by country: Key findings and comparative analysis

Country	Key Findings	Benefits Compared to 4G	Shortcomings / Challenges
South Korea	Among the first to roll out nationwide 5G, a strong user base and dense urban deployment.	Faster download speeds, better streaming quality, and enhanced industrial automation.	Limited rural coverage; high infrastructure costs.
United States	Widespread deployment in urban centers; mmWave and sub-6GHz usage.	Improved data speeds and low latency for innovative city applications.	Uneven coverage, high spectrum licensing fees.
China	Largest 5G network globally by user base and infrastructure.	Supports IoT, smart manufacturing, and public safety applications.	Concerns about surveillance and international trade restrictions.
Germany	Focused on industrial 5G for manufacturing and logistics.	Private networks for factories and automation; improved reliability.	Slower consumer rollout; spectrum licensing delays.
India	Recently launched commercial 5G services in major cities.	Potential for digital inclusion, telemedicine, and education.	Network instability; lack of affordable 5G devices.

CONCLUSION

Integration of 5G technologies into global network infrastructure is catalyzing innovation and reshaping socio-economic paradigms, (Thompson et al., 2014).

However, the global rollout of 5G technology has unfolded unevenly across different countries, shaped by infrastructure readiness, financial capacity, and regulatory support. Updated global metrics show incremental improvements in internet accessibility, digital literacy, and network readiness, underscoring regional disparities in 5G deployment capacity (ITU, 2021).

This chapter provided a detailed review of how nations advance in their 5G deployment, revealing the complexity of achieving universal next-generation connectivity. While countries such as South Korea, China, and the United States have made substantial progress, other regions face technical, logistical, and economic challenges. 5G's potential to bridge the digital divide remains significant—but only if strategic efforts continue to address disparities in deployment, spectrum access, and equipment investment (Seils, Christner, & Jin, 2010). While 5G is being rolled out globally, access remains uneven, with early-stage availability differing widely across countries, (Fieldfisher, 2020).

The implications of 5G go far beyond higher download speeds. This technology enables real-time communication, connected vehicles, smart city infrastructure, and healthcare innovation. However, this promise will remain out of reach for many unless global and local stakeholders coordinate to remove economic, technical, and policy roadblocks.

The global rollout of 5G represents a transformative leap in wireless technology, reshaping how people, devices, and industries connect and interact with one another. However, the availability of 5G varies significantly across countries, reflecting differences in infrastructure, economic development, and regulatory environments. While leading nations such as the United States, South Korea, and China have achieved substantial progress in deploying 5G networks, other regions are still grappling with implementation challenges, including limited funding, spectrum allocation complexities, and technological upgrades.

By 2023, 5 G is estimated to power over 10% of the world's mobile connections, underscoring its growing impact on global connectivity. This advancement drives innovation in healthcare, transportation, and manufacturing, while transforming everyday experiences through faster speeds and enhanced network reliability. Nevertheless, disparities in 5G availability highlight the risk of widening the global digital divide. Countries with limited infrastructure and resources often face delays in adoption, which can hinder their ability to fully leverage the benefits of this revolutionary technology.

Initiatives to bridge these gaps are underway, with governments, international organizations, and private companies collaborating to accelerate 5G deployment in underserved countries. These initiatives aim to promote equitable access and ensure that 5G's transformative potential benefits a wider range of populations. Innovations such as network sharing, cost-effective infrastructure solutions, and public-private partnerships are crucial in overcoming deployment barriers.

In conclusion, the availability of 5G by country illustrates both the opportunities and challenges of a global technological revolution. While early adopters are reaping the benefits of enhanced connectivity, the journey toward universal 5G access remains ongoing. By addressing the disparities in deployment and fostering global collaboration, stakeholders can ensure that 5G becomes a unifying force for innovation, economic growth, and societal progress across all regions.

REFERENCES

Bishen, A. (2019). *5G and the Cloud.* 5G Americas. https://www.5gamericas.org/5g-and-the-cloud

Chhabra, A. & Rogers Communications Inc. (2023). *Convergence With 5G Applications and Use Case Monetization (2023).* SCTE TechExpo. https://www.nctatechnicalpapers.com/Paper/2023/3629_Chhabra_5105_paper

Cisco. (2020). *Cisco Annual Internet Report (2018–2023) White Paper*. Cisco. https://www.cisco.com/c/en/us/solutions/collateral/executive-perspectives/annual-internet-report/white-paper-c11-741490.html

5G. co.uk. (2025). *The 5G Comparison Site*. 5g.co.uk. https://5g.co.uk/

De Looper, C. (2019). *What Is 5G? Digital Trends.* Yale University. https://archive-yaleglobal.yale.edu/content/what-5g-digital-trends

Deloitte. (2025). *5G: The chance to lead for a decade*. 5G Americas. https://www.5gamericas.org/5g-the-chance-to-lead-for-a-decade/

DIGI. (20250. *Go-To 5G Connectivity*. Digi International Inc. https://www.digi.com/solutions/by-technology/5g

Ericsson. (2019). *5G for business: a 2030 market compass. Setting a direction for 5G-powered B2B opportunities*. Ericsson.com. https://d110erj175o600.cloudfront.net/wp-content/uploads/2019/10/the-5g-for-business-a-2030-compass-report-2019.pdf

Ericsson (2023). *Ericsson Mobility Report.* Ericsson.com. https://d110erj175o600.cloudfront.net/wp-content/uploads/2023/06/21123003/Ericsson-Mobility-Report-June-2023.pdf

Ericsson. (2025). *Ericsson Mobility Report | Read the latest edition. Ericsson.com.* https://www.ericsson.com/en/mobility-report/reports/june-2020/fwa-outlook

Evans, A. (2017). *Intel accelerates the future with the world's first global 5G modem.* Microwave Journal. https://www.microwavejournal.com/articles/27684-intel-accelerates-the-future-with-worlds-first-global-5g-modem

Explained, E. M. F. (2019) *5G and EMF Explained.* AMTA. https://emfexplained.info/site/misc/emf/downloads/5G&EMF%20Explained_AMTA_23Aug_2019_20.pdf

Fieldfisher (2020). *Impact of 5G on AI and industrial IoT applications*. Fieldfisher.com. https://res.cloudinary.com/fieldfisher/image/upload/v1610106432/PDF-Files/The_impact_of_5G_on_AI_December_2020_glpw1r.pdf

Greene, B. (2018, December 9). *5G Latency - Reality Checks*. SENKI. https://www.senki.org/5g-latency-reality-checks/

Horwitz, J. (2019). *The definitive guide to 5G low, mid, and high band speeds*. VentureBeat. https://venturebeat.com/mobile/the-definitive-guide-to-5g-low-mid-and-high-band-speeds/

innovationoecd. (2017). *OECD Digital Economy Outlook 2017: Presentation at Global Parliamentary Network, 11 October 2017*. SlideShare; Slideshare. https://www.slideshare.net/slideshow/oecd-digital-economy-outlook-2017-presentation-at-global-parliamentary-network-11-october-2017/80686796

Intelligence, G. S. M. A. (2020). *The Mobile Economy 2020*. GSMA Intelligence. https://www.gsmaintelligence.com/research/the-mobile-economy-2020

Intelligence, G. S. M. A. (2023). *The Mobile Economy 2023*. GSMA Intelligence. https://www.gsma.com/solutions-and-impact/connectivity-for-good/mobile-economy/wp-content/uploads/2023/03/270223-The-Mobile-Economy-2023.pdf

ITU. (2018). *Setting the scene for 5G: Opportunities & challenges*. ITU. https://www.itu.int/en/ITU-D/Documents/ITU_5G_REPORT-2018.pdf

ITU. (2020). *Measuring digital development: Facts and figures 2020*. International Telecommunication Union. https://mpt.gov.by/en/node/6911/printable/pdf#:~:text=ITU%E2%80%99s%20Measuring%20digital%20development%3A%20Facts%20and%20figures%202020,Latest%20figures%20show%20a%20wide%20urban-rural%20connectivity%20gap

ITU. (2021). *Measuring digital development: Facts and figures 2021*. International Telecommunication Union. https://mpt.gov.by/en/node/6911/printable/pdf#:~:text=ITU%E2%80%99s%20Measuring%20digital%20development%3A%20Facts%20and%20figures%202020,Latest%20figures%20show%20a%20wide%20urban-rural%20connectivity%20gap

Nokia. (2019). *SETAR and Nokia bring 5G to Aruba in end-to-end deal*. Noki. https://www.nokia.com/newsroom/setar-and-nokia-bring-5g-to-aruba-in-end-to-end-deal/

Nokia. (2020). *Nokia 5G Readiness Report: Forecast $8 Trillion Impact on Global GDP by 2030*. Benton Foundation. https://www.benton.org/headlines/nokia-5g-readiness-report-forecast-8-trillion-impact-global-gdp-2030

OECD. (2021a). *The digital transformation of SMEs*. https://www.oecd.org/ OECD. https://www.oecd.org/en/publications/the-digital-transformation-of-smes_bdb9256a-en.html

Panetta, K. (2019, October 21). *Gartner Top 10 Strategic Technology Trends for 2020.* Gartner. https://www.gartner.com/smarterwithgartner/gartner-top-10-strategic-technology-trends-for-2020

Pirinen, P. (2014, November 1). A brief overview of 5G research activities. *IEEE Xplore*. https://doi.org/DOI: 10.4108/icst.5gu.2014.258061

Pirinen, P. (2014, November 1). A brief overview of 5G research activities. *IEEE Xplore*. https://doi.org/DOI: 10.4108/icst.5gu.2014.258061

Polese, M., Jana, R., & Zorzi, M. (2017). TCP and MP-TCP in 5G mmWave Networks. *IEEE Internet Computing*, *21*(5), 12–19. DOI: 10.1109/MIC.2017.3481348

Satariano, A., & Bayer, E. (2019). MWC 2019: Huawei Politics, Robots and Speedy 5G. *The New York Times*. https://www.nytimes.com/2019/02/27/technology/mwc-2019-5g-huawei.html

Seils, D. M., Christner, J., & Jin, R. (2010). *Deploying Cisco Wide Area Application Services* (2nd ed.). Cisco Press., https://www.ciscopress.com/store/deploying-cisco-wide-area-application-services-9781587059124

Selinis, I., Katsaros, K., Allayioti, M., Vahid, S., & Tafazolli, R. (2018). The Race to 5G Era: LTE and Wi-Fi. *IEEE Access : Practical Innovations, Open Solutions*, *6*, 56598–56636. DOI: 10.1109/ACCESS.2018.2867729

Simnovus (2024). *Sustainability in 5G RAN: A Key to Green Connectivity.* Simnovus. https://simnovus.com/sustainability-in-5g-ran-a-key-to-green-connectivity/

Strategy, A. (2021). *The Impact of 5G on the U.S. & European Economies*. Qualcomm. https://www.qualcomm.com/content/dam/qcomm-martech/dm-assets/documents/accenture_5g_economic_impact_us_eu_overview_slides.pdf

Taylor, P. (2025). *Forecast number of mobile 5G subscriptions worldwide from 2023 to 2028 (in billions)*. Statista. https://www.statista.com/statistics/760275/5g-mobile-subscriptions-worldwide/

Thompson, J., Ge, X., Wu, H.-C., Irmer, R., Jiang, H., Fettweis, G., & Alamouti, S. (2014). 5G wireless communication systems: Prospects and challenges [Guest Editorial]. *IEEE Communications Magazine*, *52*(2), 62–64. DOI: 10.1109/mcom.2014.6736744

ADDITIONAL READING

Dolcourt, J. (2019). *We ran 5G speed tests on Verizon, AT&T, EE and more: Here's what we found*. CNET. https://www.cnet.com/tech/mobile/features/we-ran-5g-speed-tests-on-verizon-at-t-ee-and-more-heres-what-we-found/

ihs.com. *(2017) The 5G Economy: How 5G Technology Will Contribute to the Global Economy*. Docslib. https://docslib.org/doc/6895067/the-5g-economy-how-5g-technology-will-contribute-to-the-global-economy

World Economic Forum. (2018). *The Future of Jobs Report 2018*. World Economic Forum. http://www3.weforum.org/docs/WEF_Future_of_Jobs_2018.pdf

Section 2
The Technology of 5G

Chapter 5
The Technological Perspective of 5G

ABSTRACT

5G is more than a mobile broadband upgrade, it's a transformative force enabling innovation across industries. With features such as ultra-reliable low-latency communication (URLLC), massive machine-type communication (mMTC), and enhanced mobile broadband (eMBB), 5G addresses the unique needs of sectors including automotive, healthcare, utilities, and finance. eMBB boosts speed, latency, and user density for seamless connectivity. mMTC supports vast networks of low-cost, battery-powered devices, enabling smart meters, IoT sensors, logistics systems, and wearables. 5G empowers both individuals and organizations with next-level connectivity and efficiency.

INTRODUCTION

The fifth generation of wireless, of the 5G technology, is far more than an incremental advancement in mobile broadband—it is a transformative enabler of innovation across industries. With its unparalleled capabilities, 5G represents a technological revolution, empowering individuals and organizations to achieve new levels of connectivity, efficiency, and creativity. Far from being limited to faster speeds, 5G introduces robust features such as ultra-reliable low-latency communications (URLLC), massive machine-type communications (MTC), and enhanced mobile broadband (eMBB), each addressing distinct needs within the digital ecosystem. These features position 5G as a game-changing technology across the automotive, healthcare, public safety, utilities, manufacturing, and finance sectors.

A recent survey revealed that 89% of industry leaders across eight diverse sectors anticipate 5G to revolutionize their respective fields. Enhanced mobile broadband (eMBB) enhances data rates, reduces latency, and increases operator density,

DOI: 10.4018/979-8-3373-0497-7.ch005

providing seamless broadband access that scales with rising consumer demands. Meanwhile, mMTC enables connectivity for vast networks of inexpensive, battery-operated devices, making it indispensable for smart meters, IoT sensors, logistics, and wearable technology applications. Finally, URLLC delivers ultra-reliable, low-latency communication, crucial for time-sensitive operations such as remote surgeries, autonomous vehicles, industrial automation, and smart grids. Popovski et al. (2018) state that ultra-reliable low-latency communication (URLLC) in 5G enables critical applications such as autonomous driving and industrial automation.

Partnerships among governments, tech companies, and global institutions are actively addressing these challenges, aiming to expand 5G access to more regions and diverse populations. According to Evercom Communication Technology Co., Ltd., 5G LTE antennas are engineered to support a wide range of frequencies (700 MHz to 3800 MHz), ensuring robust performance across diverse 5G and LTE deployment scenarios, (Evercom, 2025).

The 5G technologies lay the foundation for smarter cities, safer industries, and more efficient global operations by bridging the gap between people, devices, and machines. In this chapter, we will explore the technological aspects of 5 G, the innovations driving its capabilities, and their implications for the future of connectivity and industrial transformation.

Far more than an evolution of mobile broadband, 5G wireless entry will be a crucial enabler of the Internet of Things (IoT), empowering industries and individuals to reach new heights in innovation and efficiency. A recent survey of eight diverse sectors (automotive, media, health care, public safety, utilities, finance, web, and manufacturing) revealed that 89% of respondents expect 5G to become a game-changer in the industry.

5G wireless access gathers several broad categories:

Enhanced mobile broadband internet (eMBB), and robust machine type communications (ultra-reliable low-latency and MMTC) communications (URLLC). Advanced features of 5G, like eMBB, support real-time virtual and augmented reality experiences, (Park et al., 2020).

eMBB enhances information flow, latency, operator density, coverage, and global mobile broadband internet access capacity.

mMTC interacts with inexpensive, massive quantities of battery-driven products to assist users, such as those utilizing intelligent metering strategies and field and body sensors.

Lastly, URLLC facilitates communication between machines and devices by offering ultra-reliability, high availability, and low latency. It is critical for vehicular communication, remote surgery, factory automation, industrial control, innovative grid applications, and public safety.

Earlier 5G deployments leveraged legacy 4G networks to accelerate initial rollouts. The 5G system architecture defined in Release 17 integrates core network functionalities with service-based interfaces, enabling dynamic network slicing and edge computing.

However, much more notably, the 4G network core links most first deployments of 5G use a non-standalone (NSA) structure, where the 5G nodes (known as gNodeBs). It enables the 5G level to be anchored to 4G, thereby enhancing the complete user experience by automatically falling back to an LTE connection when coverage from the 5G level is unavailable.

Regarding deployment difficulties, 5G rollouts require the densification of cell towers in buildings and metro areas. Although it is still in the early stages, information and processing are being routed through the center toward the edge to mitigate bottlenecks. As the system's metro level becomes vital, powering these locations will be crucial from price and time perspectives. PBS: Energy, backhaul, and website acquisition have typically been critical in creating a system; this is especially true for 5 G.

Main Characteristics

The fifth-generation (5G) wireless network introduces a set of advanced technological characteristics that distinguish it from previous generations, including ultra-low latency, enhanced data throughput, massive device connectivity, and high spectrum efficiency—features that collectively enable next-generation applications across multiple industries (Chowdhury et al., 2018). The 5G has the following characteristics:

Lower Latency Connectivity

Latency refers to the time it takes for devices to respond to each other over a wireless network. For example, 3 G networks have a response time of 100 milliseconds, 4G is around 30 milliseconds, and 5G will be as low as 1 millisecond. Therefore, it is virtually instantaneous, opening a new world of connected applications.

Substantial Developments in Mobile Unit Engineering and Network Architecture

The substantial developments in mobile unit engineering and network architecture lead to lower latency.Significant advances in mobile device technology and network architecture led to lower latency with 5G technology. The 5G Network—Mobile

Network Architecture substantially changed the Core Network (Core) and the Radio Access Network (RAN) to deliver low latency, (Huawei, 2025).

Latency in a computer system describes the time it takes for an information packet to travel. Lower latency results in a higher data transmission fee. Ultra-low latency is the primary factor driving the 5G services market. The ultra-low latency of 5G is nearly 50 times lower than that of the current 4G network. The 5G engineering is anticipated to achieve one-millisecond latency, whereas the latency in 4 G is fifty milliseconds. Due to this, 5G services will significantly rely on manufacturing uses and connected IoT ecosystems, as these connections and applications require reliable and faster network services.

Ultra-low latency is a key element driving the current market for 5G solutions. Latency refers to the period an information package requires to travel through a laptop system. Reduced latency results in a greater detail transmission fee. The ultra-low latency of 5G is practically fifty times smaller compared to the present 4G network. The 5G engineering supports one-millisecond latency in 4G, whereas the latency fee is fifty milliseconds. Due to this specific need, 5G providers will play a significant role in the manufacturing sector and within IoT ecosystems, as the connections and applications require reliable and faster network products.

Millimeter-Wave Spectrum

5G technology utilizes high-frequency millimeter waves, enabling ultra-fast data transmission but with limited range, necessitating more densely placed antennas to maintain consistent coverage, (Team Roar, 2017). The mmWave enables users to access much more bandwidth than before.

For instance, the Millimeter trend is an underdeveloped spectrum band utilized in a wide range of services and products. These include high-speed, point-to-point wireless community area networks (WLANs) and high-speed broadband access. Sixty GHz millimeter waves use Unlicensed short-range data links.

The millimeter-wave (millimeter band) spectrum spans the frequency range from 30 GHz to 300 GHz.

Scientists are testing 5G wireless broadband engineering on the millimeter-wave spectrum.

Millimeter-wave, also often referred to as extremely high frequency (EHF) or very high frequency (VHF) by the International Telecommunications Union (ITU), can be utilized for high-speed wireless broadband internet communications. The millimeter trend is an undeveloped spectrum band used in extensive services and products, including high-speed, point-to-point wireless community area networks (WLANs) and broadband internet access. Furthermore, millimeter-wave technology is utilized in various products for wireless and mobile networks, enabling data rates of up to ten Gbps.

Millimeter waves have short wavelengths, ranging from 1 millimeter to 10 millimeters in length. They have very high atmospheric attenuation, absorbed by environmental fumes, reducing their range and power. Humidity and rain can affect performance and reduce signal strength, resulting in rain fade. Because of its short kilometer range, a millimeter wave travels by line of sight. Therefore, physical objects like trees and buildings obstruct its high-frequency wavelengths.

High-bandwidth point-to-point corresponds to millimeter waves ranging from 71 GHz to 76 GHz, 81 GHz to 86 GHz, and 92 GHz to 95 GHz. This requires a license from the Federal Communications Commission (FCC). *The 60 GHz millimeter waves use unlicensed short-range details backlinks*. For example, the IEEE Wi-Fi standard 802.11ad operates at a 60 GHz millimeter wave frequency.

Increased Bandwidth

Users are increasingly demanding increased bandwidth. Technological innovation, which involves much more information than previously, such as high-quality video streaming and virtual reality software programs, is a necessary component. Telecoms aim to leverage 5G to deliver high speeds, low latency, and enhanced value to an increasingly large base of mobile data users.

Virtualization of Mobile Networks

In 5G, NFV will separate a physical network into multiple virtual networks, which can support different radio access networks (RANs) or services for specific customer segments. Marriwala et al. (2020) analyzes the vulnerabilities of 5G radio access networks.

The virtualization of mobile networks is a vital component of 5 G. Network Function Virtualization (NFV) plays a crucial role in orchestrating this virtualization. Operators can utilize a software-defined network (SDN) to create a system topology with multiple hierarchies. The various hierarchies will contain radio access system (RAN) radios of numerous sizes.

Technology Response Time (Milliseconds)

The evolution from 4G to 5G is marked by dramatic improvements in response time, with 4G LTE systems averaging 20–30 milliseconds, 5G enhanced mobile broadband reducing this to 4–5 milliseconds, and 5G URLLC systems achieving latencies as low as 1 millisecond—necessitating substantial architectural changes in both the Core Network and the Radio Access Network (RAN) to support such ultra-low latency performance.

Core Network Changes and Radio Access Network Changes

A vital characteristic of the redesigned core system, signaling, and sent-out servers is moving information closer to the end user and reducing the route between products for critical programs. Excellent examples are video-on-demand streaming services, wherein it is feasible to keep a message or cache of popular information on community servers; therefore, access is quicker.

Regarding Radio Access Network changes, reducing latency requires reconfiguring the Radio Access Network (RAN) in a highly adaptable fashion and making it configurable to support the different attributes of the service types that the 5G method envisages.

The environment interface's high reliability and low latency are key characteristics of 5 G. They require newer radio strategies to minimize time delays over the air within several TTI (time transmit intervals). This reliability and low latency are combined with robustness and coding enhancements to achieve higher reliability (e.g., an email is delayed or lost every billion times).

A virtual, configurable, and dynamic RAN enables the system to achieve high throughput and low latency. It also allows the mobile network to adapt to changes in network traffic, new topology, and network fault demands.

Configuration

The brand-new structure will exist as a 4G/5G split RAN (Radio Access Network), where the person airplane (5G) and the management airplane (4G) will be different. This new structure requires separating standard goal hardware from specialized network hardware. The performance of standard goal hardware (nodes) is ideal for network-performing virtualization (NFV), where the specialized hardware in the RAN will be dynamically configurable.

Table 1. outlines the primary technological features that distinguish 5G from earlier wireless generations. These include reduced latency, expanded bandwidth, and the use of millimeter-wave frequencies, all of which support new use cases across industries. The integration of virtualization technologies such as NFV and SDN enhances network adaptability and resource efficiency. Collectively, these capabilities position 5G as a foundational platform for advanced digital services and next-generation connectivity.

Table 1. Key characteristics of 5G

Characteristic	Description
Ultra-Low Latency	5G reduces latency to about 1 millisecond, nearly 50 times lower than 4G (50 ms), enabling near-instantaneous communication. This is crucial for applications such as industrial IoT, autonomous vehicles, robotics, and remote medical services.
Millimeter-Wave Spectrum (mmWave)	5G operates in very high-frequency bands, ranging from 30 to 300 GHz, known as millimeter waves, which allow data speeds of up to 10 Gbps. However, mmWaves have a limited range and are sensitive to obstacles such as buildings, humidity, and rain, requiring a line-of-sight connection.
Increased Bandwidth	5G offers significantly higher bandwidth, which is essential for data-intensive applications such as 4K/8K video streaming, virtual reality, and real-time online gaming.
Mobile Network Virtualization (NFV & SDN)	5G leverages Network Function Virtualization (NFV) and Software-Defined Networking (SDN) to create multiple virtual networks on the same physical infrastructure. Each virtual network can be tailored for specific use cases or customer segments.
5G Network Architecture (Core & RAN)	5G introduces a redesigned network structure: – Core Network: Moves data and caching closer to the end-user to reduce the number of transmission paths. – Radio Access Network (RAN): Highly dynamic and configurable to support high throughput and low latency. Configuration: This implementation features a 4G/5G split between the user plane (5G) and the control plane (4G).

THE TRANSITION FROM 4G TO 5G

Hajlaoui et al. (2020) provide a comprehensive comparative analysis of 4G and 5G technologies, focusing on their respective capabilities in delivering Quality of Service (quality of service) within mobile environments. Their study highlights the technical advancements introduced by 5G, such as reduced latency, increased data throughput, and enhanced network efficiency, which necessitate its integration into modern communication infrastructures. The 3GPP technical specifications establish the foundational protocols and standards that enable 5G network functionality across radio and core segments.

Release 17 of the 3GPP system architecture outlines enhancements for service-based architecture, supporting scalability, network slicing, and edge deployments.

Complementing this, Usmonov, Rustamov, and Shukurillaevich (2019) trace the historical evolution of wireless technologies from 1G to 5 GG. They propose a flexible 5G architecture that allows mobile terminals to dynamically switch between various Radio Access Technologies (RATs) based on user-defined criteria. Their work also introduces forward-looking concepts such as the World-Wide Wireless

Web (WWWW) and Dynamic Adhoc Wireless Networks (DAWN), which aim to transform the future landscape of global mobile communication.

Long-Term Evolution (LTE) With WiMAX, Given the Advancements Promised by 5G

Following this, the authors Agiwal et al. (2021) provide a comprehensive survey on 4G-5G dual connectivity, highlighting it as a strategic solution to address the challenges of early 5G deployment. They explain that due to the initially sparse distribution of 5G infrastructure and high costs, operators can maintain seamless and economically viable services by interworking with existing 4G LTE networks. This dual connectivity ensures continuous coverage, even in areas with 5G coverage gaps, while optimizing the use of existing 4G infrastructure. The paper evaluates different dual connectivity models standardized by 3GPP, focusing on performance metrics, radio resource management, and spectrum sharing challenges. Moreover, it outlines research directions for future 5G standalone network migration.

Similarly, Agiwal, Roy, and Saxena (2016) address the technical and strategic aspects of migration and interworking between 4G and 5G networks. Their work focuses on providing system architectures that facilitate a smooth transition, allowing mobile operators to maximize the use of their existing infrastructure while introducing new 5G capabilities. Key challenges identified include supporting NSA (Non-Standalone) and SA (Standalone) 5G subscribers, integrating 5G radio technologies without overwhelming the legacy network, and implementing advanced features like network slicing to enhance service quality. The paper emphasizes that successful migration depends on tailoring the 5G deployment strategy according to the operator's current 3G/4G infrastructure and targeted new services.

To understand the advancements introduced by 5G, it is essential to compare it directly with its predecessor, 4 G. The transition from 4G to 5G represents a substantial improvement in mobile network capabilities. 5G not only enhances bandwidth and reduces latency but also introduces new architectural frameworks. While 4G is limited to Non-Standalone (NSA) architecture, 5G supports both NSA and Standalone (SA) modes. Table 2 summarizes the key differences that define this technological evolution.

Table 2. Comparison between 4G and 5G technologies

4G	5G
High latency	Ultra-reliable low-latency
Limited bandwidth	Enhanced bandwidth
NSA architecture	NSA and SA architecture

Long Term Evolution (LTE) with WiMAX in the Context of 5G Advancements

Agiwal et al. (2021) present a comprehensive survey on 4 G–5 G dual connectivity, identifying it as a strategic approach to support early 5G deployment. Given the initially limited availability and high deployment costs of 5G infrastructure, the authors argue that integrating 5G with existing 4G LTE networks allows operators to deliver reliable and cost-effective services. This dual connectivity model addresses 5G coverage gaps by leveraging the mature 4G infrastructure, ensuring an uninterrupted user experience. The section further explores various dual connectivity frameworks defined by 3GPP, analyzing key factors such as performance, radio resource management, and spectrum sharing. Finally, it discusses future research directions to support a smooth transition toward fully standalone 5G networks, especially in environments where WiMAX and LTE have previously played complementary roles.

Access Network Changes

The term "G" stands for "generation" in mobile communication technology. Thus, 5G refers to the fifth generation of mobile network evolution. Long-Term Evolution (LTE) is a wireless broadband communication standard developed by the 3rd Generation Partnership Project (3GPP), serving as a foundational component of 4G networks and enhancing data transmission speeds and network capacity, *While LTE brought substantial speed improvements over 3 G, it technically did not meet the full specifications of 4 G*. Nevertheless, most carriers market LTE as "4G LTE" because it offers performance that is functionally close to true 4 G.

Importantly, 5G is not designed to replace 4G LTE fully, but rather to complement it. LTE and 5G architectures will coexist in the coming years, working in tandem to meet a wide range of use cases—some of which are complex and even contradictory. The new 5G systems incorporate an internationally standardized radio access technology known as New Radio (NR), marking a significant evolution beyond existing LTE standards. The 3GPP's technical specifications provide foundational standards for 5G New Radio (NR), including spectrum usage, interface protocols, and interoperability guidelines.

5G offers a step-change improvement over LTE by delivering significantly faster data speeds, increased bandwidth, and support for a broader range of devices and services. As a foundational technology, 5G continues to evolve, with major telecom operators investing in research and upgrades to progressively expand its capabilities.

The transition from 4G to 5G is expected to unfold gradually. Early 5G deployments (as of 2020) primarily relied on existing 4G LTE infrastructure, including the network core and available spectrum. These non-standalone (NSA) configurations represent a hybrid model. The next stage involves complete standalone (SA) 5G deployments, in which 5G will operate independently of the legacy LTE systems. This dual deployment model ensures a smoother migration path and a more stable user experience during the transitional phase.

The 5G New Radio Technology

5G New Radio technology has two versions: non-standalone, based on commercial launches expected by year-end and throughout 2019, and standalone, with an entirely new core architecture.

Based on the latest 3GPP Release 15 standards, which address 5G networking, the classification for the first wave of devices and networks is Non-Standalone (NSA), referring to existing 4G infrastructures that support 5G networks.

Two primary deployment choices exist for new 5G networks: adding to existing 4G non-standalone (NSA) networks and 5 G-only, i.e., standalone (SA) networks. These deployment choices apply to 5G Core and 5G NR.

The NSA makes it more comfortable for community operators to deploy the 5G system because it can reuse the electricity from 4G facilities. Simultaneously, SA involves a 5G primary that will send super-fast, ultra-low latency, a flexible network, and decrease operational costs.

A mobile device that supports these networks is known as a two-mode 5 G-supported device.

Earlier 5G deployments could have leveraged legacy 4G networks to accelerate the initial rollouts. However, most of the first G deployments use a non-standalone (NSA) structure. The 4G network core links with the 5G nodes (geodes), anchoring the 5G level to 4 G. It enhances the user experience by falling back to an LTE connection when coverage from the 5G level is unavailable.

Figure 1. 5G vision: Use cases, spectrum, and deployments

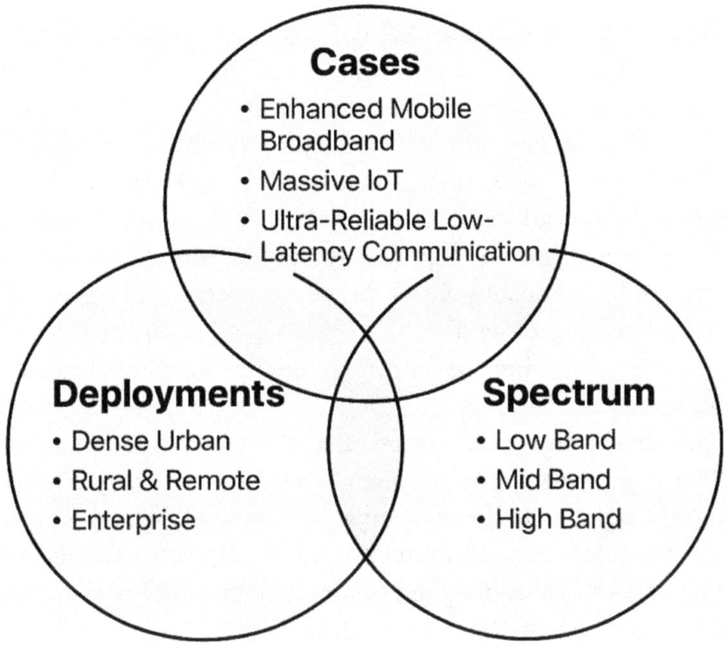

Deployment Challenges

The deployment of 5G presents a range of technical and logistical challenges that go beyond those encountered with previous mobile generations. One of the core requirements is network densification, particularly in urban environments. To meet 5G's high-capacity and low-latency demands, operators must install a large number of small cells in buildings, along streets, and across metro areas. This shift pushes data processing from centralized core networks toward the network edge, aiming to reduce latency and avoid bottlenecks.

This edge-oriented architecture requires significant investment in power supply, backhaul infrastructure, and site acquisition, which are already key cost drivers in traditional cellular deployments. For 5G, however, these concerns are even more critical, especially at the metro-cell level where dense installations must be executed efficiently and economically.

Non-Standalone (NSA) Deployment

Non-Standalone 5G New Radio (NSA 5G NR) represents the initial stage of 5G deployment. In this configuration, 5G services are delivered over the existing 4G LTE core and radio access network, with the addition of a 5G New Radio (5 G NR) carrier to support higher throughput. NSA deployments are already operational in regions such as Norway and China. By early 2019, 5 G-compatible smartphones became commercially available in select markets.

This NSA model primarily supports Enhanced Mobile Broadband (eMBB) use cases, enabling significantly faster download speeds and improved media experiences. As noted by Park et al. (2022), 5G's eMBB capabilities are a catalyst for transformative applications in immersive media, augmented reality (AR), and virtual reality (VR).

While NSA offers a faster path to market, it inherits certain limitations from LTE. For example, millimeter-wave (mmWave) bands, used in many eMBB scenarios, have limited propagation range and are more susceptible to signal interference from physical obstacles such as buildings or trees. NSA acts as a transitional phase, allowing engineers to test, refine, and optimize 5G technologies before full-scale Standalone (SA) deployment becomes widespread.

THE 4G LTE INFRASTRUCTURE IN NSA (NON-STANDALONE) DEPLOYMENTS

The initial phase of 5G rollout is built on the foundation of existing 4G LTE infrastructure, as outlined in 3GPP Release 15 /GPP 2021a), which defines the Non-Standalone (NSA) deployment model. In this configuration, 5 G-capable devices primarily connect to 5G frequencies for data transmission, while control signaling and non-data operations continue to rely on the 4G LTE network.

This approach enables faster commercialization of 5G services without requiring a complete overhaul of legacy infrastructure. For example, a 5 G-enabled smartphone in an NSA setup leverages high-speed 5G data channels while maintaining session management, mobility, and voice services over 4G LTE.

This hybrid architecture offers immediate performance benefits, including enhanced mobile broadband capabilities, while reducing the deployment burden on operators. However, it also introduces operational complexity, requiring seamless coordination between the 4G and 5G radio access networks.

Figure 2. Non-Standalone (NSA) 5G deployment

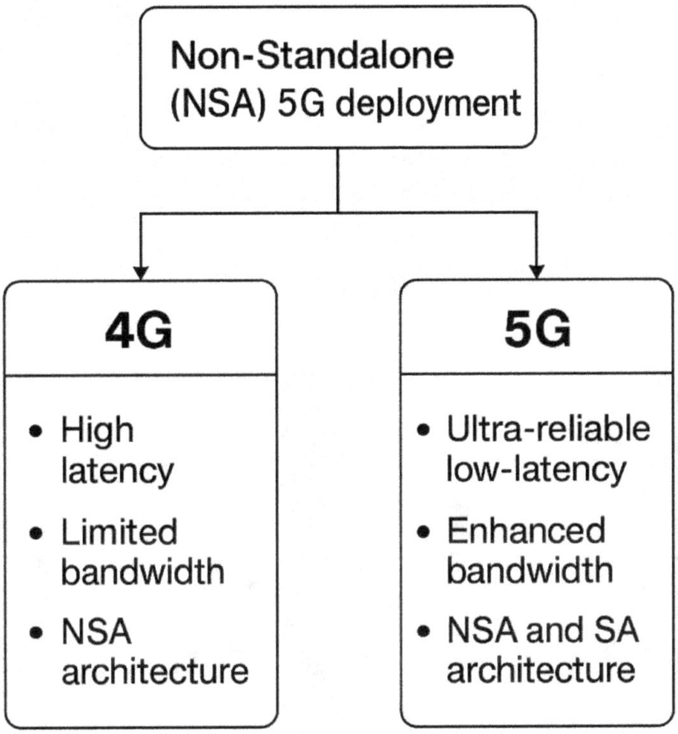

THE STANDALONE (SA) DEPLOYMENT

The 2020 timeframe leads to a standalone 5G NR, which is more significant than the preceding LTE and the non-standalone variant. Standalone 5G NR will lower operators' costs and enhance users' performance. The standalone version envisions an entirely new RAN and core system.

Like the NSA, standalone systems utilize millimeter-wave bands, mid-band, and low-frequency signals to provide complete area coverage and multi-gigabit connections.

On the other hand, SA may be a simple 5G system. In this case, the 5G community has dedicated 5G facilities to offer massive velocity upgrades and small community latency (delay). Moreover, the 5G SA system is accessible from the 4G network.

The 5G Standalone (SA), a networking and unit standard, remains under evaluation and is anticipated to be signed off by 3GPP. The benefit of Standalone is simplification and enhanced efficiency, which can reduce costs and continuously improve throughput performance with the system's advantage, while also facilitating the development of innovative cellular utilization instances, such as ultra-reliable low-latency communications (URLLC). However, after the SA standard qualifies this season, operators' final migration from 5G NSA to SA must be invisible to the person.

Figure 3. 5G Standalone 5G deployment

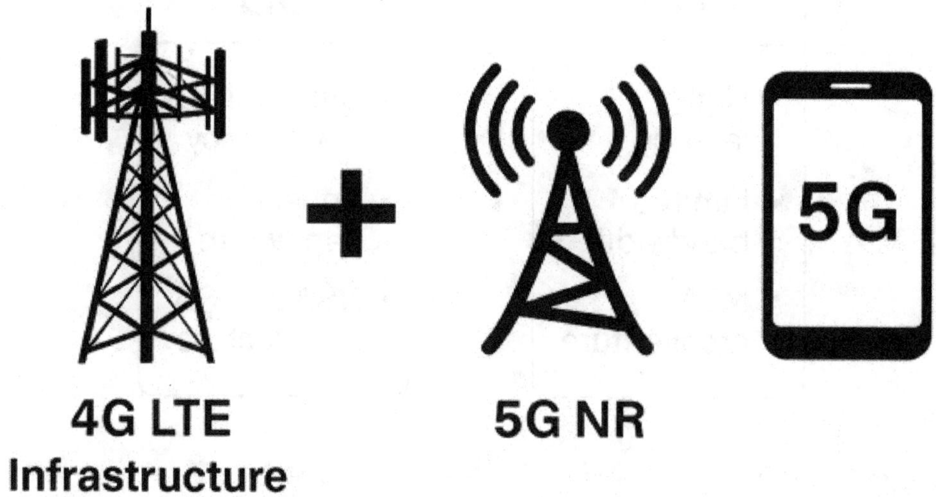

The release of fifteen 5G standards papers from 3GPP states that the 5G NSA architecture comprises the 5G stereo access system (RAN) and 5G New Radio (NR), which work in conjunction with the existing 4G LTE infrastructure and the primary system. *The 5G RAN links to the LTE core and requires a unique radio wave spectrum from 4G LTE.* The blend of 5G RAN, 5G NR, and the current 4G LTE infrastructure supports LTE services but has enhanced 5G features, such as reduced latency.

G NR will operate in the frequency range from below 1 GHz to 100 GHz with various deployments. There will generally be more coverage for every base station (macro sites) at reduced carrier frequencies and a minimal coverage area for every base station (micro and pico sites) at higher carrier frequencies. Offering higher service quality and optimum reliability requires a qualified spectrum to remain functional as the backbone of the 5G wireless network. Transmission in the unlicensed spectrum is utilized instead of incurring significantly higher data fees and increasing the capability.

THE EVOLUTION TOWARD A STANDALONE 5G ARCHITECTURE

Transitioning to Standalone 5G Infrastructure

To fully realize the potential of 5G, operators must transition from Non-Standalone (NSA) deployments to Standalone (SA) architecture. This shift requires the deployment of dedicated 5G Radio Access Networks (RAN), small cells, and a cloud-native 5G core infrastructure. As defined in 3GPP Release 15, the new 5G core is based on a Service-Based Architecture (SBA), which enables virtualization, network slicing, and dynamic resource allocation to meet various service demands (TELCOMA, 2025).

The 5G Radio Access Network (RAN)

The Radio Access Network (RAN) connects user devices to the mobile core via radio frequencies and is responsible for managing radio resources and transferring data between them. In 5G, RANs are optimized for diverse environments—rural, suburban, and urban—by leveraging sub-6 GHz and mid-band frequencies. These frequencies, although offering lower data rates than mmWave, provide better coverage and penetration, making them suitable for less densely populated areas.

Virtual RAN (vRAN) and Open RAN (O-RAN)

vRAN and O-RAN represent the next generation of radio access network (RAN) technologies. They disaggregate software functions from hardware, allowing the deployment of network functions on commercial off-the-shelf servers. This flexibility is critical for managing traffic surges and optimizing infrastructure costs. In vRAN, the user plane and control plane are decoupled, enabling features like massive MIMO, NFV, and network slicing. These innovations allow service providers to dynamically allocate resources and tailor network performance to various use cases.

Open RAN and the Role of AI/ML

Open RAN (O-RAN) introduces open interfaces and vendor-neutral hardware, promoting interoperability and faster innovation. By incorporating Artificial Intelligence (AI) and Machine Learning (ML) into the RAN Intelligent Controller (RIC), networks can perform real-time and non-real-time optimizations of radio resources. This improves performance and adaptability, especially in dynamic usage environments. According to Pipeline Magazine (2021), this openness fosters broader industry collaboration and accelerates the rollout of flexible, future-proof 5G networks.

Examples of Radio Access Technologies

Historically, various Radio Access Network (RAN) architectures have been developed to support successive generations of mobile communication technologies, including GRAN (GSM Radio Access Network), GERAN (an enhanced version of GRAN incorporating EDGE capabilities), UTRAN (UMTS Radio Access Network), and E-UTRAN (Evolved Universal Terrestrial RAN) designed for LTE.

These foundational systems have evolved into the 5G RAN, which now supports cloud-native functions and real-time adaptability through advanced radio access network (RAN) architectures.

STRATEGIC DEVELOPMENT OF 5G

Scenario studies conducted by 3GPP highlight technical and performance requirements for beyond-5G access technologies in diverse environments. Scenario analyses by 3GPP define future requirements for next-generation wireless technologies, with emphasis on ultra-low latency, mobility, and reliability.

Advanced requirements explored in next-generation access scenarios address latency, reliability, and ultra-dense network deployment needs for future 5G evolution. The 5G technology has been strategically developed to address three key usage scenarios:

- Enhanced Mobile Broadband (eMBB) – focused on high-speed data for consumer applications
- Massive Machine-Type Communications (mMTC) – supporting billions of connected IoT devices
- Ultra-Reliable Low-Latency Communications (URLLC) – enabling critical real-time services

Each of these pillars meets specific technological and societal needs, from smart cities and telemedicine to immersive media and autonomous systems.

Enhanced Mobile Broadband (eMBB)

Enhanced Mobile Broadband (eMBB) is the most commercially mature 5G application. By 2020, eMBB accounted for approximately 42.5% of the 5G market, and its influence is expected to grow significantly by 2025. eMBB supports high-speed services like 4K/8K video streaming, cloud gaming, virtual reality (VR), and augmented reality (AR).

The initial deployments of eMBB relied on Non-Standalone (NSA) 5G architecture, which pairs 5G data capabilities with 4G control signaling. This approach enabled fast market entry and supported services in densely populated urban environments, with downlink speeds reaching up to 1 Gbps indoors and 300 Mbps outdoors. Millimeter-wave (mmWave) antennas mounted on lamp posts, buildings, buses, and utility poles provide this high-speed coverage, although their limited range and line-of-sight requirements necessitate dense deployment.

In rural settings, eMBB may rely on lower-frequency omnidirectional antennas, offering more modest speeds (e.g., 50 Mbps) but broader coverage, replacing legacy 4G LTE networks with improved but cost-effective solutions.

Massive Machine-Type Communications (mMTC)

Massive Machine-Type Communications (mMTC) is designed to support large-scale Internet of Things (IoT) ecosystems, such as smart cities, smart agriculture, and logistics, with up to billions of connected devices. This communication type is crucial for applications involving smart meters, industrial sensors, and low-power connected objects, all of which require low data rates, high scalability, and energy efficiency.

MmTC technologies must operate with minimal human intervention, sending small data packets with latency under 10 milliseconds. It is distinct from traditional mobile communication as it prioritizes connection density and battery longevity over throughput. According to Popovski et al. (2018), mMTC plays a vital role in enabling ubiquitous smart infrastructure.

While many IoT devices still rely on Wi-Fi, mMTC solutions using cellular networks offer more robust coverage, especially in remote and mission-critical environments. For example, drones equipped with 4G/5G can support emergency response, while connected vehicles utilize mMTC for telemetry and diagnostics. Despite this, autonomous cars often rely on onboard systems due to potential coverage limitations in remote areas.

Ultra-Reliable and Low Latency Communications (URLLC)

URLLC is the 5G pillar focused on critical communication systems that demand extremely low latency and high reliability. It targets use cases like autonomous vehicles, remote surgery, industrial automation, and public safety systems, where latency must be reduced to 1 ms or less and reliability must exceed 99.9999%.

This level of performance is achieved through advances in the 5G NR (New Radio) specifications defined by 3GPP Release 16, which include:

- Shorter and more robust data packets
- Preemptive scheduling and retransmissions
- Packet duplication and multi-connectivity for redundancy

These improvements are necessary for real-time decision-making in scenarios where even a brief delay could lead to critical failure, for instance, in robotic surgery or preventing vehicular collisions.

In the long term, URLLC could also expand 5G's role in geolocation services, offering a potential alternative or complement to satellite-based systems, such as GPS, especially in urban areas where signal obstruction is common, (Park et al., 2022).

The strategic development of 5G is structured around three foundational use cases, each targeting specific performance needs and market demands. These use cases—eMBB, mMTC, and URLLC—demonstrate the versatility of 5G in addressing a wide range of applications, from high-speed consumer data to industrial automation and control. Table 3 summarizes their core characteristics, deployment requirements, and expected impact. Together, they define the roadmap for global 5G implementation.

Table 3. Strategic development of 5G: Key use cases

Use Case	Description
Enhanced Mobile Broadband (eMBB)	Focuses on high-speed data access for densely populated urban areas, enabling UHD video streaming, AR/VR, cloud-based gaming, and virtual meetings. Utilizes mmWave antennas and necessitates dense, small cell deployment. Provides speeds up to 1 Gbps indoors and 300 Mbps outdoors. Expected to dominate the market share until 2025.
Massive Machine-Type Communications (mMTC)	It supports a wide range of IoT devices, featuring high connection density and low energy consumption, which is crucial for smart cities, smart meters, and logistics applications. It prioritizes scalability, low cost, low power, and moderate data rates (around 100 Kbps). Latency is maintained at around 10 ms, which is crucial for enabling machine-to-machine (M2M) communication.
Ultra-Reliable Low Latency Communications (URLLC)	Designed for critical applications where low latency ($\leq$ 1 ms) and high reliability (99.9999%) are essential, such as autonomous vehicles, remote surgeries, and public safety systems. Uses advanced MAC and physical layer features to ensure minimal delays and packet loss.

THE CONVERGENCE OF WIRELESS COMMUNICATIONS

As the rollout of 4G mobile communication networks occurred, industry representatives began to examine the technical advancements of the coming generation (5G). *As a result, several study projects involving key international mobile network operators, infrastructure companies, andAcademic institutions have recently released research to establish the technical foundations of 5 G.* Nevertheless, the architecture of the upcoming 5G methods, their performance, and the provision of mobile offerings must still be clearly defined. For instance, enhanced mobile broadband facilitates high-speed cloud-based services, (Park et al., 2022). Cloud-native principles embedded in 5G networks transform operational models and architectural design through dynamic resource orchestration and edge integration (Bishen, 2019).

An Emergent Convergence Vision

We observe that advancements in network architecture are the primary driver of capacity growth. Heterogeneous networks and the convergence of information and communication technology are two critical techniques. The evolution of the air interface would enhance link and system spectrum efficiency, thereby facilitating improvements in network efficiency.

The vision for 5G is to converge evolved versions of existing cellular networks and other complementary radio access solutions. Thus, 5G is not a stereo access interface but a network of networks. The seamless integration of various air interfaces, protocols, and frequency bands requires paradigm shifts in how networks cooperate and interact to provide data rates of many gigabits per second with an end-to-end latency of a couple of milliseconds. An overview of the 5G Core Network illustrates the integration of control and user plane separation, enabling flexible service provisioning and low-latency pathways (TELCOMA, 2025).

The 5G and the Internet of Things (IoT)

5G is crucial to the Web of Things due to the demand for a faster system with a higher capacity to serve connectivity must-haves efficiently. The 5G spectrum enlarges the frequencies where digital cellular solutions will transfer information.

The Internet of Things (IoT) is yet another wide area for advancement utilizing a supercharged 5G wireless network. The Internet of Things (IoT) will connect every object, sensor, application, and device to the Internet.

IoT applications will collect necessary information from millions of sensors and devices, creating an effective network for information collection, transmission, processing, real-time analytics, and control.

5G presents a significant prospect for the Internet of Things due to the flexibility of its underutilized spectrum, accessibility, and cost-effective solutions for deployment. In addition, IoT can gain from 5G networks in numerous areas:

Another area where 5G Ultra Wideband might influence Things is the improvement of augmented and virtual reality (AR/VR). The 5G Ultra Wideband's ultra-low latency can enhance AR/VR expertise and technology choices in various fields, including business and education.

According to Verizon, 5G and Business IoT, 5 G-enabled IoT is expected to drive technological growth and create and support 22 million jobs worldwide. This job growth stems from the digitization of transportation, as well as other physical industries, manufacturing, and agriculture. Consider building sites, mines, oil derricks, and freighter fleets: these industries will significantly benefit from super-fast data transmission due to the time-sensitive nature of their output.

5G can drive advancements in intelligent machinery and smart manufacturing. On a larger scale, 5G could enable IoT to perform real-time traffic analysis, improve public safety and security, and potentially facilitate remote surgeries.

Augmented Reality and the 5G Technology: A Transformative Alliance

The integration of augmented reality (AR) and 5G marks a pivotal evolution in how we interact with digital content. With its ultra-low latency (as low as 1 millisecond) and high data transfer rates, 5G unlocks the full potential of augmented reality (AR) and virtual reality (VR), making real-time, high-definition experiences possible without lag or interruption. This enables users to overlay digital information onto the physical world with seamless precision, enriching everyday activities like shopping, education, and entertainment.

In the real estate sector, virtual platforms like Matterport enable potential buyers to explore multiple properties remotely in a single day, providing realistic 3D walkthroughs that replicate in-person visits and streamline the decision-making process, (Matterport, 2025).

In sports and entertainment, immersive VR technologies enable fans to virtually attend live games or concerts with 360-degree visual immersion and synchronized audio, replicating the in-stadium atmosphere and enhancing user engagement, (Finley, 2018).

Education will also be revolutionized—students can explore historical sites, conduct virtual science experiments, or dissect 3D anatomy models, all from the comfort of their classrooms or homes. In retail, AR mirrors can enable customers to try on clothes or visualize furniture in their living rooms before making a purchase.

Thanks to 5G, AR is no longer limited by hardware or bandwidth constraints. This powerful combination promises to reshape consumer expectations and enable industries to deliver interactive, personalized, and immersive experiences on an unprecedented scale.

COMPONENTS OF 5G INFRASTRUCTURE

The architecture of 5G builds upon decades of evolution in radio access networks (RAN), now enhanced with cutting-edge technologies such as massive MIMO (Multiple-Input, Multiple-Output), wide-spectrum bandwidths, and multi-band carrier aggregation. These innovations enable faster, more efficient, and scalable wireless communications, defining the next generation of mobile connectivity.

New Radio (NR) Standalone Architecture

5G New Radio (NR) is the newly developed air interface standard by 3GPP, purpose-built to meet the demanding requirements of next-generation networks. Unlike Non-Standalone (NSA) modes that rely on 4G LTE cores, the Standalone (SA) version of 5G NR operates independently with its dedicated 5G core. This enables advanced functionalities such as network slicing, ultra-reliable low-latency communication, user and control plane separation (UCPS), and virtualization—all of which are crucial for industrial automation, autonomous systems, and smart cities.

The adoption of 5G NR SA is projected to accelerate rapidly, particularly in the Asia-Pacific region, where governments and industry leaders are investing heavily in standalone 5G infrastructure. South Korea, China, and Japan are at the forefront, launching initiatives and pilot programs to support national digital transformation goals.

Wireless Solutions and Use Cases for 5G NR

5G NR also supports a diverse range of wireless applications, particularly in the realm of short-range Internet of Things (IoT) devices. These include use cases in healthcare, manufacturing, utilities, and energy sectors that require robust, low-latency, and high-reliability wireless connectivity. The rapid growth of IoT ecosystems is driving the demand for specialized 5G solutions that can handle dense device deployments with minimal latency.

Among wireless engineering innovations, the NR Standalone architecture stands out as a key driver of 5G's strategic development. According to the GSMA, the Asia-Pacific region is expected to lead global 5G adoption, supported by its large connected population and coordinated efforts among telecommunications providers, equipment manufacturers, and governmental agencies.

MULTIPLE OUTPUTS AND MULTIPLE INPUTS MIMOO) ANTENNAS

With NR, large-scale superior antenna methods enable state-of-the-art beam-forming. MIMO (Multiple-Input, Multiple-Output) employs valuable strategies to boost throughput, capacity, and coverage. Multi-antenna techniques will also be essential for URLLC and will be utilized to improve reliability. Additionally, the scalable numerology of NR offers excellent methods to achieve minimal latency, as a larger subcarrier spacing (SCS) reduces the transmission time interval.

Multiple-input and multiple-output (MIMO) antennas have long been a characteristic of business public wireless and Wi-Fi methods; however, 5G demands substantially higher MIMO capabilities. In addition to boosting the resiliency (signal-to-noise ratio or SNR) associated with a transmitted signal and the channel capability, a typical frequency may be steered together in several directions without raising spectrum consumption.

MIMO and Electronic Signal Processors

The thriving operation of MIMO methods involves implementing essential electronic signal processors in a world with plenty of signal interference, or "spatial diversity," which refers to a wide variety of signal paths between the receiver and the transmitter.

The range of appearance times from various obstacles, styles, and duplexing (TDD) stations could send path redundancy for identical signals or boost the channel capability by transmitting multiple areas of the modulated information. Initially conceived in the 1980s, there are variations between traditional multi-user MU MIMO and Massive MIMO. Still, a variety of antennas and a considerable number of consumers supported them. Therefore, the level of MIMO suggested the number of transmitters and the number of receivers, i.e., 4x4.

Beam Steering and Beam Switching

Beam steering alters the stage of the input signal on all radiating components. Phase-shifting enables the signal to target a particular receiver. An antenna can employ radiating components with the same frequency to guide a single beam along an absolute path. Various frequency beams additionally steer in multiple directions to serve numerous users. Massive MIMO (Multiple Input Multiple Output) technology plays a pivotal role in enhancing the capacity and efficiency of 5G networks. By deploying a large number of antenna elements at base stations, Massive MIMO enables the simultaneous transmission of multiple data streams to different users, thereby significantly increasing spectral efficiency and network throughput (Techloy, 2025).

To support this complex antenna architecture, advanced components, such as Toshiba's SPDT RF switch (model TCWA1225G), have been introduced. This switch offers high input power handling and low insertion loss within a compact footprint, addressing the spatial constraints of base stations and facilitating the implementation of Massive MIMO configurations. *This approach significantly enhances spectral efficiency and network capacity,* (Chataut & Akl, 2020).

The base station dynamically submits and estimates the signal's guidance as the endpoint moves, efficiently monitoring the user. If a beam cannot scan someone, the endpoint might change to an alternative beam.

This granular monitoring is possible because 5G base facilities must be closer to their owners than earlier generations of movable infrastructure. The infrastructure for mMTC connects vast networks of low-cost, battery-operated devices, which is crucial for wearable technologies, (Popovski et al., 2018).

Figure 4. mMTC communication architecture

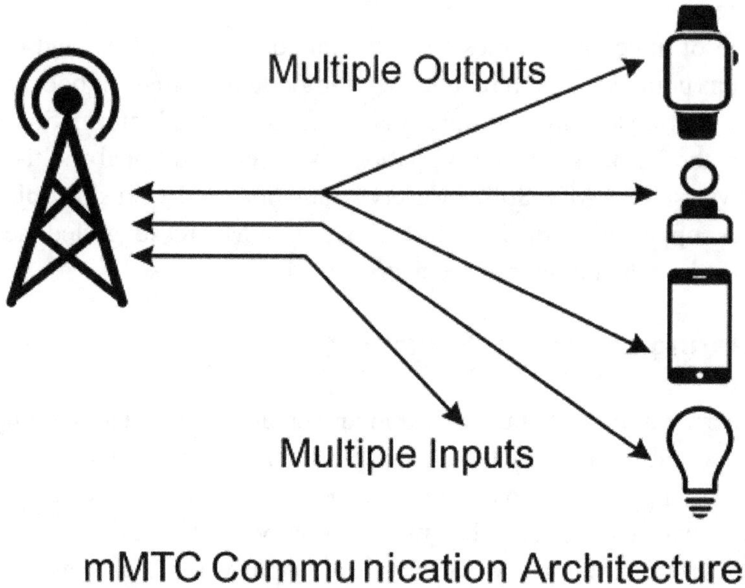

NETWORK FUNCTIONS VIRTUALIZATION VS. SOFTWARE-DEFINED NETWORKING

Network Function Virtualization (NFV) describes the virtualization of network functions. In contrast, Software-defined networking (SDN) describes a system architecture that injects programmability and automation into the system by decoupling network management and forwarding capabilities. So, NFV virtualizes network infrastructure, and SDN centralizes network management. NFV and SDN develop a plan that is operated and handled by software.

SDN) generally includes an SDN controller, northbound application program interfaces (APIs), and southbound APIs. The controller enables network administrators to monitor networking and determine policies and behaviors on the central infrastructure. In Software-Defined Networking (SDN), southbound application programming interfaces (APIs) facilitate communication between the SDN controller and the network infrastructure, enabling the controller to collect real-time data from network devices and enforce configurations to maintain optimal performance. Conversely, northbound APIs enable applications and services to communicate their resource requirements to the controller, facilitating dynamic orchestration and efficient network management, (Kerner & Ferguson, 2024).

Advantages of NFV Network

The advantages of NFV for Network operators that virtualize their networks include the ability to save money, shorten the time-to-market for new or updated services, and have a much broader scope with alternative methods available for services and applications. Some other advantages include the following.

Less Vendor Lock-In: Running virtual network functions (VNFs) on commercial off-the-shelf (COTS) hardware means locking businesses into proprietary, fixed-function solutions that require significant time, truck rolls, and labor to deploy and configure.

COTS (Software Defined Radio for 5G Development. Commercial-off-the-shelf (COTS) Software-Defined Radio (SDR) products for military radar and communication applications offer high performance and design flexibility.

Greater Energy Efficiency

A virtualized information center or other infrastructure helps run at a lower cost. For example, the information center's footprint, cooling requirements, or power consumption could remain the same, improving workload capability.

This factor is feasible because one server can run several VNFs simultaneously, so only a few servers need to perform the same amount of work. In addition, when networks require modifications, a company can upgrade its infrastructure through a program rather than conducting another rollout. It reduces company updates to their networks and data centers.

Problems Associated With NFV

Problems Associated with NFV: Challenges surrounding NFV are rooted in three key aspects of the technology: the NFV supervisor (NFVM), Virtual Network Functions (VNFs), and NFV Infrastructure (NFVI). Unfortunately, in practice, integrating the three parts rather than relying on theoretical simplification poses difficulties and complexity for network operators when deploying NFV at scale.

The Lean NFV has attempted to resolve these issues by creating a brand-new method of NFV architecture.

Critical Importance of Virtualization

Virtualization is crucial in achieving 5 G. Operators rely on software-defined networks (SDN) and network functions virtualization (NFV) to transition from LTE to 5G and rapidly scale their networks.

In 5G, NFV divides a physical network into multiple virtual networks, each supporting different radio access networks (RANs) or services for specific customer segments.

Network Functions Virtualization (NFV) would decouple network operations from proprietary hardware appliances and operate them as software in virtual devices (VMs). The various features—firewalls, traffic management, and virtual routing—are collectively referred to as equal network capabilities (VNFs).

NFV uses virtualized networking elements to create an infrastructure that is utterly independent of hardware. Conventional computing materials, storage space, and network capabilities have positioned business off-the-shelf (COTS) hardware as x86 servers.

Virtualized methods allocate resources on x86 servers to virtual machines (VMs). *Numerous virtual machines (VMs) can operate on a single server and scale to utilize the remaining free resources*. This means that resources are less idle, and information centers with virtualized infrastructure can use them more effectively. NFV could virtualize information and balance airplanes in information facilities and external networks.

Figure 5. Virtualized network functions and NFV infrastructure coordination

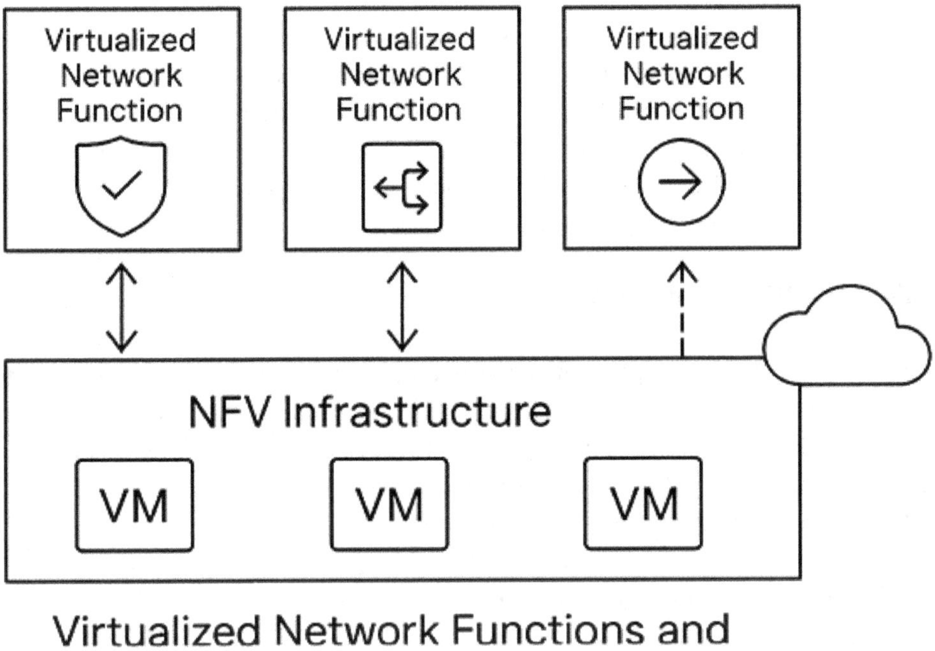

Advantages of SDN in 5G Networks

The Integration of SDN Into 5G Infrastructure Provides Several Strategic Benefits

Flexibility and Programmability: SDN allows dynamic and real-time changes to network configurations without manual reprogramming of individual devices. This adaptability is crucial in 5G, where network demands shift based on user density, application types, and service-level agreements (SLAs).

Centralized Control: A centralized SDN controller provides a global view of the network, enabling better traffic engineering, fault management, and load balancing across 5G slices and heterogeneous networks.

Cost Efficiency: By decoupling software and hardware, SDN reduces reliance on proprietary systems, resulting in lower capital expenditures (CapEx) and operational expenditures (OpEx).

Enhanced Security: SDN enables more granular security policies, dynamic threat detection, and quicker mitigation. This is especially relevant in 5G environments that support critical applications, such as autonomous vehicles and smart grids.

Support for Network Slicing: Software-defined networking (SDN) is fundamental for implementing network slicing in 5G, where different virtual networks operate on the same physical infrastructure, each customized for specific service requirements (e.g., enhanced Mobile Broadband (eMBB), Ultra-Reliable Low-Latency Communications (URLLC), and Massive Machine-Type Communications (mMTC).

SDN and NFV Integration

SDN and NFV are complementary technologies that together form the software-based backbone of 5G networks. While SDN manages the separation of control and data planes, NFV virtualizes network functions such as firewalls, load balancers, or gateways that traditionally ran on dedicated hardware.

In a 5G context, this combination allows:

- Rapid deployment of new services.
- Agile scaling of network resources.
- Isolation of network slices for different tenants or services.

SDN Use Cases in 5G

Practical implementations of SDN in 5G include:

Dynamic Traffic Steering: Adjusting data flows across the network in real time based on congestion or quality of service (quality of service) requirements.

Multi-access Edge Computing (MEC): Software-Defined Networking (SDN) helps manage resources between central and edge clouds for latency-sensitive applications.

End-to-End Network Automation: SDN enables closed-loop automation through telemetry and AI, optimizing performance without requiring human intervention.

These capabilities position SDN as a foundational pillar in building scalable, secure, and responsive 5G networks.

MULTI-ACCESS EDGE COMPUTING (MEC)

Multi-access Edge Computing (MEC) offers software content providers cloud computing resources and an IT service environment situated at the network edge, thereby enhancing performance and reducing latency for time-sensitive applications (Taleb et al., 2017). Leveraging use facilitates these developments by enabling high bandwidth, ultra-low latency, and real-time entry to radio network information.

Localized Content Delivery and Network Efficiency

MEC empowers network operators to move compute and storage resources closer to the user, at the network edge, thereby reducing latency and minimizing data transport over the core network. This architecture enables ultra-fast access to localized content and services. Examples include:

- Streaming platforms cache video segments at edge nodes to reduce buffering.
- Innovative city applications use edge processing to instantly manage traffic lights, security cameras, and public services.
- Retail analytics is delivered in real-time within stores, based on user behavior or environmental sensors.

By minimizing backhaul traffic and processing delays, MEC enhances the quality of experience (QoE) for end-users and ensures the efficient utilization of network resources.

Sensors as 5G Support

Edge computing gains further significance when integrated with sensors and Internet of Things (IoT) devices, which are foundational to industrial and smart infrastructure use cases. In a 5G environment:

- Environmental sensors provide real-time data for monitoring air quality, weather, and energy usage.
- Industrial sensors support predictive maintenance, robotics coordination, and quality control in manufacturing.
- Healthcare sensors enable remote patient monitoring, with low-latency emergency interventions facilitated by edge computing.

These sensors generate high volumes of data that require real-time processing, making MEC essential for decision-making rather than relying on centralized data centers.

Industry Opportunities and Business Models

MEC creates a new value chain and ecosystem that fosters innovation among mobile operators, third-party developers, and enterprise customers. Key opportunities include:

Edge-as-a-Service (EaaS): Operators can lease computing capacity at the edge to developers or enterprises, providing a flexible and scalable solution.
Custom Vertical Solutions: Tailored edge apps for logistics, gaming, smart agriculture, or healthcare.
Revenue-Sharing Models: Operators hosting third-party applications at the edge can share subscription or usage-based revenue with these applications.

Ultimately, the integration of Multi-access Edge Computing (MEC) with 5G networks enables real-time, high-performance services that were previously unattainable due to limitations in latency and bandwidth, marking a significant shift in IT infrastructure capabilities. It represents a shift in how mobile networks deliver value, moving from simple data transmission to intelligent, distributed computing environments.

Figure 6. Multi-Access edge computing (MEC)

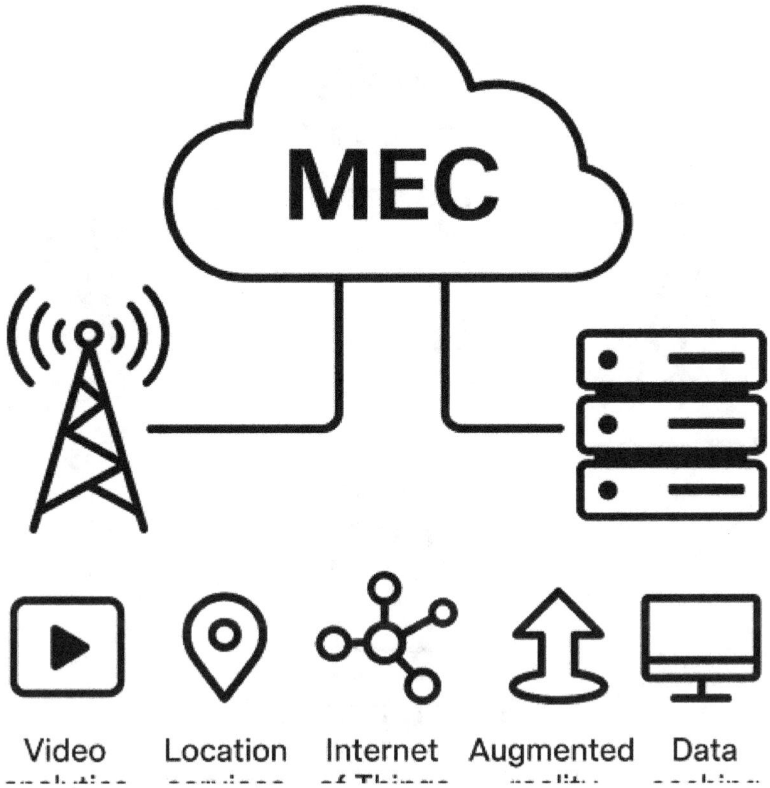

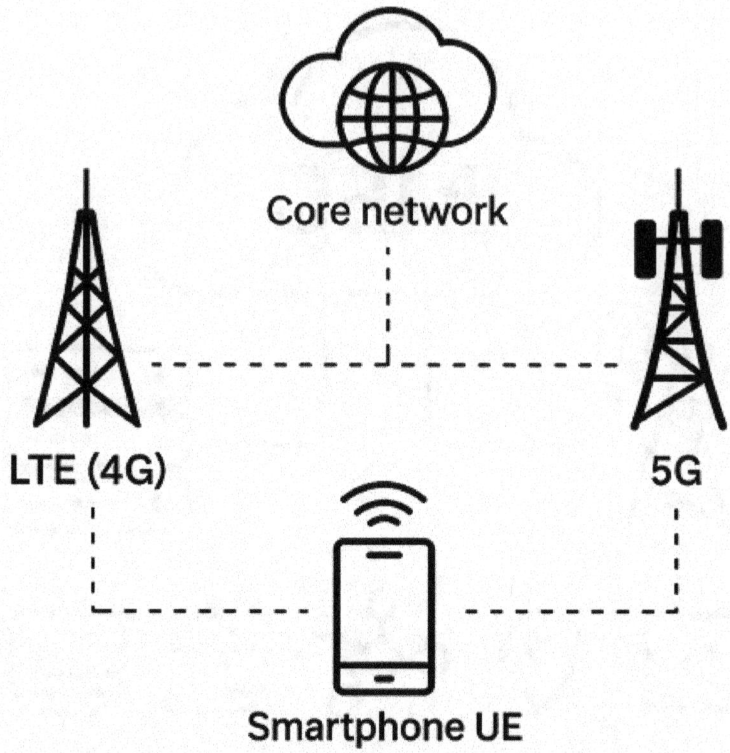

Non-Standalone (NSA) 5G Deployment

Source: Oppo

The higher-level expectations of industry actors regarding 5G are significantly less thorough than those outlined in the 3GPP recommendation. Although high-level needs are gathered from numerous sources, the first standard was ITU-R M.2083-0.

This standard was the first to describe the benefits of principal abilities in various use scenarios and explain all three critical correspondence scenarios: MMMTC, URLLC, and EMBB.

Characteristics Reveal the Main Demands

Their characteristics reveal the main demands for mass MMTC or IoT traffic, which are (i) network energy harvesting and (ii) connection density, whereas for URLLC site traffic, latency, and mobility,

Although the IoT does not have higher throughput expectations (unlike eMBB), connection density specifications are as high as one million endpoints per square kilometer. Thus, the mass MMTC is not far from broadband access specifications in a group (approximately half a million users per square kilometer).

When discussing the demands of industrial IoT programs, the need for real-time (low-latency) procedures, robust security, method interoperability, integrability, sound engineering, deployment, and practical maintenance emerges. 3GPP TS 22.104 innovation defines the demands for cyber-physical command and control in vertical domains.

The reliability of this quasi-standard is well-supported by standardized performance indicators, including latency ranging from 0.5 milliseconds to 500 milliseconds across different scenarios and ultra-high reliability levels of 99.999% or higher, (Varga et al., 2020). *Furthermore, long-range reliability (with a mean time between problems that range from one morning to ten years) should ensure the broad applicability of 5 G.*

This 3GPP TS 22.104 recommendation blankets different performance needs for periodic correspondence (as well as nondeterministic correspondence). *It included positioning performance, clock synchronization, mixed traffic, and network operation.*

For instance, advancements in 5G sensors in healthcare services elaborate on scenarios. Moreover, these advancements encompass cases that facilitate the distribution of electrical energy with partners such as affiliated hospitals and health facilities.

While vehicle-to-everything (V2X) interaction will likely become the "killer program for 5G"—and individual needs are thoroughly addressed in 3GPP TS 22.186—this area falls out of the newspaper's emphasis, targeting Industrial IoT. Likewise, though rail communications have strict requirements elaborated in 3GPP TS 22.289, this paper does not concentrate on the problems of the transportation URL.

Critical Capabilities of 5G Technology

The critical capabilities of 5G technology play a significant role in addressing diverse use cases. Enhanced Mobile Broadband (eMBB) prioritizes high peak data rates, user-experienced data rates, and local traffic capacity, catering to applications that require robust connectivity, such as streaming and augmented reality. Enhanced

continued on following page

Table 4. Continued

mobile broadband in 5G ensures higher data speeds, essential for augmented reality applications, (Park et al., 2022).

Ultra-reliable low-latency communication (URLLC) is essential for autonomous vehicles and remote surgery scenarios. It focuses on low latency and high reliability to ensure seamless real-time interactions. Thus, URLLC ensures the seamless operation of industrial automation and autonomous vehicles (Popovski et al., 2018).

Massive Machine-Type Communication (mMTC) emphasizes high connection density and network energy efficiency, supporting IoT ecosystems with large-scale device connectivity. These capacities, tailored to specific needs, ensure the versatility and transformative potential of 5G across various industries and use cases.

CONCLUSION

Chapter 5 provides a detailed exploration of the key technologies and innovations that underpin 5G, showcasing its transformative role across various sectors. 5G combines a range of engineering innovations, including network slicing, millimeter-wave spectrum, enhanced mobile broadband, ultra-reliable low-latency communication, and massive machine-type communication, to enable various industrial use cases. These technological advances form the foundation for intelligent infrastructure, real-time operations, and adaptive connectivity, all of which are essential to the future of global communications.

5G enables new levels of latency, reliability, and connection density in technology. These features are crucial in facilitating the digitalization of industries and realizing smart cities, autonomous vehicles, and scalable Internet of Things (IoT) ecosystems. The chapter also examined critical components, such as virtualized networks, open RAN, and multi-access edge computing, which contribute to a flexible, adaptive, and high-capacity mobile infrastructure. Table 4 provides an overview of the key technologies explored and their corresponding industrial implications.

Table 4. Key technologies of 5G and their industrial impacts

Technology	Technical Function	Industrial Impact
eMBB	High-speed data transmission	Supports video streaming, AR/VR, and smart cities
mMTC	Massive device connectivity	Enables IoT, logistics, and smart metering
URLLC	Ultra-low latency, high reliability	Essential for autonomous vehicles, remote surgery
Network Slicing	Virtual partitioning of networks	Custom networks for verticals like healthcare, manufacturing

Technology	Technical Function	Industrial Impact
Edge Computing	Processing near the data source	Real-time analytics reduces latency
O-RAN	Open Radio Access Network	Vendor-neutral, flexible deployment
Virtualization (NFV/ SDN)	Software-based network control	Scalable and programmable mobile architecture

The technological advancements underpinning 5G have revolutionized wireless communication, establishing it as a cornerstone of the digital era. Far beyond an evolution of mobile broadband, 5G integrates Enhanced Mobile Broadband (eMBB), Massive Machine-Type Communications (MTC), and Ultra-Reliable Low-Latency Communications (URLLC) to cater to the diverse demands of industries and individuals. These innovations transform the way humans, devices, and machines interact, driving breakthroughs in efficiency, precision, and connectivity.

According to Park et al. (2022), 5 G's enhanced mobile broadband (eMBB) capabilities drive revolutionary applications in immersive media and augmented reality.

eMBB has enhanced mobile broadband capabilities, enabling higher data speeds, reduced latency, and excellent coverage, which are essential for applications such as video streaming, augmented reality, and cloud-based services. Similarly, mMTC provides the infrastructure for connecting vast networks of low-cost, battery-operated devices, making it pivotal for IoT applications such as smart metering, logistics, and wearable technologies.

Meanwhile, URLLC offers unparalleled reliability and ultra-low latency, supporting critical use cases such as remote surgeries, autonomous vehicles, industrial automation, and public safety systems. These features collectively ensure that 5G addresses the distinct needs of both consumer and industrial applications. Popovski et al. (2018) emphasize that ultra-reliable low-latency communication (URLLC) in 5G enables critical applications such as autonomous driving and industrial automation. Ultra-reliable low-latency communication ensures seamless operation of industrial automation and autonomous vehicles (Popovski et al., 2018).

The impact of 5G extends far beyond technology; it is a transformative force across industries. 5G fosters innovation in healthcare, automotive, manufacturing, and public safety by enabling more intelligent, connected systems. It creates opportunities for businesses to achieve greater efficiency and unlock new revenue streams while improving the quality of life through enhanced connectivity and safety measures.

In conclusion, 5 G's technological perspective highlights its role as a catalyst for change in the modern world. Its ability to seamlessly integrate high-speed broadband, massive device connectivity, and ultra-reliable communication makes it indispensable for the future of industries and society as 5G evolves and expands.

REFERENCES

Agiwal, M., Kwon, H., Park, S., & Jin, H. (2021). A Survey on 4G-5G Dual Connectivity: Road to 5G Implementation. *IEEE Access: Practical Innovations, Open Solutions*, *9*, 16193–16210. DOI: 10.1109/ACCESS.2021.3052462

Agiwal, M., Roy, A., & Saxena, N. (2016). Next Generation 5G Wireless Networks: A Comprehensive Survey. *IEEE Communications Surveys and Tutorials*, *18*(3), 1617–1655. DOI: 10.1109/comst.2016.2532458

Bishen, A. (2019). *5G and the Cloud*. 5G Americas. https://www.5gamericas.org/5g-and-the-cloud/

Chataut, R., & Akl, R. (2020). *Massive MIMO Systems for 5G*. https://doi.org/DOI: 10.32545/encyclopedia202006.0016.v2

Chowdhury, M. Z., Fujii, T., Muntean, G.-M., Choi, J.-W., & Araniti, G. (2018). Emerging Small Cell Wireless Technologies for 5G: Architectures and Applications. *Wireless Communications and Mobile Computing*, *1–2*. DOI: 10.1155/2018/8969264

Evercom (2025). *5G LTE Antenna*. Evercomtech.com. https://www.evercomtech.com/5g-lte-antenna

Finley, K. (2018). *What is 5G? The Complete Guide to When, Why, and How*. Wired; WIRED. https://www.wired.com/story/wired-guide-5g/

Hajlaoui, E., Zaier, A., Khlifi, A., Ghodhbane, J., Hamed, M. B., & Sbita, L. (2020, September 1). *4G and 5G technologies: A Comparative Study*. IEEE Xplore. https://doi.org/DOI: 10.1109/ATSIP49331.2020.9231605

Huawei. (2025). *White Paper: 5G Network Architecture - A High-Level Perspective*. HuaweiTech. https://www.huawei.com/en/huaweitech/industry-insights/outlook/mobile-broadband/insights-reports/5g-network-architecture

Kerner, S. M., & Ferguson, K. (2024, December 17). *5 predictions about 5G adoption in 2023 and beyond | TechTarget*. WhatIs.com. https://www.techtarget.com/whatis/feature/5-Predictions-about-5G-Adoption-in-2021-and-Beyond

Marriwala, N. Tripathi, C. C., Kumar, D., & Jain, S. (2020). Mobile Radio Communications and 5G Networks. In *Lecture notes in networks and systems*. Springer International Publishing. https://doi.org/DOI: 10.1007/978-981-15-7130-5

Matterport (2025). Explore the business benefits of 3D tours. Matterport.

Park, Seongmin, Chopi, Bomin, Yim, Kangbin (2020). Vestiges of Past Generation: Threats to 5G Core Network. *Advances in Intelligent Systems and Computing*, 468–480. https://doi.org/DOI: 10.1007/978-3-030-50399-4_45

Popovski, P., Trillingsgaard, K. F., Simeone, O., & Durisi, G. (2018). 5G Wireless Network Slicing for eMBB, URLLC, and mMTC: A Communication-Theoretic View. *IEEE Access: Practical Innovations, Open Solutions*, *6*, 55765–55779. DOI: 10.1109/access.2018.2872781

Taleb, T., Samdanis, K., Mada, B., Flinck, H., Dutta, S., & Sabella, D. (2017). On Multi-Access Edge Computing: A Survey of the Emerging 5G Network Edge Cloud Architecture and Orchestration. *IEEE Communications Surveys and Tutorials*, *19*(3), 1657–1681. DOI: 10.1109/COMST.2017.2705720

Team Roar. (2017). *All You Need To Know About 5G*. Roar Media. https://archive.roar.media/english/tech/sponsored/all-you-need-to-know-about-5g

TELCOMA. (2025). *TELCOMA 5G Certifications and 5G Training Courses for Professionals including 6G, 4G-LTE, IoT, Cloud.* Telcoma. Telcomaglobal.com. https://www.telcomaglobal.com/p/5g-core-network-5g

Usmonov, B., Rustamov, U., & Shukurillaevich, U. (2019). *5G Technology Evolution*. 978-1-7281-2564-0/19/$31.00 ©2019 IEEE. https://www.academia.edu/43046535/5G_Technology_Evolution

Varga, P., Peto, J., Franko, A., Balla, D., Haja, D., Janky, F., Soos, G., Ficzere, D., Maliosz, M., & Toka, L. (2020). 5G support for Industrial IoT Applications— Challenges, Solutions, and Research gaps. *Sensors (Basel)*, *20*(3), 828. DOI: 10.3390/s20030828

ADDITIONAL READINGS

Abbas, N., Zhang, Y., Taherkordi, A., & Skeie, T. (2018). Mobile Edge Computing: A Survey. *IEEE Internet of Things Journal*, *5*(1), 450–465. DOI: 10.1109/jiot.2017.2750180

Accenture. (2021). *The impact of 5G on the European economy.* Accenture Strategy. https://www.accenture.com/content/dam/accenture/final/a-com-migration/manual/r3/pdf/pdf-144/Accenture-5G-WP-EU-Feb26.pdf

Al-Fuqaha, A., Guizani, M., Mohammadi, M., Aledhari, M., & Ayyash, M. (2015). Internet of Things: A Survey on Enabling Technologies, Protocols, and Applications. *IEEE Communications Surveys and Tutorials*, *17*(4), 2347–2376. DOI: 10.1109/COMST.2015.2444095

Cisco. (2017). *Cisco Visual Networking Index: Forecast and Trends, 2017–2022 White Paper*. Cisco. https://www.cisco.com/c/en/us/solutions/collateral/service-provider/visual-networking-index-vni/white-paper-c11-741490.html

Cisco. (2023, March 9). *Cisco Annual Internet Report (2018–2023) White Paper*. Cisco. https://www.cisco.com/c/en/us/solutions/collateral/executive-perspectives/annual-internet-report/white-paper-c11-741490.html

Deloitte. (2025). *Connected Edge: Converging cloud, edge computing, and advanced connectivity*. Deloitte. https://www.deloitte.com/us/en/services/consulting/5g-edge-computing-solutions.html

Dubernet, I., & Axhausen, K. W. (2019). The German value of time and value of reliability study: The survey work. *Transportation*, *47*(3), 1477–1513. DOI: 10.1007/s11116-019-10052-4

Ericsson. (2023). *Ericsson Mobility Report*. Ericsson.com. https://d110erj175o600.cloudfront.net/wp-content/uploads/2023/06/21123003/Ericsson-Mobility-Report-June-2023.pdf

Farooq, M., Waseem, M., Mazhar, S., Khairi, A., & Kamal, T. (975 C.E.). A Review on Internet of Things (IoT). *International Journal of Computer Applications, 113*(1). https://research.ijcaonline.org/volume113/number1/pxc3901571.pdf

Fatima, S., & Shri Ramtej Kondamuri. (2025). Machine Learning in Future Wireless Communications: Innovations, Applications, and Open Challenges. *Wireless Personal Communications*. DOI: 10.1007/s11277-025-11770-y

Foukas, X., Patounas, G., Elmokashfi, A., & Marina, M. K. (2017). Network Slicing in 5G: Survey and Challenges. *IEEE Communications Magazine*, *55*(5), 94–100. DOI: 10.1109/mcom.2017.1600951

GSMA. (2025). *Network Slicing*. GSMA. https://www.gsma.com/solutions-and-impact/technologies/networks/5g-network-technologies-and-solutions/5g-standalone-networks/network-slicing/

Ihs.com. (2017). *The 5G economy: How 5G technology will contribute to the global economy*. DOCSLIB.ORG. https://docslib.org/doc/6895067/the-5g-economy-how-5g-technology-will-contribute-to-the-global-economy

ITU. (2018). *Setting the Scene for 5G: Opportunities & Challenges.* ITU. https://www.itu.int/en/ITU-D/Documents/ITU_5G_REPORT-2018.pdf

Kelkkanen, V., Fiedler, M., & Lindero, D. (2021). Synchronous Remote Rendering for VR. *International Journal of Computer Games Technology, 2021*, 1–16. DOI: 10.1155/2021/6676644

Meiik Kottkamp, Anil Pandey, Raddino, D., Rosessler, A., & Reiner Stuhlfauth. (2019). *5G new radio fundamentals, procedures, testing aspects.* Rohde & Schwarz Gmbh & Co. Kg.

Lagorio, A., Cimini, C., Pinto, R., & Cavalieri, S. (2023). 5G in Logistics 4.0: Potential applications and challenges. *Procedia Computer Science, 217*, 650–659. DOI: 10.1016/j.procs.2022.12.261

Nguyen, D. C., Pathirana, P. N., Ding, M., & Seneviratne, A. (2020). Blockchain for 5G and beyond networks: A state of the art survey. *Journal of Network and Computer Applications, 166*, 102693. DOI: 10.1016/j.jnca.2020.102693

Nokia (2020). *Nokia 5G Readiness Report Data breakdown -regional findings and methodology.* Nokia. https://branden.biz/wp-content/uploads/2020/10/Nokia_5G_Readiness_Report_slideshare_Report_EN.pdf

Nwajana, A., & Ihianle, I. K. (2021). *Handbook of Research on 5G Networks and Advancements in Computing, Electronics, and Electrical Engineering.* IGI Global., DOI: 10.4018/978-1-7998-6992-4

Odida, M. (2024). The Evolution of Mobile Communication: A Comprehensive Survey on 5G Technology. *Journal of Sensor Networks and Data Communications, 4*(1), 01-11. https://doi.org/DOI: 10.33140/jsndc.04.01.06

Ratasuk, R., Vejlgaard, B., Mangalvedhe, N., & Ghosh, A. (2016, April 1). *NB-IoT system for M2M communication.* IEEE Xplore. https://doi.org/DOI: 10.1109/WCNC.2016.7564708

Sachan, S., Sharma, R., & Sehgal, A. (2023). Energy Efficiency and Scalability of 5G Networks for IoT in Mobile Wireless Sensor Networks. In: Bhushan, B., Sharma, S.K., Kumar, R., Priyadarshini, I. (eds) 5G and Beyond. Springer Tracts in Electrical and Electronics Engineering. Springer, Singapore. https://doi.org/DOI: 10.1007/978-981-99-3668-7_8

Salih, A. A., Zeebaree, S. R. M., Abdulraheem, A. S., Zebari, R. R., Sadeeq, M. A. M., & Ahmed, O. M. (2020). *Evolution of mobile wireless communication to the 5G revolution*. Technology Reports of Kansai University 62(5):2139-2151 ResearchGate. https://www.researchgate.net/publication/342549960_Evolution_of_Mobile_Wireless_Communication_to_5G_Revolution

Shafi, M., Molisch, A. F., Smith, P. J., Haustein, T., Zhu, P., De Silva, P., Tufvesson, F., Benjebbour, A., & Wunder, G. (2017). 5G: A Tutorial Overview of Standards, Trials, Challenges, Deployment, and Practice. *IEEE Journal on Selected Areas in Communications*, *35*(6), 1201–1221. DOI: 10.1109/jsac.2017.2692307

Sharma, P. (2013). Evolution of Mobile Wireless Communication Networks-1G to 5G as well as Future Prospective of Next Generation Communication Network. *IJCSMC, 2*(8), 47–53. https://citeseerx.ist.psu.edu/document?repid=rep1&type=pdf&doi=8e324078c7b0848c5e8c573861878cdbe417e89e

Taylor, P. (2025). *Forecast number of mobile 5G subscriptions worldwide from 2023 to 2028(in billions)*. Statista. https://www.statista.com/statistics/760275/5g-mobile-subscriptions-worldwide/

Varga, P., Peto, J., Franko, A., Balla, D., Haja, D., Janky, F., Soos, G., Ficzere, D., Maliosz, M., & Toka, L. (2020). 5G support for Industrial IoT Applications— Challenges, Solutions, and Research gaps. *Sensors (Basel)*, *20*(3), 828. DOI: 10.3390/s20030828

Chapter 6
Evolution of Mobile Telecommunications With 5G

ABSTRACT

Spectrum allocation is vital to 5G deployment, impacting coverage, capacity, and performance. It requires balancing low (sub-1 GHz), mid (1–6 GHz), and high-frequency (above 6 GHz) bands. Low bands offer wide rural coverage, while high-frequency millimeter waves deliver ultra-fast speeds for urban and industrial applications like AR and autonomous vehicles. Global regulators, including the ITU-R, shape spectrum frameworks—especially in the 24–86 GHz range—to support varied use cases. Their focus includes fair access, efficient allocation, and enabling both mobile operators and new industry-specific players to deploy 5G effectively.

INTRODUCTION

The evolution of mobile telecommunications has reached a pivotal milestone with the introduction of Fifth-Generation (5G) technology, marking a transformative chapter in global connectivity. The 5G technology represents a significant leap from previous generations, enabling faster, more reliable, and scalable communication networks. The evolution of 5G-compatible devices is accelerating, with Ericsson projecting substantial growth in the number and diversity of devices supporting advanced features such as standalone 5G and network slicing (Ericsson, 2024a).

As a result of extensive research into emerging demands and evolving market trends, 5G marks a significant advancement over previous generations by meeting the increasing need for faster, more reliable, and highly scalable communication networks.

DOI: 10.4018/979-8-3373-0497-7.ch006

This breakthrough goes beyond a simple upgrade, representing a transformative shift that unlocks new opportunities across industries and reshapes the fabric of modern society. Odida (2024) presents 5G as a transformative leap in mobile communication, characterized by ultra-fast data transfer, minimal latency, and the integration of advanced applications such as IoT, augmented reality, and autonomous vehicles—all of which require improved spectrum use and higher quality of service.

This chapter explores the evolution of mobile telecommunications, emphasizing the technological advancements and market trends that have shaped the emergence of 5 G. By examining its capabilities, applications, and transformative potential, this section highlights how 5G redefines modern telecommunications and sets a new standard for global connectivity.

NETWORK SLICING

At the heart of 5G innovation lies network slicing—a groundbreaking technique that enables multiple virtual networks to coexist on a single physical infrastructure. This approach provides exceptional flexibility and efficiency, allowing each slice to be tailored to the unique needs of individual users, applications, or industries. The result is a robust and adaptable framework that supports various applications, from smart cities and autonomous vehicles to advanced Internet of Things (IoT) ecosystems and mission-critical operations.

5G enhances the capabilities of autonomous robots in industrial environments. For example, Lagorio et al. (2023) argue that 5G technology has the potential to shape the logistics sector by facilitating real-time data exchange, automation, and enhanced system integration. Nevertheless, they emphasize that the successful adoption of 5G in this context depends on addressing significant challenges, including high implementation costs, substantial infrastructure requirements, and persistent data security concerns.

The implications of 5G extend beyond improved speed and connectivity; they redefine the role of telecommunications as the foundation for digital transformation. By enabling previously unattainable applications, 5G unlocks new opportunities for innovation across various sectors, including manufacturing, logistics, healthcare, and entertainment. From powering real-time analytics and remote surgeries to enhancing augmented reality experiences, 5G sets the stage for a brighter, more interconnected world.

The intense advancement of the Fifth-Generation Mobile Communications method (also known as 5G) is underway worldwide, marking a significant improvement over current mobile communication methods, based on extensive investigations into future demands and market trends for mobile communication in the 2020s.

According to Salih et al. (2020), the evolution of mobile communication from 1G to 5G marks a significant technological leap, aiming to enable more intelligent and ubiquitous connectivity by integrating IoT technologies that link countless devices across society.

Sharma (2013) describes the evolution of wireless communication from 1G to 5G as a progression focused on expanding capabilities, moving from basic voice services in 1G to high-speed, multi-user mobile broadband in 4 G. 5G is envisioned as an intelligent network integrating multiple wireless technologies simultaneously.

According to Odida (2024), the introduction of 5G addresses the limitations of previous mobile generations by aiming for exponential increases in capacity and efficiency through technologies such as Massive Multiple Input Multiple Output (MIMO). It also supports massive machine-type communication and ensures ultra-reliable connectivity.

The evolution of mobile telecommunications has reached a transformative milestone with the advent of 5G technology. Representing a significant leap from its predecessors, 5G addresses the escalating demands for faster, more reliable, and scalable connectivity to support emerging industry applications.

Central to 5G's innovation is network slicing, a revolutionary approach that allows multiple virtual networks to coexist on a shared infrastructure, each tailored to specific user requirements. The innovative feature of network slicing in 5G allows customized virtual networks to operate on shared infrastructure. *This capability ensures unparalleled flexibility, efficiency, and customization, paving the way for diverse applications, including smart cities, autonomous vehicles, and advanced Internet of Things (IoT) ecosystems.* As the cornerstone of modern telecommunications, 5G redefines connectivity and unlocks unprecedented opportunities for digital transformation.

The 5G Network Slicing

5G network slicing is a groundbreaking feature of the Fifth-Generation telecommunications system, designed to revolutionize connectivity by shaping multiple virtual networks on a shared infrastructure. This innovative approach enables each network slice to be customized for specific applications, ensuring optimal performance, scalability, and efficiency. 5G's ability to support IoT ecosystems and mission-critical applications highlights its transformative potential across industries" (Zhang & Wang, 2021). *By meeting a wide range of industry-specific requirements, 5G network slicing plays a pivotal role in enabling advanced technologies such as the Internet of Things (IoT), smart cities, and autonomous systems, positioning itself as a cornerstone of next-generation digital infrastructure*, (Sowa, 2020).

According to Sharma (2013), 5G is expected to surpass previous generations by offering faster and more efficient data access, enabling user terminals to interact with various wireless systems in real-time seamlessly, and forming a highly interconnected digital environment.

Definition

Network slicing is a virtual network structure similar to software-defined media (SDN) and network functions virtualization (NFV). *These two closely associated network virtualization solutions move contemporary networks toward software-based automation*. SDN and NFV enable greater flexibility by partitioning network architectures into virtual components. Network slicing enables the creation of multiple virtual networks on top of a standard physical infrastructure, allowing tailored services with specific performance and security requirements to coexist efficiently (Balador et al., 2018).

In this virtualized network scenario, physical elements are logical, and secondary (software-based) partitions are paramount, allocating capability to specific functions dynamically as needed. As it requires change, so could the saved resources. Network slicing permits the construction of slices dedicated to rational, self-contained, and partitioned networks.

5G network slicing is an architecture that enables the creation of virtual and independent networks on the same physical network infrastructure.

A computer network's design is its architecture. It is a framework for specifying the network's physical components, functional organization, and configuration, as well as its operational principles and procedures, and communication protocols.

A network architecture specification in telecommunications typically includes a detailed description of services and products delivered via a communications network, as well as clear rates and billing structures.

The Internet's architecture relies on the Internet Protocol Suite rather than a specific model to connect network nodes and support various hardware links.

Each network slice is an isolated and tailored end-to-end network designed to fulfill the diverse requirements of a particular application.

Figure 1. 5G network slicing

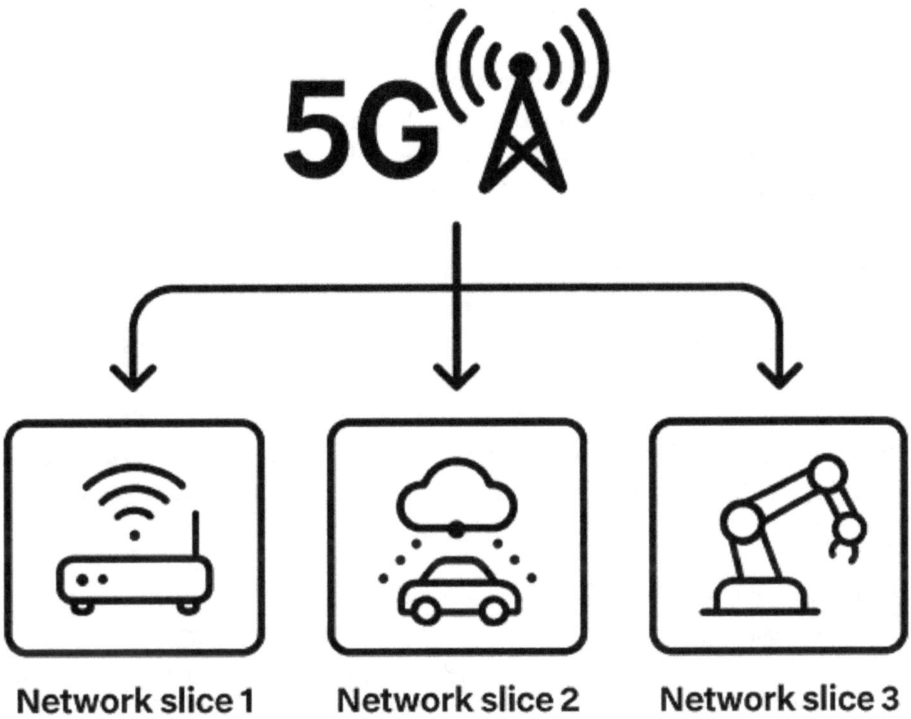

5G SMARTPHONE ARCHITECTURE: CONSIDERATION IN SMARTPHONE DESIGN

Integrating 5G technology into smartphones is not merely an upgrade of network capabilities, but a comprehensive architectural transformation. As manufacturers move toward the next generation of devices, discussions around 5G smartphone architecture must be central to the design process. This ensures optimal performance, thermal management, battery efficiency, and compatibility with global 5G standards.

The architecture of a 5G smartphone relies on the sophisticated integration of hardware and software components, engineered to meet the demands of next-generation connectivity. At its core is the 5G modem, such as the Snapdragon X60, which is capable of handling both mmWave and sub-6 GHz frequencies, supporting both Standalone (SA) and Non-Standalone (NSA) networks. Paired with an advanced antenna array and RF front-end, the system enables high-speed data transfer with beamforming and MIMO technologies. The System on Chip (SoC) unifies CPU, GPU, and AI engines, coordinating data processing and power efficiency. Thermal management systems ensure heat is dissipated during intensive 5G usage. Power management ICs regulate the phone's energy needs, striking a balance between performance and battery life. The software stack enables seamless handovers, reduces latency, and optimizes throughput. AI and machine learning enhance real-time network selection, antenna tuning, and battery conservation. Together, these elements create a fast, intelligent, and reliable 5G smartphone experience. Below is an overview of the key components that define 5G smartphone architecture.

Table 1. Key components of 5G smartphone architecture

Component	Function and Importance
5G Modem (e.g., Qualcomm Snapdragon X55/X60)	Handles 5G NR connectivity across multiple bands and supports standalone and non-standalone networks.
Antenna Array and RF Front-End	Enables mmWave and sub-6 GHz signal processing with beamforming and MIMO support.
SoC (System on Chip)	Integrates CPU, GPU, and AI engines; manages communication with modem and other system elements.
Thermal Management System	Dissipates heat from high-speed data processing and modem usage.
Battery and Power Management ICs	Supports higher power demands of 5G connectivity while maintaining energy efficiency.
Software Stack and Firmware	Optimizes network handovers, reduces latency, and enhances data throughput.
AI/ML Integration	Improves antenna tuning, power management, and network selection in real-time.

For this reason, this technology plays a vital role in supporting 5G mobile networks, which are designed to efficiently accommodate a wide range of services with varying service level requirements (SLRs). Leveraging the concepts of software-defined networking (SDN) and network function virtualization (NFV) enables the realization of a service-oriented network view, which facilitates the implementation of scalable network slices within a joint network infrastructure.

From a business model perspective, each network slice manages a mobile virtual network operator (MVNO). *The infrastructure provider (telecommunication infrastructure) rents its physical resources to MVNOs, which share the underlying physical network.* An MVNO can autonomously deploy multiple network slices customized to the various applications it provides to its users according to the availability of the assigned resources.

The NGMN Alliance divides the hierarchical structure of network slicing into three main layers: service, network function, and infrastructure. Beneath it, the network function layer oversees the core operations and functionalities necessary to meet diverse service demands. At the base, the infrastructure layer includes the radio access, transport, and core networks, forming the backbone that enables both the physical and virtual dimensions of connectivity. By coordinating these layers, the orchestrator and network slice controller ensure efficient management and dynamic resource allocation, thereby enhancing network adaptability and flexibility. This layered architecture enables the creation of tailored, independent network slices for specific use cases, thereby improving performance and optimizing resource utilization.

The service layer is designed for virtual mobile operators and third-party providers, ensuring flexible and tailored service delivery.

Figure 2. Generic slicing framework. Source: adapted from Description of Network Slicing Concept. NGMN Alliance, 2016.

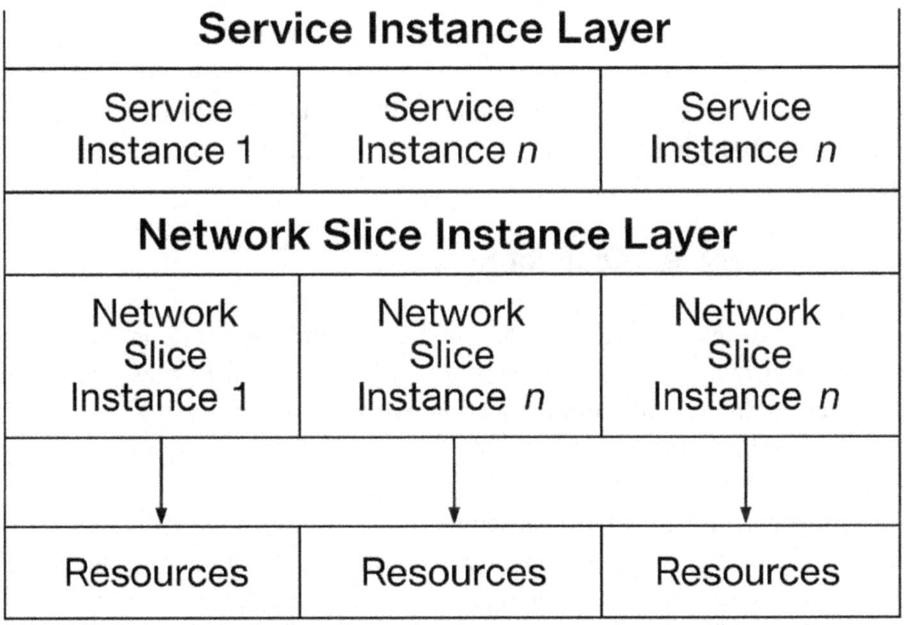

NETWORK SLICING FOR INNOVATIVE GRID SERVICES

5G network slicing apps are an excellent choice for intelligent grid services. It divides the 5G system into logically isolated networks. 5G network slicing enables the power system to flexibly privatize slices with various network capabilities and service-level agreement (SLA) assurances based on the multiple demands of different solutions within a power system. Zhang and Wang (2021) explain how 5G networks are transforming the energy sector by enabling smart grids and efficient energy management systems. Innovative grid systems benefit from 5G network slicing by achieving enhanced reliability and low-latency communication"

The 5G SA-Based Power Slicing Innovation Project

The 5G SA-based power slicing innovation project, a collaboration among China Telecom, SGCC (State Grid Corporation of China), and Huawei, was established in September 2017 through breakthroughs in confirming technology feasibility, exploring the feasibility of technology companies, and cultivating the environment.

The 5G power slicing may support differentiated service scenarios, assist power system enterprises, and considerably lower the total ownership cost (TCO) of power communication networks, yielding carriers a significant return on investment (ROI). Thus, the 5G smart grid development task has become a valuable reference point for cooperation among global enterprises and operators in the 5G era.

5G network slicing enables service providers to create multiple virtual networks tailored to specific application requirements, offering a flexible and efficient way to allocate network resources. For example, 5G enhances real-time analytics and enables groundbreaking applications such as autonomous driving and remote surgeries, (Agarwal et al., 2019).

Each Slice Serves a Distinct Purpose

Each slice serves a distinct purpose, providing high-speed internet through mobile broadband, connecting numerous devices with massive IoT slices, and enabling low-latency, highly reliable communication with mission-critical IoT slices. Unlike 4G networks, which often fail to meet future demands, 5G slicing optimizes network segments for various applications, including communication, entertainment, retail, manufacturing, and medical infrastructure.

Providers can cater to diverse industries by isolating these slices while maintaining high performance and security. This approach significantly enhances the scalability and adaptability of 5G networks, facilitating advanced applications such as autonomous vehicles, smart cities, and remote healthcare. Through network slicing, 5G technology delivers tailored end-to-end solutions for various use cases, making it a cornerstone of next-generation connectivity.

A Key Benefit of the 5G Network for Network Operators

According to 5G Americas, a key benefit of slicing the 5G network for network operators will be the ability to deploy features that help individual clients and specific industry segments. In addition, 5G network slicing offers immediate savings compared to deploying complete functionality for products that will use only a component of that functionality. A derivative advantage may be the ability to deploy 5G methods faster, allowing for a faster time to market.

Several vendors, such as Ericsson, think 5G network slicing will be the primary key to meeting the technical requirements. The evolution of 5G-compatible devices is accelerating, with Ericsson projecting substantial growth in the number and diversity of devices supporting advanced features such as standalone 5G and network slicing (Ericsson, 2024a). *The vast diversity requires the terminology of power, bandwidth, and pace to recognize the brand-new era of 5G connectivity.* According to Ericsson, the greater suppleness brought about by network slicing will enable the meeting of future flexibility requirements, improve efficiency, and reduce costs. The analysis of emerging traffic patterns demonstrates how user behavior and data consumption are actively shaping the architecture and deployment strategies of future 5G networks, (Ericsson, 2024c).

Mobile data traffic is expected to surge dramatically in the coming years, driven primarily by the proliferation of high-bandwidth 5G applications such as streaming, gaming, and industrial automation (Ericsson, 2024d).

The global number of mobile subscriptions continues to rise, with 5G subscriptions representing an increasingly dominant share of the total mobile ecosystem, (Ericsson, 2024e).

Table 2 illustrates how Network slicing is a fundamental feature of 5G that enables the creation of multiple virtual networks over a single physical infrastructure. Each slice can be customized to meet specific performance, reliability, and latency requirements, making it ideal for innovative grid applications. In the energy sector, it enables operators to allocate dedicated slices for functions such as real-time monitoring, fault detection, and automated control systems. This improves operational efficiency while maintaining high levels of security and service quality. Projects like the collaboration between China Telecom, SGCC, and Huawei showcase their potential in powering digital transformation in utilities. Network slicing also supports the integration of renewable energy and IoT devices. Overall, it enhances grid responsiveness, flexibility, and sustainability.

Table 2. Network slicing for innovative grid services

Aspect	Description
Definition and Function	Network slicing divides the 5G network into logically isolated virtual networks tailored to specific application needs, enabling flexible resource allocation.
Applications in Smart Grids	It supports intelligent grid services with differentiated service scenarios, enhancing reliability, low latency, and efficiency in energy management systems.
Example Project	China Telecom, SGCC, and Huawei launched the 5G SA-based power slicing innovation project in 2017 to test technology feasibility and promote global collaboration.
Advantages for Providers	Reduces total cost of ownership (TCO), accelerates time to market, enables faster deployment, and provides ROI through efficient resource usage and targeted service delivery.
Industry Impact	Facilitates advanced applications, such as autonomous vehicles, remote healthcare, and smart cities. Enhances scalability, flexibility, and security across industries.

NETWORK SLICING AND THE SMART GRID

The slicing of 5G networks has emerged as a transformative solution for advancing innovative grid services, providing tailored connectivity to meet the diverse and complex requirements of modern power systems. This technology divides the 5G network into logically isolated slices, each customized to serve specific applications with distinct network functions and Service Level Agreements (SLAs). These SLAs ensure that each slice meets the unique performance demands of innovative grid services, including ultra-reliable low-latency communication (URLLC), massive machine-type communication (MTC), and enhanced mobile broadband (eMBB).

Most Significant Advantages of 5G Network Slicing

One of the most significant advantages of 5G network slicing for the smart grid is its ability to provide flexibility, scalability, and connectivity. Power systems can dynamically allocate slices to critical operations such as distributed energy resource management, fault detection, and real-time grid monitoring. This customization enables utilities to deliver exceptional coverage, robust security, and high reliability while maintaining cost efficiency.

Moreover, 5G slicing differentiates network capabilities, optimizing bandwidth and latency for metering, automation, and renewable energy integration applications. The hierarchical structure of network slicing supports service differentiation and efficient resource allocation" (Agarwal et al., 2019). *5.* G technology's multi-layered approach supports diverse applications, from industrial automation to immersive entertainment experiences.

The 5G Network Slicing Deployment Offers a Less Expensive and More Straightforward Alternative

The 5G network slicing deployment offers a less expensive and more straightforward alternative to traditional fiber-based communication systems, particularly in remote and hard-to-reach locations. By leveraging 5G infrastructure, utilities can enhance grid operations without extensive investments in physical connectivity. Additionally, the improved security features of 5G slicing safeguard critical grid data, reducing the risks of cyberattacks and unauthorized access. Salih and colleagues (2020) highlight that 5G surpasses previous generations by offering higher data rates, enhanced services, and broader application support, thereby meeting the growing demand for real-time communication and comprehensive digital integration.

In conclusion, 5G network slicing is a game-changer for the smart grid, enabling power systems to benefit from tailored, efficient, and reliable connectivity. By meeting the diverse demands of modern grid operations, this technology facilitates the transition to a more intelligent, sustainable, and resilient energy ecosystem.

SLICING AND APPLICATION SCENARIOS OF SOPHISTICATED GRID

The sophisticated grid, supported by 5G network slicing, enables advanced communication systems that cater to the diverse requirements of intelligent grid scenarios. This development integrates four standard slicing approaches to address the grid's critical needs of intelligent distributed feeder automation. First, ultra-low latency communication is essential, typically within milliseconds, to ensure real-time responsiveness in fault detection and feeder reconfiguration operations. Second, high isolation is crucial as distributed feeder automation operates in sensitive areas of the power system, such as the I/II production zones, necessitating strict segregation from less critical III/IV management services.

Third, reliability is paramount, with a stringent requirement of 99.999% uptime to ensure the continuous and dependable operation of grid systems. 5G slicing applies tailored performance to each scenario, optimizing network resources for specific demands. For example, a slice dedicated to distributed feeder automation ensures ultra-low latency and high isolation, unaffected by other network activities. Moreover, these advanced communication strategies support key performance indicators

(KPIs) like efficiency, responsiveness, and fault resilience within the intelligent grid infrastructure.

These features collectively enable the integration of distributed energy resources, support predictive maintenance, and optimize grid stability. By addressing the diverse needs of intelligent grid applications, 5G slicing plays a transformative role in enhancing the operational efficiency, reliability, and scalability of modern energy systems, paving the way for a more resilient and sustainable power infrastructure.

MOBILE ACCESS TO THE NETWORK ECOSYSTEM EVOLUTIONS WITH THE 5G

5G methods address a wide range of use cases with varying data rates, required powers, densities, reliabilities, latencies, and other factors. This sizable span implies numerous trade-offs and a very scalable structure. The brand-new 5G ecosystem is complex and requires the participation of multiple stakeholders, including component manufacturers, chipset providers, network equipment suppliers, wireless device manufacturers, mobile network operators, software developers, and end-users, all overseen by various standardization entities. The evolution of mobile telecommunications has culminated in introducing 5G, redefining connectivity globally," (Won et al., 2022).

This section examines the evolution of mobile access network ecosystems with the introduction of 5 G. We identify the current leading players in the Radio Access Network and present the potential changes in their positioning within the value chain in the 5G era. We also identify new players that may play essential roles in 5G and determine their relationships with the classical players. Additionally, 5G technologies and services offer a new value chain of activities. Finally, the ecosystem evolutions highlight a net value analysis. Mobile operators occupy the center of the scene, and the current and new players surround them within four categories: customers, suppliers, competitors, and complementors.

The mobile ecosystem has evolved from a linear relationship between cellular customers and networks to a system of individual businesses involved in various phases of the value chain. The value chain is now unique, divided into sections that include articles related to services and programs, access devices, integration services, network infrastructure, a wide range of subsections, and market apps. Additionally, given the compelling nature of the telecommunications industry, it is understandable that the value chain is dynamic and constantly evolving.

Hardware Providers

Hardware providers form the backbone of the telecommunication value chain, producing essential network equipment and user devices. These providers include prominent infrastructure vendors such as Ericsson, Nokia, Huawei, and Samsung, as well as leading device companies like Apple, LG, and Nokia. Their contributions range from base stations and network controllers to end-user devices, such as mobile phones, tablets, and dongles. As critical components, base stations enable seamless network communication, acting as the foundation for any telecommunication infrastructure. User devices designed by these hardware providers serve as access points for customers to engage with the network, ensuring robust connectivity and user-friendly interfaces. These components represent the starting point for delivering innovative telecommunication services, making hardware providers indispensable in the industry's ecosystem.

Software Providers

Software providers play a crucial role in the telecommunications value chain by developing middleware and applications that enable the efficient operation of networks with user devices and infrastructure. These providers include renowned operating system designers like Apple with iOS, Google with Android, and Microsoft with its Mobile platform. Their contributions ensure seamless device operations and compatibility with network protocols. Also, software providers' middleware facilitates network monitoring, management, and troubleshooting. Base station manufacturers often include proprietary software to enhance equipment performance and optimize network operations. These software solutions are critical in heterogeneous networks, where infrastructure from multiple vendors must work cohesively. By creating robust software ecosystems, these providers enhance user experiences and support the operational efficiency of telecommunication networks.

Facility and Equipment Providers

Facility and equipment providers are essential collaborators in the telecommunications industry, offering critical infrastructure for network deployment and management. These include tower companies that provide physical sites for network equipment, urban furniture organizations that enable small-cell deployments in urban areas, and facility managers responsible for maintaining coverage and capacity extensions. These entities often work closely with Mobile Network Operators (MNOs) to ensure optimal network performance and reach. Additionally, MNOs frequently subcontract specialized companies or equipment vendors to operate and manage networks efficiently and reduce operational burdens.

These partnerships are crucial for expanding network coverage, enhancing capacity, and facilitating the ongoing development of telecommunications infrastructure.

Network Operators

Network operators, also known as Mobile Network Operators (MNOs), play a critical role in the telecommunications industry. They hold licenses to operate mobile networks and are responsible for deploying and managing infrastructure software, as well as providing connectivity services. Mobile Network Operators (MNOs) manage equipment such as base stations and network controllers, and work closely with facility and equipment managers to ensure optimal network coverage and performance. Prominent examples of network operators include AT&T, Vodafone, Verizon, and Orange, all of which have established extensive networks to deliver high-speed and reliable mobile services. Their roles extend beyond infrastructure management as they ensure seamless connectivity, customer satisfaction, and the rollout of advanced technologies such as 5G, shaping the future of communication.

Content Suppliers

Content suppliers play a crucial role in the telecommunications ecosystem by providing information and entertainment that attract and retain customers. These providers offer a wide range of digital content, such as maps, email services, news, television programming, and social media platforms, making them integral to the user experience on modern mobile networks. Examples include tech giants such as Google, which offers services like Gmail and Maps, as well as news organizations, cloud providers, and streaming platforms. While content suppliers were less critical in voice-dominated networks, their importance has grown with the rise of smartphones and data-driven networks, where user engagement heavily depends on accessible, high-quality content.

Over-the-Top (OTT) Players

Over-the-top (OTT) players have emerged as significant disruptors in the wireless telecommunications value chain. Unlike traditional telecom services, OTTs (Over-the-Top services) bypass Mobile Network Operators (MNOs) to deliver voice, video, and messaging services directly over the Internet, often at no additional cost to users. Notable examples include Viber and Skype, which have gained widespread adoption for video conferencing and voice calls. Though they are not classically part of the telecommunications value chain, OTT players have reshaped the industry by offering cost-effective alternatives to traditional communication services. Their growing influence highlights the evolving dynamics of the telecommunications sector, where digital platforms are increasingly central to user engagement and connectivity. The research by Lagorio et al. (2023) suggests that while 5G can play a pivotal role in advancing smart logistics, its widespread adoption remains contingent on addressing significant technical and financial barriers, without which its transformative potential may not be fully realized.

Service Suppliers

Service providers play a vital role in the telecommunications ecosystem. They offer end customers a wide range of wireless services, enabling seamless communication and connectivity. Their solutions encompass voice call services that cater to local, regional, national, and international communication needs. These voice services include features such as voicemail, call waiting, caller ID, and call forwarding, enhancing user convenience and flexibility in managing calls.

In addition to voice services, service providers deliver robust data services tailored to modern digital lifestyles. These include SMS messaging for quick text communication, streaming services for entertainment and education, and email services to support professional and personal correspondence. Web browsing and text alerts enable users to stay updated with the latest information, while mobile TV services provide on-the-go access to television content.

Service providers are also innovating to meet the growing demand for advanced data solutions, including real-time streaming, cloud gaming, and IoT connectivity. By enabling advanced IoT solutions, 5G contributes significantly to innovative city development and infrastructure optimization.

By integrating advanced network technologies, such as 5G, they aim to deliver ultra-fast speeds, low latency, and reliable connectivity, making mobile internet access more efficient and user-friendly. Moreover, the flexibility and scalability of these services enable providers to cater to a diverse customer base, ranging from individual consumers to businesses that rely on mobile solutions for their operations.

As wireless communication continues to evolve, service providers are strategic in driving innovation, enhancing user experiences, and enabling the digital transformation of industries. They bridge the gap between technology and everyday life, and their comprehensive offerings ensure that customers remain connected and empowered in an increasingly interconnected world.

End Users

The telecommunications value chain culminates with the end users, who are the ultimate beneficiaries of wireless services and products. These end users encompass a broad spectrum of individuals, businesses, and public administrations that rely on mobile networks for communication, connectivity, and productivity. Customers use smartphones and other wireless devices for daily activities such as messaging, streaming, and social networking, while businesses leverage telecommunications for operational efficiency, collaboration, and customer engagement. On the other hand, administrations depend on mobile networks for governance, public safety, and infrastructure management.

End users are critical in shaping the value chain, as demand drives innovation and service development. A growing user base encourages further service adoption, enhancing revenue streams for service providers and fostering network effects through their mutual influence.

For example, as more individuals and businesses adopt 5 G-enabled devices, the network ecosystem expands, enabling the delivery of advanced applications such as smart city solutions, IoT deployments, and remote work technologies.

Moreover, the behavior and preferences of end users influence pricing models, service quality, and the evolution of network standards. Providers must adapt to their needs, ensuring low latency, high data speeds, and affordable access to maintain competitiveness. The dynamic interaction between end users and service providers is a vital feedback loop that drives the continuous advancement of telecommunications technologies and ensures their alignment with societal and business needs.

The deployment of 5G is transforming the mobile network ecosystem by redefining the roles of its key stakeholders. From hardware manufacturers to content creators, each actor contributes to enabling and scaling the capabilities of next-generation connectivity. Table 3 outlines the evolving responsibilities and contributions of these ecosystem players within the 5G landscape.

Table 3. Mobile access to the network ecosystem evolutions with the 5G

Ecosystem Role	Description
Hardware Providers	This includes major vendors like Ericsson, Nokia, Huawei, and Apple. Provide essential equipment like base stations, controllers, and user devices. Critical for enabling network infrastructure and customer access.
Software Providers	Develop operating systems (e.g., iOS, Android) and middleware for network and device management. Ensure compatibility, optimize performance, and facilitate efficient troubleshooting across multi-vendor environments.
Facility and Equipment Providers	Provide infrastructure support, including towers, urban sites, and maintenance services. Collaborate with MNOs to extend network reach and capacity, ensuring stable and widespread 5G deployment.
Network Operators (MNOs)	Key players include AT&T, Vodafone, and Verizon. They own network licenses, deploy infrastructure, manage connectivity, and roll out advanced technologies like 5G, which are central to service quality and innovation.
Content Suppliers	Provide engaging digital content such as maps, emails, news, and video. Examples include Google and streaming services. Their content drives data consumption and user engagement in 5G networks.
Over-The-Top (OTT) Players	Disrupt traditional telecom by offering services like video and messaging (e.g., Skype, Viber) over the internet. Bypass MNOs and reshape the telecom value chain with cost-effective communication alternatives.
Service Providers	Deliver voice, data, SMS, and emerging 5 G-based services like cloud gaming and IoT. Focus on ultra-fast speeds, low latency, and reliability. Enable digital transformation for both consumers and industries.
End Users	Individuals, businesses, and public institutions utilize mobile networks to drive demand, influence service offerings, and enable advanced applications such as smart cities, IoT, and remote work solutions.

FOG COMPUTING (FC)

Fog computing has emerged as a pivotal architecture in the 5G era, enabling real-time data processing and low-latency services by decentralizing cloud capabilities closer to the network edge and IoT devices.

Definition

Fog computing extends cloud capabilities through IoT and node gateways, integrating computing, networking, and storage services closer to the data sources. It enables faster data processing and supports applications requiring low latency and real-time analytics, effectively bridging the gap between cloud systems and end devices. This decentralized approach enhances performance by reducing the dependency on centralized cloud data centers. Additionally, fog computing offers improved bandwidth management and energy efficiency, making it an ideal solution for IoT, autonomous vehicles, and innovative city applications. It also adds an extra layer of security by processing sensitive data locally, thereby minimizing the risks associated with transmitting data over long distances.

As mobile networks evolve with the integration of 5G, traditional cloud-based models are no longer sufficient to meet the stringent requirements of ultra-low latency and high reliability for emerging applications. This has led to fog computing, a decentralized architecture that brings computation closer to the data source. The Figure 3 illustrates how fog computing operates between IoT devices and the centralized cloud, enabling faster processing and real-time responsiveness in critical use cases.

Figure 3. Fog computing architecture in the 5G ecosystem

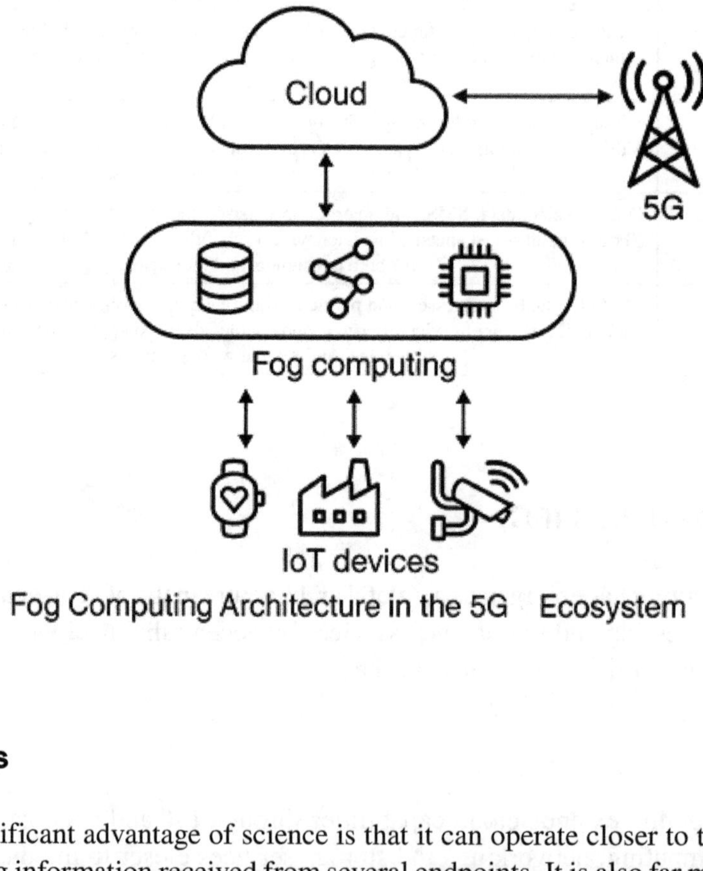

Benefits

A significant advantage of science is that it can operate closer to the end-user, processing information received from several endpoints. It is also far more scalable than edge computing if a system's public provides information through numerous data points. For instance, Nwajana and Ihianle (2021) provide multidisciplinary insights into computing advancements driven by 5 G.

5G networks promise to reduce latency to sub-millisecond levels, thereby alleviating infrastructure bottlenecks. 5G networks must support fog computing to achieve this goal.

High-frequency millimeter waves will push interaction within 5G networks. For example, a primary 5G cellular network links mobile people to a base station hooked to the core system. Even though lower-latency radio interfaces offer sub-millisecond interaction between the end-user and a base station, forwarding cloud-based program requests from the core system to the cloud would considerably increase latency. Therefore, 5G networks must move processing power closer to the end-user, which can be achieved with fog computing.

DEPLOYMENT SCENARIOS OF PRIVATE 5G NETWORKS

Private 5G networks can be deployed in two main ways: (1) a fully isolated network built on enterprise premises (5G Island), or (2) shared infrastructure using a mobile operator's public 5G resources. Isolated deployments offer maximum control, privacy, and ultra-low latency but come at higher costs and require skilled personnel. Shared deployments, such as network slicing, reduce costs by leveraging the carrier's infrastructure while still offering logical separation for enterprise use.

Netmanias identifies multiple deployment models:

- Fully isolated 5G LAN (no sharing)
- RAN sharing
- RAN + control plane sharing
- Full core and RAN sharing (end-to-end slicing)
- Local breakout options (N3/F1)

Scenario #1: Enterprise-Isolated 5G LAN

In this model, the enterprise deploys its own 5G components (gNB, 5GC, MEC) using local spectrum where permitted (e.g., Japan, Germany). *It avoids limitations of wired LANs and Wi-Fi (e.g., limited range, instability) while benefiting from 5G's low latency and strong security*. However, it demands substantial investment, fiber connectivity for backhaul, and skilled network engineers.

Table 4. Scenario #1: Isolated 5g LAN built by the enterprises (local 5g frequency, full private, no-sharing)

Aspect	Description
Deployment	The enterprise deploys a complete private 5G network (gNB, UDM, 5GC CP, UPF, MEC) locally using a non-licensed local 5G frequency, with no reliance on public mobile operator infrastructure.
Who Builds	It is typically built by the enterprise itself, though third parties or mobile operators may assist in some countries, depending on national regulatory frameworks (e.g., Japan, Germany, USA).
Advantages	It provides ultra-low latency and high connectivity, removes the limitations of wired and Wi-Fi LANs, and enables new applications or the scaling of existing ones. It is ideal for rural factories needing optical backhaul links.
Security and Privacy	Offers complete control over data, including traffic, subscription, and operational data, as the network is entirely private and isolated from public infrastructure.
Challenges	High deployment costs make it difficult for smaller enterprises. Also requires in-house personnel with 5G expertise to manage and operate the network infrastructure.
Technical Requirements	The absence of fiber infrastructure may hinder deployment; 5G ensures low latency (a few milliseconds) via URLLC services and requires backhaul solutions for sustained connectivity.

Scenario #2: Isolated 5g Lan Made by Mobile Operators (Licensed 5G Frequency, Full Private, No-Sharing)

The unique 5G network structure is the same as one. The sole variation with one) is that Mobile operators build and operate 5G LAN in the business with their own licensed 5G Frequency.

In this scenario, mobile network operators (MNOs) are responsible for deploying and managing a private 5G local area network (LAN) within an enterprise's premises. Unlike Scenario #1, where enterprises build and own the network using local, unlicensed spectrum, Scenario #2 leverages the operator's licensed spectrum. This ensures access to high-quality, interference-free frequency bands, enhancing network performance and reliability.

Mobile operators provide the full 5G stack, including gNB, UDM, 5GC CP, UPF, and MEC components. This setup enables enterprises to leverage telecom providers' expertise and scale without incurring infrastructure costs. Such deployments are particularly beneficial in highly regulated or resource-constrained environments.

Furthermore, the network remains fully isolated from public systems, preserving privacy and data integrity. This model suits businesses prioritizing guaranteed service levels, security, and fast deployment while reducing internal technical demands. The approach fosters industry-grade private connectivity with minimal risk.

Table 5. Scenario #2: Isolated 5G LAN made by mobile operators (licensed 5g frequency, full private, no-sharing)

Aspect	Description
Deployment	The enterprise deploys a complete private 5G network (gNB, UDM, 5GC CP, UPF, MEC) locally using a non-licensed local 5G frequency, with no reliance on public mobile operator infrastructure.
Who Builds	It is typically built by the enterprise itself, though third parties or mobile operators may assist in some countries, depending on national regulatory frameworks (e.g., Japan, Germany, USA).
Advantages	It provides ultra-low latency and high connectivity, removes the limitations of wired and Wi-Fi LANs, and enables new applications or the scaling of existing ones. It is ideal for rural factories needing optical backhaul links.
Security and Privacy	Offers complete control over data, including traffic, subscription, and operational data, as the network is entirely private and isolated from public infrastructure.
Challenges	High deployment costs make it difficult for smaller enterprises. Also requires in-house personnel with 5G expertise to manage and operate the network infrastructure.
Technical Requirements	The absence of fiber infrastructure may hinder deployment; 5G ensures low latency (a few milliseconds) via URLLC services and requires backhaul solutions for sustained connectivity.

Scenario #3: RAN Sharing Between the Public and Private Network

Multi-Access Edge Computing (MEC) plays a pivotal role in enhancing 5G architecture by relocating data traffic and processing closer to end-users. This proximity significantly reduces latency and supports real-time performance for bandwidth-intensive and mission-critical applications such as autonomous vehicles and smart manufacturing.

Key attributes of MEC:

- Proximity to users
- Ultra-low latency
- High bandwidth availability
- Virtualization and container support

MEC enables local processing via small-scale data centers—located in metro hubs, cell sites, or even on-premises—thus reducing reliance on centralized cloud infrastructure. It enhances responsiveness and lowers network congestion, making it ideal for applications like AR/VR, industrial IoT, and automated control systems.

Role of the 5G Core (5GC)

The 5GC introduces a modular and service-based architecture that separates the Control Plane (CP) and User Plane (UP), offering the flexibility required for network slicing, differentiated services, and agile deployments.

Core Functions of 5GC:

- Dynamic network slicing and resource orchestration
- Support for URLLC and eMBB traffic
- Secure service isolation and data management
- Seamless integration with MEC and SDN/NFV framework

Table 6. Scenario #3: RAN sharing between the public network and private network

Aspect	Description
Deployment Model	RAN (gNB) is shared between public and private networks, while UPF, UDM, 5GC CP, and MEC are deployed privately within the enterprise premises.
UPF (User Plane Function)	It handles packet processing near the network edge. It allows efficient data transmission and separates public and private network traffic within a shared RAN environment.
UDM (Unified Data Management)	Manages user data, profiles, and authentication for private devices. Operates on the control plane and is integrated into the enterprise for data privacy and local management.
MEC (Multi-Access Edge Computing)	It reduces latency by processing data near the user. This enables high-bandwidth and low-latency services such as autonomous systems, robotics, and real-time monitoring.
5GC (5G Core)	It provides secure, scalable, and high-performance core network functionalities. It also allows service differentiation and seamless control between public and private operations.
Security and Privacy	Enterprise retains control over private network data and services, isolating sensitive information from public network exposure, even with shared base station infrastructure.
Use Cases	Ideal for enterprises requiring high-speed, real-time applications like autonomous vehicles, drones, or factory automation, while leveraging shared RAN infrastructure to reduce costs.

Scenario #4: RAN and Control Plane Sharing Between a Public and Private Network

In this fourth deployment model, enterprises adopt a hybrid approach by privately hosting the User Plane Function (UPF) and Multi-access Edge Computing (MEC), while Radio Access Network (RAN) elements (gNBs) and control plane functions like the Unified Data Management (UDM) and 5G Core Control Plane (5GC CP) are shared with the mobile operator.

This scenario allows businesses to benefit from ultra-low latency for strategic applications such as autonomous vehicles and robotics, while leveraging the mobile operator's infrastructure for user authentication and mobility management. However, since subscription and operational data are stored in the operator's cloud, enterprises may face challenges in controlling their data. The following table summarizes the key features and implications of this configuration.

The enterprise makes UPF and MEC private and committed. 5G base stations (gNBs) in the enterprise and 5G CPs and UDMs in mobile operators' cloud are shared between public and private networks (RAN and Control Plane Sharing). Therefore, UDM, 5GC CP, and the gNB logically separate private and public networking, and the MEC and UPF are physically separated.

5GC UDM and CP perform balanced plane works (authentication, mobility) for private network equipment and public network products in mobile operators' networks.

The enterprise's private network systems, UPF and gNB, are integrated with and handled by the mobile operator's network (via an N2 or N4 interface). Therefore, the operation and subscription information of the private network products stored on the mobile operator's server, rather than in-house, might represent an issue.

Table 7 illustrates a fourth deployment scenario where both the RAN and control plane components are shared between public and private 5G networks. In this model, only the UPF and MEC remain fully private and deployed on enterprise premises, ensuring real-time responsiveness for critical operations. This configuration strikes a balance between cost efficiency and performance by leveraging the mobile operator's core services. However, it introduces new considerations around data privacy and integration. Table 7 outlines the key technical and operational aspects of this hybrid deployment.

Table 7. Scenario #4: RAN and control plane sharing between a public and private network

Aspect	Description
Deployment Model	RAN (gNB) and Control Plane (UDM, 5GC CP) are shared between public and private networks. UPF and MEC are deployed privately on the enterprise premises.
Data Separation	Private traffic (e.g., internal video or device control) stays within the enterprise—public traffic (e.g., speech, internet access) routes to the mobile operator's core network.
UPF and MEC	Remain physically and logically separated in the enterprise. Enable ultra-low latency for mission-critical services such as robotics, drones, and URLLC applications.
5GC CP and UDM	Shared with the mobile operator. They handle public and private device authentication and mobility, but store subscription and operational data in the operator's cloud.

continued on following page

Table 7. Continued

Aspect	Description
Integration and Interfaces	Private network UPF and gNB are integrated and managed via N2/N4 interfaces with the mobile operator's core. This tight integration may raise concerns over enterprise data control.
Security and Privacy	Sensitive traffic is retained locally, but operational data is stored externally. Logical separation supports secure operations, though complete in-house control is reduced.
Use Cases	Applicable to enterprises prioritizing URLLC while accepting shared infrastructure. Suits advanced automation, remote control systems, and cost-conscious deployments.

Scenario #5: Ran and Also Core Sharing (End-to-End Network Slicing) Between a Public and Private Network

This scenario happens when the venture deploys gNB, and the mobile operator hosts MEC and UPF in its advantage cloud. Private and public networks share "logically separated 5G RAN and also Core" (gNB, MEC, 5GC, UPF, UDM) (End-to-End Network Slicing).

Unlike in three and four, where the enterprise contains MEC and UPF, this instance includes only gNB. Thus, there is no community traffic between the private 5G equipment and intranet (LAN) products, such as local intranet servers or PCs. Therefore, the traffic should go to the UPF in the operator's advantage cloud. Then, it should grow within the enterprise through a leased line to connect to the LAN products.

Additionally, MEC, which offers 5G program offerings to 5G products in the business, is far from the equipment in the mobile operator's advantage cloud.

In this structure, network latency (RTT) is often a significant issue due to the distance between the enterprise (5G devices) and the operator's advantage cloud (UPF, MEC).

Visitors' private network products transfer OData out of the enterprise onto the mobile operator's network, and there is an issue regarding information traffic security. The movable operator slices MEC and UPF on its edge cloud to help keep our network traffic distinct from the public. Furthermore, other private network traffic concerns timid CEOs, such as the potential for internal CCTV video traffic to leak outside their enterprise.

As with situation four, it is not enjoyable for an enterprise to have available subscription information saved on a mobile operator's network instead of on Airers4You's network.

This architecture incurs the least cost to create a private 5G system for mobile operators compared to cases two, three, and four, which required UPF and 5GC CP deployment within the enterprise.

Nevertheless, the enterprise is concerned about protection (data traffic from private network terminals, operational information, and subscription information for remote network equipment) and network delay (between individual 5G equipment and MEC program servers, and between remote 5G devices and intranet/LAN products).

Table 8 presents Scenario #5, where the enterprise deploys only the gNB locally, while the mobile operator manages all core components—MEC, UPF, 5GC, and UDM—through end-to-end network slicing. In this model, public and private traffic coexist within logically separated slices, thereby minimizing on-site infrastructure requirements. However, all enterprise data must pass through the operator's infrastructure, which raises security concerns and increases latency for time-sensitive applications (Zhang & Wang, 2021; ETSI, 2017). *Despite these limitations, this architecture is ideal for cost-conscious enterprises that do not require ultra-low-latency or strict data sovereignty.*

Scenario #5 represents the most centralized model among private 5G deployment options. In this configuration, the enterprise installs only the on-site radio node (gNB), while the mobile operator manages all other functions—MEC, UPF, 5GC, and UDM—through network slicing. This architecture reduces capital expenditure and simplifies enterprise involvement in network management. However, it also introduces trade-offs related to latency, data privacy, and reliance on external infrastructure. Table 8 summarizes the technical features, benefits, and limitations of this shared architecture.

Table 8. Scenario #5: Ran and also core sharing (end-to-end network slicing) between a public and private network

Aspect	Description
Deployment Model	The enterprise deploys only gNB on-site, while the mobile operator hosts MEC, UPF, 5GC, and UDM through end-to-end network slicing.
Network Architecture	The mobile operator remotely operates the entire core and edge network (MEC, UPF, 5GC CP, UDM). Logical slices separate private and public traffic within shared infrastructure.
Latency Considerations	Increased round-trip time (RTT) due to physical distance between the enterprise and the operator's edge cloud, especially critical for latency-sensitive applications.
Security and Privacy	Enterprise traffic flows through the mobile operator's network, raising concerns about data leakage, particularly for sensitive content such as CCTV feeds and operational information.
Connectivity Limitations	The lack of local MEC and UPF means that even internal traffic must travel to the operator's cloud via leased lines, limiting direct intranet access.

continued on following page

Table 8. Continued

Aspect	Description
Cost and Scalability	This is the most cost-effective architecture for mobile operators to support private networks. No local core deployment is needed, reducing enterprises' CAPEX.
Use Cases	Best suited for cost-sensitive deployments where ultra-low latency is not a priority. May not meet the strict security or performance demands of mission-critical enterprise applications.

Scenario #6: N3 LBO (Local Breakout): Case of SK Telecom in Korea

The GNB deployment at the venture, as in case 5, N3 GTP Tunnel, requires a bridge between the UPF and the gNB whenever a device is attached, whether it is a CCTV camera or a smartphone. This requirement is necessary because these units are all public network devices.

As found in (b) above, the venture presents the MEC Data Plane (non-3GPP tools, ETSI MEC) and MEC Applications (MEC Applications). *The Mobile Edge Platform (MEP) in the mobile operator's Orchestrator sends a traffic rule on the MEC DP through the MP22 interface (if the Destination IP address is a community network).* The venture also offers private 5G products, regional wired LAN systems, a community MEC program, and a server Local Breakout.

The MEC DP examines the spot IP addresses of packets belonging to other GTP Tunnels from the gNB (GTP Decap) and routes the User IP package on the internal private network in case it is community site traffic.

Although this strategy is not a common 3GPP technique, it will separate private network traffic from public site traffic.

(This is the ETSI MEC's Bump in the Wire method.) Compared to situation five, the mobile operator's network does not transfer web traffic or route website traffic to the private network. Therefore, personal network information website managing traffic protection is as distinct as cases three and four.

Unlike cases three and four, the expense of creating a private 5G system decreases the UPF (considered the costliest gear among all 5G standard equipment) by using an inexpensive MEC DP (SDN/P4 switch) without requiring expensive UPFs for the business.

Additionally, because the MEC is present in the venture and manages the visitors, Positive Many Meanings—MEC DP breakouts —will be ready to offer ultra-low-delay program services.

Nevertheless, since MEC DP is not a 3GPP UPF, it cannot perform charging functions and mobility management for private network equipment. (It can apply several of these features since the operator can generate Positive Many Meanings, a proprietary specification that implements these capabilities.)

Like situations four and five, it is not pleasant for an enterprise to have available subscription information saved on a mobile operator's network instead of on airers4you's network.

This scenario happens when the venture deploys gNB, and the mobile operator hosts MEC and UPF in its advantage cloud. Private and public networks share "logically separated 5G RAN and also Core" (gNB, MEC, 5GC, UPF, UDM) (End-to-End Network Slicing).

Unlike in three and four, where the enterprise contains MEC and UPF, this instance includes only gNB. Thus, there is no community traffic between the private 5G equipment and intranet (LAN) products, such as local intranet servers or PCs. Therefore, the traffic should go to the UPF in the operator's advantage cloud. Then, it should grow within the enterprise through a leased line to connect to the LAN products.

Additionally, MEC, which offers 5G program offerings to 5G products in the business, is far from the equipment in the mobile operator's advantage cloud.

In this structure, network latency (RTT) is often a significant issue due to the distance between the enterprise (5G devices) and the operator's advantage cloud (UPF, MEC).

Visitors' private network products transfer of Data out of the enterprise onto the mobile operator's network, and there is an issue regarding information traffic security. The movable operator slices MEC and UPF on its edge cloud to help keep our network traffic distinct from the public. Furthermore, other private network traffic concerns timid CEOs, such as the potential for internal CCTV video traffic to leak outside their enterprise.

As with situation four, it is not enjoyable for an enterprise to have available subscription information saved on a mobile operator's network instead of on Airers4You's network.

This architecture incurs the least cost to create a private 5G system for mobile operators compared to cases two, three, and four, which require the deployment of UPF and 5GC CP within the enterprise.

Scenario #6 illustrates an innovative hybrid model using N3 Local Breakout (LBO), notably implemented by SK Telecom. In this setup, private traffic is separated locally using the MEC Data Plane, eliminating the need for a full UPF to be deployed on-site. This significantly lowers infrastructure costs while enabling ultra-low latency services within the enterprise. It combines the advantages of localized processing with centralized control for subscription data. Table 9 outlines the deployment, capabilities, and trade-offs of this efficient architecture.

Table 9. Scenario #6: N3 LBO (local breakout): Case of SK telecom in Korea

Aspect	Description
Deployment Model	Enterprise deploys gNB on-site. The mobile operator handles MEC and UPF in its cloud. Traffic is routed via an N3 GTP tunnel and broken out locally using the ETSI MEC's data plane.
Local Breakout (LBO)	Uses a non-3GPP approach ('Bump in the Wire') with MEC Data Plane to separate private and public network traffic, routing local packets internally to avoid unnecessary transit to the operator cloud.
Cost Efficiency	Reduces enterprise costs by replacing expensive UPF equipment with a cost-effective MEC DP (SDN/P4 switch), eliminating the need for a complete 5G core deployment in-house.
Latency & Performance	Provides ultra-low latency services by managing traffic locally via MEC, which resides within the enterprise and processes traffic directly near devices.
Limitations	MEC DP cannot perform full UPF functionalities such as mobility management and charging. Proprietary enhancements may be needed to bridge the gap.
Security and Privacy	Improved compared to full core-sharing models. Data remains local, and internet traffic does not route through the mobile operator's public network; however, subscription data is externally hosted.
Use Cases	Ideal for enterprises seeking secure and low-latency applications like real-time video analytics or autonomous systems, without the investment burden of deploying a full 5G UPF.

Scenario #7: F1LBOo (Local Breakout): Case of KT in Korea

This scenario is similar to the number 6 but introduces a key architectural distinction. Instead of placing the entire gNB at the enterprise site, only the Radio Unit (RU) and Distributed Unit (DU) are deployed locally.

The central unit (CU) and other core functions reside in the mobile operator's edge cloud. Private network traffic is routed via the F1 interface rather than the N3 interface, as shown in the N3 LBO. This setup places the breakout point earlier in the radio access chain, offering potential latency improvements. It enables localized traffic management while maintaining centralized control.

KT currently uses this model in Korea. While it retains the cost benefits and some security advantages of local breakout, this approach still relies on the operator's infrastructure for centralized processing. In this context, key advantages of 5G include high data rates, energy efficiency, low packet loss, and dynamic data access. However, like all mmWave-based solutions, challenges such as high path loss, interference from solid objects, and signal blockages remain key hurdles to broader adoption.

It is the same as in case six, though the real difference is that only the RU/DU in the enterprise and the CU are deployed. This deployment places the mobile network's edge cloud locally, breaking private network traffic from the F1 user interface instead of the N3 interface.

Based on Won et al. (2017), the benefits of 5G engineering are the following: globally accessible; dynamic information access; lower battery consumption; high method amount spectral efficiency; with sixth sense know-how; faster information transmission compared to the prior decades; and consistency in packet failure rates of 10

Table 10. Scenario #7: F1LBOo (local breakout): Case of KT in Korea

Aspect	Description
Deployment Model	Only RU/DU are installed at the enterprise site. The mobile operator manages CU and core components, with breakout via the F1 interface.
Breakout Interface	A local breakout occurs at the F1 interface (between DU and CU), unlike Scenario #6, where a breakout occurs at the N3 interface between the gNB and UPF.
Performance	Due to an earlier breakout in the data flow, there is potential for improved latency. This allows localized traffic handling while maintaining centralized control functions.
Security and Privacy	Enables isolation of private network traffic before it reaches the operator's core, offering moderate security benefits. Subscription data still resides with the operator.
Cost Efficiency	It minimizes on-site infrastructure costs by limiting deployment to RU/DU. It also reduces the need for expensive UPFs while providing some local processing capabilities.
Technical Challenges	Depends on robust signal quality and connectivity to the central unit (CU). Vulnerable to mmWave limitations like high path loss and signal blockages by physical obstacles.
Use Cases	Suited for enterprise environments requiring local processing with reduced latency, such as real-time monitoring, AR/VR, and localized automation solutions.

CONCLUSION

The evolution of mobile telecommunications has culminated in the groundbreaking introduction of 5G, a technology poised to redefine connectivity and reshape industries worldwide. Building on decades of innovation, 5G transcends the limitations of its predecessors by delivering unprecedented speed, reliability, and scalability. Its advanced capabilities, such as network slicing and low-latency communication, shift the network design paradigm and usage, enabling tailored solutions for diverse industries and applications.

At its core, 5G addresses the pressing demands of a rapidly digitizing world. From enhancing everyday connectivity to powering transformative technologies like autonomous vehicles, smart cities, and advanced IoT ecosystems, 5G has established itself as the cornerstone of digital transformation. Its ability to support multiple virtual networks on shared infrastructure ensures efficiency and flexibility, catering to the unique needs of various sectors while maximizing resource utilization.

The impact of 5G extends beyond technological advancements. It is a critical enabler for innovation across sectors, including healthcare, manufacturing, logistics, and entertainment. With its capacity to facilitate real-time communication, enable remote operations, and deliver immersive experiences, 5G empowers businesses and societies to unlock new levels of productivity and creativity. Furthermore, as global adoption accelerates, 5G is expected to drive economic growth, bridge digital divides, and improve the quality of service for millions of users. Therefore, 5G plays a pivotal role in bridging digital divides and fostering economic growth on a global scale" (Zhang & Wang, 2021).

In conclusion, the advent of 5G marks a transformative era in mobile telecommunications, setting new standards for connectivity and innovation. As industries and societies adapt to their capabilities, the 5G technologies will continue to shape the future of communication, fostering a more connected, intelligent, and sustainable world. This chapter has highlighted the journey of mobile telecommunications, culminating in the revolutionary potential of 5G and its role as a catalyst for global progress in the years to come.

REFERENCES

Agarwal, A., Agarwal, K., Agarwal, S., & Misra, G. (2019). 34-37 towards 5G Networks and Challenges. *American Journal of Electrical and Electronic Engineering, 7*(2), 34–37. https://doi.org/(NGMN Alliance, 2016)DOI: 10.12691/ajeee-7-2-2

Balador, A., Kouba, A., Cassioli, D., Foukalas, F., Severino, R., Stepanova, D., Agosta, G., Xie, J., Pomante, L., Mongelli, M., Pierini, P., Petersen, S., & Sukuvaara, T. (2018). Wireless Communication Technologies for Safe Cooperative Cyber Physical Systems. *Sensors (Basel), 18*(11), 4075. DOI: 10.3390/s18114075

Ericsson. (2024a). *5G device outlook*. Ericsson.com. https://www.ericsson.com/en/reports-and-papers/mobility-report/dataforecasts/5g-device-outlook

Ericsson. (2024b). *5G network coverage outlook*. Ericsson.com. https://www.ericsson.com/en/reports-and-papers/mobility-report/dataforecasts/network-coverage

Ericsson. (2024c). *Exploring how traffic patterns drive network evolution*. Ericsson.com. https://www.ericsson.com/en/reports-and-papers/mobility-report/articles/traffic-patterns-drive-network-evolution

Ericsson. (2024d). *Mobile data traffic forecast*. Ericsson.com. https://www.ericsson.com/en/reports-and-papers/mobility-report/dataforecasts/mobile-traffic-forecast

Ericsson. (2024e). *Mobile subscriptions outlook*. Ericsson.com. https://www.ericsson.com/en/reports-and-papers/mobility-report/dataforecasts/mobile-subscriptions-outlook

ETSI. (2017). *5G; Study on Scenarios and Requirements for Next Generation Access Technologies (3GPP TR 38.913 version 14.2.0 Release 14)*. 3GPP. https://www.etsi.org/deliver/etsi_tr/138900_138999/138913/14.02.00_60/tr_138913v140200p.pdf

Lagorio, A., Cimini, C., Pinto, R., & Cavalieri, S. (2023). 5G in Logistics 4.0: Potential applications and challenges. *Procedia Computer Science, 217*, 650–659. DOI: 10.1016/j.procs.2022.12.261

Nwajana, A. O., & Ihianle, I. K. (2021). *Handbook of Research on 5G Networks and Advancements in Computing*. IGI Global.

Odida, M. (2024). The Evolution of Mobile Communication: A Comprehensive Survey on 5G Technology. *Journal of Sensor Networks and Data Communications, 4*(1), 01-11. https://doi.org/DOI: 10.33140/jsndc.04.01.06

Salih, A. A., Zeebaree, S. R. M., Abdulraheem, A. S., Zebari, R. R., Sadeeq, M. A. M., & Ahmed, O. M. (2020). *Evolution of mobile wireless communication to the 5G revolution*. Technology Reports of Kansai University 62(5):2139-2151 ResearchGate.

Sharma, P. (2013). Evolution of Mobile Wireless Communication Networks-1G to 5G as well as Future Prospective of Next Generation Communication Network. *IJCSMC, 2*(8), 47–53.

Sowa, L. (2020, September 15). *Advocates for New Wireless Technologies Claim Sooner Rollout, Explicate Exciting 5G Attributes – Broadband Breakfast*. Broadband Breakfast.

Won, H., Hong, Y.-K., Choi, M., Garcia, H., Shin, D., Yoon, Y.-S., Lee, K., Xin, H., & Yeo, C.-D. (2022). Microwave absorption performance of M-type hexagonal ferrite and MXene composite in Ka and V bands (5G mmWave frequency bands). *Journal of Magnetism and Magnetic Materials*, *560*, 169523. DOI: 10.1016/j.jmmm.2022.169523

Zhang, Z., & Wang, Q. (2021). Application Status and Prospects of 5G Technology in Distribution Automation Systems. *Wireless Communications and Mobile Computing*, *2021*, 1–9. DOI: 10.1155/2021/5553159

ADDITIONAL READINGS

Agiwal, M., Roy, A., & Saxena, N. (2016). Next Generation 5G Wireless Networks: A Comprehensive Survey. *IEEE Communications Surveys and Tutorials*, *18*(3), 1617–1655. DOI: 10.1109/comst.2016.2532458

Bhawsar, V. (2023). *The Evolution of Mobile Networks from 1G to 5G and Beyond*. Cloudthat. https://www.cloudthat.com/resources/blog/the-evolution-of-mobile-networks-from-1g-to-5g-and-beyond

Campbell, K., Morelli, B., & O'neil, B. (2017). *IHS Economics / IHS Technology The 5G economy: How 5G technology will contribute to the global economy*. https://www.qualcomm.com/content/dam/qcomm-martech/dm-assets/documents/5g_economy_ihs_economic_impact_analysis_january_2017.pdf

Dongfei, Z. (2016). *On the path to 5G with 4.5G*. https://www-file.huawei.com/-/media/corporate/pdf/publications/communicate/80/03-reaching-5g-with-45g.pdf

Ericsson. (2023). *The 5G Advanced, an evolution towards 6 G*. https://www.ericsson.com/en/reports-and-papers/white-papers/5g-advanced-evolution-towards-6g

EXFO. (2021). *The evolution of 5G mobile networks: the what, why and how. EXFO Inc.* https://www.exfo.com/contentassets/331b8bd3b1c442718e03927c77204338/exfo_wpaper130_evolution-of-5g-mobile-networks_en.pdf

GSA. (2015). *The Road to 5G: Drivers, Applications, Requirements and Technical Development*. GSA. https://gsacom.com/paper/5g-device-ecosystem-report/

ITU. (2018). *Setting the Scene for 5G: Opportunities & Challenges*. ITU. https://www.itu.int/en/ITU-D/Documents/ITU_5G_REPORT-2018.pdf

ITUGSR. (2025). *Focus Group on IMT-2020*. ITU. https://www.itu.int/en/itu-t/focusgroups/imt-2020/pages/default.aspx

Moore, K. (2023, August 11). *From 1G to 5G: The Evolution of Mobile Communications*. Mpirical. https://www.mpirical.com/blog/the-evolution-of-mobile-communication

Patel, B. (2022, July 28). *The Evolution of 5G and The Backup Power It Requires*. Schneider Electric Blog. https://blog.se.com/telecommunications/2022/07/28/the-evolution-of-5g-and-the-backup-power-it-requires/

Qualcomm. (2016). Making 5G NR a reality Making 5G NR a reality Leading the technology inventions for a unifi ed, more capable 5G air interface. Qualcomm. https://www.qualcomm.com/media/documents/files/whitepaper-making-5g-nr-a-reality.pdf

Chapter 7 Spectrum, Requirements, and Trends of 5G

ABSTRACT

The spectrum allocation for 5G is a critical factor in the successful deployment of next-generation wireless networks, influencing the efficiency, coverage, and capacity of 5G services. It involves balancing sub-1 GHz, mid-range (1-6 GHz), and high-frequency millimeter-wave bands (above 6 GHz) to meet diverse application needs. While lower frequency bands provide broader coverage for rural areas, high-frequency millimeter waves enable ultra-fast data rates essential for urban and industrial use cases like augmented reality and autonomous vehicles. Global regulatory bodies, such as the ITU-R, are pivotal in defining the spectrum frameworks for 5G, focusing on the 24-86 GHz range to support various applications. These efforts ensure timely deployment and address challenges like spectrum allocation, fair access, and non-discriminatory authorization models. Regulatory bodies face the challenge of establishing equitable spectrum authorization models to ensure fair access for stakeholders" (ITU-R, 2019)

INTRODUCTION

The spectrum allocation for 5G plays a pivotal role in developing and deploying next-generation wireless networks, as it determines the efficiency, coverage, and capacity of 5G services. Identifying and optimizing appropriate spectrum bands for mobile communications remains a key regulatory challenge, requiring a coordinated effort among stakeholders to ensure the timely deployment of 5G radio access networks. It includes allocating existing mobile service bands and exploring new

DOI: 10.4018/979-8-3373-0497-7.ch007

frequency bands across a broad spectrum range—sub-1 GHz, mid-range bands between 1 and 6 GHz, and high-frequency millimeter-wave bands above 6 GHz.

Global regulatory bodies, such as the ITU-R, have focused on enabling the utilization of new spectrum bands in the 24-86 GHz range, particularly in preparation for global frameworks, including the World Radiocommunication Conference. These efforts aim to accommodate diverse 5G applications, from high-speed urban connectivity to industrial use cases in factory sites.

Regulators continue to face the challenge of developing equitable spectrum authorization frameworks that ensure all stakeholders have fair, transparent, and non-discriminatory access to the spectrum. This includes established mobile network operators (MNOs) and new market entrants aiming to support specialized verticals such as manufacturing, logistics, or industrial automation. Balancing the needs of incumbent operators with the innovation potential of emerging players is crucial to fostering competition, encouraging investment, and ensuring that 5G capabilities are widely accessible across various sectors of the economy.

Frieden (2019) analyzes the challenges faced by the United States in securing international consensus for 5G spectrum allocation during the 2015 ITU World Radio Conference, attributing the failure to underestimated timelines for consensus-building and differing national approaches to spectrum management.

To clearly illustrate how 5G bands are distributed across frequency ranges, a 5G spectrum chart should be included in Chapter 4, "Spectrum Allocation and Network Planning." This chart should show allocations for FR1 (sub-6 GHz) and FR2 (mmWave), along with example use cases such as rural coverage (low-band), urban capacity (mid-band), and ultra-high-speed access (high-band). *Highlighting global or regional allocations (e.g., n78, n258) enhances comprehension.* This visual tool helps readers understand how physical layer capabilities are mapped to real-world deployments. The figure below illustrates how the 5G spectrum is divided into low-band, mid-band, and high-band frequencies, each serving distinct performance and coverage needs within the network architecture.

Figure 1. 5G spectrum allocation across low, mid, and high frequency bands

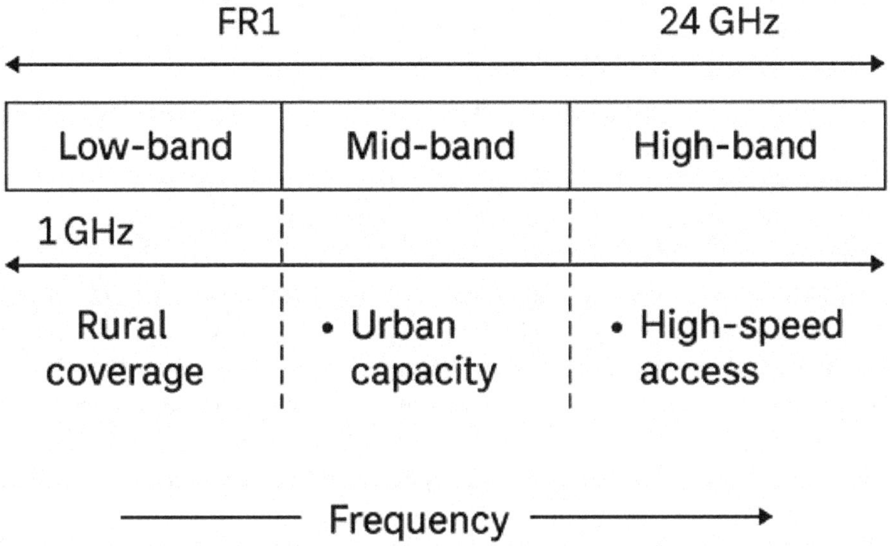

The versatility of the 5G spectrum stems from its ability to span sub-6 GHz and millimeter-wave bands, offering a balance between coverage and capacity. Lower frequency bands are ideal for rural and less densely populated areas because they can transmit data over longer distances, albeit at lower speeds. In contrast, higher frequency bands, such as millimeter waves, enable ultra-fast data rates and enhanced network capacity, making them indispensable for high-density urban environments and emerging applications like augmented reality and autonomous vehicles.

This chapter examines the key considerations surrounding 5G spectrum allocation, the technical requirements for effective deployment, and emerging trends in the field of 5 G. Examining the interplay between spectrum management and technological advancements highlights the foundation of the global 5G ecosystem.

SPECTRUM CONSIDERATIONS

According to Jhon (2025), the primary regulatory challenge in 5G deployment involves identifying and allocating suitable spectrum bands for mobile communications, ensuring timely availability for stakeholders to facilitate the efficient rollout of 5G radio access networks.

Throughout the 5G rollout, this involves obtaining options in bands currently allocated to the mobile service and different opportunity bands covering a range of frequencies, including those below 1 GHz, between 1 and 6 GHz, and above 6 GHz. In addition, identifying a completely new 5G spectrum at the global level, as outlined in ITU-R, focuses on the frequency range of 24.86 GHz, which supports local tasks in planning for the World Radiocommunication Conference in 2019.

Additionally, regulators must improve spectrum authorization models to grant entry rights to utilize all candidates' 5G spectrums objectively, non-discriminatorily, transparently, and proportionately. In 5G, applicants include existing mobile network operators (MNOs) and new entrants seeking to deploy 5G networks in specific areas, such as manufacturing companies at factory sites.

The 5G spectrum selects radio wavelengths in the sub-6 GHz range and the millimeter-wave frequency range, which extends from 24.25 GHz and above. It describes the radio frequencies with consumer equipment (UE) information to cellular-based facilities at the data endpoint. LTE networks utilize frequencies in the sub-6 GHz range and share the area with 5G traffic. The lower frequency bands are used for less densely populated areas, as information can travel more slowly, albeit more efficiently, on them.

LTE VS. 5G SPECTRUMS BASED

LTE vs. 5G Spectrums Based on the requirements frame, 3GPP, the radio frequency spectrum utilized by LTE networks ranges from 700 MHz to 2.7 GHz. In concept, LTE peak data fees are seventy-five Mb/s for the uplink and 300 Mb/s for the downlink when utilizing a multiple-input multiple-output (MIMO) antenna process called spatial multiplexing.

The frequency bands for 5G networks are available in 2 sets:

- Frequency range one (FR1) is from 450 MHz to 6 GHz, including the LTE frequency range.
- Frequency range two (FR2) is from 24.25 GHz to 52.6 GHz.

The sub-6 GHz range is often referred to as FR1, and the millimeter wave (mmWave) spectrum is often referred to as FR2. In concept, the peak data rates for a fully mature 5G network are 20 Gb/s for downlinks and 10 Gb/s for uplinks, as standardized by the International Telecommunications Union.

- Latency And Reliability Of Spectrum

3GPP began standardizing NR in April 2016 to make it commercially available before 2020. 3GPP is taking a phased approach to defining the 5G specs.

Although NR does not need to be backward-compatible with LTE, its initial release should suit its upcoming evolution. Moreover, since NR should support a broad range of use cases, many of which are yet to be defined, advanced compatibility is essential.

The evolution of reliability and end-to-end latency for different technologies highlights the differences between the unlicensed and licensed spectrums, as well as LTE (4G) and NR technology. 5G NR (New Radio) is a new radio access technology (RAT) designed for the 5G (fifth-generation) mobile network.

LATENCY AND RELIABILITY CHARACTERISTICS

Their latency and reliability characteristics can categorize wireless communication technologies. Non-real-time applications, such as asset monitoring and wireless sensors, typically require lower reliability and higher latency. Real-time applications, such as mobile robots and automated guided vehicles, require moderate reliability and low latency. Challenging real-time applications, such as time-critical closed-loop controls, require extremely high reliability and ultra-low latency. Licensed spectrums, such as LTE and NR, are a better fit for applications that require high reliability and low latency. In contrast, people commonly use unlicensed spectrums, such as Wi-Fi, for non-real-time or less critical applications. This classification enables the development of tailored solutions for diverse industrial and technological needs.

The first standardization stage with a restricted NR function was completed by 2018. This standardization stage will fulfill all the demands of IMT-2020 (the next generation of mobile communication methods specified by ITU-R in 2019. NR will likely continue to develop beyond 2020 with a series of releases, including extra features and functionalities. The International Telecommunication Union's IMT-2020 framework sets benchmarks for ultra-reliable coverage and low latency in 5G networks" (ITU-R, 2015).

Although NR does not need to be backward-compatible with LTE, its initial release should suit its upcoming evolution. Furthermore, advanced computability is essential since NR should support numerous use cases, many of which are still to be defined.

Qualified Spectrum for Interference Control

The accessibility of spectrum methods is essential for conference needs regarding capability, latency, and bitrates. Careful handling of spectrum resources offers reliable and predictable service levels on the factory floor. The achievable performance will depend on a few factors:

- The quantity of spectrum available
- The spectrum needed - decreased band (below 2GHz), the mid-band (high band/mmWave or 2-5GHz) (26GHz and above)
- The licensing regime
- the spectrum as TDD or FDD
- The stereo access technology used in the coexistence scenarios applies to the range.

Table 1 presents a comparative overview of LTE (4G) and 5G New Radio (NR) spectrums, highlighting key distinctions in frequency bands, data rates, latency, and use cases. With the rapid evolution of mobile networks, understanding the spectrum landscape is crucial for designing efficient communication systems. LTE operated primarily in lower bands, while 5G expands into both sub-6 GHz and high-frequency mmWave bands. These developments enable transformative use cases such as IoT, real-time automation, and immersive technologies. This comparison lays the groundwork for evaluating 5G's capabilities compared to legacy systems.

Table 1. LTE vs. 5G spectrums based

Aspect	LTE (4G)	5G (NR)
Frequency Range	700 MHz to 2.7 GHz	FR1: 450 MHz to 6 GHz (includes LTE) FR2: 24.25 GHz to 52.6 GHz (mmWave)
Peak Data Rates	Uplink: 75 Mb/s Downlink: 300 Mb/s (using MIMO spatial multiplexing)	Uplink: 10 Gb/s Downlink: 20 Gb/s (per ITU-R IMT-2020)
Latency and Reliability	Higher latency and lower reliability for real-time critical applications	Ultra-low latency and ultra-high reliability (suitable for time-critical applications)

continued on following page

Table 1. Continued

Aspect	LTE (4G)	5G (NR)
Standardization	3GPP LTE standards finalized before NR, backward-compatible with older tech	Started in April 2016, phased releases from 2018, designed for IMT-2020 and future use cases
Use Cases	Mainly broadband and mobile communication	Supports broadband, real-time automation, IoT, robotics, and more
Spectrum Licensing	Licensed spectrum: better for high-reliability and low-latency needs	Licensed and specialized spectrums to manage interference and ensure performance
Spectrum Types	Primarily FDD and TDD below 3 GHz	Uses TDD and FDD, includes sub-6 GHz and mmWave above 24 GHz

The transition from LTE to 5G introduces a paradigm shift in wireless communication. 5G significantly enhances data rates, reduces latency, and expands frequency utilization, offering unprecedented performance for diverse applications. While LTE remains relevant in existing networks, 5G's spectrum strategy enables scalability and innovation across industries. The expanded use of mmWave and sub-6 GHz bands positions 5G as a foundational technology for future digital ecosystems. As deployment accelerates, spectrum management becomes a critical factor in maximizing 5G's potential.

ESTIMATES OF SPECTRUM REQUIREMENTS

Estimates of spectrum requirements range from tens to plenty of megahertz. Nearly all new mid-band spectrums use TDD, while a significant portion of the range is probably allocated to mobile operators for FDD. The latency of a DDD product is inherently less than that of a corresponding TDD process.

The Mid-band spectrum is well-designed for interior deployments. Its propagation characteristics make it easy to provide good coverage with a small set of transmission areas.

Estimates of spectrum requirements range from tens to plenty of megahertz. Nearly all new mid-band spectrums use TDD, while a significant portion of the range is probably allocated to mobile operators for FDD. The latency of a DDD product is inherently less than that of a corresponding TDD process. The Mid-band spectrum is well-designed for interior deployments. Its propagation characteristics make it easy to provide good coverage with a small set of transmission areas. Coverage at mmWave is commonly spottier, needing denser stereo deployment, but mmWave is nonetheless an excellent complement to the mid-band for in-factory deployments since it enables:

Coverage at mmWave is commonly spottier, needing denser stereo deployment, but mmWave is nonetheless an excellent complement to the mid-band for in-factory deployments since it enables:

Higher Method Capability

Higher method capability is one of the defining features of advanced 5G systems, leveraging larger bandwidths, innovative antenna technologies, and beamforming techniques to deliver exceptional performance. These enhancements enable a compact and efficient form factor, ideal for deployment in indoor environments, such as offices, shopping malls, and public spaces. Beamforming, in particular, focuses signal strength toward specific devices, enhancing efficiency and minimizing interference. Load balancing technology plays a crucial role in strengthening overall network capacity and reliability, particularly in densely populated environments where demand peaks can compromise service availability.

Integrating millimeter-wave (mmWave) radio signals with existing networks further amplifies this capability. By coordinating internal and external mobile networks, mmWave offers drastically reduced latency, even in time-division duplexing (TDD) spectrums, and ensures shorter transmission times. This synchronization allows for seamless data transfer across devices and networks, making 5G a powerful enabler of applications that require high-speed, low-latency communication, such as augmented reality (AR), virtual reality (VR), and industrial automation. These advancements significantly enhance user experience, support innovative IoT solutions, and open avenues for new business models, highlighting the critical importance of higher method capability in the evolution of wireless technology.

Guarantees Against Wild Interference

A qualified spectrum ensures reliable, low-latency communication for mission-critical applications. Unlike unlicensed technologies like Wi-Fi and MulteFire, which operate in shared bands and cannot guarantee consistent performance, licensed spectrums provide bounded latency and superior reliability. Unlicensed technologies face limitations due to their reliance on listen-before-talk mechanisms, which can lead to back-off delays in high-interference scenarios. This limitation becomes particularly problematic in environments with heavy traffic, where uncontrolled interference can disrupt communication and degrade network performance.

Unlicensed spectrums may offer viable solutions for less critical applications, such as providing connectivity for consumer-grade IoT devices or temporary network setups. However, mission-critical operations, such as autonomous driving, telemedicine, and industrial automation, require the guarantees provided by licensed spectrums. These spectrums ensure stable performance by mitigating wild interference and maintaining consistent service levels, even in high-load scenarios. This distinction highlights the importance of strategic spectrum allocation and the prioritization of licensed technologies in deploying 5G networks to support a diverse range of applications.

Apruzzese et al.'s study (2023) examines the transformative potential of 5G and its associated spectrum, as well as technologies such as IoT, AI, and blockchain, in revolutionizing logistics and supply chain operations. Focusing on the transportation and logistics sector, the research highlights how these technologies can enhance real-time data exchange, automation, and connectivity. However, it also addresses challenges like infrastructure requirements, data security, and the need for substantial investments.

Table 2 outlines the various spectrum requirements essential for supporting diverse 5G applications and services. As 5G expands, the demand for spectrum grows significantly, ranging from tens to hundreds of MHz, especially in the mid-band and mmWave ranges. Each spectrum type brings unique trade-offs in terms of latency, deployment complexity, and interference control. Efficient spectrum allocation is crucial for enabling emerging use cases, such as industrial automation, AR/VR, and autonomous systems. This table helps clarify how spectrum planning shapes network performance and strategic 5G deployment.

Table 2. Estimates of spectrum requirements

Aspect	Details
Estimated Requirements	It ranges from tens to hundreds of MHz; the mid-band spectrum uses mainly TDD and some FDD allocation.
Latency Differences	FDD has inherently lower latency than TDD.
Mid-Band Spectrum	Well-suited for indoor deployment with fewer transmission points needed.
mmWave Spectrum	Requires denser deployment but complements mid-band with higher performance.
Higher Method Capability	Larger bandwidth, beamforming, and advanced antennas enable efficient high-capacity indoor networks.
Beamforming	Focus signal to specific devices, reducing interference and increasing efficiency.
mmWave Integration	Reduces latency and enhances data transfer in both internal and external networks.
Application Examples	Enables AR/VR, industrial automation, IoT, and other high-demand applications.
Unlicensed vs. Licensed Spectrum	Licensed spectrum ensures low-latency, reliable communication; unlicensed spectrum (such as Wi-Fi and MulteFire) is subject to interference.
Interference Control	Licensed spectrum mitigates interference and ensures performance in mission-critical environments.

continued on following page

Table 2. Continued

Aspect	Details
Critical vs. Non-Critical Use	Unlicensed is suitable for consumer IoT; licensed is necessary for applications such as telemedicine, automation, and autonomous driving.
Research Insights	Apruzzese et al.'s (2023) study found that 5G, combined with IoT, AI, and blockchain, can transform logistics; however, infrastructure and security challenges persist.

THE TRANSFORMATIVE POTENTIAL OF 5G

The transformative potential of 5G lies in its ability to significantly enhance the capabilities of current mobile generations, addressing the escalating demands for connectivity in a digital-first world. With 5G, data volume in each area is expected to increase by 1,000 times, accommodating the explosive growth in user data traffic and connected devices, which are projected to rise by 10 to 100 times. This remarkable scalability ensures that 5G networks can support the diverse and growing needs of industries, consumers, and IoT ecosystems.

Key performance benchmarks set by the International Telecommunication Union (ITU) within the IMT-2020 framework illustrate these advancements. Ultra-reliable coverage enables the support of 10 to 100 times more devices per unit area, ensuring seamless connectivity in densely populated or industrial environments. The minimum user data rate is pegged at 100 Mbps, providing consistent high-speed access even with lower service quality. Moreover, 5G networks promise a capacity increase of 10,000 times that of previous generations, supporting a surge in applications such as autonomous vehicles, telemedicine, and smart cities. Advanced services like smart cities and autonomous vehicles rely on 5G's ability to integrate heterogeneous nodes seamlessly," (Hossain & Hasan, 2015).

Latency, a critical metric for real-time applications, is reduced to under 1 millisecond, enabling instantaneous communication essential for mission-critical tasks. Furthermore, improvements in the energy efficiency of IoT devices extend battery life to 10 years, addressing concerns about sustainability and operational costs. With ultra-low deployment and operation costs, 5G networks promise to democratize access to high-speed connectivity. These advancements collectively define the 5G era as unparalleled speed, reliability, and efficiency, setting the stage for innovative applications and global connectivity.

The International Telecommunication Union (ITU), which has created a global R&D program on International Mobile Telecommunications for 2020 and past values, has created a 5G framework illustrated in the following table.

- Ultra-reliability Coverage: 10-100 more devices
- User Data rate: 100 Mbps low-end data rate
- Capacity: 10000 more traffic
- Latency: <1ms latency
- Cost: M" M ultra-low cost
- Energy Consumption: 10 years of battery

The requirements for 5G networks, as outlined in Table 1, showcase the ambitious goals set to meet the requests of the evolving digital ecosystem. Data rates are expected to range between 1 and 10 Gbps, significantly outpacing previous generations, enabling applications such as ultra-high-definition streaming, real-time virtual reality, and advanced cloud services. Data volume targets are equally impressive, supporting 9 GB per hour during busy periods and up to 500 GB per user per month, ensuring robust capacity for high-traffic environments and data-intensive applications.

Latency, a critical factor for real-time operations, is set to be less than 5 milliseconds, enabling applications such as remote surgeries, autonomous driving, and industrial automation. This reduction in response time ensures seamless and efficient operations in mission-critical scenarios. Battery life enhancements are another cornerstone of 5G requirements, intended to achieve a decade-long operational lifespan for connected devices. This milestone addresses the sustainability needs of IoT devices, minimizing maintenance costs and ensuring long-term functionality.

The ability to connect 300,000 devices per access point demonstrates the scalability of 5G networks, enabling massive IoT deployments and accommodating dense urban environments. Reliability, pegged at an unprecedented 99.999%, ensures consistent service quality, even in challenging conditions, providing the robustness required for applications in public safety, healthcare, and industrial systems.

These requirements reflect the vision of 5G as a transformative technology that will enable unparalleled connectivity, efficiency, and innovation across industries.

5G TECHNICAL REQUIREMENTS

The deployment of 5G represents a decisive leap in the evolution of wireless communications, setting unprecedented standards for speed, connectivity, and reliability. The next-generation mobile network must meet stringent technical requirements to support emerging applications such as autonomous vehicles, industrial automation, and immersive virtual environments.

The FP7 ICT METIS project ("Mobile and Wireless Communications Enablers for the Twenty-Twenty Information Society") outlined these performance benchmarks, providing a comprehensive vision of the capabilities needed to meet future societal and industrial demands. The project specifies high data rates ranging from 1 to 10 Gbps, the ability to handle massive data volumes (up to 500 GB per user per month), and ultra-low latency of less than 5 milliseconds, enabling real-time responsiveness.

Moreover, the vision includes a battery life of up to ten years for IoT devices, support for 300,000 connected devices per access point, and a reliability level of 99.999%, essential for mission-critical applications. These requirements underscore the transformative potential of 5G, positioning it not only as a technological upgrade but also as the foundation of the 21st-century digital economy.

The technical performance benchmarks for 5G are designed to support a wide range of high-demand use cases, including autonomous vehicles and large-scale Internet of Things (IoT) applications. Table 3 presents the core performance targets expected from next-generation networks, including ultrafast data rates, ultra-low latency, and massive device connectivity. These values are not arbitrary; they reflect the necessary capabilities to meet the needs of both consumers and industries by 2030. Achieving such specifications requires coordinated advances in spectrum, infrastructure, and software systems. These metrics act as reference points for network designers, vendors, and policymakers. The following table summarizes the core technical requirements that define the promise of 5G networks.

Table 3. 5G technical requirements

Requirements	Desired Value
Data Rate	1 to 10 Gbps
Data Volume	9 GB/h (busy period) and 500GB/month/user
Latency	Less than 5 ms
Battery Life	One decade
Connected devices	300000 devices/AP
Reliability	99.999%

Svensson et al., 2020

Fulfilling the technical benchmarks outlined in Table 3 is essential to leverage the full transformative potential of 5G networks. With latency reduced to under 5 milliseconds, near-perfect reliability, and the ability to connect hundreds of thousands of devices per access point, 5G redefines the expectations for mobile communications. These capabilities are foundational for enabling the next generation of smart cities, advanced industrial automation, and immersive real-time digital experience; however, their deployment hinges on coordinated global efforts in research, infrastructure investment, and regulatory alignment. Aligning technological innovation with these stringent performance targets is key to a successful and scalable 5G rollout.

In practical terms, 5G could power urban applications such as smart metering, intelligent lighting, environmental monitoring, and dynamic traffic control through dense wireless sensor networks (WSNs). *These systems, built on very low-cost and energy-efficient devices, require a unified communication architecture that can support a diverse range of technologies.* 5G's flexibility and performance make it an ideal enabler. Additionally, mobile video surveillance—a growing use case—relies on 5G for high-speed data transmission and real-time analytics. While this demands slightly higher energy use, it requires high reliability and robust security, both of which 5G is designed to support.

SECURITY STRUCTURE REQUIREMENTS

The typical Security structure and methods for the 5G system are excellent illustrations of these standards. This security structure was developed in collaboration with the 3GPP criteria bodies and partnership with the Internet Engineering Task Force (IETF). *Additionally, the International Telecommunication Union (ITU) has outlined several key requirements for 5G in these areas.* Still, it does not possess a comparable paper establishing rules for 5G protection on the same level as 3GPP's publications.

The 3rd Generation Partnership Project (3GPP) is the business consortium that sets standards for 5 G. It describes some methods utilizing the 5G NR (5G New Radio) application as "5G," a definition that became common by late 2018.

The International Telecommunications Union (ITU) establishes Minimum standards. Previously, several reserved the phrase" 5" G for devices that provide download speeds of 20 Gbit/,s as specified in the ITU's IMT-20200 document.

Initially, the term referred to the International Telecommunication Union's IMT-2020 standard format, which required a theoretical download speed of 20 gigabits per second and a publish speed of 10 gigabits per second, along with other specifications. Then, the industry standards group 3GPP chose the 5G NR (New Radio) standard, combined with LTE, as its proposal for submission to the IMT-2020 standard.

The requirements for 5G networks represent a leap in connectivity standards. Data rates of 1 to 10 Gbps enable high-speed applications while handling 9 GB/hour during peak times, and 500 GB/month per user ensures robust data capacity.

Latency under 5 milliseconds supports real-time operations, which are critical for applications such as autonomous driving. A decade-long battery life enhances sustainability for IoT devices, and connecting 300,000 devices per access point demonstrates the scalability of the solution. Reliability at 99.999% guarantees dependable performance for mission-critical services.

The first stage of the 3GPP 5G specifications in Release 15 was finished by 2019. Therefore, the next stage of Release-16 will be completed in 2020.

5G NR can include lower frequencies (FR1) under 6 GHz and higher frequencies (FR2) above 24 GHz. Nevertheless, the pace and latency in initial FR1 deployments using a 5G NR application on 4G hardware (non-standalone) are significantly better than those of brand-new 4G methods, which are estimated to be 15-50% better.

TECHNICAL REQUIREMENTS FOR USERS' EQUIPMENT, CHIEFLY SMARTPHONES AND TABLETS

5G requirements are demands for users' equipment, chiefly smartphones and tablets, a nd the foundation. Positive Many Meanings - facilities in a 5G network. Some other standards include different functions within the 5G program. There is a focus on confidentiality and integrity, followed by replay protection in the 5G security requirements outlined in the SDxCentral post, "Data Security in the Cloud Best Practices." Discover much more about integrity and privacy. "Replay protections are utilized to avoid replay strikes. A malicious actor will intercept a message, extract the included credentials, and send a similar, altered note to the exact location. The effect is a response delivered to the initial sender, rather than towards the enemy.

Beyond pace improvement, 5G anticipates the unleashing of a tremendous IoT (Internet of Things) ecosystem whereby networks can work communication requirements for vast amounts of connected devices, with the proper trade-offs between speed, latency, and price, with the following criteria:

- As much as 10 Gbps information rate - ten to 100x improvement over 4G and 4.5G networks
- 1-millisecond latency
- 1000x bandwidth in every device area
- As many as 100 times more devices are connected to a unit area (compared with 4G LTE), with 99.999% availability.
- 100% coverage
- 90% reduction in network energy usage
- Up to 10-year battery life for lower-power IoT devices

MAIN APPROACHES TO HANDLING REQUIREMENTS

The requirements for 5G networks demand innovative technologies and approaches to achieve the unprecedented performance and scalability needed for modern connectivity. One of the most impactful strategies is the densification of existing cellular networks. It involves incorporating a vast number of small cells to facilitate peer-to-peer (P2P) interactions, including device-to-device (D2D) and machine-to-machine (M2M) communication. These multi-tier heterogeneous networks enhance coverage, capacity, and efficiency, making them essential for high-density urban areas and industrial applications. Network densification, including heterogeneous and multi-tier networks, is pivotal for achieving high data capacity in 5G systems," (Hossain & Hasan, 2015).

Simultaneous Transmission and Reception Through Full-Duplex (FD) Communication

Another critical advancement is simultaneous transmission and reception through full-duplex (FD) communication. It enables devices to send and receive data simultaneously on the same frequency. FD communication significantly improves spectral efficiency. It reduces latency, which is essential for real-time applications such as autonomous driving and remote medical procedures. "Full-duplex communication substantially improves spectral efficiency and reduces latency, critical for real-time applications" (Hossain & Hasan, 2015).

Massive multiple-input multiple-output (MIMO) and millimeter-wave (mmWave) technologies are pivotal in meeting 5G's high-speed and high-capacity demands. Massive MIMO utilizes an extensive array of antennas to maximize spatial multiplexing. At the same time, mmWave communications provide access to high-bandwidth spectrum, ensuring faster data transfer rates for applications such as augmented reality and cloud gaming. Millimeter waves enable ultra-fast data rates essential for applications like augmented reality and autonomous vehicles," (ITU-R, 2015).

Energy-Aware Communication and Energy Harvesting Techniques

Energy-aware communication and energy harvesting techniques aim to enhance energy efficiency. These methods minimize power consumption, extend the operational life of IoT devices, and support sustainable network operations. Energy harvesting techniques in 5G networks aim to reduce power consumption and extend the operational life of IoT devices," (ITU-R, 2015).

Cloud-based radio access networks (C-RAN) further enhance 5G by centralizing network resources, enabling efficient load balancing and seamless network management. Wireless information virtualization complements C-RAN. It creates flexible, software-defined network architectures that adapt to dynamic user demands.

These approaches collectively form the foundation of 5G's ability to deliver ultra-reliable, low-latency, and high-capacity connectivity, shaping the future of communication and enabling transformative applications across industries.

Approaches and Technologies to Handle the Requirements

The main approaches and technologies to handle the requirements for 5G could be categorized as follows:

- The densification of present cellular networks, with the extensive inclusion of small cells for peer-to-peer (P2P) interaction (e.g., device-to-device (D2D) and machine-to-machine (M2M) communication), has enabled multi-tier heterogeneous networks.
- Simultaneous transmission and reception, such as full-duplex (FD) communication, massive multiple-input multiple-output (MIMO), and millimeter-wave (mm-wave) communications technologies, contribute to improved energy efficiency through energy-aware communication and energy harvesting.
- Cloud-based radio access system (C-RAN).
- Virtualization of wireless information.

SEAMLESS INTEGRATION OF 5G WITH INDUSTRIAL ETHERNET LANS FOR SMART MANUFACTURING

The integration of 5G networks with existing industrial Ethernet Local Area Networks (LANs) marks a pivotal advancement in the digital transformation of manufacturing environments. This seamless convergence enables manufacturers to bridge legacy systems with next-generation wireless capabilities, allowing for uninterrupted data flow between machines, control units, and cloud platforms. Integrating 5G technology into existing Ethernet-based systems offers significant advantages for medical manufacturing facilities. As many factories already rely on Ethernet for deterministic communication, incorporating 5G ensures backward compatibility without necessitating a complete overhaul of the existing infrastructure. This seamless integration enables enhanced connectivity and real-time data exchange, which is crucial for optimizing manufacturing processes in the pharmaceutical industry, (Sadek, 2024).

One of the key advantages of this integration is the ability to perform real-time performance monitoring. 5G's ultra-low latency and high reliability enable precise tracking of machine operations, production cycles, and maintenance needs. Manufacturers can now make data-driven decisions in real time, minimizing downtime and increasing operational efficiency. This is especially critical in environments where milliseconds matter, such as robotics or precision machining.

Data integrity and privacy are also enhanced through 5G's advanced security protocols. Industrial networks deal with sensitive intellectual property, production data, and customer information, which must be protected from cyber threats. 5G provides end-to-end encryption, secure authentication, and enhanced threat detection mechanisms, making it a trustworthy partner for industrial environments.

Furthermore, 5G's precise positioning capabilities support applications that require tight time synchronization, such as collaborative robotics (cobots), automated guided vehicles (AGVs), and digital twins. Devices connected via 5G and Ethernet can operate in perfect coordination across production floors, increasing productivity and safety.

A key enabler of this integration is network slicing, which allows manufacturers to dedicate virtual network segments to different applications. For instance, a slice can be reserved for latency-sensitive robotic arms, while another handles less critical sensor data. This ensures that each process receives the network performance it requires, without interference or congestion.

In summary, the fusion of 5G and industrial Ethernet LANs is not just a technical upgrade—it is a strategic move toward innovative manufacturing ecosystems. It brings together speed, precision, security, and flexibility, making factories more intelligent, agile, and future-ready.

THE 5G NETWORK COOPERATION FOR REQUIREMENTS AND STRUCTURE

As 5G aims to unify diverse use cases—from enhanced mobile broadband (eMBB) to ultra-reliable low-latency communications (URLLC) and massive machine-type communications (mMTC)—network cooperation becomes not only a technical challenge but also a policy and regulatory imperative. To accommodate such heterogeneous demands, 5G networks must function as open ecosystems where legacy systems (such as LTE) and emerging technologies (including AR, AI, and blockchain) coexist under a common architectural framework.

This coexistence requires harmonized spectrum allocation across countries and continents, preventing fragmentation that could stifle global scalability. Collaborative efforts between governments, standardization bodies (e.g., 3GPP, ITU), and telecom operators are crucial to ensuring seamless interoperability. Without such alignment, innovations in connected healthcare, intelligent transportation, and the autonomous industry could be delayed or remain siloed.

Infrastructure sharing, including the co-location of small cells, fiber, and cloud resources, becomes essential in both densely populated urban environments and rural areas where economic viability is challenging. Multi-operator core networks (MOCNs) and RAN sharing enable more efficient investments and faster rollouts, thereby promoting digital equity.

Furthermore, 5G requires intelligent orchestration of macro, micro, and pico base stations. Software-defined networking (SDN) and network function virtualization (NFV) enable the dynamic allocation of resources to meet the quality-of-service (QoS) needs of diverse verticals. Edge computing nodes must also cooperate to minimize latency and facilitate local data processing for critical applications, such as smart grids or autonomous vehicles.

In this new era, cross-sector collaboration is equally vital. Industrial manufacturers, transportation agencies, healthcare institutions, and telecom vendors must co-design and co-deploy tailored solutions that align with the capabilities of 5 G.

Ultimately, 5G is not a standalone technology—it is a platform for convergence. Its success depends on tightly coordinated technical architectures and globally aligned deployment strategies. As Frieden (2019) highlights, without cooperative spectrum policies and inclusive design approaches, the full potential of 5G will remain unattainable.

STRUCTURE OF 5G NETWORK

The structure of 5G networks introduces a multi-tier architecture to meet the demands of high connectivity, low latency, and increased data capacity. This architecture comprises nodes and cells with diverse qualities and capacities, including macrocells, small cells (such as picocells and femtocells), and device-to-device (D2D) user equipment (UE). These heterogeneous nodes enable seamless and efficient communication, thereby enhancing network performance across diverse environments.

The multi-tier architecture enables greater flexibility and scalability by allowing the system nodes to self-organize. This self-organization addresses the increasing complexity of network coordination and management among different tiers. Tasks such as load balancing, power adaptation, spectrum allocation, and interference minimization are handled autonomously, ensuring optimal network operation and resource utilization.

A unique feature of 5G networks is the capability of user equipment (UE) to establish simultaneous connections with multiple base stations (BS) or access points (AP), potentially using different radio access technologies (RATs). This multi-connectivity enhances reliability, reduces latency, and supports seamless handovers, which are crucial for autonomous vehicles and industrial automation applications.

Integrating heterogeneous nodes, such as wearable devices, intelligent machines, base stations, and user equipment, into a unified system ensures seamless connectivity across all devices. This integration, often managed through cloud-based systems, supports dynamic and efficient network management. By combining these elements, the 5G network structure establishes a robust foundation for advanced communication systems that support applications such as smart cities, IoT ecosystems, and augmented reality.

CORRESPONDENCE EFFECTIVENESS WITHIN 5G METHODS

The correspondence effectiveness in 5G networks has been significantly improved by adopting advanced techniques, ensuring seamless and efficient communication across diverse scenarios. One key enhancement is Coordinated Multipoint (CoMP) reception and transmission, which synchronizes data processing across multiple network points to reduce interference and optimize signal quality. Network-assisted interference cancellation and suppression enhance communication by mitigating noise and signal overlap, which is crucial for densely populated areas with high device usage. Non-orthogonal multiple access (NOMA) technology enables efficient spectrum reuse, allowing users to share the same frequency band without compromising performance. Spectrum reuse and the adoption of millimeter-wave bands are essential for achieving the expected 1,000-fold throughput improvement over 4G," (Hossain & Hasan, 2015).

Additionally, three-dimensional and full-dimensional MIMO (Multiple-Input, Multiple-Output) technologies utilize spatial multiplexing to enhance network capacity and maintain high signal quality. These innovations ensure that 5G networks facilitate the growing demand for faster and more reliable communication.

USAGE OF A LOT OF REMOTE RADIO HEADS (RRHS)

The deployment of multiple Remote Radio Heads (RRHs) connected to central processing nodes has revolutionized the architecture of 5G networks. These Remote Radio Heads, often managed through cloud-based systems, optimize the distribution of network resources by facilitating high-speed backhaul and front-haul connections to the core network. This setup ensures low latency and high throughput, which are crucial for applications such as IoT, autonomous vehicles, and smart cities. RRHs are essential in reducing network congestion and enhancing coverage, particularly in densely populated urban areas. Their modular nature allows for scalable deployment, enabling network operators to expand capacity and reach as demand grows. By separating the radio and processing functions, RRHs enhance network flexibility, support efficient load balancing, and enable dynamic resource allocation. This architecture ensures robust and reliable communication, catering to the complex requirements of modern wireless networks.

AN EVOLVING COMPLEXITY

The architecture of 5G networks introduces significant complexity due to advancements like energy-aware full-duplex (FD) communication and spectrum virtualization. These technologies enable devices to operate across multiple spectrum bands, from traditional radio frequencies (RF) to millimeter waves, while maintaining backward compatibility with older technologies, such as 3G and 4 G. Energy efficiency is critical, as power-hungry multimedia applications demand prolonged battery life and sustainable operation. 5G equipment integrates energy harvesting technologies that capture and utilize ambient energy to power devices, addressing this issue.

However, supporting multiple radio access technologies (RATs) and implementing FD communication capabilities present substantial challenges for designing and testing internal chips and front-end modules. These evolving demands require sophisticated engineering solutions to ensure that 5G networks can deliver on their promises of enhanced performance, scalability, and energy efficiency, paving the way for a more connected future.

Dense Heterogeneous Networks

The 5G cellular network is a multitier heterogeneous network comprising macrocells, along with many low-power nodes (e.g., small cells, relays, remote radio heads (RRHs), and provisions for P2P (such as D2D and M2M) communication. Deploying heterogeneous nodes inside 5G methods has a substantially higher density than modern regular single-tier (e.g., macrocell) networks. The heterogeneity of various classes of base stations (e.g., small cells and macrocells) provides adaptable coverage areas and increases spectral efficiency. Decreasing the cell's dimensions improves the spot's spectral efficiency, allowing for better spectrum reuse. Furthermore, small indoor areas (such as houses, public vehicles, and office buildings) improve coverage. Moreover, the underlying cellular architecture can substantially enhance the system's total energy efficiency, spectrum effectiveness, and wireless peer-to-peer (P2P) interaction (e.g., D2D/M2M interaction among autonomous sensors, actuators, and user equipment (UE).

Full-Duplex Communication

An FD transceiver can transmit a similar frequency in an FD correspondence pattern in a comparable period. However, the wireless node (e.g., a base station or a user equipment (UE)) can decode a received signal while simultaneously transmitting on the same frequency band due to inner interference between the transmitter and receiver circuits, known as self-interference (SI).

Nevertheless, developing in-band FD radios is feasible with the latest advancements in antenna and electronic baseband technologies, as well as RF interference cancellation strategies. Furthermore, FD interaction can double the spectral effectiveness of the physical coating by removing its frequency band or time slot for each downlink and uplink transmission.

Current studies suggest that FD methods are achievable and offer substantially greater data rates than traditional half-duplex (HD) communication devices. FD technology may also resolve issues in existing wireless networks, such as hidden terminals, throughput loss due to congestion, and long end-to-end waiting times.

One possible method of reducing SI found in FD radios is to combine the antenna cancellation, RF interference cancellation, and electronic interference cancellation methods.

Energy-Aware Communication and Energy Harvesting

One of the primary difficulties in 5G networks is improving the power efficiency of battery-constrained wireless devices. Harvesting electricity from energy sources might be a stylish solution to prolong battery life and enhance the system's overall energy efficiency. For instance, the UE can harvest green energy (e.g., solar and wind power).

Nevertheless, due to the stochastic dynamics of green resources, available energy may differ considerably across locations and water conditions. Thus, harvesting power from these resources might not be practical for quality-of-service and reliability (quality of service)- constrained wireless uses.

Conversely, background radio signals (e.g., RF) harness the power of 5 G. For example, in an RF-powered energy harvesting network (RF EHN), the UE can gather power from the hybrid access point (HAP), utilizing RF signals for information processing and transmission.

Defining Features and Trends of 5G Networks

The defining features and trends of 5G networks showcase the transformative advancements expected from this next-generation technology. (Hossain & Hasan, 2015) One of the primary features is the improvement in capacity and throughput, with data rates projected to be approximately 1,000 times higher than those of 4G, achieving cell data rates of up to 10 Gbps. Trends such as spectrum reuse, millimeter-wave bands (e.g., 28 GHz and 38 GHz), and technologies like massive MIMO and device-to-device (D2D) communication are proposed to meet these demands.

Another crucial aspect is reduced latency, targeted at 2 to 5 milliseconds for end-to-end communication. Full-duplex communication, Cloud Radio Access Networks (C-RAN), and Device-to-Device (D2D) technologies are critical enablers of this low-latency performance, ensuring real-time responsiveness for applications such as autonomous driving and remote surgeries.

Network densification, a hallmark of 5G, aims to support up to 1,000 times more mobile data per unit area and up to 10,000 times more connected devices per unit area. Heterogeneous and multi-tier networks are key trends addressing these requirements, ensuring seamless connectivity in dense urban environments.

Innovations like network virtualization and Cloud Radio Access Network (C-RAN facilitate advanced services and applications, such as smart cities and machine-to-machine (M2M) communication. These developments enable service-oriented communication tailored to specific user needs.

Wireless charging and harvesting technologies support improved energy efficiency, including up to 10 times longer battery life, making 5G ideal for IoT and autonomous applications. Additionally, self-organizing and cognitive networks enhance communication and network management, providing resilience and adaptability.

The features and trends of 5G collectively position it as a cornerstone for future connectivity, enabling unprecedented performance, efficiency, and diversity of applications.

Table 3 highlights the core features and technological trends that distinguish 5G from previous generations. With expectations of massive data throughput, ultra-low latency, and high device density, 5G is engineered to meet the demands of modern digital ecosystems. The evolution includes breakthroughs in spectrum usage, massive MIMO, full-duplex systems, and network virtualization. These capabilities are foundational for emerging applications, such as smart cities, the Internet of Things (IoT), and autonomous systems. This table summarizes the enabling technologies and strategic directions shaping the future of 5G deployment.

Table 3. Features and trends of 5G networks

5g Expectations and Features	Trends/Proposals
Capacity and throughput improvement, high data rate (~1000x of throughput improvement over 4G, cell data rate ~10 Gb/s, signaling loads less than 1~100%)	Spectrum reuse and use of different bands (e.g., mm-wave communication using 28~GHz and 38~GHz bands), multi-tier network, D2D communication-RAN, massive-MIMO
Reduced latency (2~5 milliseconds end-to-end latencies)	Full-duplex communication, Cloud Radio Access Network (C-RAN, D2D communication
Network densification (~1000x higher mobiledataperunitarea,100~10000x higher number of connecting devices)	Heterogeneous and multi-tier network
Advanced services and applications (e.g., smart city, service-oriented communication)	C-RAN, network virtualization, M2M communication
Improved energy efficiency (~10x prolonged battery life)	Wireless charging, energy harvesting
Autonomous applications and network management, Internet of Things	Communication, self-organizing, and cognitive network

(Hossain & Hasan, 2015)

CHALLENGES FOR 5G DEVELOPMENT

The development of 5G networks presents significant challenges that span technological innovations, network architecture, and the capabilities of telecom operators. As the demands for advanced connectivity grow, 5G unit architectures become increasingly complex compared to their 4G predecessors. Products in 5G networks must effectively operate across various spectrum bands, including radio frequencies (RF) and millimeter waves (mmWave), while maintaining backward compatibility with earlier technologies such as 3G and 4G. This requirement necessitates sophisticated engineering solutions to ensure seamless integration and reliable performance.

Challenges of Technology

Technological advancements in 5G face challenges, particularly in terms of energy efficiency and component design. Power-intensive multimedia applications demand sustainable solutions, making energy-harvesting technologies critical for extending device battery life. Full-duplex (FD) communication, which enables simultaneous transmission and reception, introduces an additional layer of complexity, necessitating innovative designs for internal chips and front-end modules. Additionally, supporting multiple Radio Access Technologies (RATs) across diverse spectrum bands calls for advanced testing and design strategies to ensure interoperability and robust functionality.

Challenges With the Structure of the 5G Network

The multi-tier architecture of 5G networks introduces new structural challenges. Integrating heterogeneous nodes such as macrocells, picocells, and femtocells requires sophisticated network management to optimize performance. Self-organizing capabilities must handle real-time tasks like load balancing, power adaptation, and interference mitigation. Moreover, ensuring seamless connectivity across various network layers while accommodating a growing number of connected devices presents logistical and operational difficulties.

Challenges Gap Capabilities of Telecom Operators

Telecom operators face capability gaps in adapting to the demands of 5 G. Deploying infrastructure that supports advanced features like network slicing and ultra-reliable low-latency communication (URLLC) requires significant investments in technology and workforce training. Operators must also address gaps in spectrum allocation and licensing, which can delay the rollout of comprehensive 5G services. Bridging these gaps is crucial for realizing the full potential of 5G and maintaining a competitive edge in the global market.

These challenges collectively underscore the complexity of 5G development, necessitating collaborative efforts among industry stakeholders, regulatory bodies, and technology innovators to overcome barriers and realize the transformative potential of this next-generation network.

- Adequate Medium Access Control: A reasonable medium access control requires the thick deployment of entry points and operator terminals. As a result, user throughput and latency are low, and hotspots will not be proficient in cellular technology to offer higher throughput. In addition, research and development in various disciplines integrate and facilitate the development of new functions.
- Traffic Management: Compared to the standard human-to-human visitors in cellular networks, an astounding number of machine-to-machine (M2M) products in a cell could cause severe method conflicts. These, i.e., stereo access system (RAN) difficulties, will trigger congestion and overload.

Challenges With the Structure of the 5G Network

The rollout of 5G networks presents several critical challenges related to technology, infrastructure, and governance. These challenges must be addressed to fully realize the potential of 5 G.

Support for Multiple Services

Unlike traditional network generations, 5G must accommodate a wide range of services, devices, and technologies across diverse geographic regions. The challenge lies in creating standardized, robust, user-centric, and high-capacity wireless services that can meet the broad and growing expectations of users.

Infrastructure Complexity

- Researchers and engineers continue to face technical challenges in implementing and standardizing 5G infrastructure. The dense deployment of base stations and edge computing facilities further complicates the situation.
- Communication, Navigation, and Sensing Integration: 5G is expected to manage vast volumes of data from heterogeneous sources. These services rely heavily on spectrum availability and demand robust computational infrastructure to handle real-time data transmission, navigation, and sensing tasks efficiently.

Security and Privacy

With increased data exchange and connectivity, protecting individual and organizational data becomes paramount. 5G must address growing concerns related to cybersecurity, data privacy, trust, and secure network operations.

Cyberlaw and Regulation

The high speed and ubiquity of 5G could inadvertently fuel cybercrime and digital fraud. Therefore, robust cyberlaw frameworks and international cooperation are crucial for managing legal, political, and governance risks at both national and global levels.

Radio Access Network (RAN) Complexity

RAN design remains one of the most significant challenges due to its central role in ensuring seamless connectivity, efficient resource allocation, and maintaining service quality in heterogeneous environments.

The 5G Methods Require a Crucial Change in Cellular Structure and Critical Wireless Solutions

The deployment of 5G networks demands a fundamental transformation of cellular architecture and the adoption of advanced wireless technologies. The unprecedented need for network capacity and enhanced user experience necessitates innovation across multiple technical dimensions. Among these, the Radio Access Network (RAN) remains a central challenge.

One of the most pressing needs is increased bandwidth, which necessitates more efficient spectrum utilization. This includes not only optimizing existing frequency bands but also leveraging unlicensed spectrum, exploring new frequency ranges, and, in particular, migrating toward millimeter-wave (mmWave) bands. To support this shift, new waveforms and technologies, such as Massive Multiple Input Multiple Output (MIMO), become essential.

Network densification—the deployment of a high number of small cells—and traffic offloading are also vital to boosting capacity. These efforts require the integration of cloud RAN, context-aware multi-connectivity, and device-to-device (D2D) multi-hop communications.

To enhance overall network efficiency, advanced techniques such as interference management, improved radio link protocols, and resource optimization must be adopted. At the same time, spectrum expansion strategies—such as carrier aggregation, Authorized Shared Access (ASA), and Licensed Shared Access (LSA)—are crucial to meeting the growing capacity demands.

These evolutions reinforce the necessity for virtualization and network notarization, which together form the backbone of next-generation 5G infrastructure. The convergence of these technologies enables scalable, flexible, and high-performance mobile networks. As highlighted by ZTE (2014) and recent insights from Ordóñez de

Pablos and Zhang (2023), the path forward lies in harmonizing 5G with innovations such as AI, advanced IoT, and emerging 6G paradigms.

The development of 5G networks presents a wide range of challenges across technological, architectural, regulatory, and operational domains. As outlined in Table 4, these issues span energy efficiency, real-time network management, infrastructure readiness, and spectrum expansion. Operators must navigate the complexity of deploying dense, multi-tier networks while ensuring low latency and high throughput. Meanwhile, legal and cybersecurity concerns demand new frameworks to protect users and systems. These overlapping challenges underscore the need for coordinated global efforts, innovative engineering, and robust policy support to realize the full potential of 5 G.

Table 4. Challenges for 5G development

Challenge Category	Description
Technological Challenges	Energy efficiency, full-duplex communication, complex component design, and support for multiple RATs.
Network Architecture	Multi-tier structure (macrocells, picocells, femtocells), self-organizing networks, real-time management (load, power, interference).
Operator Capability Gaps	Infrastructure investments, training, spectrum allocation, and licensing limitations.
Medium Access Control	Requires dense deployment of terminals; hotspots face limits in throughput and latency.
Traffic Management	High volume of M2M devices causes congestion and RAN overload.
Multiple Services & Standardization	Support for diverse technologies across regions; standardization to ensure robust and universal service.
Infrastructure	Implementation and standardization challenges in deploying 5G services.
Communication, Navigation, Sensing	Requires spectrum access and powerful processing of vast, heterogeneous data.
Security & Privacy	Trust, cybersecurity, and personal data protection concerns.
Cyberlaw Legislation	Rapid 5G adoption may increase cybercrime; global legal frameworks are needed.
Radio Access Network (RAN)	Bandwidth demands require new waveforms, spectrum efficiency, massive MIMO, and virtualization.
Densification & Efficiency	More sites, cloud radio access, multi-connections, D2D communication, and interference control are needed.
Spectrum Expansion	Utilization of mmWave bands, carrier aggregation, ASA/LSA techniques to increase capacity.

CHALLENGES OF GAP CAPABILITIES FOR TELECOM OPERATORS

According to the GSMA, mobile operators' attempts to optimize 5G system economics will depend on their specific capabilities and assets across massive 5G expense owners and optimizers.

The GSM Association (commonly known as the GSMA or the Global System for Mobile Communications, initially GroupSpécialal Mobile) is an industry organization representing the interests of mobile network operators worldwide. Over 750 mobile operators are GSMA participants, and 400 vendors in the broader mobile ecosystem are associate members. The GSMA represents its members through business programs, working groups, and industry advocacy initiatives.

Significantly, the speed of 5G era price optimization will depend on the gap between the present ('as is') and demanded ('to be') abilities of their 5G assets, which will differ among operators. Figure eight outlines a generic 5G era gap analysis between the present and required 5G era abilities for an average European older market mobile operator.

Several gaps in the operators' 5G assets and abilities. First, there are a few gaps in the operator system's sharing capabilities and assets, which will remain a crucial expense optimizer for 5 G. Next, you will discover small-to-medium, easier-to-bridge spaces in the necessary 5G era features in RAN, backhaul, and vitality, which are essential 5G era price accelerators.

Lastly, we find moderate gaps, which may challenge several operators, particularly in seller support, in a crucial group of areas related to primary system infrastructure, automation, AI, and MEC.

This paper examines the current assets and capabilities, required advancements, and gap analysis for various 5G network infrastructure components. It highlights radio network infrastructure, backhaul, core network infrastructure, energy, and automation, emphasizing the technological and operational transitions needed. Significant gaps exist in densification, energy optimization, and automation, underscoring the challenges operators face in adopting and scaling 5G networks. This analysis is a roadmap for aligning existing assets with emerging 5G requirements.

Table 5 continues by analyzing the transformation needed in 5G infrastructure components to meet future demands. As shown, significant gaps remain between existing and required capabilities, particularly in areas such as the 5G core network, energy systems, and automation. These high-severity gaps reflect the need for investment in next-generation infrastructure, including cloud-native core systems, AI-driven orchestration tools, and real-time edge processing. Medium-level gaps, such as those in backhaul and RAN, still require densification and modernization to achieve optimal 5G performance. This gap analysis identifies where stakeholders should focus their resources to build a scalable, efficient, and intelligent 5G ecosystem.

Table 5. 5G network assets, capabilities, and gap analysis

Infrastructure Component	Current Capabilities (As-Is)	Required Capabilities (To-Be)	Gap Severity
Radio Access Network (RAN)	4G/LTE infrastructure with limited 5G upgrades	Densified 5G NR deployments with massive MIMO	Medium
Backhaul	Legacy fiber and microwave links	High-capacity and ultra-low latency backhaul	Medium
Core Network Infrastructure	Centralized 4G EPC with limited 5G functions	Service-based, cloud-native 5G core	High
Energy Systems	Basic energy efficiency measures	Advanced power management and optimization	High
Automation & AI	Low automation, manual configuration	AI-driven automated network orchestration	High
Multi-access Edge Computing (MEC)	Minimal or no MEC deployments	Distributed MEC with real-time data processing	Medium to High
System Sharing Capabilities	Limited or no active sharing agreements	Full active and passive sharing mechanisms	Low to Medium

Adapted from GSMA

CONCLUSION

Chapter 7 highlights the 5G Network Assets, Capabilities, and Gap Analysis, emphasizing the complexity and strategic importance of managing spectrum allocation, meeting technical requirements, and monitoring emerging trends in 5G development. From sub-1 GHz bands for long-range coverage to high-frequency mmWave bands for ultra-fast communication, the versatility of the 5G spectrum lies at the heart of its transformative power. Global regulatory efforts, particularly those led by the ITU-R, have sought to harmonize spectrum usage across regions, encouraging innovation while ensuring fair access. Simultaneously, industry players must address diverse technical and economic requirements, including low-latency performance, battery longevity, and high device density.

The deployment of 5G is also closely tied to its evolving architecture. Densification through small cells, the adoption of full-duplex communication, and the integration of energy-aware technologies enable the network to meet demands for speed and efficiency. Moreover, techniques such as network slicing, cloud RAN, and device-to-device communication illustrate the depth of technological coordination required to scale 5G effectively. The table below summarizes the key requirements and technological advancements defining 5G performance benchmarks.

Table 6. Summary of 5G requirements and enabling technologies

Requirement	Target/Benchmark	Enabling Technology
Data Rate	1–10 Gbps	mmWave, Massive MIMO, Carrier Aggregation
Latency	<1–5 ms	Full-Duplex Communication, Network Slicing
Connected Devices	Up to 300,000 per AP	Densification, IoT Optimization
Reliability	99.999%	Licensed Spectrum, Beamforming
Battery Life	Up to 10 years	Energy Harvesting, Efficient Protocols
Network Efficiency	90% less energy usage	Cloud RAN, Energy-Aware Design

Spectrum allocation and management for 5G are critical components in shaping the success of this transformative technology. Integrating a wide range of frequency bands—from sub-1 GHz for broad coverage in rural areas to millimeter waves for ultra-fast speeds in urban environments—5G demonstrates remarkable versatility in meeting diverse application needs. The versatility of the 5G spectrum stems from its ability to span sub-6 GHz and millimeter-wave bands. It is offering a balance between coverage and capacity" (Hossain & Hasan, 2015). This dynamic approach to spectrum utilization highlights the careful balance between ensuring extensive network reach and achieving the high speeds and low latencies necessary for advanced use cases.

Regulators and industry stakeholders play pivotal roles in navigating the complexities of spectrum allocation. The efforts to harmonize global frequency bands, led by organizations such as ITU-R, have laid the groundwork for consistent deployment across regions. Simultaneously, the challenge of designing fair and transparent spectrum authorization models underscores the importance of inclusivity, enabling new entrants, specialized industries, and traditional mobile network operators (MNOs) to access and leverage 5G capabilities. The spectrum allocation for 5G is a critical factor influencing next-generation wireless networks' efficiency, coverage, and capacity" (Hossain & Hasan, 2015).

To ensure that 5G's transformative potential reaches densely populated urban centers and underserved rural and industrial areas, this collaborative approach is essential.

The trends in 5G spectrum usage, particularly the integration of millimeter waves with sub-6 GHz frequencies, have enabled unparalleled flexibility in addressing the unique demands of modern connectivity. Applications ranging from consumer broadband to IoT-driven industrial automation benefit from this adaptability, paving the way for innovations in smart cities, autonomous vehicles, and immersive virtual experiences. As more regions adopt 5G and refine their spectrum strategies, the global digital ecosystem is expected to witness accelerated growth and innovation.

In conclusion, the effective management of the 5G spectrum is foundational to unlocking the full potential of next-generation wireless networks. It requires ongoing collaboration between regulators, industry players, and global organizations to ensure equitable access, optimized performance, and widespread adoption. As the world embraces 5G, the strategic deployment and utilization of spectrum will continue to drive technological advancements, bridge digital divides, and transform how we connect, interact, and innovate.

REFERENCES

Apruzzese, G., Colajanni, M., Ferretti, L., & Marchetti, M. (2023). 5G and Beyond: The Role of Emerging Technologies in Logistics and Supply Chain Management. *Sustainability*, *15*(15), 11846. DOI: 10.3390/su151511846

de Pablos, O., & Patricia, Z. Xi (2023). 5G, Artificial Intelligence, and Next Generation Internet of Things: Digital Innovation for Green and Sustainable Economies. IGI Global.

Frieden, R. (2019). The evolving 5G case study in spectrum management and industrial policy. *Telecommunications Policy*, *43*(9), 101826. DOI: 10.1016/j.telpol.2019.04.001

Hossain, E., & Hasan, M. (2015). 5G cellular: Key enabling technologies and research challenges. *IEEE Instrumentation & Measurement Magazine*, *18*(3), 11–21. DOI: 10.1109/MIM.2015.7108393

ITU-R. (2015). *IMT Vision – Framework and overall objectives of the future development of IMT for 2020 and beyond.* International Telecommunication Union. https://www.itu.int/en/publications/Documents

Jhon, R. (2025). *What are the frequency bands for FR1 and FR2?* Teletopix.org. https://teletopix.org/what-are-the-frequency-bands-for-fr1-and-fr2/#gsc.tab=0

Sadek, A. (2024). *AI and ERP are powerful together in the medical manufacturing industry.* Expert ERP Solutions - MEBSCO - [ERP Solutions] Your Trusted Digital Transformation Partner. https://mebsco.com/2024/11/07/ai-and-erp/

Svensson, T., Li, J., Wanlu, S., Lakshamana, T. R., Ström, E., & Yutao, S. (2020) *Information Society (METIS).* (2020). Research.chalmers.se. https://research.chalmers.se/en/project/5261

ADDITIONAL READING

Cognizant. (2019). *Driving the convergence of the physical and digital world.* Cognizant. https://www.cognizant.com/en_us/insights/documents/driving-the-convergence-of-the-physical-and-digital-worlds-codex3530.pdf

GSMA. (2018). *Road to 5G: Introduction and Migration.* GSMA. https://www.gsma.com/futurenetworks/wp-content/uploads/2018/04/Road-to-5G-Introduction-and-Migration_FINAL.pdf

GSMA. (2022). *Maximising the socio-economic value of spectrum*. GSMA. https://www.gsma.com/connectivity-for-good/spectrum/wp-content/uploads/2022/01/mobile-spectrum-maximising-socio-economic-value.pdf

GSMA. (2023). *Spectrum Policy Trends 2023*. GSMA. https://www.gsma.com/spectrum/wp-content/uploads/2023/02/Spectrum-Policy-Trends-2023-1.pdf

GSMA Association. (2021). *Estimating the mid-band spectrum needs in the 2025-2030 time frame*. GSMA https://www.gsma.com/spectrum/wp-content/uploads/2021/07/Estimating-Mid-Band-Spectrum-Needs.pdf

Huawei. (2018). *Huawei 5G Wireless Network Planning Solution White Paper*. Huawei Technologies Co. https://www-file.huawei.com/-/media/corporate/pdf/white%20paper/2018/5g_wireless_network_planing_solution_en.pdf

Huawei (2019). *5 G-Oriented Edge Data Center Facility White Paper*. Huawei Technologies Co. https://digitalpower.huawei.com/upload-pro/index/index/5G-oriented-Edge-Data-Center-Facility-White-Paper.pdf

Huawei. (2022). *Striding Towards the Intelligent World: Wireless Network White Paper*. Huawei Technologies Co. https://www-file.huawei.com/-/media/corp2020/pdf/giv/striding-towards-the-intelligent-world/the_intelligent_world_wireless_network_en.pdf

ITU. (2017). *Forging paths to 5 G. ITU News MAGAZINE*. https://www.itu.int/en/itunews/Documents/2017/2017-02/2017_ITUNews02-en.pdf

Markit, I. H. S. (2019). *The 5G Economy. How 5G will contribute to the global economy*. IHS Markithttps://www.qualcomm.com/content/dam/qcomm-martech/dm-assets/documents/the_ihs_5g_economy_-_2019.pdf

Qualcomm. (2023). *G spectrum update and innovations for future wireless systems. QCOMResearch*. https://www.qualcomm.com/content/dam/qcomm-martech/dm-assets/documents/global-5g-spectrum-status-and-innovations-for-future-wireless-systems.pdf

Section 3
Applications

Chapter 8
Promising Applications and Services for the 5G Market

ABSTRACT

5G technology has ushered in a transformative era in mobile communication, redefining connectivity through unprecedented speed, ultra-low latency, and exceptional reliability. Designed to meet the stringent demands of diverse Internet of Things (IoT) use cases, 5G enables gigabit-level data rates, massive connection densities, and real-time operational capabilities. By seamlessly connecting devices, machines, and systems, 5G facilitates innovation and efficiency across industries, paving the way for advanced applications that need to be more attainable with previous network generations. Recent analyses, including those conducted by IoT Analytics, predict IoT-driven use cases will contribute to 76 million 5G connections by 2025. These use cases span critical sectors like smart cities, industrial automation, healthcare, automotive systems, and public safety, showcasing the versatility and scalability of 5G networks.

INTRODUCTION

The advent of 5G technology marks a transformative era in mobile communication, characterized by its unparalleled speed, ultra-low latency, and exceptional reliability. As the fifth generation of wireless communication, 5G has been meticulously designed to address the stringent requirements of diverse Internet of Things (IoT) use cases. These include gigabit-level data rates, support for massive connection densities, and the ability to meet real-time operational demands across industries. With its ability to seamlessly connect devices, machines, and systems, 5G stands

DOI: 10.4018/979-8-3373-0497-7.ch008

poised to revolutionize how industries operate and innovate. The authors, Nam Tuan Le et al. (2016), provide a comprehensive survey of key enabling technologies for 5G networks, emphasizing the roles of cloud computing, Internet of Things (IoT), and software-defined networking (SDN) in addressing challenges related to data traffic, high-speed connectivity, and Quality of Service (quality of service), while proposing new architectures to support flexible and reliable next-generation mobile systems.

Recent insights from IoT Analytics underscore the expansive potential of 5 G. They project that IoT-driven use cases will account for 76 million 5G connections by 2025. These use cases span an array of verticals, including smart cities, industrial automation, healthcare, automotive systems, and public safety, reflecting the diverse applications of 5G technology. Each domain is poised to benefit immensely from the enhanced performance and scalability of 5G networks, enabling previously unattainable solutions that were not possible with earlier generations of wireless technology.

The promising applications of 5G are rooted in its ability to adapt to the distinct needs of various industries. 5G technology ushers in transformative changes in mobile communication, enabling high-speed, low-latency connectivity crucial for IoT applications" (IoT Analytics, 2025). Whether by supporting high data throughput for enhanced mobile broadband (eMBB), facilitating massive Internet of Things (IoT) connectivity with enormous machine-type communication (MTC), or ensuring ultra-reliable low-latency communication (URLLC) for mission-critical applications, 5G provides a versatile and robust framework. This adaptability positions 5G as a key driver of digital transformation across industries. Enhanced mobile broadband (eMBB) and massive machine-type communication (mMTC) are fundamental aspects of 5G's versatile framework," (Hossain & Hasan, 2015).

This chapter examines the most promising applications and services enabled by 5G, exploring their potential to transform industries, boost efficiency, and unlock new opportunities. By examining use cases with significant market potential, this section highlights the transformative impact of 5 G on the global economy and its role in shaping the future of connectivity.

APPLICATION CASES #1: IMPROVED INDUSTRIAL IOT WITH 5G

5G technology has opened new avenues for enhancing industrial IoT (Internet of Things) applications by providing high-speed, low-latency, and reliable connectivity. According to Gartner, 57% of enterprises identify 5G as a critical enabler of IoT interactions and actively pursue its integration into their operations. This transformation is driven by the unique capabilities of 5G, including its ability to handle massive numbers of device connections and support real-time data transmission across various Internet of Things (IoT) systems. These features empower industries to optimize workflows, automate processes, and improve operational efficiency.

One of the most promising aspects of 5 G-enabled industrial IoT is its integration with advanced technologies, such as machine learning algorithms and big data analytics. This synergy enables the real-time analysis of vast amounts of data generated by IoT devices, facilitating predictive maintenance, automated decision-making, and enhanced supply chain management. For example, in manufacturing, 5G networks can support the seamless operation of interconnected machinery, reducing downtime and increasing productivity. Similarly, 5G can enhance asset tracking and route optimization in logistics, resulting in faster and more cost-effective deliveries.

The benefits of 5G for industrial IoT also extend to energy and resource management. Smart grids powered by 5G can optimize energy distribution based on real-time consumption patterns, reducing waste and ensuring sustainability. Smart grids powered by 5G improve energy distribution and reduce waste through real-time monitoring and analytics" (Deutsche Telekom, 2019). Additionally,5 G-enabled IoT solutions can enhance safety in hazardous environments by enabling remote monitoring and control of equipment. These advancements collectively position 5G as a transformative force in industrial IoT, driving innovation and creating new business opportunities across various sectors. As industries adopt 5G technology, its impact on productivity, sustainability, and automation is expected to grow exponentially.

A variety of high-potential application cases as an outcome of this are as follows:

Wireless Manufacturing Command

The deployment of 5G technology enables wireless manufacturing commands in environments previously restricted due to latency constraints. The manufacturing sector will significantly benefit from 5G, contributing nearly 25% of the 5G ultra-low latency market revenue by 2028" (ABI Research, 2025). 5G enables real-time communication, which is crucial for seamless operations on connected assembly lines, by providing ultra-low latency and reliable connectivity. These lines traditionally relied on hardwired connections, which now operate flexibly, allowing movable robots to work collaboratively with human teams. This capability increases operational efficiency and opens manufacturing opportunities in remote or previously inaccessible areas. Wireless manufacturing commands supported by 5G reduce the complexity and costs associated with wiring while ensuring consistent performance, even in demanding industrial settings. Furthermore, this approach enhances adaptability, enabling manufacturers to reconfigure assembly lines quickly to meet changing demands or introduce new products without significant downtime.

Sensible Industrial Facilities

The emergence of 5G has transformed industrial facilities by enabling the robust communication architectures necessary for interconnected, semi-automated, and highly structured operations. IoT systems, robotics, and human operators can seamlessly exchange data in these facilities, fostering greater collaboration and efficiency. Previously reliant on wired networks, these devices now benefit from 5G's low-latency, high-capacity wireless connectivity, which supports real-time data transfer and decision-making. This innovation allows factories to integrate flexible and adaptive systems, making production lines more responsive to market demands.

Additionally, 5G's ability to handle massive device connections ensures that industrial facilities can scale their operations without compromising performance. The ability to support enormous device connections makes 5G indispensable for large-scale IoT deployments in healthcare and automotive industries. This interconnected infrastructure reduces downtime, streamlines maintenance processes, and enhances overall productivity, positioning 5G as a cornerstone of brilliant industrial facilities.

Wearable Technology Interaction

Wearable technology, including smart glasses, helmets, and other devices, has become increasingly prevalent on manufacturing floors, providing workers with critical information and enhancing both safety and efficiency. With the advent of 5G, these wearables can now operate with greater independence and functionality. The high-speed mobile data offered by 5G ensures seamless communication and service flow, even when devices are out of range of traditional networks. This capability transforms wearable technology from supplementary tools into essential standalone devices capable of real-time data processing and communication. 5G enhances storage and data delivery speeds, enabling wearables to support complex applications in a broader Internet of Things (IoT) landscape, like augmented reality (AR) for training, maintenance, and design visualization. This technological advancement enhances productivity and situational awareness while reducing the reliance on stationary systems, marking a significant evolution in manufacturing technology.

APPLICATION CASES #2: VEHICLE TELEMATICS IN AUTOMOTIVE & TRANSPORTATION

The automotive and transportation sectors are emerging as leaders in adopting 5G technologies for IoT applications, driven by the transformative potential of vehicle telematics. AT&T, a prominent player in this domain, has established one of the most mature IoT ecosystems, connecting millions of automobiles through telematics applications. These include vehicle diagnostics, user-based insurance models, and location tracking, which enhance operational efficiency, safety, and user convenience. As 5G becomes more widely implemented, telematics integration into automotive systems is expected to grow exponentially, enabling even more advanced capabilities.

By 2022, it is projected that 100% of new automobiles manufactured by Original Equipment Manufacturers (OEMs) will be equipped with intelligent telematics systems. These systems incorporate various features, including real-time vehicle diagnostics, connectivity tools, and precise location tracking. They enable data-driven insights for predictive maintenance, personalized insurance policies based on driving behavior, and optimized navigation systems. These advancements mark a significant shift toward more intelligent, more interconnected transportation networks.

Current telematics applications primarily transmit modest volumes of transient data, which traditional cellular networks (2G, 3 G, and 4G) have been sufficient to support. However, introducing 5G unlocks new possibilities, offering ultra-low latency, higher bandwidth, and excellent reliability. These improvements are crucial for more complex applications, such as autonomous vehicles, real-time traffic management, and fleet optimization. 5G also supports vehicle-to-everything (V2X) communication, which enhances safety through real-time alerts and collaborative driving technologies.

As 5G continues to evolve, its role in transforming the automotive and transportation sectors will become increasingly significant, paving the way for safer, more efficient, and intelligent transportation systems. This application case highlights the significance of 5G in shaping the future of mobility and the broader Internet of Things (IoT) landscape. Shrivastava, Rana, and Taneja (2025) emphasize the pivotal role of 5G in driving future Internet of Things (IoT) frameworks.

Vehicle Telematics: Enabling Intelligent Fleet Management

Vehicle telematics is a transformative technology integrating advanced communication and data analytics to enhance vehicle monitoring and management. At its core, a telematics system comprises a tracking device installed in a vehicle, enabling the collection, transmission, and storage of telemetry data. This device connects to the vehicle's onboard diagnostics (OBDII) or CAN-BUS port using a SIM card and onboard modem to facilitate communication via wireless networks. These networks may include GPRS, 4G mobile data, or satellite connections, ensuring seamless data transmission to a centralized server.

The system captures vehicle-specific data, including GPS-based location, speed, idling time, fuel consumption, and events involving harsh acceleration or braking. It transmits this information to a server, processes it, and presents it to end-users through secure web interfaces or mobile applications optimized for smartphones and tablets. The accessible, real-time insights allow fleet operators and individual users to make data-driven decisions, optimize operations, and enhance overall safety and efficiency.

Telematics has an incredibly impactful role in engagement, providing intensive insights across an entire fleet. By analyzing patterns and events, operators can improve route planning, reduce fuel costs, and proactively address vehicle maintenance issues. Additionally, telematics supports compliance with regulatory requirements, such as electronic logging for hours of service in commercial transport.

Beyond fleet operations, vehicle telematics plays a vital role in user-based insurance, real-time vehicle diagnostics, and connected mobility ecosystems. As telematics technology evolves and integrates with 5G networks, it is poised to revolutionize the automotive and transportation sectors, enabling smarter, safer, and more efficient mobility solutions.

According to Verizonconnect.com, Telematics operates through a vehicle tracking device that captures location and diagnostic data from the vehicle. This information is transmitted via cellular or satellite networks to centralized servers. The servers process and interpret the data, making it accessible through secure platforms. End-users can view vehicle updates in near real-time using desktop or mobile applications, enabling efficient monitoring and management of fleet operations. The process ensures seamless connectivity and actionable insights for improved decision-making.

Figure 1. How telematics works

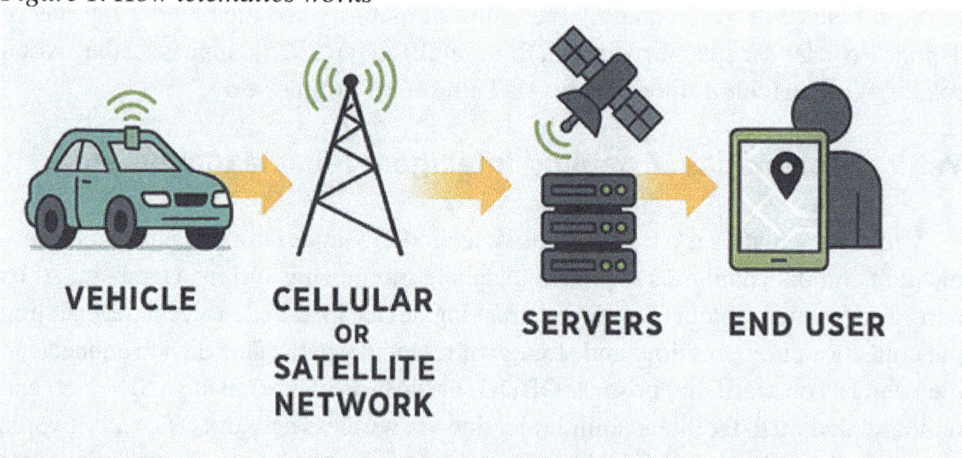

In-Vehicle Infotainment (IV) Systems

In-vehicle infotainment (IV) systems transform the driving experience by combining hardware and software to deliver entertainment and real-time information to drivers and passengers. These systems provide seamless access to audio, video, and multimedia content, including music and video streaming, navigation aids, and updates on traffic and weather conditions.

With the advent of 5G technology, the capabilities of IV systems have undergone significant improvements. 5 G offers faster and more reliable connectivity than LTE networks. Its ultra-low latency and high-speed data transmission enable advanced applications, such as augmented reality (AR) and virtual reality (VR)-based navigation and entertainment services.

These advancements cater to the growing demand for personalized and interactive in-car experiences, such as real-time traffic analysis, location-based services, and cloud gaming. Furthermore, 5G connectivity enables the integration of in-car voice assistants and connected services, allowing drivers to control functions such as climate, lighting, and infotainment features through voice commands. Automakers are leveraging 5 G-powered IV systems to differentiate their offerings, creating opportunities for innovation in autonomous driving support, seamless over-the-air updates, and improved vehicle-to-everything (V2X) communication. These systems also open new revenue streams for automakers by enabling partnerships with streaming platforms and app developers. As in-vehicle infotainment evolves, it is poised to redefine the automotive experience, making vehicles not just modes of transport but hubs of connectivity and entertainment, fostering both convenience and enjoyment for users.

APPLICATION CASES #3: SMART GRID AUTOMATION

Intelligent grid automation represents a transformative shift in the energy sector. It leverages advanced technologies and processes to optimize the traditional electric grid.

Driven by the exponential increase in global electricity demand—projected to grow by 75% by 2030—the smart grid integrates innovative devices, software, and applications to enhance grid reliability, efficiency, and sustainability. It utilizes real-time data analytics and communication technologies to ensure balanced energy distribution, minimize transmission losses, and improve outage management. This evolution reflects broader advancements in connectivity, including the Internet, cellular, and wireless technologies, which enable seamless communication between grid components.

The deployment of 5G plays a pivotal role in intelligent grid automation. It provides ultra-reliable, low-latency communication that supports critical grid applications, including demand response, distributed energy resource management, and automated fault detection and isolation. Moreover, 5 G-enabled smart grids empower consumers with greater control over their energy usage through real-time monitoring tools, cost-saving energy management systems, and dynamic pricing models.

As energy costs rise and consumer expectations grow, the smart grid meets these demands by ensuring uninterrupted energy availability while incorporating renewable energy sources by integrating systems into an intelligent, responsive, and sustainable ecosystem, paving the way for smart grid is r a gr through its more efficient energy future ration of IoT devices, machine learning, and cloud-based systems, a more efficient energy future.

Power and Utility Industry Suitable for Real-Time Management

In the power and utility industry, integrating 5G technology transforms how energy systems are managed, offering real-time control and automation to optimize grid operations. With its ultra-reliable low-latency communication (URLLC) and massive machine-type communication (MMTC) capabilities, 5G enables faster fault detection and response, ensuring seamless operation and reduced downtime for critical energy infrastructure. Addressing the growing challenges posed by integrating Distributed Energy Resources (DER), such as renewable energy sources like solar and wind, into the power grid is vital. These advancements ensure compliance with increasingly stringent regulations and enhance grid resilience against evolving cybersecurity threats.

By enabling predictive maintenance, 5G facilitates continuous monitoring of equipment, such as transformers and substations, allowing for the identification of potential failures before they occur. It reduces maintenance costs, enhances reliability, and minimizes disruptions to the energy supply. Furthermore, 5G enables the dynamic control of energy flow, allowing utilities to balance demand and supply while efficiently incorporating decentralized energy generation.

Rana and Smith (2022) highlight the role of 5G in enhancing Industrial Internet of Things (IoT) applications, which are critical for real-time data acquisition, predictive analytics, and system optimization in the energy sector. The technology also empowers utilities to implement advanced cybersecurity protocols, safeguarding operations against emerging threats. Overall, the power and utility industry's adoption of 5G marks a significant leap toward creating a more innovative, reliable, and efficient energy ecosystem equipped to meet the demands of a rapidly evolving energy landscape.

The intelligent grid system infrastructure integrates various energy sources, such as offshore wind farms, solar plants, and fossil fuel power plants, into a centralized data center. This network is interconnected through intelligent nodes, facilitating real-time communication for efficient energy distribution. Key components, including intelligent buildings, electric vehicles (EVs), and smart homes, are seamlessly connected to the grid, ensuring optimal energy usage and distribution. The system features a robust communication network that connects energy generation sources with transmission and distribution systems, enhancing reliability and efficiency. This interconnected framework underscores the potential of smart grids in modernizing energy networks to support sustainability and technological advancements.

Figure 2. Example of a smart grid and its communication network (Adapted from Deutsche Telekom)

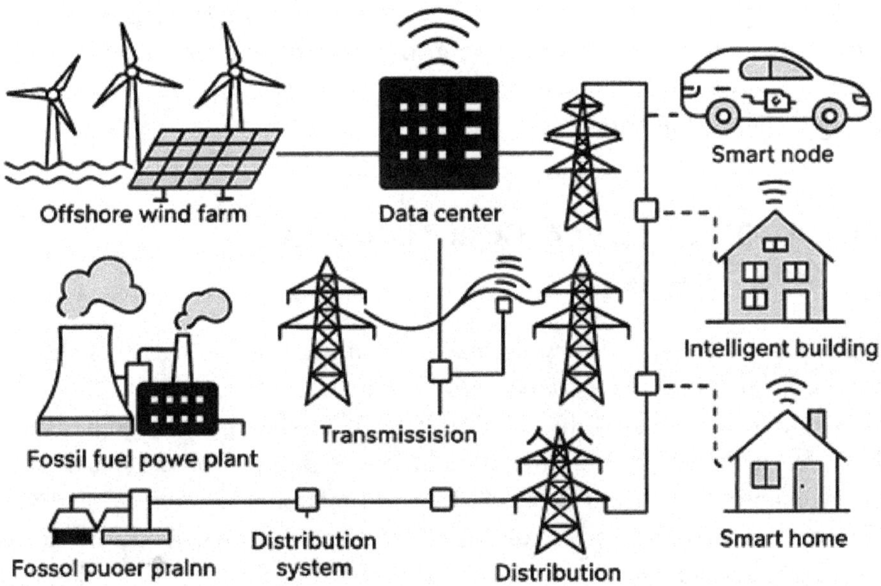

Higher Deployment Flexibility Than Wired Alternatives

The flexibility of 5G deployments compared to wired alternatives marks a significant advancement for energy operators seeking to modernize their power grid infrastructure. While reliable, traditional wired networks, such as Power Line Communications (PLC) and optical fibers, require extensive physical infrastructure and higher costs for installation and maintenance. These limitations often hinder rapid upgrades and expansions, especially in remote or challenging environments. On the other hand, 5G networks offer a wireless alternative that reduces dependency on physical cabling, allowing energy operators to implement new systems quickly and cost-effectively.

With the rising demand for grid automation, 5G enables the deployment of advanced sensors and management applications that enhance operational efficiency and effectiveness. These wireless systems can be easily integrated with renewable energy sources, such as solar and wind, facilitating the smooth adoption of Distributed Energy Resources (DERs) into existing grids. Furthermore, 5G's capabilities in ultra-low latency, high reliability, and massive IoT connectivity ensure performance comparable to wired networks, making it suitable for mission-critical applications.

The reduced deployment cost and increased adaptability make 5G an attractive option for rural and underdeveloped areas where laying physical cables is either economically unviable or technically challenging. Additionally, 5G networks give operators the agility to scale operations in response to dynamic energy demands. As utilities continue to invest in innovative grid initiatives, the enhanced deployment flexibility of 5G over wired alternatives is anticipated to play a pivotal role in driving the energy sector's digital transformation.

APPLICATION CASES #4: COLLABORATIVE AND MOBILE ROBOTS

In the manufacturing industry, there is a prevailing opinion that 5G is best suited for connecting manufacturing robots with varying degrees of mobility, ranging from fixed collaborative robots to fully mobile robots, such as autonomous guided vehicles (AGVs). 5G technology offers transformative capabilities for the manufacturing sector by enabling enhanced connectivity and functionality for fixed and mobile robotics systems. Collaborative robots, often referred to as cobots, are designed to work alongside humans on manufacturing floors, performing tasks that require precision and speed. With 5G's ultra-low latency and high reliability, cobots can achieve real-time responsiveness, enabling seamless interaction with human workers and improving workplace safety and efficiency.

5G enables advanced navigation and coordination for mobile robots, including autonomous guided vehicles (AGVs). These robots rely on consistent, high-speed communication to process real-time data from sensors, cameras, and control systems. The expansive bandwidth and fast data transfer rates of 5G networks allow AGVs to make instantaneous decisions, avoid obstacles, and optimize routes in dynamic industrial environments.

Additionally, 5G's ability to manage multiple device connections simultaneously enables efficient operation in complex and densely populated manufacturing settings where numerous robots, IoT devices, and sensors are active. This connectivity enables the fleet management of AGVs, allowing them to operate collaboratively on tasks such as material transport, inventory management, and warehouse optimization.

The integration of 5G also enhances the deployment of edge computing, enabling robots to process data locally while maintaining a connection to centralized cloud systems for analytics and control. It reduces latency and enhances the accuracy of decision-making. By leveraging 5G's network slicing capabilities, manufacturers can dedicate specific network resources to robotics operations, ensuring consistent performance even during peak usage. As industries continue to embrace automation, 5G will play a critical role in revolutionizing collaborative and mobile robotics applications.

Sensors for Situational Awareness

Modern robots rely on various advanced sensors to gain situational awareness and interact effectively with their environment. These sensors include LiDAR, cameras, ultrasonic, infrared, and proximity sensors that enable the robots to detect obstacles, navigate spaces, and identify objects or people around them. As technology advances, these robots become increasingly intelligent, capable of performing complex tasks autonomously, such as transporting goods within a warehouse, assisting on assembly lines, or even conducting predictive maintenance on industrial equipment. This enhanced intelligence comes at the cost of processing vast amounts of data in real time, presenting significant challenges to existing communication and processing frameworks.

Accommodating Adequate Processing Capabilities

The demands of incorporating sufficient processing power into robots can increase their cost and complexity, especially for more miniature robots designed to perform precise, repetitive tasks. Providing these machines with the computational resources they need to process and analyze sensor data locally adds to the burden of hardware design and costs. However, 5G technology offers a solution by enabling robots to offload some of their processing to edge or cloud-based servers.

With 5G's low latency and high reliability, data from sensors can be transmitted to remote servers in real time, where advanced analytics and decision-making processes occur. This setup simplifies the robot's hardware requirements and significantly reduces production costs, making robotics solutions more accessible to various industries. Additionally, 5G's robust connectivity ensures uninterrupted communication between the robot and the server, allowing robots to execute time-sensitive tasks precisely and efficiently. This paradigm of offloading computational tasks to external servers is revolutionizing the capabilities and affordability of robots equipped with situational awareness technologies.

Accommodating Adequate Processing Capabilities With 5G

Integrating advanced processing capabilities into increasingly smaller robots has historically posed significant challenges due to the high costs and complexities of local computational hardware. These robots, often designed for specific industrial tasks such as assembly, transportation, or inspection, require real-time decision-making based on sensor data. This demand for instant processing adds to their hardware burden, increasing costs and limiting scalability, especially in smaller units with restricted space for hardware components.

The advent of 5G technology offers a transformative solution to these challenges. By leveraging 5G's low-latency and high-reliability communication, robots can offload a significant portion of their data processing tasks to edge or cloud-based servers. This offloading reduces the robots' hardware complexity, making them more compact and cost-effective. Instead of embedding powerful processors within each robot, manufacturers can utilize centralized or distributed computing resources enabled by 5G to handle data analysis and decision-making in real time.

Moreover, 5G ensures seamless, high-speed data transmission between robots and cloud servers, enabling them to perform complex tasks without delays. This capability also allows the deployment of sophisticated algorithms, such as artificial intelligence (AI) and machine learning, to optimize robot performance. As a result, businesses can deploy more intelligent, more efficient robots at a fraction of the cost, driving innovation and adoption in the manufacturing, logistics, and healthcare industries. This approach reduces upfront investment and allows easier maintenance and upgrades, such as bulk.

Benefits of Mobile Robots

Your collaborative robot has some clear benefits. You can probably think of specific services tailored to your applications.

The most evident benefit is that the robot can move around the workspace alone. For example, we have seen distinct groups use their cobots (Control Objectives for Information and related Technology) to move between workstations, thereby boosting uptime. Adding a mobile base enables the robot to perform this task autonomously.

Reduce The Active Time Needed For Changeovers. Added autonomy means that your team members do not waste time switching between tasks on the robot. Instead, they can focus on the more value-added assignments.

Reduce the Time Between Tasks—One of the huge drawbacks of physically moving the robot is that it causes. The robot can sit inactive for quite a while until someone is ready to move it. A mobile base implies that this downtime does not happen.

Improve Performance For Picking Assignments — A mainstream application for mobile robots is picking and material handling. The robot moves between locations and identifies items, such as those on shelves.

Better Use of Space — Robots use significantly less space than traditional industrial robots because they do not require safety fencing. However, they still require a fixed workspace. On the other hand, mobile robots can be stored in one location and moved to different areas as needed.

APPLICATION CASES #5: VIDEO SURVEILLANCE

Video surveillance represents a transformative application of 5G connectivity, poised to deliver unparalleled enhancements to public safety and security measures. The global rise in security risks has prompted many government sectors and urban areas to invest in public surveillance systems, with this trend expected to continue accelerating. These systems aim to comprehensively monitor critical areas, ensuring rapid response to security threats and fostering public safety.

Acquisition of Wireless Communications

Traditional video surveillance methods rely heavily on wired networks, which can be costly and time-consuming. However, the growing adoption of wireless communications, such as Wi-Fi and cellular networks, offers faster deployment, greater flexibility, and significant cost reductions. These advantages contribute to the profitability and scalability of surveillance projects, making them accessible to a broader range of organizations and municipalities.

Functionality Improvement for Video Content Analytics

While current LTE networks provide sufficient support for live and recorded HD video streaming, the advent of 5G significantly enhances real-time video analytics. This advancement enables the deployment of advanced features, such as facial recognition and anomaly detection, which support public safety efforts with actionable intelligence. Additionally, 5G enables the simultaneous operation of large-scale camera networks with minimal latency.

Video Surveillance Usage

Typical video surveillance monitors high-traffic public areas such as stadium entrances, train platforms, airports, and secure facility perimeters. These systems are crucial for identifying potential threats, ensuring event security, and protecting critical infrastructure.

Video Surveillance Process

The surveillance process involves identifying key areas for monitoring and strategically deploying cameras to ensure adequate coverage. By analyzing live footage and recorded images during high-risk periods, security personnel can swiftly detect and address improper activities, enhancing overall situational awareness and safety.

APPLICATION CASES #6: NETWORK INTELLIGENCE WITH V2X

The transformative power of 5G enables advanced Vehicle-to-Everything (V2X) communication, revolutionizing traffic management and road safety. This cutting-edge technology leverages 5G's ultra-low latency and high-speed data transmission to facilitate the real-time sharing of information among vehicles, infrastructure, and pedestrians. Integrating road users into a unified network, V2X ensures a more thoughtful, safer, and efficient transportation ecosystem. Core applications include delivering early warnings for jaywalking, detecting potential blind spot accidents, and providing alerts for road hazards like potholes or traffic congestion. These innovations enhance proactive measures, reducing accidents and improving overall road safety.

Architecture of V2x (Vehicle-to-Everything)

The V2X architecture is built upon a 5 G-based framework encompassing roadside sensors, V2X base stations, and 5G communication devices designed for vehicles and other road users, such as cyclists and pedestrians. This robust infrastructure supports seamless interaction across all mobility modes, creating a cohesive and intelligent transportation system.

Real-Time Traffic Monitoring

V2X systems monitor traffic data collected from roadside infrastructure and moving vehicles in real time and provide critical warnings to drivers about hazardous road conditions, traffic congestion, and imminent safety concerns. These alerts enable drivers to make informed decisions, thereby improving traffic flow and reducing the likelihood of accidents. Intelligent Transportation Systems (ITS), powered by Vehicle-to-Everything (V2X) technology, play a significant role in enhancing road safety and operational efficiency.

Enhanced Connectivity in Smart Cities

5G enhances connectivity among distributed sensors within innovative city ecosystems, allowing rapid and precise information exchange. By integrating AI and machine learning, 5G facilitates real-time data analytics, improving operational efficiency in smart cities and manufacturing" (Jones & Ng, 2021). This connectivity supports specialized solutions tailored to dynamic traffic conditions, enabling cities to implement more brilliant transportation strategies. By seamlessly linking infrastructure, vehicles, and road users, 5 G-powered V2X communication paves the way for a safer and more sustainable future in urban mobility.

Applications Potential

The fifth-generation mobile communication network (5G) is faster than the current 4G and is expected to revolutionize other sectors, including production, automotive, healthcare, and energy.

The marketplace is segmented into the following categories: Utilities and Energy, Automotive, Healthcare, IT and Telecommunication, Retail, Industrial, and Consumer Electronics. IT and telecommunication are the most significant vertical segments, while manufacturing is probably the fastest-growing sector. brilliant manufacturing benefits significantly from 5G's ultra-reliable low-latency communication.

With 5G deployments underway in nearly all geographies, several studies are examining the business potential of 5G digitalization. This analysis reveals that the ICT market has gained a deeper understanding of its operations, with notable shifts apparent in the redistribution between the marketplace projections for business-to-customer (B2C) and small business-to-business (B2B) services, as well as modifications in the B2B utilization case clusters studied.

Expectations of Considerable Development of 5G for B2B Services

According to Ericsson, the expectations of considerable development of 5G for B2B services establish an addressable industry projection for the future; there is currently a broader context for these observations. For example, throughout 200+ use cases, ten industries, and a pair of 5G-enabled pick and case clusters, 18% of the total ICT addressable business digitalization worth and 47% of the 5 G-enabled service providers could resolve B2B worth.

Service providers play unique roles in various ecosystems, from connectivity provisioning to B2B service delivery and end-user application creation. There is considerable opportunity to gather information across all functions, and service providers will likely play a crucial role in business digitalization.

The secret to leveraging the strength of usage case clusters: When entering completely new 5G B2B marketplaces, addressing industries individually and considering a single utilized case is not ideal for competitively producing synergies. Consequently, Ericsson has grouped use cases with similar attributes to analyze opportunities using case sorting or clustering, rather than just by business. Ericsson discovered that grouping usage cases into nine utilization case clusters covers approximately 9% of the market opportunity for service providers.

The journey to record the importance of 5G beyond mobile broadband begins today. Digitalization in the industry enables service providers to create and expand their businesses beyond connectivity. The 5G landscape offers enormous potential but is complex to navigate, requiring an in-depth understanding of the various driving forces and barriers affecting multiple industries in focus. With the market's structure, the probability of capturing this opportunity is significantly higher within the next 5-7 years. Addressing these changes may enable service providers to uncover additional revenue streams of up to 35% beyond the current range of small businesses by 2030.

Because of this development, the moment to act is in 4 G's first 5G deployments to construct the impetus. It is crucial to invest in 4G to succeed in 5G and beyond; private 5G networks represent an essential first step and a foundation for creating momentum in serving a broad array of 5 G-enabled use cases.

Significant Growth Across Various Industries

The adoption of 5G technology is projected to drive significant growth across various industries by 2030, with key sectors expected to experience substantial compound annual growth rates (CAGRs) by then. Industries like media and entertainment, agriculture, and public safety are expected to see the highest opportunities for digital transformation enabled by 5G connectivity. Manufacturing, healthcare, and energy utilities are also likely to benefit significantly from the enhanced capabilities of 5G, including low latency, high reliability, and scalability. These advancements will enable new business models, improve operational efficiencies, and support innovative applications in diverse fields. This transformative potential highlights the strategic role of 5G in shaping the future of industry and commerce.

APPLICATION CASES #6: NETWORK INTELLIGENCE WITH V2X

The transformative power of 5G enables advanced Vehicle-to-Everything (V2X) communication, revolutionizing traffic management and road safety. This cutting-edge technology leverages 5G's ultra-low latency and high-speed data transmission to facilitate the real-time sharing of information among vehicles, infrastructure, and pedestrians. Integrating road users into a unified network, V2X ensures a more thoughtful, safer, and efficient transportation ecosystem. Core applications include delivering early warnings for jaywalking, detecting potential blind spot accidents, and providing alerts for road hazards like potholes or traffic congestion. These innovations enhance proactive measures, reducing accidents and improving overall road safety.

Architecture of V2x (Vehicle-to-Everything)

The V2X architecture is built upon a 5 G-based framework encompassing roadside sensors, V2X base stations, and 5G communication devices designed for vehicles and other road users, such as cyclists and pedestrians. This robust infrastructure supports seamless interaction across all mobility modes, creating a cohesive and intelligent transportation system.. 5G supports ultra-reliable low-latency communication (URLLC), a critical requirement for autonomous vehicles and remote surgeries" (Ericsson, 2020).

Real-Time Traffic Monitoring

Through real-time monitoring of traffic data collected from roadside infrastructure and moving vehicles, V2X systems provide critical warnings to drivers about hazardous road conditions, traffic congestion, and imminent safety concerns. These alerts enable drivers to make informed decisions, thereby improving traffic flow and reducing the likelihood of accidents. Intelligent Transportation Systems (ITS), powered by Vehicle-to-Everything (V2X) technology, play a significant role in enhancing road safety and operational efficiency.

Enhanced Connectivity in Smart Cities

5G enhances connectivity among distributed sensors within innovative city ecosystems, allowing rapid and precise information exchange. This connectivity supports specialized solutions tailored to dynamic traffic conditions, enabling cities to implement more brilliant transportation strategies. By seamlessly linking infrastructure, vehicles, and road users, 5 G-powered V2X communication paves the way for a safer and more sustainable future in urban mobility.

Some popular applications of smart homes are:

- Smart Locks and garage-door openers allow the homeowner to grant access to friends or visitors
- Smart TVs connect to the Internet and enable access to content through applications like on-demand video and music.
- Smart Security Cameras enable homeowners to monitor their homes when they are away or on vacation. Intelligent motion sensors can also distinguish between residents, visitors, pets, and burglars and notify authorities if suspicious behavior is detected.
- Smart Blinds can adjust to maintain the same light level using sun-tracking.

Smart thermostats intelligently manage home temperatures, allowing users to schedule, monitor, and adjust settings remotely. These advanced devices go beyond simple programming, learning household routines, and adapting automatically to create an optimal balance of comfort and energy efficiency.

- Smart Lighting adjusts the light intensity and color according to the lightness of the room, time, and human presence.,
- Smart Irrigation can automate the irrigation process by analyzing soil moisture and climate conditions.

The 5G Network Slicing for Smart Home

Mobile telecommunications and operator vendors are preparing themselves because of the development of 5G networks worldwide. 5G networks intend to become adaptable and programmable for end-to-end connect-and-compute infrastructures: application-aware, service-aware, time--, location-, and context-aware. 5G networks demonstrate a revolution in the capabilities, effectiveness, and spectrum utilization of previous-generation 4G networks.

Advantages of 5G N Gt Network Slicing

5G network slicing offers numerous advantages, transforming how networks operate and catering to diverse use cases. Slicing enhances flexibility and adaptability by enabling a single network to provide services tailored to specific user demands. This approach ensures that network resources are utilized efficiently, with operators allocating the resources required for each slice. For example, one slice may prioritize ultra-low latency for critical applications, while another focuses on high throughput for data-intensive tasks.

Another critical advantage is the improvement in operational efficiency, as slicing allows operators to manage and scale resources dynamically according to real-time demands. Furthermore, it overcomes the limitations of traditional "DiffServ" (Differentiated Services), a mainstream quality-of-service mechanism, by providing a more sophisticated and reliable solution for prioritizing network traffic.

Advantages of 5G Network Slicing

Therefore, Several of the advantages of 5G network slicing are:

- An individual network can offer various services tailored to the user's demands and other use cases.
- Network operators can allocate the correct number of necessary resources per network slice, enabling efficient and effective resource utilization. For instance, one network slice can deliver very low latency and minimal data price, while the opposite configuration can provide high throughput.
- It can help network operators reduce both operating expenses (OPEX) and capital expenditures (CAPEX).
- It significantly improves operational effectiveness and the time to market for 5G network service distribution.
- It mitigates all the downsides of "DiffServ," the most widely used quality of service arrangement.

Smart Cities

The development of Smart Cities is now a strong global trend. In the past, telecommunications solutions have been essential components of urban infrastructure. As 5G becomes a ubiquitous technology in all cities, telecommunication technologies are enhancing Smart Cities and serving as a key driver of change for most cities. Research on 5G mobile networks is steadily growing, with its implementation in

Global Business and Smart Cities. Artificial Intelligence and the Internet of Things (IoT) will also be incorporated, resulting in immediate advancements in smart cities.

Applications that require delay-tolerant information require a specific design of CMTC (cloud-enabled). The essential nature of the information guarantees an accurate transmission to the destination.

Examples of CMTC apps include unmanned applications, such as autonomous vehicles, electric grid control, traffic safety control, and remote healthcare. As you can observe, these services align with the perception of electronically automating cities.

This cloud-based service enables real-time, high-throughput, and low-latency access to applications that are inherently intolerant of latency. For conventional, centralized network architectures, these latencies cause visitors to traverse the entire network to reach the main point and then to the end-user equipment.

Smart City Applications

With the deployment and commercialization of 5G networks, empowering vertical ventures with 5G—one of its most significant application scenarios—has become a hot topic in innovative city development.

By advancing specialized assets to achieve V2X, 5G technology enables integrated "cloud-application-terminal" networks for communication, supervision, decision-making, and new transportation architecture, paving the way for innovative applications.

5G and smart security can enable city-wide security checking frameworks and proficiently protect the city anywhere, anytime.

Traditional checking techniques only partially serve environmental administration because they require more regional coverage and frequency. Thus, as 5G technologies develop and are applied, new environmental protection and regulations emerge.

5G technology can meet the needs of innovative medical services and vigorously promote fair, accessible, and inclusive medical reform.

5G and other new data technologies can help the government's response to achieve innovative governance.

THE 5G APPLICATIONS FOR MANUFACTURING

5G technology revolutionizes the manufacturing industry by enabling more flexible, cost-efficient, and intelligent production processes while significantly reducing maintenance expenses. According to ABI Research, by 2028, manufacturing is expected to contribute nearly 25% of the total revenue generated from 5G ultra-low latency use cases, underscoring its transformative impact on the sector.

5G enables manufacturing production operations to be more adaptable and economical, while enhancing efficiency and reducing maintenance expenses. ABI Research forecasts that by 2028, manufacturing will account for nearly a quarter of the total revenue generated in the 5G ultra-low latency use cases market.

The Road to 5G Manufacturing

5G technologies provide the system characteristics necessary for manufacturing. High reliability and low latency help in critical applications. Excessive bandwidth, link density, secure, and ubiquitous connectivity. These requirements, which manufacturers currently depend on, are based on fixed-line networks. The movable 5G technology enables greater flexibility and is less expensive, with shorter lead times for reconfiguring factory floor output, making format modifications, and implementing improvements.

According to the determination, the best essential manufacturing utilizes case categories, which 5G enables operators to use. These include manufacturing automation and control systems, planning and design methods, and area products.

Industrial automation, control methods, automation, and the influence of robots, intelligent logistics systems, and factories
Planning and design method, simulation of factory process, and training support
Field product applications are used to gather and monitor data.

The path to manufacturing 5G technologies provides the network characteristics vital for manufacturing. Low latency and high reliability are essential for supporting critical applications. High bandwidth and connection density make secure universal connectivity. These are requirements that manufacturers currently rely on fixed-line networks. The mobile 5G technology enables higher flexibility, lower costs, and shorter lead times for reconfiguring factory floor production, making design changes, and implementing adjustments.

Utilization of Real-Time Granular Location Information

A prime example is where real-time, granular location information throughout the supply chain can significantly increase agility and efficiency, as well as lower operational risks. The widespread adoption of 5G allows manufacturers to monitor and track the location and velocity of components and materials throughout the supply chain. This technology enables them to control that understanding and correct each aspect of the supply chain.

Increased Usage of Sensor Data to Record Data

Another benefit is the dramatically increased usage of sensor data to record data about a particular characteristic. That could be the heat and humidity for a production facility location, a car, the pace of a filler top in a box procedure, and the chemical structure from a lab device. This intelligence enables more accurate real-time conclusions and predictions about trends and performance, allowing companies to continually monitor and fine-tune their procedures and react to events before they have a material impact on their operations.

Capability for Devices to Act Intelligently

Instead of the usual client/server strategy, which remains foundational in cloud computing, it generates an interconnected web of machines, receptors, and products that cooperate to enhance complex tasks in real-time. As a result, 5G becomes vital to making that unit work, as do the emerging new Wi-Fi requirements.

Ultimately, there is a legitimate use case for leveraging real-time data access and analysis (either automatically or manually) to reduce costs, maximize value, and mitigate risk, thereby enabling advancements in agility, efficiency, and performance.

THE EXAMPLE FROM ERICSSON

In the following, we examine three examples coming from Ericsson for applications and service development:

Augmented Reality for Troubleshooting

Ericsson's Tallinn processing plant is the first to utilize Augmented Reality for troubleshooting. It helps moderate the expense of breakdowns—extra components, material, labor, and cradles—and reduce production downtime. We utilize AR to identify operational shortcomings, including inadequate maintenance planning, ineffective failure diagnostics, and insufficient training. To date, we have achieved time savings of up to 50%.

The Connected Screwdriver

China Mobile and Ericsson are enabling automation by applying cellular IoT technology. The world's first cellular IoT-based trial for connected tools, such as screwdrivers, took place at Ericsson's radio product manufacturing site in Nanjing. Approximately 1,000 high-precision screwdrivers in the factory require routine calibration and lubrication based on utilization times, a manual procedure performed periodically and documented in handwritten logs.

High-accuracy devices feature real-time motion sensors that are attached to NB-IoT modules. The information runs through a cellular IoT network over the organization's private cloud and back-end systems, automating calculations and intelligent analyses of the collected data. With connected screwdrivers, the factory replaces the manual tracking of hardware utilization information with an automated solution, reducing manual work by 50%. As a result, the cost per device decreases, and manufacturers gradually phase out manual monitoring completely.

CONCLUSION

5G technology enables a wide range of innovative applications across multiple sectors. Each use case leverages 5G's strengths, such as low latency, massive connectivity, and high throughput. However, these benefits also come with challenges and deployment considerations, which are summarized in the table below.

Table 1. 5G application cases – benefits and shortcomings

Application Case	Benefits	Shortcomings
Industrial IoT	Predictive maintenance, workflow optimization, and automation	High infrastructure cost, cybersecurity risks
Vehicle Telematics	Real-time tracking, predictive diagnostics, personalized insurance	Privacy concerns, high data volume management
Smart Grids	Efficient energy distribution, fault detection, and consumer empowerment	Integration complexity, reliance on latency-sensitive networks
Collaborative Robots	Enhanced factory flexibility, reduced downtime, and cost-effective automation	Dependence on a stable 5G network, initial investment costs
Video Surveillance	Real-time analytics, enhanced public safety, and broad deployment flexibility	Data privacy issues, storage, and bandwidth demands
V2X (Vehicle-to-Everything)	Accident prevention, traffic efficiency, and proactive hazard alerts	Coverage limitations, infrastructure deployment cost

continued on following page

Table 1. Continued

Application Case	Benefits	Shortcomings
Smart Cities	Environmental monitoring, intelligent transportation, and innovative governance	Interoperability challenges, high deployment costs
In-Vehicle Infotainment	AR/VR applications, voice-controlled interfaces, enhanced user experience	Driver distraction risk, continuous connectivity needed

5G technology enables a wide range of innovative applications across multiple sectors. Each use case leverages 5G's strengths, such as low latency, massive connectivity, and high throughput. However, these benefits also come with challenges and deployment considerations, which are summarized in the table below.

By 2025, IoT-driven applications are expected to comprise a substantial share of the 5G market, with millions of devices leveraging this next-generation network to improve efficiency, safety, and scalability. This wide-ranging impact underscores 5G's potential to serve as the backbone of modern digital infrastructure, fostering innovation and enabling industries to meet the demands of an increasingly connected world. Whether in logistics, public safety, or environmental monitoring, deploying 5G applications demonstrates its ability to address critical challenges and optimize performance.

Despite its transformative potential, realizing 5 G's full benefits requires addressing challenges such as infrastructure investment, spectrum management, and regulatory compliance. Collaboration among stakeholders—including governments, network operators, and technology providers—is essential to overcoming these barriers and ensuring equitable access to 5 G's benefits. Continued technological advancements and strategic partnerships will further accelerate the adoption of 5G services, making them a cornerstone of global economic growth and societal progress.

In conclusion, 5G is more than a technological upgrade; it catalyzes change across industries and societies. The promising applications and services enabled by 5G are paving the way for a future defined by unprecedented connectivity, efficiency, and innovation. As the deployment of 5G continues to expand, its impact will resonate far beyond telecommunications, shaping a more intelligent, connected, and sustainable world.

REFERENCES

Deutsche Telekom. (2019). *Highlights - Deutsche Telekom AG Interim Report 2019*. Deutsche Telekom. https://report.telekom.com/interim-report-q2-2019/at-a-glance/highlights.html

Hossain, E., & Hasan, M. (2015). 5G cellular: Key enabling technologies and research challenges. *IEEE Instrumentation & Measurement Magazine*, *18*(3), 11–21. DOI: 10.1109/MIM.2015.7108393

Le, N. T., Hossain, M. A., Islam, A., Kim, D., Choi, Y.-J., & Jang, Y. M. (2016). Survey of Promising Technologies for 5G Networks. *Mobile Information Systems*, *2016*, 1–25. DOI: 10.1155/2016/2676589

Lema, M. A., Laya, A., Mahmoodi, T., Cuevas, M., Sachs, J., Markendahl, J., & Dohler, M. (2017). Business Case and Technology Analysis for 5G Low Latency Applications. *IEEE Access : Practical Innovations, Open Solutions*, 1–1. DOI: 10.1109/access.2017.2685687

Research, A. B. I. (2025). *Top industries/applications driving 5G market growth from 2025 to 2030*. Abiresearch.com. https://www.abiresearch.com/blog/5g-market-top-industries-applications-2025-to-2030

Shrivastava, S., Rana, A., & Taneja, A. (2025). *Driving Next Generation IoT with 5G and Beyond*. 33–64. https://doi.org/DOI: 10.1002/9781394256075.ch3

Tahir, M., Habaebi, M. H., Dabbagh, M., Mughees, A., Ahad, A., & Ahmed, K. I. (2020). A Review on Application of Blockchain in 5G and Beyond Networks: Taxonomy, Field-Trials, Challenges and Opportunities. *IEEE Access : Practical Innovations, Open Solutions*, *8*, 115876–115904. DOI: 10.1109/access.2020.3003020

ADDITIONAL READING

Ericsson (2019). *5G for business: A 2030 market compass*. Ericsson.com. https://d110erj175o600.cloudfront.net/wp-content/uploads/2019/10/the-5g-for-business-a-2030-compass-report-2019.pdf

Finley, K. (2018, December 13). *What is 5G? The Complete Guide to When, Why, and How*. Wired; WIRED. https://www.wired.com/story/wired-guide-5g/

Fitzpatrick, A. (2019). *Everything you need to know about 5 G*. Time. https://time.com/5594360/what-to-know-about-5g/

GII Global Information. (2024). *5G Services Market by Communication Type, Application, End-User - Global Forecast 2025-2030.* 360iResearch. https://www.giiresearch.com/report/ires1611927-5g-services-market-by-communication-type.html

GSA. (2015). *The road to 5G: Drivers, applications, requirements, and technical development.* GSA. https://gsacom.com/content/uploads/2015/11/151118-GSA-the-Road-to-5G.pdf

GSA. (2025). *5G market snapshot February 2025.* GSA. https://gsacom.com/paper/5g-market-snapshot-february-2025/

Market Research Reports. (2015, February 5). *5G Wireless Ecosystem Forecast Report: 2015-2025.* Prweb.com. https://www.prweb.com/releases/5g_wireless_ecosystem_forecast_report_2015_2025/prweb12497540.htm

Reportlinker. (2015, January 28). *The 5G Wireless Ecosystem: 2015 - 2025 - Technologies, Applications, Verticals, Strategies & Forecasts.* Prnewswire.com; Cision PR Newswire. https://www.prnewswire.com/news-releases/the-5g-wireless-ecosystem-2015---2025---technologies-applications-verticals-strategies--forecasts-300027177.html

ReportsnReports. (2016, March). *5G Wireless Market Worth $250 Billion by 2025: $6 Billion Spend Forecast on R&D for 2015-2020.* Prnewswire.com; Cision PR Newswire. https://www.prnewswire.com/news-releases/5g-wireless-market-worth-250-billion-by-2025-6-billion-spend-forecast-on-rd-for-2015-2020-570611171.html

Vyn, Z. Research. (2025). *5G services market growth, demand, report 2025–2030.* VynZ Research. https://www.vynzresearch.com/ict-media/5g-services-market

Wills, J. (2019). *5G Technology: Which Country Will Be the First to Adapt*? Investopedia. https://www.investopedia.com/articles/markets-economy/090916/5g-technology-which-country-will-be-first-adapt.asp

Chapter 9
Risk and Challenges in Application Development

ABSTRACT

5G is ushering in a new era of mobile communication, offering unmatched speed, ultra-low latency, and high reliability. Designed for the complex demands of the Internet of Things (IoT), 5G enables gigabit speeds, massive device connectivity, and real-time operations. It connects machines, devices, and systems, driving innovation across industries. IoT Analytics projects 76 million 5G IoT connections by 2025. These use cases span smart cities, industrial automation, healthcare, automotive, and public safety—demonstrating 5G's versatility and scalability for mission-critical applications previously out of reach with earlier networks.

INTRODUCTION

The advent of fifth-generation (5G) telecommunications networks has led to a transformative era for the digital economy, enabling a new wave of applications that demand ultra-fast communication on an industrial scale. The capabilities of 5G extend far beyond enhanced speed, offering unparalleled opportunities through innovations like network slicing. This feature enables operators to allocate resources to specific verticals, such as healthcare and energy, ensuring tailored solutions that meet industry-specific requirements. Singh et al. (2024) examine the security issues associated with 5G network slicing, highlighting key risks such as inadequate isolation, authentication flaws, and vulnerabilities in inter-slice communication. They stress the importance of implementing adaptive, slice-specific security strategies to maintain the integrity and reliability of services within the flexible 5G framework.

Agrawal et al. (2023) explore the transformative potential of 5G technology in healthcare, highlighting its benefits, including enhanced telemedicine, real-time monitoring, and precision medicine. They also address cybersecurity risks, system

DOI: 10.4018/979-8-3373-0497-7.ch009

failures, and privacy issues, emphasizing the need for robust regulatory frameworks and ethical standards.

Applications leveraging 5G include driverless cars, telemedicine, factory automation, intelligent electric grids, and smart cities. These advancements integrate seamlessly with artificial intelligence (AI), creating an ecosystem where AI-enabled 5G networks enhance performance and efficiency. For instance, telemedicine can enable real-time remote surgeries with ultra-low latency, while intelligent grids optimize energy distribution using real-time data analytics. The transformative potential of 5G became evident as it began its global rollout in 2020, offering scalable solutions for industries worldwide.

Despite its promising opportunities, 5 G deployment is not without challenges. The authors, Dutta and Hammad (2020), examine the security challenges and opportunities of 5G networks from a system-level perspective, emphasizing the need for security-by-design across all layers of 5G architecture, and highlighting how the integration of SDN, NFV, and edge cloud extends both the risks and potential for enhanced resilience and trustworthiness in next-generation wireless networks. Wazid et al. (2021), investigate the critical security and privacy challenges in 5G-enabled Internet of Things (IoT) communications, identifying a wide range of potential attacks and emphasizing the need for robust protocols in key management, authentication, access control, and intrusion detection to secure diverse applications like remote surgery, autonomous vehicles, and IoT drones. For instance, Narayanan et al. (2024) explore relaying solutions framed within the 3GPP standards for 5G-IoT applications.

READINESS BY MARKETS AND INDUSTRIES

The required infrastructure demands significant investment, and the technological complexity of its setup can introduce risks, including cybersecurity threats. Moreover, the regulatory frameworks and spectrum management policies in different regions complicate its implementation. Nevertheless, 5G remains a pivotal driver of digital transformation, promising to reshape industries and economies while addressing the intricate challenges of its deployment.

Readiness by Markets

The global rollout of 5G wireless networks marks a transformative shift in telecommunications, promising unprecedented speed, ultra-low latency, and connectivity capabilities. However, the readiness of markets to adopt 5G technology varies significantly across regions, shaped by economic, infrastructural, and regulatory factors. Assessing 5G readiness involves evaluating the preparedness of markets to navigate the complexities of its deployment, including spectrum availability, infrastructure upgrades, and policy frameworks. Developed nations, such as the United States, South Korea, and China, lead the 5G race with substantial investments and established infrastructure. At the same time, emerging economies face challenges due to limited resources and inconsistent regulations. Furthermore, the market readiness for 5G encompasses the ability of industries such as healthcare, automotive, and manufacturing to capitalize on its transformative potential. By examining the opportunities and addressing the risks, such as cybersecurity threats and high deployment costs, this analysis highlights the global disparities and progress in achieving 5G readiness.

The 5G readiness of markets can help examine the risks and challenges associated with the upcoming global rollout of new 5G wireless networks.

Without a doubt, 5G is helping set the stage for notable change. However, the landscape of 5G markets remains complex, with diverse and sometimes conflicting interpretations of their structure. This confusion impacts consumers and complicates the industry's potential to measure itself against a standard set of 5G expectations and requirements.

New 5G technical standards ultimately enable the development of applications that can open new opportunities, inform business models, and transform everyday routines for several industries and users worldwide. Nevertheless, many of those standards were absent in the primary 5G rollouts, but they were to be applied in subsequent releases within the following years. Every technology provides new opportunities and challenges for the wireless market, and each market is one that the brand-new use cases envision. Service providers must tackle new application development to recognize these opportunities before they are entirely released.

Readiness Across Industries: A Diverse Landscape

The readiness for 5G applications reveals significant variations across industries, highlighting distinct levels of preparedness and investment in this groundbreaking technology. Sectors such as telecommunications and manufacturing exhibit higher readiness scores, driven by their commitment to harnessing 5G for innovation, operational excellence, and competitive advantage. The authors, Ding and Janssen (2018), explore the diverse requirements and challenges of 5G-enabled applications across vertical sectors such as smart mobility, energy, industrial IoT, and AR/VR, emphasizing the need for adaptable network architectures that can handle varying demands in latency, bandwidth, and security, while integrating emerging technologies like edge computing, blockchain, and virtualization. Tahir et al. (2020), present an in-depth review of Blockchain's application in 5G and beyond networks, proposing a taxonomy of its uses and demonstrating how Blockchain can address key challenges like decentralization, security, and interoperability, particularly in supporting enabling technologies such as SDN, NFV, and edge computing across front-haul, edge, and core network layers.

These industries have adopted 5 G-enabled solutions, including automation, Internet of Things (IoT) applications, and advanced analytics, underscoring their strategic focus on future-proofing operations and unlocking new opportunities. Chen et al. (2022) illustrate that 5G's ultra-low latency and real-time communication capabilities are revolutionizing telemedicine, enabling remote surgeries and facilitating access to healthcare in underserved areas.

In contrast, sectors such as agriculture and education are gradually integrating 5G into their ecosystems, constrained by infrastructural and financial limitations. The disparity reflects the challenges some industries face in overcoming barriers to entry, such as high deployment costs, technical complexities, and regulatory hurdles. Nevertheless, the transformative potential of 5G across diverse verticals is undeniable. 5G supports seamless telehealth consultations and advanced medical imaging in healthcare, enhancing patient outcomes. Similarly, 5G enables autonomous vehicles and vehicle-to-everything (V2X) communication in the automotive sector, ensuring safer and more efficient transportation systems.

This variance in 5G readiness emphasizes the need for tailored strategies and investments across industries. Governments and private stakeholders must collaborate to foster equitable access to 5G technologies, bridging the gap between high- and low-readiness sectors. By addressing these disparities, 5G can fulfill its promise of driving innovation and global economic growth. Qualcomm (2020) projects that 5G technology could generate trillions in economic output over the next decade through innovation and productivity gains.

For example, the Healthcare and energy industries are also progressively aligning their infrastructure and processes to benefit from 5G's low latency and better reliability. The automotive and logistics sectors are preparing to utilize 5G for advancements in automation and connected systems, while consumer-focused industries are gearing up to deliver enhanced experiences. This readiness underscores the diverse potential of 5G across various economic sectors.

The adoption of 5G optimizes both short-term and long-term establishments within the ecosystem and markets. 5G enhances existing services and enables new use cases, such as autonomous vehicles, uninterrupted video, zero-latency virtual reality, immersive entertainment, and latency-free gaming. On the manufacturing front, 5G is crucial to realizing and expanding the full potential of intelligent technologies, such as the Internet of Things (IoT), with its influence on smart homes, innovative industries, and smart cities. Zanella et al. (2018) highlight that smart cities benefit significantly from 5G's high-speed and low-latency capabilities, enabling efficient resource management and enhanced citizen services.

Sticking to first-quarter sales of 37 million first-generation 5G smartphones this season, with shipments just commencing, global shipments surged to 120 million units in 2020. This rollout is the fastest for a brand-new wireless generation, generating six times more device shipments than the earlier record holder, LTE, which is more than in a similar timeframe. With significant market momentum and positioning, 5G smartphone shipments are expected to surge to over 525 million units in 2023.

Table 1 highlights the varying degrees of 5G readiness across different markets and industries. It outlines the infrastructure and investment requirements necessary for full deployment, as well as disparities between developed and emerging economies. The table also examines industry-specific preparedness, showcasing leaders in sectors such as telecom and automotive, while identifying lagging sectors, including agriculture and education. Additionally, it explores the role of technological standards, integration with emerging technologies, and the impact of consumer adoption. This table provides an overview of the global 5G landscape and the collaborative efforts needed for inclusive digital transformation.

Table 1. 5G readiness across markets and industries

Area of Readiness	Description
Infrastructure & Investment Needs	High deployment costs and complex setups introduce risks, including cybersecurity threats. Regional regulatory differences complicate implementation.
Market Readiness by Country	Developed countries, such as the United States, South Korea, and China, lead with robust infrastructure. Emerging economies face challenges in accessing spectrum and navigating regulations.
Industry Readiness	The telecom, manufacturing, and automotive sectors demonstrate high preparedness. Healthcare also advances with the integration of telemedicine and remote operations.
Challenges in Low-Readiness Sectors	Sectors such as agriculture and education lag due to financial and infrastructural constraints.
Technological Standards & Evolution	Initial rollouts lacked full standards. Subsequent releases introduce capabilities to support new applications, shaping business models and services.
Integration with Emerging Tech	Blockchain, edge computing, Software-Defined Networking (SDN), and Network Function Virtualization (NFV) are key enablers for decentralized, secure, and interoperable 5G networks.
Consumer Impact & Adoption	5G enables immersive entertainment, VR/AR, and real-time gaming. 5G smartphone shipments surged from 37 million in early rollouts to over 525 million by 2023.
Government & Private Sector Role	Public-private partnerships are crucial for achieving equitable access and bridging readiness gaps across sectors and regions.
Long-Term Ecosystem Potential	5G drives smart cities, smart homes, and the Industrial Internet of Things (IIoT). Key to digital transformation across verticals.

THE CHALLENGES AND OPPORTUNITIES IN 5 G DEPLOYMENT

The deployment of 5G networks is sparking optimism and concern among industry leaders as they navigate the significant challenges and opportunities it presents. Deutsche Telekom's Uwe Janssen, Vice President of Research and Innovation, highlighted the substantial infrastructure investments required for 5G deployment. He emphasized that while shareholders demand improved returns on capital investments, the industry's ability to deliver satisfactory earnings remains under scrutiny. It underscores the financial strain on operators to justify the heavy expenditures associated with 5G technology.

Similarly, Berit Svendsen, CEO of Telenor Norway, noted that the telecommunications sector has been experiencing stagnant profits for years. Svendsen stressed the importance of identifying profitable business models and practical use cases for 5G to navigate this economic challenge. Telenor Norway has taken a proactive approach, focusing on innovative applications such as autonomous vehicles, fish farming, and remote surgery, which have the potential to drive revenue and create new market opportunities.

Vodafone Group's Director of Technology for Architecture and Innovation, Matt Beal, highlighted the urgent need to capitalize on 5 G-enabled IoT solutions. He argued that the industry must move beyond mere anticipation and implement practical applications, particularly those that leverage 5 G's high-speed, low-latency capabilities. This shift is critical for demonstrating the technology's tangible benefits and addressing skepticism surrounding its economic viability.

Despite the challenges, major telecom vendors like Cisco, Ericsson, Huawei, Nokia, ZTE, and Samsung are actively working to unlock the profitability of 5G applications globally. By transitioning from LTE networks to virtualized platforms with purpose-built operating systems, these companies are paving the way for the next evolutionary step in telecommunications. Their efforts aim to demonstrate that 5G is a technological advancement and a transformative tool with immense economic potential.

The Global Mobile Suppliers Association (GSA), a not-for-profit organization representing key stakeholders in the mobile ecosystem, plays a vital role in advancing 5G adoption. While the GSA has faced criticism for slow LTE adoption, it has significantly expanded its product offerings. By July 2017, the association reported 8,623 LTE products available worldwide, signaling progress in meeting market demands. As 5G adoption accelerates, organizations like the GSA play a crucial role in fostering collaboration and addressing the technical and economic challenges that lie ahead.

In summary, the successful deployment of 5G will depend on the industry's ability to balance substantial infrastructure investments with innovative business models and applications. By addressing profitability concerns and demonstrating 5G's transformative potential, the telecommunications sector transforms challenges into opportunities for laying the foundation of a more connected 5G future.

Challenges of Monetization of the 5G Applications

The path to monetizing 5G applications is riddled with significant challenges, demanding operators to think beyond traditional revenue generation models. Unlike previous generations of mobile networks, simply charging consumers more for faster data may not suffice to achieve a satisfactory return on the extensive investments in 5G infrastructure. Although fixed wireless access (FWA) broadband opportunities exist in some scenarios, the scale of investment required for widespread 5G deployment makes achieving profitability particularly complex. Miller (2020) highlights that 35% of operators' roles are either entirely new or significantly upgraded with the advent of 5G applications, emphasizing the need for innovative strategies to unlock the economic potential of this transformative technology.

A key challenge is adapting 5G's unique capabilities to generate new revenue streams. It includes developing business models that leverage network slicing, IoT integrations, and ultra-low latency for industry-specific use cases, such as smart factories, telemedicine, and autonomous vehicles. The monetization process also requires robust mechanisms for effectively pricing these services and convincing enterprises of their value. Moreover, operators must navigate the complexities of tailoring services to meet the diverse needs and expectations of various industries.

Another hurdle is the competition from over-the-top (OTT) players and tech giants, who often capture a significant portion of the value generated by advanced connectivity. Operators must work to create differentiated offerings that go beyond connectivity, delivering end-to-end solutions for industries while managing capital and organizational investments effectively. Successful 5G monetization will depend on an operator's ability to innovate, form strategic partnerships, and implement dynamic pricing models that align with customer demands and market trends. Without this, the potential of 5G may remain underutilized, leaving operators struggling to justify their investments.

Architectured for Monetization

As technology companies prepare to capitalize on 5 G's transformative potential, the monetization framework is built on key performance components to unlock diverse revenue streams. Enhanced mobile broadband (eMBB) provides near-fiber-like experiences over wireless radio connections, enabling high-demand applications such as ultra-high-definition video, 3D content delivery, and fixed-wireless broadband. These capabilities cater to a growing consumer appetite for immersive media experiences, presenting operators with significant monetization opportunities.

Ultra-reliable, low-latency communications (URLLC) are integral for use cases that demand minimal latency and high reliability. The ecosystem relies on Applications such as augmented reality (AR), virtual reality (VR), and online gaming, which thrive, as do mission-critical services like vehicle-to-vehicle communication and remote surgeries. The ecosystem's applications require not only low latency but also a high degree of reliability to ensure safety and effectiveness. Integrating mobile edge computing amplifies these use cases, enabling real-time processing and reducing dependency on centralized cloud infrastructures. The integration of cloud infrastructure is transforming how 5G networks are architected and operated.

The new emergence of Massive connectivity, a cornerstone of 5G, supports an IoT-driven world where over a million devices can connect seamlessly within a single square kilometer. This capability fosters the proliferation of Industry 4.0 applications, such as advanced manufacturing systems, autonomous logistics, and intelligent urban planning. Applications such as Smart cities can benefit significantly from this densified network, enabling efficient resource management and improving residents' quality of life. Li and Dai (2021) highlight that scalable IoT ecosystems hinge on 5G’s robust connectivity, while Kaur and Singh (2021) emphasize 5G’s pivotal role in advancing machine-to-machine communication in industrial contexts.

Essentially, 5 G's architecture addresses various use cases, providing tailored solutions that drive revenue generation across sectors. The ability to adapt these capabilities to diverse industries ensures that 5G remains at the forefront of technological and economic evolution.

Although monetization is a critical component of 5G's adaptation potential, the 5G system's structural attributes empower specialist organizations to enhance its execution. First, however, tap into the enormous opportunities for adaptation. More than in past courses, 5G has evolved to adopt the computerized economy and move specialist communities away from providing access to the advanced environment and focusing on the value of the incentives in an automated economy. For this, three principles are most commonly used by 5G help specialist organizations in their adaptation methodologies.

Challenges in Monetizing 5G Applications: A Focus on High-Value Use Cases

The monetization of 5G applications has predominantly centered on high-value services, including augmented reality (AR), virtual reality (VR), robotics, and autonomous driving. While these areas highlight the transformative potential of 5G, this emphasis on premium applications can inadvertently hinder its broader adoption across mass-market industries. A notable disparity exists between the high-cost infrastructure investments and the affordability required for widespread connectivity solutions. As of the first quarter of 2017, Ovum reported 2.16 billion LTE subscriptions globally, underscoring the significant market penetration of previous-generation technologies compared to the nascent adoption of 5 G.

A critical factor in this challenge is the definition of 5G technology itself. Early 5G deployments, such as fixed wireless access using millimeter-wave spectrum bands in the United States, raise questions about whether these services truly embody the full promise of 5G or merely represent an evolution of point-to-point microwave technology. This ambiguity in defining 5G's scope complicates efforts to market and scale the technology effectively, as stakeholders may hesitate to invest in an unclear value proposition.

Additionally, the competition between LTE and 5G technologies poses another barrier to adoption. Advanced LTE technologies, including Narrowband IoT (NB-IoT) and Category M (Cat M), continue to capture market share in the IoT space, offering cost-effective solutions for applications that 5G also targets. Similarly, gigabit LTE solutions provide robust connectivity, further narrowing the market for 5G in specific sectors. These factors collectively contribute to a slower-than-expected rollout of 5G in mass-market use cases.

The 5G market development trajectory underscores the need for continuous monitoring and validation of its capabilities and use cases. Organizations like the Advanced Composites Group (ACG) play a crucial role in evaluating the deployment of mobile-related technologies, including 5G, LTE-Advanced, Small Cells, and Multi-Access Edge Computing (MEC). By addressing these challenges and aligning 5G applications with broader market needs, the industry can achieve a more balanced approach to monetization, ensuring both high-value and mass-market use cases are effectively served.

Challenges of Safety Issues

5G technology poses a heightened security threat because adversaries have more vectors to attack. In addition, technological development opens the door to a substantial number of connected devices, collectively referred to as the Internet of Things (IoT), and challenges related to safety issues.

This challenge becomes critical with any data-driven technology. However, the 5G rollout must deal with sophisticated and standard cybersecurity threats. For example, although 5G falls under the Key Agreement and Authentication (KAA) system, which builds trust between networks, observing individuals nearby using their phones would be easy. They may even eavesdrop on live telephone calls.

The onus is currently on carriers and network consortia to provide a digital backup for customers; user complacency may be the only challenge.

Connectivity improves, with data speeds likely to be magnitudes faster than current levels. This forces cloud-based and data virtualization offerings to be as airtight as possible to protect consumer data and privacy. Similarly, users must be cautious and aware as data stewards.

Mobile network operators have a business incentive to concentrate on security. For example, automation and manufacturing digitization are expected to generate approximately $619 billion in new revenue opportunities (a 36% growth in addressable revenue) for operators by 2026, according to Ericsson, a 5G network supplier. According to Margaris (2019), enabling network security features for industries is a fundamental differentiator for these services in the rush to implement these changes.

He predicted that organizations focus intensely on the system security that provides and connects to the 5G applications as a crucial element in selecting their service or slice provider. According to Margaris, operators can become secure digital business enablers in the 5G era by offering real value beyond connectivity.

The safety issues originate from the capability to supply on-demand service creation and self-service; instead of developing a fixed selection of network slices, operators can hand over to businesses to fashion their service, such as security features. Margaris said differentiated 5G networks are critical in recording this industry opportunity. Security is a crucial differentiator. Subsequent evolutions in network slicing require additional security features and greater flexibility in how various protection features can be attached to a specific slice.

The bottom line is that 5G is on the horizon and promises revolutionary change for the wireless industry. However, security must still be developed.

Figure 1. Challenges of monetization of the 5G applications

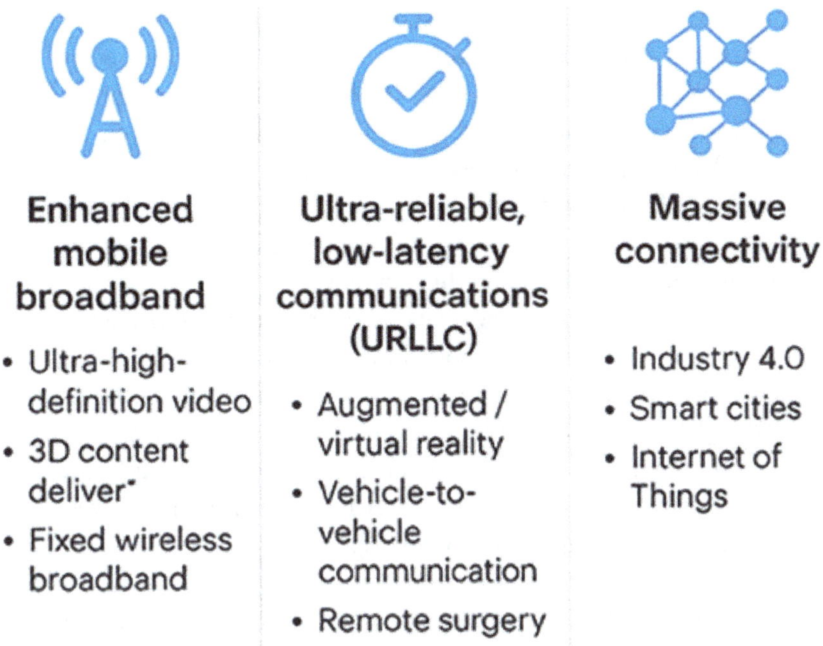

VIRTUALIZATION AND SYSTEM CUTTING

This section explores the role of virtualization and network slicing technologies in enabling flexible, scalable, and efficient 5G network architectures, which are critical to meeting the diverse demands of modern applications

Virtualization for Modern Telecommunications

Virtualization has emerged as a transformative force in modern telecommunications. It revolutionizes service delivery models by enabling a product-driven approach with exceptional speed and flexibility. Service providers can transition from traditional hardware-dependent systems to dynamic, software-based frameworks through virtualization, unlocking unprecedented agility and scalability. This shift allows them to adopt a "fail-fast" methodology, where rapid innovation and iterative testing are prioritized with minimal risk. It supported an environment of continuous improvement and swift adaptability.

One critical advantage of virtualization is its ability to optimize operational costs by reducing reliance on proprietary hardware and enabling resource sharing across multiple virtual networks. This cost-effectiveness is complemented by enhanced scalability, as virtualized systems can easily accommodate fluctuations in demand without requiring substantial infrastructure overhauls. Moreover, virtualization accelerates the time to market for new services, enabling providers to dynamically launch, modify, and refine offerings in response to evolving consumer needs and market trends. Dixit and Rani (2025) analyze trends and integration challenges between 5G and IoT technologies in this context.

Virtualization's flexibility extends beyond operational efficiencies. It empowers service providers to diversify their portfolios by introducing innovative solutions tailored to specific market segments, such as IoT applications, edge computing services, and high-reliability communication systems. By integrating virtualization into their core strategies, providers can leverage their agility to explore new revenue streams, enhance customer satisfaction, and maintain a competitive edge in the rapidly evolving telecommunications landscape.

In summary, virtualization serves as a cornerstone of digital transformation in telecommunications, redefining how services are conceptualized, developed, and delivered. Streamlining processes, optimizing resources, and fostering innovation highlight its pivotal role in shaping the future of the 5G era and beyond.

System Cutting: Revolutionizing Network Efficiency and Applications

System cutting, also known as network slicing, is a groundbreaking technological innovation. It enables partitioning a single network framework into multiple virtual instances to maximize system capabilities. This approach allows network operators to create tailored slices for specific applications or user groups, each optimized to meet distinct performance and service requirements. This method can enhance service delivery across various parameters, including latency, bandwidth, reliability, and security, thereby ensuring customized experiences that meet diverse user demands.

A core advantage of system cutting is its adaptability to various applications and requirements. Each network slice operates as an independent virtual network, allowing service providers to allocate resources dynamically based on different applications or customers' unique needs. For instance, a slice designed for IoT devices may prioritize low power consumption and connectivity for massive devices, while another focused on augmented reality applications would emphasize ultra-low latency and high data rates. This flexibility optimizes resource utilization and aligns network performance with the specific demands of various industries.

System cutting also opens new avenues for monetization through the "System Slice as a Service" model. This approach enables service providers to offer customized network slices as premium solutions for enterprise clients, catering to the healthcare, manufacturing, and autonomous transportation industries. Additionally, consumer-facing services can benefit from differentiated network slices, which offer enhanced user experiences and provide a competitive advantage in the market.

The integration of system cutting into 5G networks also supports advanced use cases, such as smart cities, remote surgeries, and real-time analytics, by ensuring that each slice delivers the precise performance metrics required. This enables the seamless coexistence of diverse applications on a single network infrastructure, promoting operational efficiency and facilitating the rapid deployment of emerging technologies.

In conclusion, system cutting represents a transformative approach to network management. It empowers providers to deliver highly specialized services, improve operational efficiency, and unlock new revenue streams. As the telecommunications market continues to evolve, this 5G technology stands as a cornerstone for enabling the full potential of 5G and beyond.

NAVIGATING THE CHALLENGES IN 5G APPLICATION DEVELOPMENT

The advancement of 5G applications is confronted with complex technical, operational, and economic challenges. One of the primary obstacles is the effective implementation of network slicing—a fundamental element of 5G—which must meet a wide range of Quality of Service (quality of service) demands.

They are tailored to support various industry verticals, such as IoT, enhanced mobile broadband (eMBB), and ultra-reliable low-latency communications (URLLC). Designing and managing these slices demands advanced orchestration tools and real-time resource allocation for optimal performance and scalability.

The exponential increase in connected devices and data volumes further compounds the complexity of 5G application development. Managing this growth requires robust network scalability and adaptive resource allocation frameworks to handle the dynamic demands of users and devices. These challenges highlight the need for cutting-edge technologies, such as AI-driven network optimization and edge computing, to ensure seamless functionality across a dense and diverse ecosystem.

Economic barriers also significantly limit 5G adoption. Small and medium-sized enterprises (SMEs), often constrained by budgetary limitations, may struggle to invest in the necessary infrastructure and applications. Bridging this gap requires innovative pricing models and support mechanisms to make 5G technology accessible across various operational scales.

Moreover, expanding the 5G ecosystem introduces heightened cybersecurity and data privacy risks. The interconnected nature of 5G networks increases the potential attack surface, necessitating the implementation of advanced security measures to protect sensitive data and maintain user trust. Robust cybersecurity frameworks and strict regulatory compliance are crucial for mitigating these vulnerabilities.

Addressing these multifaceted challenges is critical to fully realizing the transformative potential of 5G technology. Strategic innovation and stakeholder collaboration can surmount these obstacles, paving the way for the widespread adoption and integration of 5G applications across industries.

The 5G Revolution in the Media Industry

Integrating 5G technology into the media business represents a transformative leap, reshaping how content is created, distributed, and consumed. Intel's analysis projects that 5 G-enabled capabilities could drive annual industry revenues to an impressive $420 billion by 2028. Service providers who embrace this innovation gain a critical competitive edge, leveraging 5G's advancements to redefine the media and entertainment landscape.

One of the most groundbreaking aspects of 5G's impact is its facilitation of immersive media experiences through augmented reality (AR) and virtual reality (VR) applications. These technologies, empowered by 5G's ultra-low latency and high-speed connectivity, are projected to generate over $140 billion and cumulative revenues between 2021 and 2028. By enabling real-time interactions with virtual environments, 5G allows for a new level of consumer engagement, particularly in sectors such as AR gaming, live events, and interactive storytelling. Lee (2024) highlight how 5G opens doors to innovative AR applications, enhancing user experiences with seamless, high-quality performance.

Moreover, augmented reality offers a new way for individuals to interact with media, blending virtual characters, objects, and informative overlays into real-world settings. This integration requires producing high-quality 3D content and creating platforms that can effectively deliver and manage this content. Media companies face the challenge of adapting their infrastructure to handle the increased complexity and bandwidth demands of such applications.

In essence, 5G's role in the media industry is a technological advancement and a paradigm shift. It enables a dynamic and interactive media ecosystem to meet the demands of modern consumers. By empowering these capabilities, media organizations can unlock unprecedented opportunities for growth and innovation.

The Transformation of the Advertising Market

The evolution of mobile advertising is reaching new heights with the advent of 5G technology, poised to redefine the $178 billion market forecasted for 2028. By enabling unprecedented connectivity and performance, 5G is paving the way for innovative and engaging advertising experiences that surpass traditional formats. This 5G transition marks a shift from conventional display ads on social and general media platforms to interactive, immersive campaigns that leverage the unique capabilities of 5G networks.

One of 5 G's most significant advertising impacts is its influence on video content. The enhanced network capabilities of 5G are driving an 85% compound annual growth rate (CAGR) in video segment revenues between 2021 and 2028. This surge creates vast opportunities for television networks, streaming platforms, and service providers to compete directly with IPTV, cable, and satellite TV. The ability to stream high-quality, low-latency video transforms how advertisers engage with audiences, offering seamless integration of video ads into various media.

In the business-to-business (B2B) space, 5G networks enable the expansion of mobile media by providing over-the-top (OTT) service providers with enhanced capabilities. The 5G By delivering faster speeds, lower latency, and better reliability, 5G facilitates advanced advertising formats, such as virtual reality (VR) campaigns and augmented reality (AR), creating deeply engaging consumer experiences. This evolution also introduces tactile elements to media, incorporating haptic feedback in VR applications. As 5G offers more sophisticated immersive experiences, 5 G-enabled VR advertising is projected to generate over $5 billion annually by 2028.

The gaming industry exemplifies the transformative potential of 5G, with online gaming platforms leveraging reduced latency and enhanced streaming speeds to deliver an improved gaming experience. The Augmented reality games are expected to dominate the 5G gaming market, accounting for 90% of revenue by 2028, equivalent to $35.7 billion globally. Cloud gaming is expected to further capitalize on 5G advancements, surpassing $100 billion annually by 2028, as gamers increasingly demand faster, more responsive, and visually rich experiences.

Ultimately, 5G's role in advertising extends beyond technological improvements. The 5 Gt offers a dynamic framework for innovation and engagement. By facilitating new forms of interaction and enriching the user experience, 5G is transforming advertising and shaping the future of media consumption and entertainment.

New Immersive Experiences With 5G: Transforming Consumer Engagement

5G technology heralds a transformative era for immersive experiences. It unlocks industry opportunities and sets a foundation for applications expected to generate $67 billion in annual revenue by 2028. This figure aligns with the revenue of the entire global mobile media market, encompassing video, music, and gaming, as of 2017, highlighting the unprecedented economic potential of 5 G-enabled advancements.

The 5 G-driven immersive experiences facilitate retail and emerge as critical beneficiaries. According to Miller (2020), augmented reality (AR) shopping is projected to engage 100 million consumers by next year. Retailers are increasingly integrating AR into their physical and e-commerce platforms, offering shoppers an enhanced, interactive experience. AR technology overlays digital imagery in practice, allowing customers to access virtual products in their environment or try on items virtually through advanced graphical simulation. Nearly half of retailers plan to implement AR or VR technology soon, reflecting a shift toward personalization and innovation in consumer engagement, (Mitra & Das, 2020).

Beyond retail, 5 G-powered immersive experiences extend into entertainment, education, and healthcare. Augmented and virtual reality applications, enabled by 5 G's high-speed and low-latency capabilities, facilitate hyper-realistic simulations, ranging from virtual showrooms to remote medical consultations. These 5G applications leverage 5G's ability to support the real-time rendering of complex visuals, enhancing their utility and accessibility across sectors. For example, 5G's bandwidth ensures seamless streaming of AR experiences, creating opportunities for businesses to provide tailored services and build deeper connections with consumers.

In the gaming industry, 5G technology amplifies the integration of AR and VR, enabling players to engage in rich, interactive environments with minimal latency. Similarly, educational platforms can leverage immersive experiences to create dynamic learning environments, such as virtual field trips and hands-on simulations, thereby bridging the gap between theoretical and practical knowledge.

While 5G drives these technological breakthroughs, its success relies on the infrastructure and investment to support widespread adoption. The convergence of AR, VR, and 5G lays a pathway for new consumer experiences that redefine engagement, stimulate economic growth, and expand the boundaries of what digital interactions can achieve. This shift transforms industries and sets a precedent for the future of immersive technology in daily life.

Bandwidth and Low-Latency Communications in 5G: Transforming Media and Connectivity

The evolution of mobile advertising is reaching new heights with the advent of 5G technology, poised to redefine the $178 billion market forecasted for 2028. The 5G By enabling unprecedented connectivity and performance, 5G is paving the way for innovative and engaging advertising experiences that surpass traditional formats. This transition marks a shift from conventional display ads on social and general media platforms to interactive, immersive campaigns that leverage the unique capabilities of 5G networks.

5G can offer considerably greater detail bandwidth and low-latency communications. These support the increased need for high-resolution video more than mobile networks, enabling more immersive modalities such as 360-degree video. However, advanced video compression algorithms are needed to use the increased bandwidth in the new 5G networks.

One of 5 G's most significant advertising benefits is its influence on video content. 5 G's enhanced network capabilities are driving an 85% compound annual growth rate (CAGR) in video segment revenues between 2021 and 2028. This surge creates vast opportunities for television networks, streaming platforms, and service providers to compete directly with IPTV, cable, and satellite TV. The ability to stream high-quality, low-latency video transforms how advertisers engage with audiences, offering seamless integration of video ads into various media.

In the business-to-business (B2B) space, 5G networks enable the expansion of mobile media by providing over-the-top (OTT) service providers with enhanced capabilities. By delivering faster speeds, lower latency, and improved reliability, 5G enables advanced advertising formats and innovations, such as augmented reality (AR) and virtual reality (VR), creating deeply engaging consumer experiences. This evolution also introduces tactile elements to media, incorporating haptic feedback in VR applications. As 5G technology evolves, immersive experiences become more sophisticated, and 5 G-enabled VR advertising is projected to generate over $5 billion annually by 2028.

The gaming industry exemplifies the transformative potential of 5G, with online gaming platforms leveraging reduced latency and enhanced streaming speeds to deliver an improved gaming experience. Augmented reality games are expected to dominate the 5G gaming market, accounting for 90% of the revenue by 2028, which is equivalent to $35.7 billion globally. Cloud gaming is expected to further capitalize on 5G advancements, surpassing $100 billion annually by 2028, as gamers increasingly demand faster, more responsive, and visually rich experiences.

Ultimately, 5G's role in advertising extends beyond technological improvements. It offers a dynamic framework for innovation and engagement. By facilitating new forms of interaction and enriching the user experience, 5G is transforming advertising and shaping the future of media consumption and entertainment.

State-of-the-Art Video Compression Algorithms: Enabling High-Resolution Media in 5G Networks

Among the critical drivers for next-generation mobile communications, 5G can support the deployment of the Internet of Things (IoT), enabling the connection of vast numbers of objects to the web with very low latency. For instance, 5G technology allows the realization of smart cities, big data applications, and intelligent environments. The terrestrial solutions in the 5G framework augment the latest High Throughput Satellite (HTS) devices and mega Low Earth Orbit (LEO) satellite constellations. In this section, we examine the integration of 5G technology and IoT using an aerial element composed of satellites and drones in a rural setting. 5G enhances operational efficiency by integrating industry devices and systems.

State-of-the-art video compression algorithms can help reduce the bandwidth required for high-resolution video streams at lower nominal rates. They can be the primary elements of 5G video-over-mobile services. The improved bandwidth capability and reduced latency of 5G networks enable more access to complex immersive audiovisual experiences than mobile networks. The accessibility of commercial-grade head-mounted displays, combined with the help from mobile phone and tablet products, has created a thriving marketplace, a massive customer audience for virtual and augmented reality, and 360-degree clip content, surpassing mobile networks.

The proposed system can offer enhanced services like fire alarm detection, smart agriculture, animal tracking, and plant disease control.

Smart Farming With Drones

The prospects of using uncrewed aerial vehicles (UAVs) or drones for smart farming are immense. UAVs are primarily controlled by remote controllers that use radio waves. Several technologies, such as Wi-Fi or Zigbee, are used to control UAVs. However, Smart Bluetooth is a wireless technology that transfers data over short distances. Moreover, Smart Bluetooth is more cost-effective than other technologies and is available on nearly every smartphone. Therefore, farmers can use smartphones to operate their respective UAVs and Bluetooth Smart-enabled agricultural sensors in the future.

The UAV monitoring system integrates various elements, including a UAV operator, ground control station, and satellite communication, to coordinate activities across military, agricultural, surveillance, and aerial operations. These UAVs rely on data collected from diverse sensors, transmitted through base stations and the internet, to facilitate real-time analysis and response. The system ensures connectivity with rescue agencies, enabling efficient communication and coordination during emergencies. Through this networked infrastructure, UAVs enhance situational awareness and operational efficiency. The diagram illustrates the interconnected flow of data and control, highlighting the multi-purpose applications of UAVs (Agrawal & Cleland-Huang, 2021).

UAVs' main advantages for innovative agricultural applications include their versatility in variable climate conditions and ability to capture high-resolution pictures from various ranges (the average range is 50 to 100 meters). They can also determine and monitor crop quality and attacks by pests, weeds, and animals. Furthermore, farmers and other partners can access the information gathered by UAVs from cloud-based platforms remotely through applications on their smart devices, which can help predict crop yields and determine the requirements for pesticides, fertilizers, and seed sowing.

THE CASE OF 5 G AND DRONES

Governmental and commercial uses include aerial digital photography, construction, energy, agriculture, security, safety, and strategic businesses, whose possibilities are often additionally exploited by Beyond Visual Line of Sight (BVLOS) technology.

Nations worldwide are amending drone policies to permit uncrewed aerial vehicles (UAVs) to fly beyond the Visual Line of Sight (VLOS) for optimum effectiveness. Unlike VLOS flights operating on the pilot's line of sight, BVLOS flights fly beyond the visible range. As a result, BVLOS capabilities allow drones to cover much greater distances.

Beyond Visual Line of Sight (BVLOS) has several cost-effective applications. It enables service providers to perform complex drone operations and facilitates the autonomous flight of drones without human intervention.

Nevertheless, drones must become' airworthy' to make this happen. BVLOS flights typically utilize infrastructure inventory to inspect significant areas and generate orthophotomaps. One obstacle and reason for strict laws on BVLOS flights is that the pilot should always intervene when things go entirely wrong. This law can only be accomplished if you think there is a reliable and permanent link via the drone pilot. Promising know-how to devise solutions for the challenges faced by the cellular network, particularly the new 5G model.

The graphic below illustrates the evolution of technologies that link the drone to the pilot. At the start of the 2000s, municipal worker bees generally had an uplink interconnection. However, customer requests for improving the real-time handheld remote control and video clip streaming (up and downlink) were considered before worker bees were considered for business use.

With increasing flight endurance, the scale of information sets, and video streaming capabilities, alternative command abilities have been required to overcome today's RF limits. In 2015, the first application of the 4G movable details system, as a regulatory and information transmission boundary, nevertheless restricted the exploitation of other drone features until the 2010s.

The evolution of drone connectivity marks a transition from fundamental radio frequency (RF) control to advanced 5G networks. Early drones relied on RF for uplink-only communication and Wi-Fi for bidirectional data transmission, including video and remote control. The introduction of 4G/LTE enabled cellular network connectivity, providing higher data rates and improved reliability. With 5G, drones experience a bitrate up to 100 times higher and five times lower latency than 4G/LTE, allowing for seamless communication and enhanced real-time operations. This advancement positions 5G as a critical enabler for the next generation of drone applications, (Schroth, 2020).

Drone Applications

Initially known for military use, drones are now helping large companies, SMEs, and individual entrepreneurs accomplish other tasks. According to Juniper's investigation, by 2020, several industrial market regulations will have purchased 16 million drones, and the world will have fully embraced drone technologies. Below are ten unique applications of drone technologies.

The potential applications for drones powered by 5G connectivity are vast and transformative. Enhanced aerial photography and entertainment experiences can be achieved through real-time data streaming and high-definition imaging capabilities.

Drones can revolutionize logistics and delivery, ensuring faster and more efficient services. Geographic mapping and disaster management applications benefit from precise and instantaneous data transmission, enabling the development of effective response strategies. Drones enable precision farming by providing real-time insights into crop health and resource distribution in agriculture. monitoring

The following figure illustrates how state-of-the-art video compression algorithms, combined with 5G infrastructure, enable the real-time delivery of high-resolution media while integrating IoT technologies and drones in connected environments, such as innovative farming and emergency management.

Figure 2. The case of a UAV monitoring system in agriculture

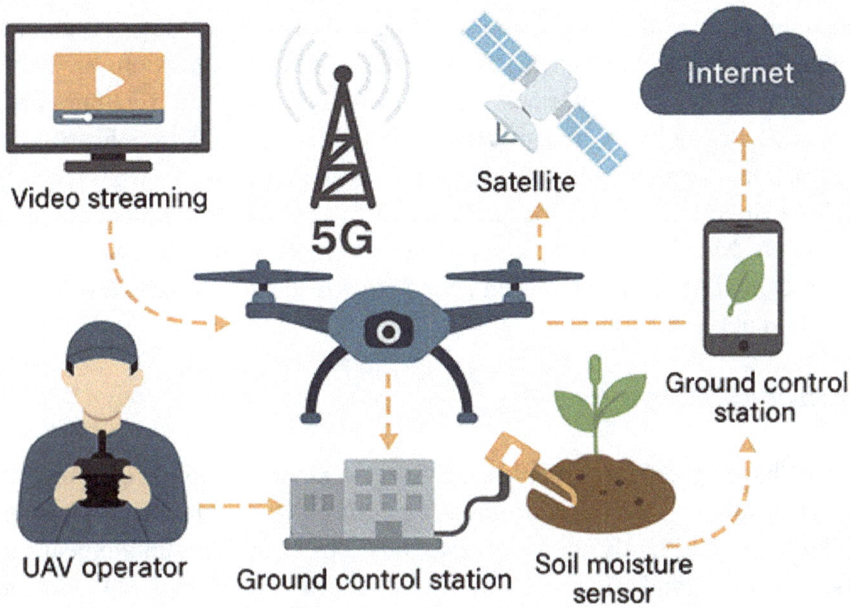

The ability of drones to locate individuals quickly in challenging terrains bolsters the effectiveness of search and rescue missions.

Additionally, drones provide accurate and timely data to support law enforcement, wildlife monitoring, and weather forecasting. These advancements underscore the pivotal role of 5G in enabling smarter, faster, and more versatile drone applications (Source: Allerin.com).

Aerial Photography

Drones have revolutionized aerial photography by offering a cost-effective alternative to helicopters and cranes, which have traditionally been used for capturing footage. The Aerial photography ability to maneuver through tight spaces and achieve unique angles has made them indispensable in filming action-packed scenes and science fiction movies. In real estate, drones are widely used to showcase properties from dynamic perspectives, enhancing marketing efforts. Similarly, sports increasingly utilize them to capture live-action from unconventional viewpoints. Journalists utilize drones to capture real-time footage during live broadcasts, providing audiences with a broader perspective of events. These flying devices offer unmatched flexibility and creativity in visual storytelling across industries.

Shipping and Delivery

The integration of drones into the shipping and delivery industry is spearheaded by companies such as Amazon, UPS, and DHL. These organizations are exploring drone technology to save labor costs and reduce road congestion. Drones are ideal for short-distance delivery, transporting lightweight items like medications, food, beverages, documents, and other small packages. Their ability to bypass road traffic and deliver directly to a location makes them particularly advantageous in urban environments. Moreover, drones can operate in remote or inaccessible areas, ensuring that essential goods reach those in need quickly and efficiently, revolutionizing logistics and delivery systems.

Geographic Mapping

Drones have become invaluable tools for geographic mapping, enabling professionals and enthusiasts to capture high-resolution data in previously inaccessible areas. They can efficiently survey remote locations such as mountain peaks, dense forests, coastlines, and islands. Drones facilitate the creation of 3D maps, aiding urban planning, environmental monitoring, and disaster management. Additionally, their use in crowdsourced mapping projects helps build comprehensive global maps. Their versatility and precision have made drones a cornerstone for geospatial analysis, benefiting fields like archaeology, agriculture, and conservation.

Disaster Management

Drones are highly effective in disaster management, offering swift access to areas affected by natural or artificial catastrophes. Equipped with high-definition cameras, radars, and sensors, they assist rescue teams in locating injured victims by navigating through rubble and debris. Their small size allows them to operate in areas inaccessible or hazardous for larger aerial vehicles, such as helicopters. Using drones reduces the reliance on human-crewed operations, conserving resources and enhancing efficiency. Ali and Khan (2021) highlight that 5G technology further supports these mission-critical applications, enhancing drone capabilities for public safety and disaster response.

Precision Agriculture

Drones are revolutionizing agriculture by providing cost-effective solutions for crop monitoring. Their infrared sensors help identify crop health issues, enabling farmers to apply targeted remedies, such as insecticides or fertilizers, as needed. This precision reduces waste, improves crop management, and increases yields. Experts predict that drones will dominate nearly 80% of the agricultural sector in the coming years, becoming indispensable tools for modern farming practices.

Search and Rescue

Drones equipped with thermal sensors are crucial for search and rescue missions, particularly in challenging terrain or adverse weather conditions. Their night vision capabilities enable them to locate missing individuals or disaster victims quickly and efficiently. Additionally, drones deliver essential supplies, such as food, water, and communication devices, to stranded individuals before rescue teams arrive, thereby significantly increasing their chances of survival in remote or inaccessible areas.

Weather Forecast

Drones are emerging as critical tools in weather forecasting. They enable scientists to monitor and analyze hazardous weather phenomena, such as hurricanes and tornadoes. Their affordability allows them to gather valuable data on environmental parameters and storm trajectories, contributing to more accurate predictions and minimizing disaster-related risks.

Wildlife Monitoring

Drones are instrumental in wildlife conservation, particularly in deterring poachers and monitoring endangered species such as rhinos and elephants. With thermal sensors and high-resolution cameras, drones offer round-the-clock surveillance that minimizes disturbance to wildlife. They also provide valuable insights into animal behavior, habitat patterns, and migration routes, thereby aiding conservation efforts.

Law Enforcement

Law enforcement agencies use drones to monitor crowds, ensure public safety, and detect criminal activities. They are particularly effective in border patrols, tracking illegal activities like drug smuggling, and investigating fires. Drones offer a strategic advantage in ensuring security and responding swiftly to threats.

Entertainment

In the entertainment industry, drones are used for innovative applications, including aerial videography and photography. By capturing dynamic visuals, they add excitement to events such as nightclubs and live performances. Drones are also gaining popularity in gaming and virtual reality experiences, offering immersive and interactive entertainment options.

As drone technology advances, its applications expand across various industries, becoming increasingly indispensable for enhancing efficiency and driving innovation. Organizations must embrace this technology, develop the necessary infrastructure, and explore its potential to stay competitive in the evolving landscape.

CONCLUSION

Chapter 9 examined the multifaceted risks and challenges associated with developing and deploying 5G applications as 5G transitions from conceptual potential to operational reality, developers and stakeholders must navigate various technical, economic, regulatory, and security hurdles. Innovations such as network slicing, virtualization, and ultra-low-latency communication present significant opportunities but require precise orchestration and substantial investment. This chapter highlighted that balancing these advancements with robust cybersecurity frameworks, clear monetization strategies, and equitable market readiness is key to unlocking 5G's transformative power.

Developing immersive applications, autonomous systems, and IoT ecosystems also reveals disparities in industrial readiness and resource allocation. ITU (2022) highlights significant disparities in digital development across regions, emphasizing the urgent need for inclusive 5G deployment. From smart farming and drone navigation to the media and healthcare sectors, 5G's versatility introduces unique benefits and vulnerabilities. The table below synthesizes the critical challenges and strategic responses identified across industries and technologies in this chapter.

The following table outlines the key challenges faced in the development and deployment of 5G applications, along with the corresponding strategic responses adopted by the industry and policymakers. As 5G technologies expand across various sectors, they encounter a diverse range of technical, economic, and regulatory challenges. These challenges impact deployment speed, user trust, and overall system performance. Strategic interventions are thus essential to ensure secure, efficient, and scalable implementation. The table categorizes these issues and presents targeted solutions aligned with the global 5G development goals.

Table 2. Key challenges and strategic responses in 5G application development

Challenge Area	Identified Risk	Strategic Response
Security & Privacy	Expanded attack surface, unauthorized access	Security-by-design, encryption, and compliance with regulatory standards
Infrastructure Costs	High deployment and operational expenses	Public-private partnerships, modular deployments
Industry Readiness	Uneven adoption across sectors	Tailored business models and technology transfers
Monetization	Unclear ROI for mass-market applications	Dynamic pricing models, service-based architecture
Cyber Threats	Data breaches and AI exploitation	Edge security frameworks, intrusion detection systems
Regulatory Issues	Lack of harmonized global policies	Cross-border regulatory alignment and spectrum coordination

5G has led to one of the most significant digital transformations since World War II. Moreover, it has the potential to transform our digital environment.

To sum up, the digital transformation of 5G continues, with breakneck data transfer speeds, reduced latency, enhanced network capability, and improved energy efficiency, which will benefit 5G applications, wearables, and IoT devices.

Furthermore, Big Data and IoT applications require future wireless systems to integrate different quality-of-service requirements.

Recent research on 5G has revealed 44 diverse IoT use cases, estimating that these cases will account for 76 million 5G connections by 2025. Thus, several promising 5G use cases have the most significant potential in the industry.

There is a prevailing opinion in the manufacturing industry that 5G is best suited for connecting manufacturing robots with varying degrees of mobility, ranging from fixed collaborative robots to fully mobile robots, such as autonomous guided vehicles. Shayea et al. (2020) provide a comprehensive survey of mobility management challenges in 5G networks, identifying critical issues such as high path loss from millimeter waves, dense deployment of small base stations, and inefficient handover mechanisms, while discussing advanced techniques and optimization strategies aimed at ensuring seamless connectivity for users in highly dynamic environments.

LTE networks offer adequate performance for remote access to live and recorded HD video. However, adopting 5G enables the functionality improvement necessary for more real-time video content analytics and even the deployment of significant cameras.

Enhancing technical resources to achieve V2X, 5G technology allows for integrating "cloud-application-terminal" networks for communication, supervision, decision-making, and a new transport architecture, paving the way for innovative applications.

As utilities and energy companies leverage 5G technologies, essential infrastructure can produce more innovative solutions in usage, distribution, transmission, energy production, and the next wave of innovative grid functions and efficiency.

Huawei's petroleum and gasoline treatments are used in 45 countries and regions worldwide, helping 70% of the world's leading 20 petroleum and gasoline companies. In addition, 5G can boost revenue and efficiencies in the medical industry, enabling healthcare providers to establish networks more quickly and effectively to handle large volumes of data.

5G addresses numerous issues in logistics and transportation, impacting logistics and supply chain management for freight treatment buyers, cargo owners, and others in different ways, including benefits such as Improved Location Tracking and Route Optimization. For example, 5G enables businesses to track their valuable cargo across all areas by optimizing coverage in previously “dead zones.” In addition, improved geolocation technology permits greater visibility into unforeseen travel circumstances and delays, ultimately improving location intelligence and mitigating ensuing problems.

Future generation system technology in 5G networks may equip U.S. first responders with new resources to drive faster, more beneficial emergency treatment results, such as apps for paramedics. Current healthcare providers design apps that enable paramedics to accurately alert the emergency department of their imminent arrival. Choudhury et al. (2020) explain that the integration of 5 G with IoT devices transforms healthcare, particularly in remote patient monitoring. Chen et al. (2021) discuss how 5G allows for the seamless integration of AI and machine learning for advanced data analytics in healthcare.

Throughout 200 use cases, across ten various industries, and with a pair of G-enabled pick case clusters, 18% of the total ICT addressable business digitalization value, and 47% of the 5 G-enabled service providers could resolve B2B value.

REFERENCES

Agrawal, A., & Cleland-Huang, J. (2021). *RescueAR: Augmented Reality Supported Collaboration for UAV Driven Emergency Response Systems.* arXiv preprint arXiv:2110.00180.

Agrawal, V., Agrawal, S., Bomanwar, A., Dubey, T., & Jaiswal, A. (2023). Exploring the risks, benefits, advances, and challenges in internet integration in medicine with the advent of 5G technology: A comprehensive review. *Cureus, 15*(11), e48767. DOI: 10.7759/cureus.48767

Chen, Y., Zhang, H., Chen, J., Guo, Y., Jiang, P., Gao, F., Bao, H., & Huang, X. (2022). Thermally Conductive but Electrically Insulating Polybenzazole Nanofiber/Boron Nitride Nanosheets Nanocomposite Paper for Heat Dissipation of 5G Base Stations and Transformers. *ACS Nano, 16*(9), 14323–14333. DOI: 10.1021/acsnano.2c04534

Ding, A. Y., & Janssen, M. (2018). *Opportunities for Applications Using 5G Networks: Requirements, Challenges, and Outlook.* In *ICTRS '18: Proceedings of the Seventh International Conference on Telecommunications and Remote Sensing* (pp. 27–34). https://doi.org/DOI: 10.1145/3278161.3278166

Dixit, E., & Rani, S. (2025). *Integrated Trends, Opportunities, and Challenges of 5G and Internet of Things.* In *Current and Future Cellular Systems*, IEEE, 139–152. https://doi.org/DOI: 10.1002/9781394256075.ch7

Dutta, A., & Hammad, E. (2020). *5G security challenges and opportunities: A system approach.* In 2020 IEEE 3rd 5G World Forum (5GWF) (pp. 109–114). IEEE. https://doi.org/DOI: 10.1109/5GWF49715.2020.9221122

ITU. (2022). *5G - Fifth generation of mobile technologies.* Www.itu.int. https://www.itu.int/en/mediacentre/backgrounders/Pages/5G-fifth-generation-of-mobile-technologies.aspx

Lee, T. (2024). *Advanced Packaging for 5G in RF and Analog Mixed Signal.* IEEE. https://eps.ieee.org/images/files/HIR_Presentations/HIR2024_5G_Lee_Final.pdf

Margaris, P. (2019, January 8). *Differentiating With 5G Security: How Mobile Service Providers Can Become Secure Business Enablers.* Palo Alto Networks Blog. https://www.paloaltonetworks.com/blog/2019/01/differentiating-5g-security-mobile-service-providers-can-become-secure-business-enablers/

Miller, M. J. (2020, October 22). *Gartner's Top Strategic Technology Trends for 2021*. PCMAG; PCMag. https://www.pcmag.com/news/gartners-top-strategic-technology-trends-for-2021

Narayanan, S., Liberg, O., Höglund, A., Tsolkas, D., Alonso, L., Passas, N., & Merakos, L. (2024). Relaying Solutions For 5G-IoT Applications: A 3GPP Perspective. *IEEE Communications Standards Magazine*, *8*(2), 28–35. DOI: 10.1109/mcomstd.0005.2300019

Qualcomm (2020). *Global update on spectrum for 4G & 5G*. Qualcomm Technologies, Inc. https://www.qualcomm.com/content/dam/qcomm-martech/dm-assets/documents/messaging_presentation_-_4g_5g_spectrums_-_december_2020.pdf

Schroth, L. (2020). *5G and Drones: What is 5G?* Drone Industry Insights. https://droneii.com/5g-and-drones-part-1-what-is-5g

Shayea, I., Ergen, M., Hadri Azmi, M., Aldirmaz Çolak, S., Nordin, R., & Daradkeh, Y. I. (2020). Key challenges, drivers, and solutions for mobility management in 5G networks: A survey. *IEEE Access : Practical Innovations, Open Solutions*, *8*, 172534–172552. DOI: 10.1109/ACCESS.2020.3023802

Singh, V. P., Singh, M. P., Hegde, S., & Gupta, M. (2024). Security in 5G Network Slices: Concerns and Opportunities. *IEEE Access : Practical Innovations, Open Solutions*, *12*, 52727–52743. DOI: 10.1109/ACCESS.2024.3386632

Tahir, M., Habaebi, M. H., Dabbagh, M., Mughees, A., Ahad, A., & Ahmed, K. I. (2020). A review of the application of Blockchain in 5G and beyond networks: Taxonomy, field trials, challenges and opportunities. *IEEE Access : Practical Innovations, Open Solutions*, *8*, 115876–115904. DOI: 10.1109/ACCESS.2020.3003020

Wazid, M., Das, A. K., Shetty, S., Gope, P., & Rodrigues, J. J. P. C. (2021). Security in 5 G-enabled Internet of Things communication: Issues, challenges, and future research roadmap. *IEEE Access : Practical Innovations, Open Solutions*, *9*, 4466–4489. DOI: 10.1109/ACCESS.2020.3047895

Zanella, A., Bui, N., Castellani, A., Vangelista, L., & Zorzi, M. (2018). Internet of Things for Smart Cities. *IEEE Internet of Things Journal*, *1*(1), 22–32. https://ieeexplore.ieee.org/document/6740844

ADDITIONAL READING

Advices, H. (2025). Huawei Themes App Archive. https://huaweiadvices.com/tag/huawei-themes-app/

Dubernet, I., & Axhausen, K. W. (2019). The German value of time and value of reliability study: The survey work. *Transportation*, *47*(3), 1477–1513. DOI: 10.1007/s11116-019-10052-4

GAO. (2019). *Agencies Need to Develop Modernization Plans for Critical Legacy Systems*. GAO. https://www.gao.gov/assets/gao-19-471.pdf

Media, R. (2017). All You Need To Know About 5 G. Team Roar. https://roar.media/english/tech/sponsored/all-you-need-to-know-about-5g 5G Networks. (2020). Massive MIMO for 5 G. 5G Networks. https://www.5g-networks.net/tag/massive-mimo-for-5g/

NIST. (2014). *Framework for Improving Critical Infrastructure Cybersecurity*. https://www.nist.gov/system/files/documents/cyberframework/cybersecurity-framework-021214.pdf

OECD. (2019). *Going Digital: Shaping Policies, Improving Lives*. OECD. https://www.oecd.org/en/publications/going-digital-shaping-policies-improving-lives_9789264312012-en.html

Sála Strečno. (2025). *5G Will Benefit the Logistics Industry*. Sála Strečno. https://www.salastrecno.sk/2019/06/16/5g-will-benefit-the-logistics-industry/

Varga, P., Peto, J., Franko, A., Balla, D., Haja, D., Janky, F., Soos, G., Ficzere, D., Maliosz, M., & Toka, L. (2020). 5G support for Industrial IoT Applications— Challenges, Solutions, and Research gaps. *Sensors (Basel)*, *20*(3), 828. DOI: 10.3390/s20030828

World Bank Group. (2025). *Digital Economy for Africa Initiative*. World Bank Group. https://www.worldbank.org/en/programs/all-africa-digital-transformation

ADDITIONAL READING

Chapter 10
5G Applications for Logistics, Energy, and Financial Services

ABSTRACT

5G is transforming logistics and shipping by boosting efficiency, connectivity, and innovation. Its high-speed data transfer, low latency, and massive connectivity enable real-time communication across transportation systems. Applications like vehicle-to-vehicle (V2V) communication support autonomous vehicle testing, fleet coordination, and route optimization. 5G enables seamless data exchange between vehicles, infrastructure, and central systems, resulting in safer and smarter logistics. When combined with IoT, 5G enhances supply chain visibility, allowing real-time tracking of location, temperature, and condition, while predictive analytics help prevent delays and disruptions.

INTRODUCTION

The logistics and shipping industries are on the brink of a technological transformation driven by the adoption of 5G networks. The study by Lagorio et al. (2023) critically examines the role of 5G technology in advancing Logistics 4.0, focusing on identifying specific domains within logistics operations where 5G implementation can yield significant improvements. The authors underscore the transformative capabilities of 5G in enabling real-time data transmission, fostering automation, and enhancing overall connectivity across logistics networks. Despite these advancements, the paper highlights several impediments to effective deployment, including the substantial infrastructural demands, concerns regarding data security, and the high financial investments required. The authors conclude that

DOI: 10.4018/979-8-3373-0497-7.ch010

although 5G presents considerable potential for revolutionizing logistics, a strategic approach that addresses these limitations is imperative to ensure its successful adoption (Lagorio et al., 2023).

As a cornerstone of next-generation connectivity, 5G offers ultra-reliable, high-speed data transfer and low latency, providing the foundation for innovative applications that can revolutionize transportation and logistics. Slamnik-Kriještorac et al. (2022) present the concept of Network Applications (NetApps) as a key enabler for harnessing 5G in the Transport and Logistics (T&L) sector, which demands ultra-low latency, high data rates, and extreme reliability.

The study underscores how NetApps serve as modular, adaptable components that simplify the deployment and management of complex T&L services by abstracting network complexities and aligning 5G capabilities with sector-specific needs. The paper demonstrates the practical value of NetApps through the VITAL-5G project, specifically focusing on a case study at the port of Galati, where NetApps improved the speed and safety of port operations by integrating vertical service requirements with 5G network slices and core functionalities, (Slamnik-Kriještorac et al., 2022).

From enhancing vehicle-to-vehicle (V2V) communications to improving operational visibility and efficiency, 5G enables organizations to address long-standing challenges in managing fleets, optimizing supply chains, and integrating smart infrastructure. 5G connectivity in logistics could improve visibility and optimize operations through real-time data exchange, (Goasduff, 2021). Apruzzese et al. (2023) examine how 5G and complementary technologies, such as IoT, AI, and blockchain, are reshaping business models in logistics and supply chains, emphasizing enhanced device density, low latency, network slicing, and improved supply chain visibility as key drivers of innovation and sustainable growth.

5 G's most significant advantage is its ability to support real-time communication between vehicles, infrastructure, and central management systems. Shehab et al. (2022) provide a comprehensive review of recent advancements in 5G networks, focusing on their role in fostering innovative and sustainable cities. The paper highlights the technical strengths of 5G, including high data rates, ultra-reliable low-latency communication, and dense connectivity, which are essential for supporting applications such as enhanced mobile broadband (eMBB), ultra-reliable low-latency communication (uRLLC), and massive machine-type communication (mMTC). The authors explore how these capabilities can drive sustainability across environmental, social, and economic domains within urban environments.

Specifically, they analyze sustainability metrics, including energy efficiency, power consumption, carbon emissions, cost, health, and security. Their review reveals that 42% of existing research focuses on environmental sustainability, followed by 37% on economic aspects, and 21% on social considerations, with particular attention to energy efficiency (20%) and power consumption (17%) as critical areas of study in 5G-enabled smart cities, (Shehab et al., 2022).

This capability is particularly impactful in advancing autonomous vehicle testing and deployment, as well as optimizing route planning and fleet management. By enabling seamless data sharing, 5G facilitates better coordination, reduces downtime, and enhances safety within logistics operations. For cities and municipalities, 5 G-powered networks provide unprecedented access to transportation system data, helping to design more innovative urban transit systems and reduce congestion.

Additionally, integrating 5G with IoT technologies in logistics enables precise tracking and monitoring of goods across the supply chain. Whether tracking shipments in real time, managing warehouse operations, or predicting delays, 5G's speed and reliability ensure that logistics providers can deliver superior service while reducing costs. Furthermore, the environmental benefits of optimized logistics processes, including reduced fuel consumption and lower emissions, align with the industry's emphasis on sustainable practices.

This chapter explores the transformative impact of 5G on logistics and shipping, analyzing how its applications enhance efficiency, improve safety, and drive innovation across the transportation sector. By exploring the potential of 5 G-enabled solutions, this section highlights the crucial role of advanced connectivity in shaping the future of global logistics and shipping.

Transportation logistics, ranging from public buses to private logistics fleets, improve visibility and balance as 5G enables the synchronization of IoT devices with AI to streamline transportation and logistics operations. In addition, 5G enables improved vehicle-to-vehicle communications, allowing much more self-driving automobile testing. These networks help cities gain access to significantly more data about their transportation systems.

Short-Term Impact

In the short term, a high-stakes race is underway to build 5G wireless networks nationwide. That requires a massive and rapid rollout of bodily technology infrastructure. Global Tranz, for instance, provides transportation and logistics solutions to large telecommunications companies involved in the rollout of 5 G.

Building out a nationwide 5G system with the broadest coverage in the market requires considerable effort from skilled employees and specialists. The gear is shipped to facilities precisely when needed. Construction stalls, costly labor idles, and missing deadlines could be easily avoided without the proper equipment system, and competitors surge ahead.

The critical element of this rollout is the behind-the-scenes logistical work to orchestrate and thoroughly transport advanced, high-value technology to the appropriate location at the correct time, unscathed, so it can continue to be built out.

Long-Term Impact

5G influences supply logistics and chain by enabling more details to be transferred faster in real-time, making improved exposure throughout the supply chain attainable. As more products through the supply chain and process become part of the "Internet of Things," 5G technology is poised to revolutionize the transportation and logistics industry by enabling real-time data exchange, enhancing supply chain visibility, and facilitating the adoption of autonomous vehicles and drones.

They produce a vibrant information stream that sends real-time signals to trigger various events. Patel and Zhang (2023) argue that 5G networks are essential for deploying advanced robotic process automation in industries.

For instance, using a 5G network, a component tote could communicate that it is 80% depleted because of this SKU, triggering a reorder of the necessary parts. This tote may trigger throughout the supply chain, leading to warehouse moves, consolidation, maybe LTL, and delivery and resupplies. The critical element to leveraging this insightful information is applying multiple technology platforms to process it into insights that generate source chain effectiveness.

To comprehend 5 G's genuine promise, we must picture a really "connected" supply chain, where the gear that manufactures the item is connected to the pickup truck that picks it up. The local store shelf, as shown, delivers real-time signals crucial in driving generation and transportation pursuits. Moor Insights & Strategy reports that 90% of logistics providers face significant challenges due to limited supply chain visibility, which 5G can address, (Sag, Townsend, & Moorhead, 2025).

For instance, a stockout in the merchant's inventory might signal to the producer how much of the product is required, where it can be produced, and when it is needed. It might also trigger the dispatch of a pickup truck. This leverages improved information regarding the offered capacity, the pickup truck, road conditions, and the car owner's performance history.

Achieving the critical mass necessary to create the 5G truth will require more than the proliferation of low-cost 5G units. It will also necessitate the development of standards that guide development and govern the interaction of these devices.

As the case with blockchain continues, these requirements can take time to create and implement. While we in transportation tend to think about enhanced visibility into trucks' movement and how that impacts capacity, pickup and delivery times, and pricing, 5G significantly impacts the supply chain. Benefits for the Logistics Industry

5G has a drastic impact on some industries, but the logistics business is the segment that benefits most from the improved connectivity and increased pace. The strategies are the lower frequency band providing more comprehensive coverage in rural and suburban areas, and the higher frequency band providing much better content in higher-density urbanized areas. Finally, the supply chain business is equipped to supply constant end-to-end coverage for checking, monitoring, and theft detection.

LONG-TERM IMPACT OF 5G ON LOGISTICS AND SUPPLY CHAIN

Historically, many issues surrounding lost indicators and inadequate coverage in outlying areas have plagued the market, resulting in bottlenecks and revenue leakage for re-establishing inventory checks. 5G solves a lot of these issues, impacting logistics and supply chain management for freight treatment buyers, cargo owners, and some in different ways, including:

Better Location Tracking and Route Optimization

5G enables businesses to track their valuable cargo across all areas by optimizing coverage in previously "dead zones." Improved geolocation technology permits greater visibility into unforeseen travel circumstances and delays, ultimately improving location intelligence and mitigating ensuing problems. 5G enables businesses to enhance their routes through improved visibility, avoiding unnecessary excursions and inefficiencies. Along with enhanced monitoring of remote countryside areas, highly trafficked areas benefit from implementing 5G networks. A strained system does not slow tracking capabilities as shipments traverse populated towns and cities. Ultimately, 5G enables logistics supervisors to account for cargo from end to end during almost all trip points.

Constant Tracking With IoT Sensors

According to a survey from Moor Insights, 90% of strategies and delivery providers consider the absence of supply chain visibility among the most significant problems in the market today. With 5G, discount sensors enable better individual item tagging and tracking. Also, near-edge computing on small-footprint infrastructure enables faster inventory checking. 5G helps provide end-to-end visibility, not only at an aggregate level but also within each item, thereby assisting in preventing revenue leakage through theft and loss.

5G enhances traceability by observing and collecting details at each step, including whether a pickup truck switches mass (which may indicate a potential problem). This development enables cargo owners to have remarkably high visibility into the delivery process, even before it becomes inaccessible or nonexistent.

Faster and Safer Port Operations

An intelligent transportation framework integrates sensors, cameras, and devices into a cohesive network, forming an advanced communications ecosystem. 5G enhances mobile connectivity for IoT-enabled logistics, ensuring seamless tracking and monitoring of assets. Within this framework, autonomous ground vehicles play a pivotal role by performing tasks such as automatically loading and unloading cargo and broadcasting inventory data to access-controlled areas. These vehicles operate seamlessly thanks to integrating into a more extensive, connected system.

The backbone of such a system relies on high-speed, reliable, and high-capacity communication, made possible only through the unparalleled capabilities of 5G technology. With its ultra-low latency and immense data transfer speeds, 5G ensures that all system components can communicate in real time, facilitating precision and efficiency in operations.

This digital transformation unlocks numerous possibilities across the transportation and logistics sectors. As demonstrated in port operations, synchronizing IoT devices and artificial intelligence with robust 5G connectivity fosters a streamlined, secure, and efficient operational environment. This convergence of technologies, from real-time cargo tracking to predictive machinery maintenance, drives greater productivity and operational control.

Moreover, 5 G-powered intelligent transportation systems pave the way for more innovative and responsive infrastructure, where interconnected devices and autonomous systems collaborate to optimize traffic, enhance safety, and reduce environmental impact. IoT, AI, and 5G connectivity improve transportation systems and revolutionize logistics by enabling lean, agile, and intelligent operations.

Enhanced Communication and Repairs Through VR and AR

The 5G network setup allows improved virtual reality (VR) to help with on-the-road backup (and different assists). For example, a remote mechanic can use VR to provide roadside guidance, reducing unforeseen downtime. In addition, augmented reality (AR) can be incorporated into cars to provide better endpoint recognition and improve car owners' security. These technological updates reduce the time allocated for maintenance and save companies money, as in-house mechanics are available to address any issues during transit.

In the transport and logistics sector, 68% of chief information officers (CIOs) prioritize improving business process efficiency, compared to 54% across all industries, reflecting a strong focus on operational optimization. Enhancing the customer experience is another key goal for 61% of logistics CIOs, closely aligning with the 60% reported across other sectors. Additionally, 58% emphasize growing revenue from existing customers, surpassing the 52% average among all CIOs. Interestingly, only 42% of logistics CIOs prioritize attracting new customers, significantly lower than the 55% in other industries. Meanwhile, 39% of logistics CIOs focus on developing new products and services, compared to 49% in different fields, highlighting differing strategic priorities.

The following table highlights the transformative impact of 5G on logistics and supply chain operations. With ultra-low latency and high data throughput, 5G enables real-time visibility, automation, and precision across all logistics functions. From connected supply chains to smart port operations, the technology supports faster, safer, and more efficient processes. As industries adopt IoT, AR/VR, and autonomous systems, 5G becomes a foundational enabler of next-generation logistics. This table outlines the key areas where 5G reshapes logistics strategy, operations, and value delivery.

Table 1. Long-term impact of 5G on logistics and supply chain

Area of Impact	Description
Real-Time Data & Supply Chain Automation	5G enables real-time tracking and triggers for inventory, improving visibility and automation in the supply chain.
Connected Supply Chains	Integration through real-time signal exchange enhances synchronization and responsiveness.
Location Tracking & Route Optimization	Improved geolocation and route management, even in areas with poor signal, reduce inefficiencies and delays.
IoT-Based Constant Tracking	Utilizing 5 G-powered IoT sensors, enhanced visibility enables item-level tracking, loss prevention, and improved inventory accuracy.
Smart Port Operations	Autonomous vehicles and real-time communications enable faster cargo handling, reducing delays and increasing safety.
AR/VR for Maintenance	Remote assistance via VR and AR reduces downtime and enhances roadside support and vehicle maintenance.
Strategic CIO Focus in Logistics	CIOs prioritize efficiency (68%), customer experience (61%), and revenue growth from existing clients (58%), indicating a shift toward internal optimization.

continued on following page

Table 1. Continued

Area of Impact	Description
Infrastructure and Security Considerations	Implementation requires significant investment, robust cybersecurity measures, and consistent adherence to regulatory standards.
Transformative Potential	5G drives smart logistics, enhances productivity, and supports digital transformation across global supply chains.

DETAILS OF THE IMPACTS OF 5G ON LOGISTICS

While geopolitical debates persist, particularly surrounding China's leadership ambitions in 5G development, the global rollout of this technology is inevitable. For North American logistics and flatbed transport sectors, 5G represents not just an upgrade but a disruptive leap forward. Offering speeds up to 100 times faster than 4G, 5G radically transforms operational capabilities across connected systems in vehicles, warehouses, and logistics platforms. Its integration into flatbed transport and other trucking operations opens the door to unprecedented levels of automation, real-time communication, and data-driven compliance.

Automation and Autonomous Operations

5G significantly accelerates the development and deployment of autonomous vehicles by enabling real-time responsiveness, high-definition mapping, and ultra-reliable low-latency communication (URLLC). Fleet managers benefit from advanced driver-assist technologies and centralized monitoring systems that will allow split-second decisions, thereby increasing safety, reducing fuel consumption, and enhancing operational efficiency. Onboard safety features powered by 5G, such as fatigue detection, lane departure warnings, and automated braking, help mitigate risks related to human error and long-haul driving fatigue.

Enhanced Communication and Supply Chain Visibility

One of 5G's most immediate impacts is the improvement of highway communication and fleet coordination. By reducing dead zones and enhancing connectivity quality, fleets can transmit real-time updates on cargo status, weather conditions, and delivery timelines without disruption. This level of instant, high-fidelity communication enhances end-to-end supply chain synchronization, enabling dynamic rerouting or rescheduling in response to delays, hazards, or sudden shifts in demand.

Regulatory Compliance and Asset Tracking

5G also modernizes regulatory adherence through improved Electronic Logging Devices (ELDs) and telematics systems. As mandates evolve, fleets using outdated 4G hardware may face compatibility and compliance issues. 5 G-enabled systems offer precise GPS tracking, reliable data logging, and cloud integration, making it easier to meet Department of Transportation (DOT) and Federal Motor Carrier Safety Administration (FMCSA) standards. Additionally, the extended coverage of 5G enables uninterrupted tracking of vehicles and assets, even in previously inaccessible rural or mountainous regions.

In summary, 5G's role in logistics goes beyond connectivity—it redefines responsiveness, efficiency, and control. For flatbed carriers and broader freight logistics, adopting 5G is not just a competitive advantage; it is a strategic necessity for navigating the increasingly digitalized future of transportation.

APPLICATIONS FOR LOGISTICS

The logistics industry stands to benefit significantly from the implementation of 5G technology, which enables real-time data transmission, greater automation, and improved tracking capabilities across the supply chain.

Real-Time Location Tracking for Shipments

5G IoT devices are not limited to transmitting related data and location information. Items could be eco-monitored at the item level by committed IoT sensors. Subsequently, real-time data on temperature, light levels, humidity, gas amounts, and other areas may influence the quality or safety of vulnerable items. With much more detailed item monitoring, LSPs (Logistic Service Providers) are going to be ready to:

- Ensure the freshness of foods along with other perishable items.
- Help manage the security of products that may be vulnerable, such as raw materials or chemicals, to ensure their protection and safety.
- Provide items of the quality the buyer expects.
- Download the realities of making a 5 G-ready world.
- Incorporated Supply Chains.

Comprehensive Product Monitoring When in Storage or Transport

Due to their in-depth knowledge of inbound goods, every logistics service provider (LSP) strives to optimize its operations. 5G enables LSPs to incorporate with downstream and upstream supply chain companies to help make the handover of things faster and more precise, empowering LSPs with the following:

- The trust is established or maintained due to a comprehensive listing of products backed by IoT product-level tracking.
- There has been a quicker turnaround of raw materials, components, and goods because vendors and companies are better equipped to predict and prepare for receiving goods.
- Better resource management is achieved throughout organizations by using accurate and timely data.
- Synchronization occurs across and between departments and organizations within a supply chain business.

Another location that may help with information handoff involves communications networks. For instance, 5G can work with various communications protocols, such as Wi-Fi, to permit the transfer of information to other methods. Chen and Zhang (2022) describe the transformative potential of 5G in creating more resilient and efficient logistics networks.

Self-driving vehicles communicate with one another, although genuinely autonomous cars are not standard on the highways. However, it is just a situation of your time.

One area where this will make the most considerable difference is design for the logistics business. Self-driving trucks will alleviate most of the issues associated with driver scheduling and regulations.

5G is crucial to a fully functioning, automated fleet. Vehicles must communicate to optimize safety and ensure instant attention to other traffic, road conditions, and other relevant factors.

Augmented Reality for Warehouse Operations

The most thrilling application of 5G is augmented reality (AR), where users can interact with computer-generated imagery projected onto their interface.

Examples of places in which this technology could be helpful include:

- Directions for shelving to store and retrieve results more comfortably and efficiently, facilitating faster storage, distribution, and retrieval.
- Remote functioning of forklifts and choosing products through instantaneous "haptic" and feedback technology
- Inter-equipment communications permit robots and other automatic machinery to optimize routing near a warehouse.

APPLICATIONS FOR RETAILING

Over the last several years, retailers have invested millions in intelligent technologies to help clients shop more efficiently and check out quicker while collecting more customer experience information. From in-store analytics to visual recognition-driven shelf monitoring, all rely upon or take advantage of the ability to transmit enormous amounts of information and access high-throughput connections, which is why 5G technology has a massive impact on how retailers operate.

According to CB Insights, over 100 million Americans made a purchase on their smartphones in 2018, the article noted. Hence, the shift to mobile shopping is primarily driven by the rise of 4G/LTE. The faster speeds of 5G technology enable a new list of possibilities, including virtual reality (VR) dressing rooms.

The addition of 5G technology to an existing network architecture can help power new technologies that many retailers are beginning to utilize, such as:

- Personalized digital signage
- Augmented reality
- Virtual reality
- Video and pattern recognition of shoppers
- Interactive mobile applications

Smart Shelves and Shopping Experience

The 5G is transforming the shopping experience. For instance, intelligent shelves incorporating RFID technology could enable an entrepreneur to compare product pickups with sales and display total prices. With 5G technology, shelves equipped with sensors can identify reduced inventory on a product, notify a distribution facility to replenish its stock, and dynamically track the shipment's progress. The data required to move within the mobile network is too much for the existing infrastructure.

Nowadays, companies like Sephora utilize virtual try-on technology to help in-store clients visualize how cosmetics will look on them before making a purchase. However, information streaming limits constrain the item. 5G technology removes such limitations—we may one day consider our clothes in augmented reality with that precision, which is currently challenging to achieve.

The 5G can create entirely new shopping experiences that would be impossible with present-day technology. For instance, when a consumer enters a shop, an augmented reality program on their smartphone triggers and guides them directly to the shelf where they can purchase the items of their choice.

The bodily groundwork for these experiences is already underway with cashier-less stores (e.g., Amazon Go). Based on Adobe Digital Insights, enhancements in connectivity resulting from 5G technology may increase list revenue by 1dolar1 12B yearly by 2021.

Retailing Applications for Augmented Reality

The launch of 5G is not just an incremental enhancement of telecoms technology. The possibility of achieving rates of 110 Gbps with pervasive connectivity and minimal latency opens up a range of options for merchants and their IT teams. 5G enables them to "supercharge" revolutionary customer experiences and deliver highly personalized engagement to customers, driven by seamless connectivity and rapid response times.

The primary instances where we imagine merchants and reputable customer goods companies applying 5G are smart shelves, real-time sales messaging and promotions, and personalized digital signage in in-store and mobile apps. In addition, he states that virtual reality (VR) and augmented reality (AR)- enabled solutions, which reflect adaptive secrets, would have affected dress retailers.

Tyagi adds that the capability to observe and track shop inventories is available in real-time. Enabling customers to seamlessly order online for home delivery if their choice is not available in the store "is another where the quick response times powered by 5G make a big difference for consumers seeking great retail experiences". Kim et al. (2020) describe how 5G enhances smart home systems by supporting interconnected devices with ultra-low latency.

Nokia-Telia 5G Network for Shopping Centers

One of the more recently available 5G network deployments within the list sector has brought the site to deep Helsinki, Finland, with a collaboration between Nokia Corporation, the Finnish global revolutionary globe head in 5G networks and buyer electronics company, and Telia, a Swedish international telecommunications company and mobile network operator with a presence in Sweden, Estonia, Lithuania, Denmark, Norway, Finland, and Latvia.

Nokia runs Telia's 5G system in the brand-new Mall of Tripla in Pasila, a district in Helsinki, Finland. The Mall opened on October 17 as part of the Helsinki smart city project.

This 5G network provides next-generation connectivity for customers, 250 stores, companies, and entrepreneurs within the shopping center. Triple is a tremendous three-block urbanized business and shopping facility with a retail space and 50,000 square meters of office space.

The 5G network in Tripla is powered by Nokia's 5G base station and small cell technology, providing all users with high capacity, low latency, and dependable connectivity. As a result, entrepreneurs and retailers can develop, test, and implement innovative retail applications and lifestyle experiences, taking a step forward in the future of retail. Furthermore, these innovative retail applications enable retailers and advertisers to implement emerging technologies, such as those featured in futuristic movies like Minority Report.

Figure 1. Applications for retailing

APPLICATIONS FOR UTILITIES AND ENERGY

Essential infrastructure, enabled by 5G technologies, can produce more innovative solutions in usage, distribution, transmission, energy production, and the next wave of innovative grid functions and efficiency. The China National Energy Administration (NEA) outlines how 5G enhances innovative grid systems by enabling real-time monitoring and fault detection.

Information from the Power System

An intelligent power grid is a modernized power system that utilizes information and communication technologies to gather data from the power system. This info is helpful when setting the production and distribution of power or adjusting power usage to save electricity, reduce losses, and improve the power system's reliability. In a typical power system, products are monitored onsite. These units can be monitored and measured remotely using intelligent grids to determine, adjust instantly, and balance power usage. These make smart grids smart. Thus, connecting these units to the communications system is essential for building a sensible grid.

A power grid system comprises 5 phases: energy generation, distribution, transformation, transmission, and usage. Communication networks across power generation, transmission, and transformation are mature, utilizing the optical system. Nevertheless, the substantial quantity of nodes spreads distribution in the distribution and power consumption phases, and the communications system's coverage is minimal. It is the last five km problem and is the main bottleneck for smart grid growth. According to the China National Energy Administration (NEA), 95% of blackouts happen in the final 5 kilometers of the power grid.

Creating Terminal Entry Networks

Creating terminal entry networks within the last five km of power distribution and consumption is essential to controlling and communicating in an intelligent grid. (NEA forecasted that in 2025, 9.63 million square kilometers discussed would have 20.13 million communication modules for China's entire energy market.) Moreover, wireless cellular networks, present-day cellular technologies, and the emerging 5G technology benefit the intelligent grid.

The 5G networks offer suitable providers for timely and critical applications in various aspects of the smart grid. In addition, 5G provides business models that power providers can utilize in current electricity markets and enhances the communication of energy with power grid systems.

This covers utilities, energy applications, current wireless technologies, and architectural modifications. Moreover, a comprehensive evaluation of 5G architecture and the power grid perspectives recognizes the user-based analyses in an intelligent grid. Finally, the current state of 5G networks in a smart grid, with a distinct assessment of electricity efficiency, clarifies the specific focus of this work.

Furthermore, we emphasize the use of reliable and intelligent grid interaction to address future challenges and develop effective roadmaps. These applications offer an in-depth understanding of networks related to future smart grids, including extensive analysis.

Thus, an intelligent energy last-mile system represents an excellent vertical for substantial 5G deployment, where various applications with

Various specifications must be managed for Smart Grid applications:

- Supervisory
- Checking (cyber monitoring and physical/aerial)
- Surveillance, fault localization, isolation, and self-healing
- Power re-routing requires much more stringent measures
- Latency, highest security, and availability

Advanced Metering Apps

Advanced Metering Apps represent critical applications for utilities and energy industries. They enable the seamless and lock-in-free integration of end-users and infrastructure, requiring stricter capacity and security. For example, Smart Electric Vehicle Charging utilizes 5G technology, incorporating and addressing capacity and latency requirements that are even more stringent.

Integration Application of 5G and Smart Grid

With the advancement of intelligent grids, existing power networks have fallen short of meeting the demanding needs of manufacturing applications, particularly in terms of bandwidth, end-to-end latency, and reliability.

Combining vertical businesses is a significant advancement in 5G mobile communication technology. In contrast, its adequate bandwidth, low latency, higher reliability, and substantial connection, coupled with the innovative grid applications program, meet its needs. The best way to use 5G in an intelligent grid is attractive. It first evaluates the upgrade of 5G bodily level technologies to determine the corresponding needs of 5G application scenarios. Then, it introduces the primary vital technologies, such as 5G portable advantage computing, computational caching, and network slicing. That is the foundation for 5G to enable vertical multi-service networks and meet diverse service demands. Several network slicing reasons, including the access system, transmission network, and primary network, are coupled with power applications. For instance, Carrillo et al. (2023) present smart RAN slicing as a solution to optimize 5G in innovative grid environments.

APPLICATIONS FOR OIL AND GAS

What Does 5G Mean for the Oil and Gas Industries?

Effectiveness and agility in oil exploration and production (E&P) are becoming increasingly necessary, driving ambitious digital initiatives within the E&P segment of the oil value chain. These include digital oilfield initiatives, IT and OT integration, digital well-site operations management, and unmanaged production platform management.

The 5G technology within the oil and gas industry is likely to drive development in oil and gas marketplaces over the next ten years, with gasoline use growing at a rate exceeding that of petroleum consumption.

Oil and Gas's Improved Needs

To meet the evolving needs of the oil and gas industry, exploration is expected to push even further from shore, enhancing isolation and reducing dependency on network marketing communications for field personnel. In addition, offshore exploration requires massive data transfer, supply, and logistics for gear, which must be received more quickly and reliably.

Existing techniques and exploration of oil and gas require communication over costly satellite links. According to a particular satellite, each boat receives its backlinks, and every website link costs approximately $15,000 per month. Likewise, the backlinks offer speech correspondence, excluding the range for significant details transfer, video clips, and web connectivity, and still have a limited selection.

Other 5G Advantages

In addition to the higher speeds, 5 G offers advantages such as ease of use and versatility. Whether or not 5G connectivity is eventually presented through cellular towers and satellites, each indicates more or less ubiquitous, blazing-fast connectivity worldwide, typically without the headache of installing Wi-Fi.

5G allows each website feature to communicate more reliably and at a much faster speed. It also produces more reliable IoT (Internet of Things) components that profit from improving and securing high-risk offshore gas and Oil tasks.

Huawei 5G Solution for the Oil and Gas Industry

Regarding the oil and gas business, wireless LTE technology is now a highly effective way to interact, enabling mobile inspection and real-time monitoring of devices and systems, particularly in hazardous and remote locations. The new LTE solution addresses the challenge of insufficient committed spectrum resources near the planet by leveraging 4.5G expertise to aggregate standard VHF/UHF narrowband discrete spectrum into broadband internet spectrum online resources.

The solution supports a latency of twenty milliseconds and a maximum of 4,000 contacts per cellular device, enabling IoT solutions within the engine oil and gasoline industry. This solution can help petroleum and gasoline businesses operate efficiently, effectively, and seamlessly, providing clients with new support and services to achieve energy cost savings.

As a prominent global provider of Communications and Information Technology (ICT) strategies, Huawei drives "Leading New ICT, Higher Efficiency" and Safety within the Oil and Gas industry. Huawei collaborates with business partners to develop ICT solutions that address the upstream, midstream, and downstream sectors, combining electronic generation with reliability management to enhance efficiency. These revolutionary 5G strategies apply to gas and oil, electronic pipelines, and high-performance computing (HPC) for function management and intelligent division.

Jerry Ji, President of the Energy Sector at Huawei Enterprise Business Group, stated, "Huawei is dedicated to providing electronic solutions to every prominent energy provider." Huawei is focused on ICT infrastructure and leverages its technical advantages in deep cloud computing, IoT, Big Data, and Artificial Intelligence (AI) to accelerate digital transformation in the Oil and gas industry. We supply clients with an open, flexible, and secure ICT infrastructure framework to support the device-pipe-cloud effort, streamlining the transition to digital transformation in the immense energy industry. We also help environment associates attain win-win outcomes, minimize electronic transformation expenses, and collectively bring value to customers.

Huawei collaborates with business associates to showcase revolutionary converged ICT solutions, including innovative HPC devices for seismic processing and reservoir simulation, IoT for gas and oil, electronic pipelines, and cloud computing, as well as Big Data and petroleum applications.

Thus far, Huawei's petroleum and gasoline treatments have been applied in 45 countries and areas worldwide, helping 70% of the world's 20 leading petroleum and gasoline companies.

Transition From Industry 3.0 to Industry 4.0 in the Oil and Gas Sector

The transition from Industry 3.0 to Industry 4.0 in the oil and gas sector is shifting from traditional automation pyramids to highly interconnected automation pillars. Lagorio et al. (2023) analyze the integration of 5G technology within Logistics

4.0, focusing on its potential applications and challenges. The study identifies key areas in logistics where 5G can be effectively implemented, including real-time data exchange, automation, and enhanced connectivity, which are expected to improve operational efficiency. However, the authors also highlight significant challenges, including the need for substantial infrastructure investments and concerns about data security. They conclude that while 5G holds considerable promise for transforming logistics operations, carefully considering these challenges is essential for successful implementation.

Key elements such as intelligent sensors, smart gateways, and adaptive security provide the foundation for advanced industrial automation. Integrating 5G connectivity introduces low-latency, highly deterministic networking, enabling real-time data exchange and enhanced operational control. Advanced analytics and digital platforms further empower the sector by leveraging industrial applications and inter-cloud connectivity to optimize processes, improve safety, and boost overall efficiency. These innovations underline the transformative potential of 5G in modernizing oil and gas operations.

APPLICATIONS FOR HEALTHCARE

5G can boost revenue and efficiencies in the medical industry, enabling healthcare methods to produce more quickly and effective networks to match the massive quantities of information involved. Medical science can also enable remote monitoring products to enhance health outcomes. 5G infrastructure is pivotal in the digital transformation of industries such as healthcare.

The healthcare vertical is poised for a paradigm shift, driven by advancements in sensing technology, the evolution of telemedicine, and innovative technology. This paper examines the various ecosystems, growing adoption, and technology of several original use cases that have led to 5 G. Several critical motivation factors influence technology evolution, including bandwidth, essential services, real-time response, and wearables. For instance, Sharma et al. (2021) explain how 5G enables wearable technology to deliver better health monitoring and fitness tracking.

Cutting-edge technologies are being combined to address these challenges, including the Internet of Things (IoT), big data, artificial intelligence, and 5G wireless transmission technology, to enhance the individual experience and improve the quality of healthcare services. This technique is not an impractical ideal, as these emerging technologies are already subtly influencing and reconstructing healthcare. Although the technologies mentioned previously integrate, this detailed review provides a brief examination of situations that focus on utilizing 5G wireless transmission technology in healthcare. However, we could highlight the potential pitfalls to the accessibility of 5G solutions.

Transformation of Healthcare With 5 G

Several of the world's largest economies, including China and the U.S., are poised to make significant investments in healthcare facilities. For example, the National Health Expenditure Account (NHEA) of the U.S. expects total healthcare spending in the U.S. to reach approximately USD 6.0 trillion by 2027, exhibiting a compound annual growth rate (CAGR) of nearly 5.5% from 2018 to 2027. Contemporary healthcare emphasizes the administration of technology-driven therapies to patients, driving the industry's demand for 5G products within the forecast period.

5G technology transforms healthcare by enabling high-speed, low-latency connectivity that integrates with cloud computing to support advanced medical applications. Wearable devices and innovative medical technologies benefit from 5G by providing real-time health monitoring and data sharing, enhancing patient care. Remote diagnosis and virtual consultations are made seamless, ensuring accessibility to medical expertise regardless of location. In smart hospital settings, 5G facilitates the efficient operation of connected devices and systems, improving service delivery and patient outcomes. Additionally, 5 G-powered virtual reality (VR) applications offer innovative approaches to medical training and treatment, while enhanced transportation systems ensure timely emergency responses.

5G technology is poised to transform healthcare delivery by enabling real-time, high-bandwidth communication that supports remote diagnostics, robotic surgery, and continuous patient monitoring, (Li, 2019).

Self-Monitoring Medicine

In terms of self-monitoring medicine, 5G is not just one technology. Instead, it mixes cutting-edge technologies, including big data, cloud computing, and artificial intelligence. In the 5G era, with computer systems' better capability to process substantial information, self-determination medication is becoming a reality.

Instead of the traditional, doctor-centered, and often fashionably presented treatment and diagnosis model, the self-determination medical-related model significantly enhances the patient's autonomy, allowing the affected person to participate throughout the treatment process.

Unlike individualized or personalized medication, self-determined medication relies on algorithms and high-speed active data. The diagnosis and treatment plan should be regular, robust, and functional, allowing for specific feedback concerning lifestyle components, behavioral elements, and therapy results to help individuals become more autonomous from medical services. During this procedure, professionals can build and grow a private overall health ecosystem using secure interactions and an interactive social network. Low and ultra-reliable latency 5G networks will play a crucial role.

Categories of Services in Healthcare With 5G

The next generation of mobile network engineering, 5G, is revolutionizing the healthcare industry just as Wi-Fi did. However, 5G improves healthcare in distinct ways. We will explore 5G's potential to enable more powerful home healthcare. We could then focus on why 5G is suited to a hospital's requirements for its connected know-how to be highly safe and protect patients' health records and information.

Home Healthcare

There is a significant gap in hygiene between patients' experiences at a clinic and those in home healthcare. Many individuals want to be treated at home rather than forced to cope with long wait times and high hospital costs. In most instances, patients are released from the hospital and arrive weeks or days later.

Throughout the United States, the readmission rate for individuals enrolled in Medicare with Medicare Part B is 14.9%, with a few states reaching up to 15.7%. As healthcare has become more accessible in recent years, viable alternatives to traditional clinic care have emerged. 5G networks support healthcare and are inching closer to that reality.

5G can improve home healthcare services, such as video chats with caregivers, more than a secure 5G mobile network. Additionally, 5G enables many wearables and other remote monitoring and rehabilitative devices, which are crucial to making home healthcare a primary treatment and recovery option. The lower latency of 5G services enables doctors and nurses to monitor patients in real-time.

Furthermore, emergencies or abnormalities are reported without delay. 5G networks can also provide caregivers with immediate, high-quality backlinks to assess individuals, lessening the likelihood of additional complications or injury.

They are unlocking the extra opportunity of home healthcare, which enables higher levels of care for older people, allowing them to remain in the convenience of their homes for as long as possible before transitioning to an assisted living or senior care facility. In addition, unleashing the multitude of monitoring, rehabilitative, and healing solutions in the house is a reality with 5G, further driving better treatment outcomes and reducing rising healthcare costs.

With 5G, unleashing the multitude of monitoring, rehabilitative, and healing services in the home is a reality, further driving better care outcomes and reducing soaring healthcare costs.

Remote Emergency and Medicine Care

Delivering effective and timely triage treatment is crucial in crises, such as those involving individuals with lengthy cardiac arrest, wherein brain injury can begin at 4 minutes. The U.S. model has typically been a patient-to-hospital train, but 5G connectivity can bring the clinic to the client, accelerating therapy. Future generation system technology, such as 5G networks, may equip U.S. first responders with a new set of resources to drive faster and more beneficial emergency treatment outcomes.

Apps for Paramedics

Current healthcare providers design apps that enable paramedics to accurately alert the emergency department of an imminent arrival. These apps provide the ambulance's GPS information and deliver the ECG, private details, pictures, and other trauma information that help ER staff get patients with the correct equipment and personnel.

4k Body Cameras for Doctors

Moreover, 5G networks support 4K body cameras with a Livelink feature, allowing ER doctors or experts to guide an EMT through a process in the area, or even evaluate, handle, and prepare for the transportation of injured patients. The network engineering required to support these high-definition (HD) streaming video feeds necessitates low latency and high capacity, which only 5G can provide.

Public safety dispatch officials often require rapid response times to gain situational awareness and provide the necessary equipment and resources. Nevertheless, other applications that access HD traffic monitoring cameras and drones equipped with 4K video can relay information to public safety officials. These applications provide real-time information about the emergency, enabling the evaluation and dispatch of the appropriate emergency response resources.

Mobile Clinics With 5G

Furthermore, 5G-enabled ambulances or mobile clinics in outlying areas could enable AR-assisted operations to enhance the capabilities of EMTs (Emergency

Medical Technicians) or even practitioners by providing guided steps for performing critical care in the field.

In rural or remote areas where health specialists may not be accessible, a generalist today has a valuable tool to identify, treat, or even perform necessary procedures using AI-assisted Augmented Reality applications.

Medical Training, Therapeutic, and Rehabilitative VR

Virtual reality is usually associated with the gaming industry, but it has also created numerous medical opportunities. Clinics can use it to immerse individuals and physicians in their chosen environment.

Virtual Reality therapy is now an effective way for clinics to utilize VR to benefit individuals and help them overcome persistent pain and injuries. For example, VR reduces chronic pain by 25% while improving the individual's well-being and saving them money by decreasing the time they must spend in a medical facility.

Furthermore, medical students and doctors can use VR to simulate functions that usually restrict the operation's staff. However, hospitals and therapeutic centers have limited their use of VR features due to concerns about expense, accessibility, and operational support. Nowadays, genuinely good VR-based healthcare instruction modules require a wired VR package to generate realism, which should be beneficial.

The latest top-quality VR engineering is currently limited to wired laptops due to the reduced latency required to develop high-speed broadband applications. Nowadays, inexpensive wireless VR works through compression and other techniques, but falls short of the intensive, immersive experiences resulting from the more significant latency of the network, producing lag that computer users experience while wearing a headset. With 5G, top VR enables cordless mobility, making the lives of both doctors and patients more comfortable. Mobile VR is poised to reach a wider audience of individuals, including those who are confined to their beds or rooms, and a wired VR setup comes at a fraction of the cost.

Figure 2. Applications for healthcare

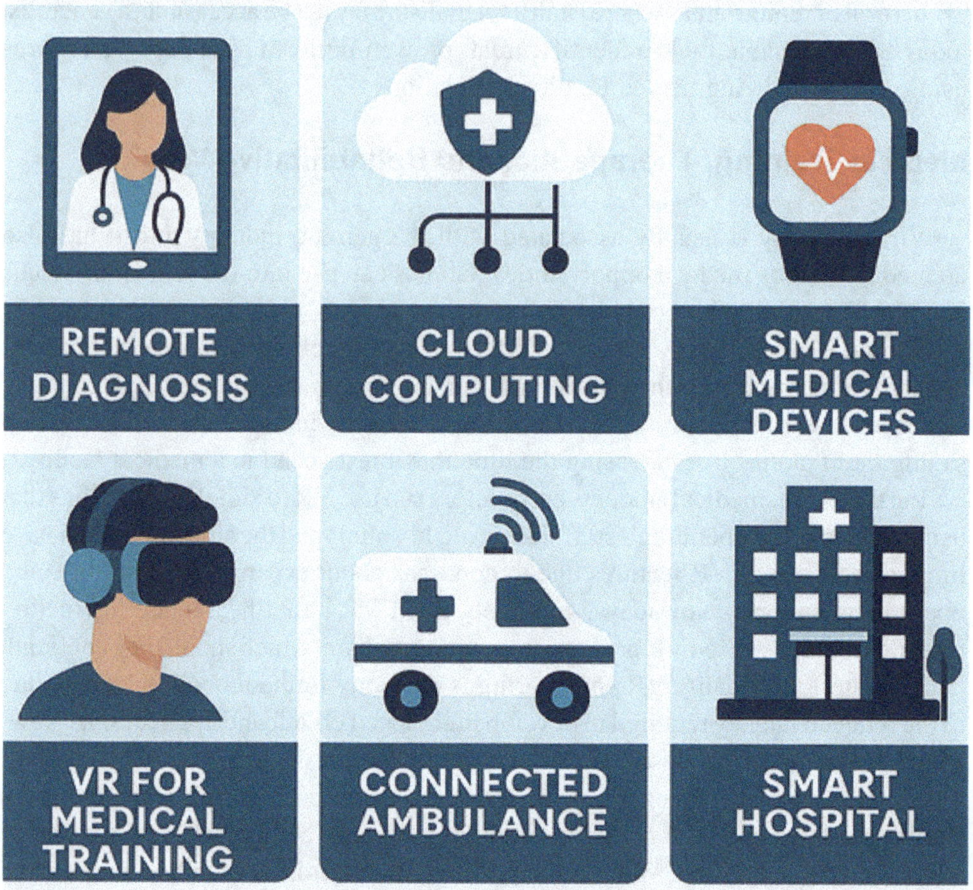

APPLICATIONS FOR FINANCIAL SERVICES

For financial institutions, a proven way to leverage these opportunities for transformation is through 5 G. Industry experts estimate that over 80 billion terminal products will connect to the 5G network within the next five years. Tian et al. (2019) explore how 5G technology drives innovation in financial services by enhancing mobile banking security, reducing transaction delays, and enabling real-time, high-frequency trading, while highlighting the potential of 5G to boost productivity, customer satisfaction, and risk management in the financial sector. In June 2020, the Commercial and Industrial Bank of China and China Mobile jointly released the white paper "Banking Innovation in the 5G Era," focusing on the revolutionary applications of 5G banking. Lee and Park (2022) emphasize that 5G is crucial in ensuring secure and high-speed connections for financial transactions.

The article stated that 5G accelerates the digitization of financial institutions, including internal operations and customer support. Additionally, enhanced speed and security enable users to complete transactions promptly, allowing remote tellers to participate.

While several banks have begun exploring consumer experience layouts using 5G, most people still have reservations about the transformation that 5G would bring to bank services and customer experiences. For example, what new economic requirements will arise in the 5G era? Do many people still need to go to a bank branch? What food do departments have to offer to draw in clients? SLD develops three predictions that Destin can materialize in the 5G era to work on several strategies and design tasks within the banking business.

5G technology enables numerous key fintech features, such as high volume, security, stability, and instant response. 5G enables primary fintech development, including high volume, instant responses, robustness, and security. 5G mobile banking also handles multiple transactions and high-speed financial requirements in city centers, malls, festivals, and football stadiums. Suppose a bank proactively incorporates 5G in the client experience. In that case, it will be better prepared to tackle competition from tech companies (such as Facebook) that aspire to enter the banking business.

Exponential Growth in Mobile Banking With 5G

5G offers growth opportunities for mobile banking. The various components of 5G that enable smartphone banking include digital deposit apps, photo and bill payment apps, Smartphone Credit Card Scanners, Electronic Meetings, Advanced ATM Apps, and Mobile Payment Apps.

5 G-enabled smartphone banking is revolutionizing the financial sector by introducing faster, more secure, and innovative user services. Key applications include advanced ATMs for seamless transactions, mobile payments with reduced latency, and digital deposits that ensure real-time fund transfers. Features like smartphone-based credit cards enhance payment flexibility, while photo bill payments simplify bill processing through AI-enabled recognition. Electronic meetings powered by 5G connectivity also facilitate secure and efficient communication for banking professionals and customers. This 5G technological advancement represents a significant leap forward, as research highlights the substantial growth of banking technology enabled by 5G advancements.

Digital Deposit App

Internet technology, wireless mobile communication standards, and mobile apps have made depositing a cheque quite simple compared to an old-fashioned paper cheque. For example, in Digital Deposit Apps using smartphones, users can deposit a cheque and send two sides or a digitally scanned image to the bank for processing. As for deposits, checks are processed at the bank.

Photo Bill Payment Apps

In Photo Bill Payment Apps, bill information is transmitted to the respective customer account after taking a digital photograph of the bill. The bill's details include the payment amount, due date, and the company issuing the invoice, collected from the user. The bill amount is paid electronically by setting a reasonable due date and a regular bill payment schedule using these apps.

Smartphone Credit Card Scanners

Typically, the credit card is scanned by a scanner or reader owned by the seller, which increases the likelihood of the password being hacked. Smartphone Credit Card Scanner: A customer's credit card is scanned by the reader attached to a smartphone. The scanner reads the card, and a specified amount of money is transferred from the user's account to the payee's history, reducing the chance of someone hacking the password.

Electronic Meeting

Teleconferencing with a bank teller or loan officer is a new concept in electronic meetings that will soon be introduced. The customer or user must obtain prior permission to make an appointment with a bank teller or loan officer. The user can open a new account or app for a loan or mortgage through this service.

Advanced ATM Apps

Banking institutions should implement a more robust security mechanism to prevent fraudulent transactions, ensure users maintain customer trust and faith, and attract new customers through ATM transactions. In this advanced ATM App, as soon as the user

When standing in front of the ATM counter, facial recognition is captured through a facial recognition scanning and identification system attached to ATMs.

Facial recognition is sent to the bank's server and matched against the stored facial recognition data. If it matches, the user's smartphone will receive an OTP. When users enter the OTP in their apps through wireless communication, the OTP is sent to the bank, which then matches it with the already generated OTP. If it matches, the ATM is opened for the user's transaction.

Mobile Payment Apps

Mobile Payment Apps allow users to enter their credit or debit card information. For example, users may have multiple cards from different banks; all the card information is loaded into mobile apps. Then, out of various cards, customers select any one card from the menu on their Smartphone and then touch the phone to an in-store reader. After this process, the payment process is like regular credit or debit card transactions.

Micro-Banks Are Everywhere

Because 5G technology can handle vast numbers of visitors and innovations, we will see many more integrated micro and mobile banking devices in crowded locations, such as event spaces. Linking to the 5G network, next-generation intelligent ATM-like products have significantly incorporated advanced capabilities, utilizing technologies such as AI-assisted self-service, facial recognition, contactless interfaces, and e-signatures. They merge scenario-specific requirements with fiscal services and rapidly occupy previously unexplored areas where physical banks were previously unlikely to be present.

Figure 3. Applications for financial services

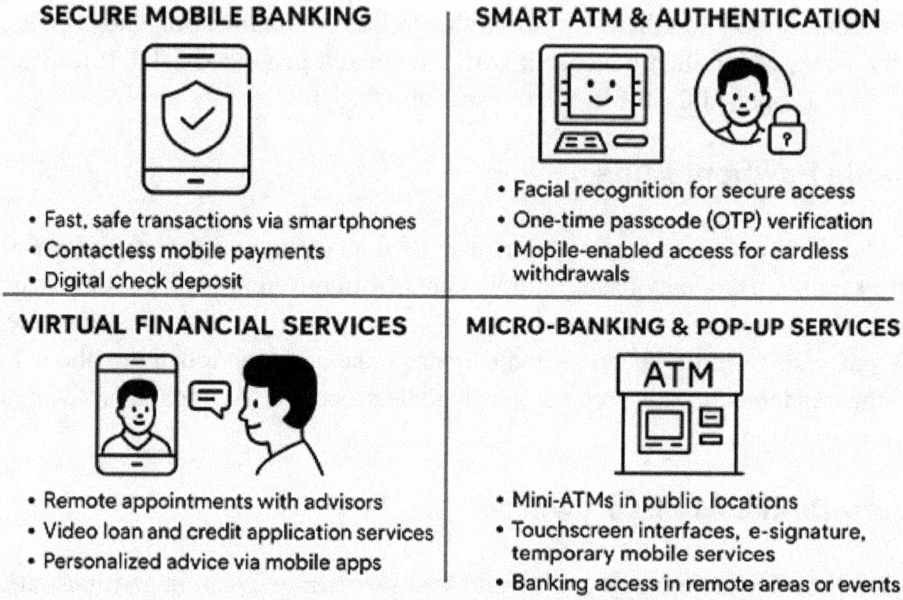

AR/VR REVOLUTIONIZES CUSTOMER EXPERIENCE

The combination of 5G high-speed transmission and cloud computing has a groundbreaking influence on AR/VR applications. It is conceivable that branches in remote areas can utilize VR to project images of customer care representatives. While this would alter the nature of several banking services, providing quality service is vital in delivering inclusive banking expertise.

The marriage of technical innovation and business models can give rise to a new ecology that produces new, previously unmet requirements. Banks that embrace and proactively deploy innovation can drive transformation, add value to their brand, and foster financial growth.

Media Ailment

5G brings new possibilities for TV, home broadband, mobile advertising, and active mobile technologies, including virtual reality (VR) and augmented reality

(AR). Improvements in 5G innovation are increasingly integrating with the broadcast communications sector, as evident in the use of cell phones and tablets.

It is an essential field of application; however, it is certainly not the only one. Many parts are poised to benefit from the various advantages offered by access to a faster, more stable, and more secure system and framework.

For instance, both the media and entertainment segments are preparing to adopt 5G to develop new capabilities or, more broadly, enhance their content creation and distribution.

Recordings, games, music, promotions, AR, and VR all undergo significant changes because 5G aims to decrease the separation between content and the general population.

Even with these, we must consider new media, such as vehicle diversion, 3D holographic showcases, live encounters in the arena, and comparative settings.

SMART AGRICULTURE

The article noted that farmers worldwide use IoT know-how to enhance crop monitoring, livestock safety, fertigation, and water management. 5G could enable real-time data collection, allowing growers to monitor and automate farming methods, thereby boosting security, efficiency, and profitability.

Drone Operations

Drones require high-speed internet connectivity to realize their full potential. They handle projects such as supplying our parcels and providing high-quality, real-time footage during health emergencies. At the same time, this is feasible to a certain degree; current communication technology has its limitations. High-speed online connectivity is a bottleneck, but this issue is expected to disappear with the massive rollout of 5G wireless connectivity. Azari (2022) notes that 5G enables highly dynamic applications such as drones and remote sensors by providing enhanced mobile connectivity.

The 5G rollout is an excellent enabler for improved drone utility for all drone technology workers. 5G is the fifth generation of electronic cellular networks. 5G network slicing leverages the physical resources of multiple virtual networks, making them more valuable and efficient.

This 5G network slicing enables drones to access advanced communication technology. It offers decreased latency (the delay before an information transfer begins), availability, capacity, and greater throughput. In addition, this capability enables drone activity to control and command (with precise and intuitive remote drone management), conduct independent flights, and share media.

One task currently open is the Horizon 2020-funded 5G DIVE initiative. The 5 G-DIVE project aims to conduct end-to-end 5G trials to demonstrate the technical advantages and business value of 5G technologies through two vertical pilot applications: Industry 4.0 and Autonomous Drone Scout. This specific effort involves a pilot examination designed to enable autonomous drone flight. Similarly, the automotive industry is exploring 5 G-enabled autonomous driving, and drone designers are excited about the prospect of independent flight.

In the US, Verizon, a telecommunications company, recognizes the significance of 5G's impact on drone technologies. As a result, it expects to become the first mobile operator to connect one million drone flights. To achieve that goal, Verizon acquired Skyward's drone procedure business.

High-Definition Footage

5G enables workers to transmit high-definition footage. The development of wireless communication standards from 4G suggests enhanced latency and data speeds that are 100 times faster than those of 4 G.

The technological rollout enables infrastructure safety inspections, which may involve high-definition video inspections of hard-to-access infrastructure, such as suspension bridges or wind generators. Since the 4 G standard would not be achievable in that specific utilization situation, the operator must be able to examine comprehensive HD pictures of the infrastructure.

This concept is apt to modify the dynamics of policing. For example, the New Zealand Police wish to utilize HD video streams from 5G networked devices in their policing efforts. The application relays footage to charge centers. With a total rollout over time, it is sensible to expect this departure to impact how police forces choose police. The knock-on outcome might be many fewer police officers on patrol. They could be displaced by officers manning worker bees in a command and control center.

Effectiveness Gains

Reaching a 5G standard implies that one-third of technology harnesses Artificial Intelligence. After the drone interaction standard shifts from 4G to 5G, it will be easier to operate AI. AI technology can be utilized onboard a drone to recognize patterns using laptop vision in a warehouse, which is a valuable efficiency advancement tool.

AI may also help locate natural disaster victims using live stream footage from 5 G-enabled drones. For example, in a wildfire scenario, firefighters can now survey land areas more efficiently, eliminating the need to place themselves in danger by doing so manually.

Many drone-based services are already set up and facilitated via 4G communication. Nevertheless, an improvement to 5G is likely to enhance their service offerings and quality levels.

"5G coverage would allow us to process some wealthy information sets outside of the drone with really low latency, basically providing us supercomputing processing powers on the drone," according to Alan Wicks, CTO of Manna.

Manna is an Irish B2B drone delivery-as-a-service business that plans to launch food deliveries in the coming weeks. Manna CTO Alan Wicks told 150sec that "4G presently allows Manna to run at scale and with much safety." (Cawfield, 2016).

Even with that, the rollout of 5G facilitates the organization's ability to do much more. For example, Wicks mentioned that "currently, due to the boundaries of 4G, real-time processing occurs on the drone."

"5G coverage would allow us to process some wealthy information sets outside of the drone with really low latency, basically providing us supercomputing processing powers on the drone" of ours, he adds.

Fleet Tracking

5 G's lower latency facilitates the tracking of drone fleets with pinpoint accuracy. Improved search capabilities prevent drones from entering restricted areas and minimize the risk of collisions.

Correct positioning can also lead to greater unit density. This technology changes the assortment of applications and use cases. For example, it might result in a scenario in which thousands of nano-drones are deployed in a specific location to capture sensor or video footage measurements.

Two-way block. It is not just a case of what 5G can do for drone technologies. Swedish telecoms and marketing company Ericsson suggests that drones could play a role in interference management on 5G networks. In a research blog post, airers4you is considering how radio scanner worker bees may contribute to maintaining network efficiency. Whereas drive-by and walk-by testing are time-intensive, drones can quickly fulfill this role. Additionally, they have greater ease of access, enabling them to calculate and evaluate outside radio signal interference from individual indoor 5G networks.

The growth of one engineer is strongly linked with that of another. Thus, drone engineering continues to establish itself as a distinct field of study. Nevertheless, the strength associated with a 5G networked environment flourishes when applied.

CONCLUSION

Chapter 10 highlights the profound impact of 5G technology across three key industries: logistics, energy, and financial services. The logistics sector benefits from enhanced visibility, real-time tracking, and automation, supported by the integration of IoT and AI, which enables predictive maintenance and streamlined operations. Innovative grid implementations and intelligent metering systems are revolutionizing consumption efficiency and grid resilience for the energy sector. 5G accelerates digital transformation in finance through secure, low-latency networks that support high-frequency trading, mobile banking, and augmented reality (AR) and virtual reality (VR)-enhanced customer experiences.

The convergence of 5G with complementary technologies such as blockchain, augmented reality, and edge computing enables greater operational agility, customer responsiveness, and sustainability. Despite infrastructure and regulatory hurdles, the trajectory of 5G adoption indicates a shift toward hyper-connected ecosystems. The table below synthesizes the principal advantages and ongoing challenges of 5G across logistics, energy, and financial services.

Table 2. Comparative analysis of 5G applications in key industries

Industry	Key Applications	Benefits	Challenges
Logistics	V2V, IoT sensors, drone delivery	Real-time tracking, optimized routing	Infrastructure cost, regulatory complexity
Energy	Smart grids, remote fault detection	Efficient energy use, predictive maintenance	Coverage gaps, latency concerns
Finance	Mobile banking, VR customer service	Faster transactions, enhanced UX	Cybersecurity, integration with legacy systems

Integrating 5G technology into logistics and shipping marks a transformative shift that redefines how goods and transportation systems are managed in an increasingly interconnected world. With its unparalleled speed, ultra-low latency, and ability to support massive connectivity, 5G offers a robust platform for real-time communication, enhanced data-driven decision-making, and optimized operations. The capability of 5G to enable seamless vehicle-to-vehicle (V2V) communication has already laid the groundwork for advancing autonomous vehicles and intelligent transportation networks, offering unprecedented opportunities for innovation.

Beyond vehicles, 5G is pivotal in improving end-to-end visibility across supply chains. Logistics providers can track shipments in real-time, predict potential delays, and enhance efficiency. By integrating IoT devices with 5G networks, logistics stakeholders can monitor critical parameters such as location, temperature, and condition of goods with greater precision, ensuring superior service quality. Additionally, harnessing predictive analytics through 5 G-powered data processing enables proactive responses to disruptions, further enhancing the resilience of supply chains.

The environmental implications of 5G in logistics are equally significant. Optimized fleet management and more intelligent transportation systems reduce consumption and carbon emissions, contributing to global sustainability goals. As cities and businesses adopt 5G to enhance their logistics and shipping operations, the benefits extend beyond efficiency, including safer roads, improved urban planning, and reduced environmental impact.

In conclusion, 5G technology represents a pivotal enabler of innovation in logistics and shipping, transforming challenges into opportunities for growth and sustainability. By fostering smarter, faster, and more efficient systems, 5G drives progress within the logistics sector and sets a new standard for global connectivity in transportation. As adoption accelerates, the full potential of 5G in logistics and shipping will continue to unfold, shaping the future of these critical industries.

REFERENCES

Apruzzese, M., Bruni, M. E., Musso, S., & Perboli, G. (2023). 5G and companion technologies boost new logistics and supply chain business models. *Sustainability*, *15*(15), 11846. DOI: 10.3390/su151511846

Azari, M. (2022). *Cellular, Wide-Area, and Non-Terrestrial IoT: A Survey on 5G Advances and the Road Toward 6G*. IEEE Communications Surveys & Tutorials; Institute of Electrical and Electronics Engineers (IEEE). https://www.academia.edu/100346983/Cellular_Wide_Area_and_Non_Terrestrial_IoT_A_Survey_on_5G_Advances_and_the_Road_To

Carrillo, D., Kalalas, C., Raussi, P., Michalopoulos, D. S., Rodriguez, D. Z., Kokkoniemi-Tarkkanen, H., Ahola, K., Nardelli, P. H. J., Fraidenraich, G., & Popovski, P. (2023). Boosting 5G on Smart Grid Communication: A Smart RAN Slicing Approach. *IEEE Wireless Communications*, *30*(5), 170–178. DOI: 10.1109/MWC.004.2200079

Cawfield, D. (2016) *Manager Insight: Patience pays off in Manulife balancing act*. Morningstar Canada. https://www.investmentexecutive.com/news/products/manager-insight-patience-pays-off-in-manulife-balancing-act/

Goasduff, L. (2021). *Gartner Forecasts Worldwide 5G Network Infrastructure Revenue to Grow 39% in 2021*. Gartner. https://www.gartner.com/en/newsroom/press-releases/2021-08-04-gartner-forecasts-worldwide-5g-network-infrastrucutre-revenue-to-grow-39pc-in-2021

Kim, E.-S., Sung, M., Lee, J. H., Lee, J. K., Cho, S.-H., & Kim, J. (2020). Coverage Extension of Indoor 5G Network Using RoF-Based Distributed Antenna System. *IEEE Access : Practical Innovations, Open Solutions*, *8*, 194992–194999. DOI: 10.1109/access.2020.3033592

Lagorio, A., Cimini, C., Pinto, R., & Cavalieri, S. (2023). 5G in Logistics 4.0: Potential Applications and Challenges. *Procedia Computer Science*, *217*, 650–659. DOI: 10.1016/j.procs.2022.12.261

Li, D. (2019). 5G and Intelligence Medicine—How the Next Generation of Wireless Technology Will Reconstruct Healthcare? *Precision Clinical Medicine*, *2*(4). Advance online publication. DOI: 10.1093/pcmedi/pbz020

Sag, A. Townsend, W., & Moorhead, P. (2025). *The G2 on 5G Podcast - Episode 219: New Intel CEO, 5G Advanced, Converged Connectivity*. Moor Insights & Strategy. https://moorinsightsstrategy.com/g2-on-5g/the-g2-on-5g-episode-219-new-intel-ceo-5g-advanced-converged-connectivity/

Sharma, P., Jain, S., Gupta, S., & Chamola, V. (2021). Role of machine learning and deep learning in securing 5G-driven industrial IoT applications. *Ad Hoc Networks*, *123*, 102685. DOI: 10.1016/j.adhoc.2021.102685

Shehab, M. J., Kassem, I., Kutty, A. A., Kucukvar, M., Onat, N., & Khattab, T. (2022). 5G networks towards smart and sustainable cities: A review of recent developments, applications and future perspectives. *IEEE Access : Practical Innovations, Open Solutions*, *10*, 2987–3006. DOI: 10.1109/ACCESS.2021.3139436

Tian, M.-W., Wang, L., Yan, S.-R., Tian, X.-X., Liu, Z.-Q., & Rodrigues, J. J. P. C. (2019). Research on financial technology innovation and application based on 5G network. *IEEE Access : Practical Innovations, Open Solutions*, *7*, 138614–138623. DOI: 10.1109/ACCESS.2019.2936860

Zhang, Z., & Patel, K. (2023, July 9). *ANUP Implementation in 5G with BGP Signaling*. Ietf.org. https://www.ietf.org/archive/id/draft-zzhang-dmm-anup5g-signaling-00.html

ADDITIONAL READING

Slamnik-Krijestorac, N., Landi, G., & Brenes, J. Alexandru Vulpe, Suciu, G., Carlan, V., Konstantinos Trichias, Ilias Kotinas, Municio, E., Athina Ropodi, & Marquez-Barja, J. M. (2022). *Network Applications (NetApps) as a 5G booster for Transport & Logistics (T&L) Services: The VITAL-5G approach*. In 2022 Joint European Conference on Networks and Communications & 6G Summit (EuCNC/6G Summit) (pp. 279–284). IEEE. https://doi.org/DOI: 10.1109/EuCNC/6GSummit54941.2022.9815830

Section 4
Research

Chapter 11
Emerging Research for 5G

ABSTRACT

This chapter examines cutting-edge research that is shaping the evolution of 5G networks, with a focus on advanced technologies that address their complex demands. Key innovations include the deployment of small cells and peer-to-peer communications (D2D, M2M) in heterogeneous networks. Enabling technologies such as full-duplex communication, massive MIMO, and millimeter-wave systems are highlighted for their role in enhancing network performance. Energy efficiency emerges as a core priority, with energy-aware communication and energy-harvesting strategies being crucial to the user experience. Additionally, cloud-based radio access networks (C-RAN) and spectrum virtualization are presented as essential for managing resources and ensuring network scalability.

INTRODUCTION

This chapter provides an overview of emerging research and new technologies for 5 G. The primary approaches and technologies to address the demands of 5G systems include densifying current cellular networks through the widespread deployment of base stations and small cells, as well as supporting peer-to-peer (P2P) interactions such as device-to-device (D2D) and machine-to-machine (M2M) communication within multi-tier heterogeneous networks (Andrews et al., 2014). Chin, Fan, and Haines (2014) examine the foundational technologies driving the evolution of 5G wireless networks, highlighting current developments and future research challenges. The article discusses how specific innovations, such as Massive MIMO, interference management techniques, and architectural enhancements, are beginning to influence standards like 3Gwithin electronic instrumentation systems PP LTE, while others remain at the research stage. The authors emphasize that while these emerging

DOI: 10.4018/979-8-3373-0497-7.ch011

technologies offer significant potential to increase throughput, improve network efficiency, and support advanced services, they also introduce complex problems related to implementation, interference, and system integration that demand further investigation (Chin, Fan, & Haines, 2014).

Amali and Ramachandran (2018) conduct an extensive review of the enabling technologies critical to the development of 5G networks, highlighting architectural innovations such as Cloud-RAN (C-RAN), Software-Defined Networking (SDN), and Network Function Virtualization (NFV). These technologies enhance network scalability, flexibility, and resource efficiency. The authors also emphasize the significance of advanced modulation schemes in fulfilling the high data rate requirements of 5G networks. Despite these technical advancements, several critical challenges remain, including the management of ultra-dense network environments, maintaining consistent Quality of Service (QoS), enhancing energy efficiency, and minimizing deployment costs. The paper concludes that effectively addressing these challenges is critical for the successful implementation and long-term optimization of 5G networks (Amali & Ramachandran, 2018).

Sun, Hu, and Qian (2023) discuss unresolved challenges and research directions in 5G and beyond networks. Jaber et al. (2016) extensively survey the critical challenges associated with 5G backhaul networks, identifying them as a significant bottleneck in deploying ultra-dense 5G infrastructures. Li, Tang, and Luo (2022) highlight critical infrastructure challenges in emerging 5G regions.

While research efforts have primarily focused on enhancing the radio access segment through technologies such as massive MIMO, network densification, and carrier aggregation, the authors emphasize that backhaul systems must also meet stringent requirements for capacity, latency, reliability, and energy efficiency. The paper critically evaluates legacy and emerging backhaul solutions, proposing an integrated framework that considers radio access and backhaul perspectives. The study argues that no single backhaul solution is sufficient to meet the complex demands of 5 G. It highlights the need for a hybrid approach to support future high-capacity networks effectively. Key insights are provided into the strengths and limitations of various backhaul technologies when considered in the broader context of 5G network architecture (Jaber et al., 2016).

Gupta and Jha (2015) present a comprehensive survey of 5G network architecture and emerging technologies designed to meet critical future demands, including increased capacity, higher data rates, reduced latency, and enhanced Quality of Service (QoS). The study focuses on key architectural advancements, including massive Multiple Input Multiple Output (MIMO) and tools such as Device-to-Device (D2D) communication, as central components of 5G networks. Additionally, the paper explores other essential innovations, including interference management, spectrum sharing via cognitive radio, ultra-dense networks, multi-radio access, full-duplex communication, millimeter-wave solutions, and the integration of cloud computing with Software-Defined Networking (SDN) for efficient radio access. A 5G architecture is proposed, incorporating small cells, D2D links, network cloud infrastructure, and the Internet of Things (IoT) as foundational elements. The authors also review global research initiatives that are actively pursuing these advancements, underscoring the collaborative effort required for 5G deployment (Gupta & Jha, 2015).

Simultaneous transmission and reception, such as full-duplex (FD) communication, massive multiple-input multiple-output (MIMO), and millimeter-wave (mm-wave) communication technologies, play a crucial role in modern wireless systems.

It has improved energy efficiency through energy-aware communication and effort harvesting.

- Cloud-based radio access system (C-RAN) and virtualization of wireless information.

With the rise of energy-aware CAN FD communication and the implementation of spectrum virtualization, the architecture of 5G units has become significantly more complex than that of 4G systems. CAN FD (Controller Area Network Flexible Data-Rate) facilitates the broadcasting of sensor data and control signals over two-wire connections within electronic instrumentation systems (Robert Bosch GmbH, 2012). This protocol is used in modern high-performance vehicles.

- Spectrum visualization is a technique that utilizes diagrams, images, or animations to illustrate the spectrum, thereby facilitating easier understanding.

Applications in 5G networks must be practical for running across multiple spectrum bands, ranging from radio frequency (RF) to millimeter-wave (mm-wave), while remaining backward compatible (e.g., with existing technologies like 3 G and 4G).

Due to the power-hungry nature of multimedia applications, energy efficiency is crucial for the 5G user experience. Therefore, it is appealing that the equipment integrates energy harvesting technologies to support multiple RATs, energy, and F.D. communication harvesting.

The integration of 5G in manufacturing relies on a suite of enabling technologies that collectively enhance efficiency, responsiveness, and security. Each of these innovations plays a critical role in modernizing production environments to support smart manufacturing objectives. The following table, adapted from Hossain and Hasan (2015), highlights the key technological trends driving this transformation. These innovations include connectivity, computing, intelligence, and protection layers that work in synergy. Understanding their functions helps stakeholders align technological investments with operational goals. The following table, adapted from Hossain and Hasan (2015), summarizes some of the topics discussed in this section.

Table 1. Enabling technologies and trends

Enabling Technology	Adapted Role/Trend in Manufacturing
5G Connectivity	Supports ultra-reliable low-latency communication for automation and real-time control.
IoT Devices	Enable seamless interconnection of sensors and machines across the production floor.
Edge Computing	Facilitates rapid data processing close to the source, reducing latency.
AI & Machine Learning	Enhances predictive maintenance and operational optimization.
Network Slicing	Allows for dedicated virtual networks tailored to specific industrial needs.
Cybersecurity Protocols	Ensure secure communication and protect critical infrastructure from threats.

The industry faces significant challenges in addressing increased bandwidth, latency requirements, and the processing of big data generated by more connected 5G equipment, as well as the data processing required on the factory floor. For example, deploying 5G immersive media services in cloud-enabled small cell networks, utilizing a neutral host business model, offers a viable solution for providing high-quality services during crowded events (Paglierani et al., 2019).

This section presents a comprehensive techno-economic analysis of deploying 5G immersive media services in venues such as stadiums, utilizing cloud-enabled small cell networks managed by a neutral host. Paglierani et al. (2019) conducted a comprehensive techno-economic analysis of deploying 5G immersive media services in venues such as stadiums, utilizing cloud-enabled small cell networks managed by a neutral host. The study developed a planning model to estimate the necessary compute, storage, and radio resources, taking into account factors like service penetration and price evolution. The economic assessment employed metrics including net present value, internal rate of return, and expected payback period, demonstrating the investment's viability for delivering innovative 5G services during crowded events.

The deployment of fifth-generation (5G) wireless systems presents several challenges in measuring and modeling network propagation channels. As Rapaport et al. (2013) highlight, the primary difficulties stem from the use of higher frequencies, such as those in the millimeter-wave (mm-wave) range, wider bandwidths, and the incorporation of massive antenna arrays. These technical advancements increase system complexity and necessitate more detailed spatial domain channel models. Additionally, the rapid increase in wireless data traffic demands more efficient use of spectrum and infrastructure, which large-scale MIMO technology supports by densifying network nodes. Consequently, this leads to intensified requirements for measurement tools and techniques that can capture the nuanced behavior of signal propagation in these dense and high-frequency environments.

Moreover, accurate testing of 5G systems must account for diverse environmental conditions, mobility scenarios, and hardware imperfections that affect signal quality and latency, (Heath et al., 2016). The measurement frameworks must integrate real-time capabilities and adaptive signal processing to cope with the dynamic characteristics of 5G channels. This includes challenges in modeling non-stationary behavior, frequency selectivity, and spatial correlation within urban and indoor environments. Furthermore, multi-user interference and the variability introduced by beamforming techniques further complicate empirical validation of theoretical models (Haneda et al., 2016). Therefore, developing standardized, scalable, and reproducible testing methodologies is essential for ensuring performance reliability and interoperability across vendors.

MEASUREMENT AND TEST CHALLENGES FOR 5G SYSTEMS

This research examines the modeling and measurement of network channels, arguing that the core challenge lies in utilizing higher frequencies, broader bandwidths, and significantly larger antenna arrays (Rapaport et al., 2013).In addition, expanding wireless website traffic requires additional spectrums (e.g., higher frequencies in the mm-wave range) and enabling technologies like large-scale MIMO, which densifies network nodes and forces new demands for channel measurement and modeling in the spatial domain. The primary challenges and considerations for measuring and modeling 5G propagation stations have several characteristics.

Efficient as Well as Practical Measurement

Measurement information is vital for extending or modifying free propagation versions. Therefore, the measurement strategy must record several frequency ranges, spatial consistency, 3D (e.g., spherical waves and elevation), and new communication models, such as small cell communications and d2d/m2m. Additionally, feasible and practical use cases (such as crowded places and vehicle-to-vehicle/roadside communications) consider measurements for mm-wave (e.g., 60 GHz and above) in both indoor and outdoor situations.

Spatial Mobility and Distributions

The current channel is drop-based; for example, the scattering surroundings are arbitrarily generated for every link. Nevertheless, since the density of backlinks is likely to increase in 5G methods, it is essential to model these backlinks regularly. It could also support the heterogeneous mobility of various network nodes.

To accurately reflect 5G environments, spatial consistency in modeling is necessary to capture realistic user mobility and signal propagation. This includes accounting for transitions between Line-of-Sight (LoS) and Non-Line-of-Sight (NLoS) conditions as users move. Furthermore, modeling should integrate correlated shadowing effects and dynamically changing urban topologies. Spatial mobility modeling also aids in designing predictive handover mechanisms and load balancing strategies. Overall, consistent spatial distributions are crucial for evaluating and optimizing 5G network performance under real-world mobility patterns.

Large-Scale Antenna Arrays, as Well as Mm-Wave Correspondence

The 5G technology uses vast-scale antenna arrays for direct communication. Directive antennas present channel designs that need corresponding angular resolution, subpath amplitude, and division enhancement. Large-scale antenna arrays necessitate the use of a spherical wave model rather than the widely employed plane wave approximation (Han et al., 2015). The utilization of mm-wave frequencies (particularly at 60 GHz) enables the achievement of spatial multiplexing gain and spectrum efficiency. Nevertheless, too-resolved angular qualities and non-line-of-sight path loss are not recognized and require additional measurement studies.

To optimize performance, accurate channel modeling must capture the beamforming characteristics of large-scale MIMO systems. Mm-wave propagation faces challenges such as high attenuation, sensitivity to blockage, and limited diffraction, especially in urban or indoor environments. Advanced ray-tracing simulations and empirical datasets are crucial for a deeper understanding and more accurate prediction of mm-wave behavior. Hybrid beamforming techniques are also being explored to strike a balance between hardware complexity and performance. Ultimately, robust channel models tailored to large-scale arrays and millimeter-wave frequencies are fundamental for the success of 5G deployment.

Testing 5G Systems

The utilization of significant channel bandwidths, high detail requirements, rapid response times, more complex antenna configurations, and support for advanced stereo access technology (RAT) pose considerable challenges for the next generation's improvement of B.S. products. Testing 5G systems is not limited to hardware; it also necessitates the validation of new communication algorithms and approaches.

From the viewpoint of an RF engineer, several of the primary critical measurement and design issues for 5G methods are as follows:

- The necessity to deal with multiple channel bandwidths (e.g., 1.4 to 20 MHz for LTE-A) and operability in higher bands (e.g., 60 GHz for mm-wave).
- (The utilization of various transmission modes in LTE A (e.g., orthogonal frequency division much access [OFDMA] for uplink, single carrier frequency division multiple access [SC FDMA] for downlink) as well as provisioning for equally transmission modes (e.g., time/frequency division duplexing [TDD/FDD]).
- (Support for multi-RAT (e.g., because of multi-standard stereo transmission requirements) along with simultaneous transmission/reception of the command, data, and energy.
- The spectral, energy, and time variations (due to heterogeneous network visitors and node density) must be considered when implementing multi-antenna strategies (e.g., spatial multiplexing, transmitter diversity, and beamforming).

The Challenges Regarding Testing and Measurement

The challenges regarding testing and measurement of the receiver radios for the 5G methods can include the following:

(i) Verifying the R.F. and baseband receiver (e.g., autonomic gain control, baseband demodulation, and hybrid car automated repeat request [HARQ] functional assessment).
(ii) Receiver effectiveness under impaired problems (e.g., phase interference and preservative, white-colored Gaussian interference [AWGN] impairments).
(iii) Receiver general performance testing (e.g., testing under channel propagation impairments).
(iv) Testing the MIMO-enabled methods (e.g., baseband coding assessment, receiver verification problems like isolation of indicators, and performance tests under fixed and fading channels).

The deployment of 5G networks presents significant challenges in testing and measurement due to its reliance on advanced technologies, including massive MIMO, millimeter waves, and ultra-dense networks. These innovations increase complexity in evaluating performance, coverage, and reliability. Hossain and Hasan (2015) identify key dimensions where measurement techniques must evolve to match the demands of 5G infrastructure.

Additionally, signal variability caused by mobility, environmental conditions, and device heterogeneity adds another layer of difficulty in evaluating receiver performance. Accurate modeling of signal propagation across diverse urban, suburban, and indoor scenarios is critical. Instrumentation must also be adapted to test over a wide frequency range, including sub-6 GHz and mmWave bands, which demand different test setups and calibration routines.

Moreover, as 5G systems use beamforming, precise alignment and beam quality assessment become necessary during receiver validation. Synchronization between multiple input/output channels under real-time traffic also represents a significant test burden. Tools must support both simulated and live network conditions to facilitate comprehensive performance analysis and evaluation.

Security testing is also increasingly vital, as 5G introduces network slicing and virtualized components that could affect receiver integrity. Engineers must ensure data confidentiality and low-latency response while handling a massive number of simultaneous connections.

Future test environments must incorporate AI-driven analytics to handle large datasets from test outputs, providing predictive insights into failures and performance bottlenecks. As a result, the evolution of test and measurement methodologies will be central to the robust deployment of 5G technologies.

CHALLENGES IN WIRELESS COMMUNICATION NETWORKS

The development of 5G wireless communication networks introduces a range of critical challenges that must be addressed to ensure successful deployment and adoption. According to Haider et al. (2011), one of the primary difficulties is spectrum scarcity, particularly in the lower-frequency bands, which are already heavily utilized by existing technologies. Another significant issue is interference management, as ultra-dense network deployments increase the likelihood of cross-cell interference.

Additionally, 5G systems must accommodate heterogeneous network environments that include macro-, micro-, and pico-cells, requiring seamless handovers and integration. Energy efficiency also emerges as a key concern, given the expected growth in connected devices and the corresponding rise in power consumption. Security and privacy must be reimagined for billions of IoT devices operating under variable trust conditions. Latency reduction and high reliability are crucial for mission-critical applications, such as autonomous vehicles and remote surgery. Furthermore, the backhaul infrastructure must be upgraded to support the enormous data throughput demands. Finally, the adoption of new technologies such as massive MIMO and millimeter-wave communications introduces complexity in hardware, signal processing, and network optimization.

According to Haider et al. (2011), the main challenges of the 5G wireless communication networks are the following:

Optimizing Performance Metrics

Due to their higher complexity, wireless communication network analyses typically consider only one or two performance metrics while neglecting others. A fair and complete assessment of 5G wireless methods requires additional performance metrics. These include spectrum efficiency, energy efficiency, delay, reliability, consumer fairness, implementation complexity, and quality of service. Consequently, a comprehensive framework is created to assess the functionality of 5G wireless methods, considering as many performance metrics as possible from different perspectives.

Realistic Channel Models for 5G Wireless

Realistic channel versions with accurate and complex trade-offs are essential for common 5G scenarios, such as massive MIMO routes and high-mobility channels (e.g., high-speed train and vehicle-to-vehicle scenarios). Standard MIMO channel models cannot be exclusively applied to massive MIMO routes, in which different antennas may observe unique sets of clusters. Huge MIMO channel designs should consider qualities that distinguish them from conventional MIMO channels, such as the spherical wavefront and non-stationary properties.

Interference Management for Cognitive Radio (Cr) Networks

A significant challenge in interference-tolerant Cognitive Radio (CR) networks for 5G is determining the most effective way to manage CR's mutual interference reliably and practically using primary methods. Power is essential for any C.R. phone system to coexist with other licensed forms to regulate transmission. An interference temperature model is created for this function to characterize the interference from the C.R. to the licensed networks. Mitigation of the interference at cognitive radio (C.R.) receivers must also be implemented to develop effective interference cancellation techniques, (Akyildiz et al., 2006).

An additional problem in interference-tolerant C.R. networks is that a feedback mechanism is crucial for periodically informing the C.R. system about the current interference status in the licensed phone system. A practical solution is that the interference declares that information may be delivered from suitable methods and collected by the primary device (or a third-party system).

CHALLENGES FROM THE RADIO RESOURCE MANAGEMENT PERSPECTIVE (RRM)

In 5G networks, RRM must dynamically adapt to rapid variations in user mobility, traffic patterns, and service requirements, (Zhang, Zhou, & Hu, 2021). One of the significant challenges lies in ensuring fair resource allocation among users with heterogeneous quality-of-service (QoS) demands while maximizing overall spectral efficiency. The introduction of network slicing further complicates RRM, as it requires the isolation and customization of resources per slice. Additionally, inter-cell interference coordination (ICIC) becomes more complex in ultra-dense environments where small cells overlap, making it challenging to manage interference. Beamforming techniques, while enhancing spatial reuse, require sophisticated scheduling algorithms to avoid beam conflicts and optimize directional transmission. Another challenge involves managing energy efficiency without compromising latency or throughput, particularly in machine-type communications (mMTC). The integration of AI-based predictive models into RRM systems introduces computational overhead and demands accurate real-time data. Moreover, maintaining load balancing across heterogeneous cells and Radio Access Technologies (RATs) is essential to avoid bottlenecks. Mobility management in high-speed scenarios, such as vehicular networks, places additional strain on radio resource management (RRM) algorithms. Addressing these challenges will require intelligent, distributed, and context-aware RRM strategies that can operate autonomously at the edge of the network.

Radio Resource Management (RRM) in Radio Access Networks (RANs) is a large-scale control problem that encompasses multiple network functionalities operating simultaneously, spanning from sub-milliseconds to seconds.

The evolution to 5G introduces new levels of complexity in managing wireless resources efficiently across ultra-dense, heterogeneous networks. As diverse services—from autonomous vehicles to industrial automation—compete for limited radio resources, the role of Radio Resource Management (RRM) becomes increasingly critical. RRM governs the allocation and optimization of spectrum, time, power, and spatial resources across multiple layers of the network. It must address strict requirements for low latency, high reliability, and massive connectivity. The following section outlines key challenges that must be addressed to ensure 5G networks deliver on their performance promises.

Interference Management Within Heterogeneous Networks

A challenge is interference management within heterogeneous networks. Due to the widespread deployment of heterogeneous nodes within 5G networks, one method of enhancing resource utilization is to utilize readily available information in a spectrum underlay manner. Nevertheless, for the underlying 5G networks, interference management is among the primary vital obstacles. One of the primary challenges for multi-tier heterogeneous networks is mitigating inter- and intra-tier, as well as inter-cell, interference. Along with the dense deployment and heterogeneity of wireless devices, coverage and targeted traffic load imbalance resulting from altering the transmit powers of various base stations (BSs) exacerbate interference management and resource allocation challenges significantly more than in standard single-tier methods.

Full-Duplex Communication

In the context of full duplex (F.D.) communication, SI (self-interference) limits the theoretical gain of the F.D. transmission. Despite many S.I. cancellation mechanisms, a specific amount of S.I. stays in the product. The F.D. transmission performance is an effective S.I. cancellation method that can eliminate recurring S.I. You can find new inter-cell and intra-cell interference sources in F.D. interaction due to simultaneous transmission and reception.

Additionally, due to the higher transmit strength of base stations (BSs), downlink-to-uplink interference becomes more intense than uplink-to-downlink interference, (Sabharwal et al., 2014).

Thus, to achieve maximal spectral efficiency, sophisticated channel estimation, interference management, and power allocation mechanisms (utilizing advanced methods such as transceiver beamforming and distributed space-time coding) are expected to consider new solutions for managing interference. The F.D. transceivers can also choose the transmission modes (e.g., HD/FD) based on the system dynamics.

Wireless Network Virtualization (WNV) encompasses spectrum sharing and infrastructure sharing, and it requires effective resource utilization, interslice isolation, and customizable slice resource allocation. The resource allocation algorithms must be created with the interests of heterogeneous and finite resources (e.g., spectrum, infrastructure, computing), the source needs of MVNOs, and fairness criteria.

Additionally, boot and control signaling stamping problems must be resolved for spectrum virtualization. Finally, the issues related to mobility management, pricing-based allocation, isolation, and resource discovery (e.g., seamless handover to various MVNOs/InPs) opened up new research opportunities.

Cloud Run

Cloud-RAN (CRAN) is a cutting-edge radio access network (RAN) technology, as outlined above [xxxviii]. In Cloud-RAN, operators can deploy mobile networks more quickly by utilizing various access technologies while sharing the same infrastructure. In conclusion, the lower deployment price and reduced system resources result in lower maintenance costs.

Cloud-RAN (C-RAN) represents a fundamental shift in how radio access networks are architected and managed in the 5G era. By centralizing baseband processing in a cloud-based infrastructure, C-RAN decouples the radio units from the computational core, allowing for more flexible, scalable, and cost-effective deployments. This architecture enhances spectral efficiency, simplifies network upgrades, and enables real-time coordination between cells. It also supports emerging technologies such as Massive MIMO and network slicing, making it ideal for dense urban environments and high-demand use cases. The following section explores the strategic benefits and implementation considerations of C-RAN in next-generation networks.

Wireless Network Virtualization

An entirely beneficial Wireless Network Virtualization (WNV) framework requires effective resource utilization, interslice isolation, and customizable slice resource allocation. The resource allocation algorithms must be created with the interests of heterogeneous and finite resources (e.g., spectrum, infrastructure, computing), the source needs of MVNOs (Mobile Virtual Network Operators), and fairness criteria. Additionally, boot and control signaling stamping problems were resolved for spectrum virtualization. The issues related to mobility management, pricing-based allocation, isolation, and resource discovery (e.g., seamless handover to various MVNOs/InPs) opened up new research opportunities. For example, Sun et al. (2022) illustrate how reinforcement learning and fog computing enhance mobile VR over 5G

Energy-Harvesting Communication

For R.F. Energy Harvesting Communication (EHNs), effective wireless charging and channel, energy, and time adaptation policies are examined. Standard multi-user scheduling schemes (e.g., greedy, opportunistic, and round-robin) do not consider the quantity of harvested energy or power specifications. Therefore, they could result in power (or transmit) outages or overflows in RF EHNs. Thus, effective multi-user booking schemes are necessary for RF EHNs. Also, powerful CSI acquisition techniques are needed for HAPs to enhance the transmit strength or the harvesting duration of the effort to transfer directional power based on the channel conditions.

Table 2 provides an overview of key emerging technologies that underpin the development of 5G cellular networks. These innovations are crucial for achieving the ambitious performance targets of 5G, including ultra-low latency, enhanced data rates, and massive connectivity. Each technology plays a distinct role, from enabling efficient spectrum usage to supporting direct device communication and reducing energy consumption. Their integration not only optimizes network performance but also supports a wide range of use cases across various industries. The following table highlights the core contributions of each technology within the 5G ecosystem.

Table 2. Summary of emerging technologies for 5G cellular networks

Emerging Technology	Role in 5G Networks
Millimeter Wave Communication	Enables high data rate transmission over short distances.
Massive MIMO	Enhances spectral efficiency and network capacity by utilizing large antenna arrays.
Full Duplex Radios	Supports simultaneous reception and transmission on the same frequency.
Device-to-Device Communication	Facilitates direct communication between nearby devices.
Cloud Radio Access Networks (C-RAN)	Centralizes baseband processing to enhance network flexibility and scalability.
Energy-Efficient Network Design	Reduces power consumption while maintaining performance.
Heterogeneous Networks (HetNets)	Combines macro and small cells to improve coverage and capacity.

Adapted from Hossain and Hasan (2015)

CHALLENGES FOR 5G BUSINESS SUPPORT SYSTEM EVALUATION

This section examines the significant challenges associated with evaluating Business Support Systems (BSS) within the context of 5G network deployments. As 5G introduces unprecedented levels of complexity, scale, and service diversity, traditional BSS architectures struggle to meet the evolving demands. These systems must now support dynamic service provisioning, real-time billing, and integration with advanced technologies such as AI and edge computing. Evaluating their performance and adaptability is vital to ensure seamless customer experiences and efficient network operations. The following section outlines the key issues and considerations for a future-ready 5G BSS infrastructure. In telecommunications, Business Support Systems (BSS and OSS/BSS refer to operations support systems and business support).

5G Technologies Can Remake Several Historical Infrastructures

5G technologies can remake several historical infrastructures in the current core system. Nevertheless, the 5G air-interface specifications include selecting new concepts and features, as well as requiring an effective development plan. These characteristics can consist of substantial MIMO, low-latency HARQ methods, high-order modulation schemes (e.g., 256 quadrature amplitude modulation [QAM]), and a wide range of combinations or configurations of RF bands and channel bandwidths to provide spectrum flexibility, assessment F.D. radios, and power-harvesting capabilities.

Instead of standard methods, the next generation of systems (including small cell systems and relays) must also deal with stereo source management (including admission management, load balancing, and stereo mobility decisions), as well as classic Layer One and Layer Two functionalities. Additionally, the anxiety assessment allows BSs, clouds, and primary networks to verify the overall performance against more network nodes.

The Assessment Problems for 5G

The assessment problems for 5G UEs resemble those for standard methods, which utilize measurement matrices such as optimum output power, receiver sensitivity, and power control. Nevertheless, OFDMA (Orthogonal Frequency-Division Multiple Access) leads to a multi-user version of the popular orthogonal frequency-division multiplexing (OFDM) digital modulation scheme. OFDMA assigns a variety of subsets of subcarriers to individual users, achieving multiple access. It represents transmission patterns in the downlink and SC-FDMA in the uplink within LTE-A/LTE-B-based 5G methods. At the same time, support for simultaneous communication and energy harvesting capabilities necessitates distinct measurement principles to enable the required performance assessments (Gupta & Jha, 2015).

The 5G Business Support Systems to Handle the Multidimensional Service Orchestration

Another fundamental challenge in evaluating 5G Business Support Systems is their ability to handle the multidimensional service orchestration required by network slicing. Each network slice, customized for a specific use case (e.g., autonomous vehicles, industrial IoT, or telemedicine), demands differentiated quality of service (QoS), security, and billing mechanisms. This multiplicity complicates BSS design and performance validation, especially as operators must manage concurrent slices with distinct lifecycles and dynamic workloads. To address this, BSS platforms must adopt flexible service catalogs, policy-driven charging systems, and AI-assisted analytics to monitor and optimize performance in real time.

Interoperability Across Heterogeneous Platforms

Additionally, interoperability across heterogeneous platforms presents a significant challenge for BSS testing. With the growing presence of third-party APIs, virtualized network functions (VNFs), and software-defined networking (SDN) layers, BSS components must interface with a complex web of legacy and next-gen systems. Ensuring seamless integration while maintaining low latency, high reliability, and end-to-end visibility becomes essential. Automated test frameworks using digital twins and continuous integration/continuous deployment (CI/CD) pipelines are becoming increasingly important in this context. These tools enable real-time emulation, rapid regression testing, and adaptive BSS behavior that aligns with evolving 5G service demands.

For appropriate R.F. (Radio Frequency) dimensions, the examination instrument must add a signaling protocol that works instantly with consumer-definable details (such as channel number). The purposeful tests for the UEs must include testing of the signaling protocol, handover testing, and end-to-end throughput. Among the principal difficulties for screening 5G, U.E.s ensure that the state modification to requirements is achieved. For instance, the unit is in a low-energy usage state for the idle mode, confirming excellent battery life, and occasionally checks for incoming emails. On the other hand, if the unit is connected (e.g., information transmission is to be scheduled), the team should awaken and quickly synchronize the uplink and downlink of its (or both, in the case of F.D. communication). Method development cycles perform Battery drain tests in all phases (e.g., product development, design validation, and program style).

A Range of Emerging Technologies

The deployment of 5G systems introduces a range of emerging technologies that each present specific challenges in terms of measurement and testing. Accurate performance evaluation is crucial for ensuring network reliability, efficiency, and scalability. From millimeter wave communications to massive MIMO and full-duplex radios, each technology brings unique requirements for signal quality assessment, interference management, and system validation. Furthermore, as 5G networks become increasingly complex with the integration of C-RAN, HetNets, and energy-efficient designs, traditional testing methods must evolve. The table below summarizes the key technologies involved and highlights their respective measurement and testing considerations in the context of 5G implementation.

The following table summarizes the problems that complicate the measurement and testing of 5G methods.

Table 3. Measurement and test issues for 5G systems

Emerging Technology	Role in 5G Networks
Millimeter Wave Communication	Enables high data rate transmission over short distances.
Massive MIMO	Enhances spectral efficiency and network capacity by utilizing large antenna arrays.
Full Duplex Radios	Supports simultaneous reception and transmission on the same frequency.
Device-to-Device Communication	Facilitates direct communication between nearby devices.
Cloud Radio Access Networks (C-RAN)	Centralizes baseband to process and enhance network flexibility.
Energy-Efficient Network Design	Reduces power consumption while maintaining performance.
Heterogeneous Networks (HetNets)	Combines macro and small cells to improve coverage and capacity.

Hossain and Hasan (2015)

CONCLUSION

Chapter 11 has illustrated the scope and depth of emerging research in the 5G domain, identifying key technological enablers and persistent challenges. Innovations such as heterogeneous multi-tier networks, full-duplex communication, and energy-harvesting systems have emerged as transformative pillars of 5 G. The chapter also emphasized the significance of wireless network virtualization and cloud radio access networks (C-RAN) as strategies to optimize resources, reduce latency, and improve scalability. Measurement and testing challenges were also highlighted, underlining the need for efficient modeling, mobility analysis, and RF system validation to ensure reliable performance across complex 5G infrastructures.

Although key enabling solutions significantly improve overall network efficiency and end-user Quality of Experience (QoE), several research challenges remain—from system and device design and testing to network management—to fully realize 5G capabilities. These challenges are generally categorized into two groups: (i) measurement and evaluation issues for emerging 5G systems, and (ii) challenges in the efficient management of wireless resources (Gupta & Jha, 2015).

The following provides a brief overview of the primary critical challenges regarding measuring, testing, and validating the functionality of 5G system components.

These developments indicate that the success of 5G depends not only on deploying advanced technologies but also on how these innovations are tested, integrated, and managed in real-world applications. The following table summarizes the emerging technologies and research topics covered, outlining their applications, advantages, and the challenges they aim to overcome.

Table 4. Summary of emerging technologies and research challenges in 5G

Technology / Research Area	**Benefits / Features**	**Applicability**	**Research Challenges**
Heterogeneous Multi-Tier Networks	Higher throughput, better spectrum use	Small cell, IoT, D2D/M2M	Interference, mode selection, and power control
Full-Duplex Communication	Low latency, spectrum efficiency	Cognitive radio, multi-hop relaying	SI reduction, MAC design
Energy Harvesting	Energy-aware, green networks	Small cells, D2D/M2M	Scheduling, energy prediction, SWIPT
Cloud RAN	Cost-effective, scalable	Heterogeneous networks	Front-haul resource allocation, BBU clustering
Wireless Network Virtualization	Flexibility, CAPEX/OPEX savings	Everything-as-a-service	Isolation, fairness, mobility, pricing
Measurement & Testing	Accuracy, multi-band validation	UEs, BSs, RF modules	Propagation modeling, channel impairments

We have introduced numerous emerging technologies for 5G cellular wireless networks. In addition, several open exploration issues, including those related to measuring and testing 5G methods, have already been identified.

Potential 5G cellular wireless networks undoubtedly involve various enabling solutions. Nevertheless, the most considerable challenge is integrating all the enabling technologies and combining them with the technologies described above—solutions like mm-wave and massive MIMO significantly influence the layout and development of 5G networks.

This chapter explains the following key points:

The main approaches and technologies to handle the demands for 5G methods can be categorized as follows: densification of present cellular networks with the enormous inclusion of a provision and small cells for peer-to-peer (P2P) interaction (e.g., device-to-device (D2D) and machine-to-machine (M2M) communication enabled multi-tier heterogeneous networks), Simultaneous transmission as well as reception (e.g., full-duplex (F.D.) communication), Massive multiple-input multiple-output (Massive MIMO) and millimeter-wave (mm-wave communications technologies, Improved energy efficiency by energy-aware communication and effort harvesting, Cloud-based radio access system (C-RAN), and also virtualization of wireless information.

The dense deployment and heterogeneity of wireless devices, coverage, and targeted traffic load imbalance resulting from altering transmit powers of various BSs help make interference management and resource allocation problems much more complicated than in standard single-tier methods.

Problems related to mobility management, pricing-based allocation, isolation, and resource discovery (e.g., seamless handover to various MVNOs/InPs) opened new research opportunities.

Cognitive Radio Networks: The C.R. system is a cutting-edge software-defined radio strategy regarded as one of the promising technologies to enhance the utilization of the congested RF spectrum.

Cloud technologies are evolving toward 5G mobile networks, characterized by an increasing number of wireless devices, enhanced device and service capabilities, and the need to access mobile services everywhere.

According to the IEEE Computer Society (Alkhatib et al., n.d.), four of the 20+ emerging technologies are Cloud Computing (Cloud), Big Data Analytics (Big Data), the Internet of Things (IoT), and Software-Defined Networks (SDN).

REFERENCES

Akyildiz, I. F., Lee, W.-Y., Vuran, M. C., & Mohanty, S. (2006). Next generation/dynamic spectrum access/cognitive radio wireless networks: A survey. *Computer Networks*, *50*(13), 2127–2159. DOI: 10.1016/j.comnet.2006.05.001

Alkhatib, H., Faraboschi, P., Frachtenberg, E., Kasahara, H., Lange, D., Laplante, P., Merchant, A., Milojicic, D., Schwan, K., Alquraishi, M., Burgess, A., Forsyth, D., Iwata, H., Mcgeer, R., & Walz, J. (n.d.). *IEEE CS 2022 Report read on*. https://ieeecs-media.computer.org/assets/pdf/2022Report.pdf

Amali, C., & Ramachandran, B. (2018). Enabling Key Technologies and Emerging Research Challenges Ahead of 5G Networks: An Extensive Survey. *JOIV: International Journal on Informatics Visualization*, *2*(3), 133. DOI: 10.30630/joiv.2.3.128

Andrews, J. G., Buzzi, S., Choi, W., Hanly, S. V., Lozano, A., Soong, A. C. K., & Zhang, J. C. (2014). What will 5G be? *IEEE Journal on Selected Areas in Communications*, *32*(6), 1065–1082. DOI: 10.1109/JSAC.2014.2328098

Chin, W. H., Fan, Z., & Haines, R. (2014). Emerging technologies and research challenges for 5G wireless networks. *IEEE Wireless Communications*, *21*(2), 106–112. DOI: 10.1109/MWC.2014.6812298

Gupta, A., & Jha, R. K. (2015). A survey of 5G network: Architecture and emerging technologies. *IEEE Access : Practical Innovations, Open Solutions*, *3*, 1206–1232. DOI: 10.1109/ACCESS.2015.2461602

Haider, F., Wang, C.-X., Haas, H., Yuan, N. D., Wang, N. H., Gao, N. X., You, N. X.-H., & Erol Hepsaydir. (2011). Spectral efficiency analysis of mobile femtocell-based cellular systems. In 2011 *IEEE 13th International Conference on Communication Technology* (pp. 347–351). https://doi.org/DOI: 10.1109/ICCT.2011.6157895

Han, S., I., C. L., Xu, Z., & Rowell, C. (2015). Large-scale antenna systems with hybrid analog and digital beamforming for millimeter-wave 5 G. *IEEE Communications Magazine*, *53*(1), 186–194. DOI: 10.1109/mcom.2015.7010533

Haneda, K., Tian, L., Asplund, H., Li, J., Wang, Y., Steer, D., Li, C., Balercia, T., Lee, S., Kim, Y., Ghosh, A., Thomas, T., Nakamura, T., Kakishima, Y., Imai, T., Papadopoulas, H., Rappaport, T., Maccartney, G., Samimi, M., & Sun, S. (2016). *Indoor 5G 3GPP-like Channel Models for Office and Shopping Mall Environments*. https://wides.usc.edu/Updated_pdf/haneda2016indoor.pdf

Heath, R. W., Gonzalez-Prelcic, N., Rangan, S., Roh, W., & Sayeed, A. M. (2016). An Overview of Signal Processing Techniques for Millimeter Wave MIMO Systems. *IEEE Journal of Selected Topics in Signal Processing*, *10*(3), 436–453. DOI: 10.1109/jstsp.2016.2523924

Hossain, E., & Hasan, M. (2015). 5G cellular: Key enabling technologies and research challenges. *IEEE Instrumentation & Measurement Magazine*, *18*(3), 11–21. DOI: 10.1109/MIM.2015.7108393

Jaber, M., Imran, M. A., Tafazolli, R., & Tukmanov, A. (2016). 5G backhaul challenges and emerging research directions: A survey. *IEEE Access : Practical Innovations, Open Solutions*, *4*, 1743–1766. DOI: 10.1109/ACCESS.2016.2556011

Li, C., Tang, Y., & Luo, F. (2022). Research on Internet Regional Security Based on 5G Wireless Communication Technology. *2022 International Conference on Smart City and Green Energy (ICSCGE)*. https://doi.org/DOI: 10.1109/ICSCGE58152.2022.00008

O'Connell, E., Moore, D., & Newe, T. (2020). Challenges Associated with Implementing 5G in Manufacturing. *Telecom, 1*(1), 48–67. MDPI. https://doi.org/DOI: 10.3390/telecom1010005

Paglierani, P., Neokosmidis, I., Rokkas, T., Meani, C., Nasr, K. M., Moessner, K., & Sayyad Khodashenas, P. (2019). Techno-economic analysis of 5G immersive media services in cloud-enabled small cell networks: The neutral host business model. *Transactions on Emerging Telecommunications Technologies*, *31*(2), e3746. DOI: 10.1002/ett.3746

Rapaport, T. S., Sun, S., Mayzus, R., Zhao, H., Azar, Y., Wang, K., & Samimi, M. (2013). Millimeter wave wireless communications for 5G cellular: It will work! *IEEE Access : Practical Innovations, Open Solutions*, *1*, 335–349. DOI: 10.1109/access.2013.2260813

Robert Bosch Gmb, H. (2012). *CAN with Flexible Data-Rate Specification Version 1.0*. Robert Bosch GmbH. https://www.bosch-semiconductors.com/media/ip_modules/pdf_2/can/can_fd_spec.pdf

Sabharwal, A., Schniter, P., Guo, D., Bliss, D., Rangarajan, S., & Wichman, R. (2014). In-band full-duplex wireless: Challenges and opportunities. *IEEE Journal on Selected Areas in Communications*, *32*(9), 1637–1652.

Sun, H., Hu, R. Q., & Qian, Y. (2023). Open Issues and Future Directions for 5G and Beyond Wireless Networks. In 5G and Beyond Wireless Communication Networks, *IEEE*, 155–156. https://doi.org/DOI: 10.1002/9781119089469.ch12

Sun, Y., Chen, J., Wang, Z., Peng, M., & Mao, S. (2022). Enabling Mobile Virtual Reality with Open 5G, Fog Computing, and Reinforcement Learning. *IEEE Network*, *36*(6), 142–149. DOI: 10.1109/MNET.010.2100481

Zhang, R., Zhou, W., & Hu, H. (2021). Towards 5G Security Analysis against Null Security Algorithms Used in Normal Communication. *Security and Communication Networks*, *2021*, 1–15. DOI: 10.1155/2021/4498324

ADDITIONAL READING

Dai, L., Wang, B., Yuan, Y., Han, S., Chih-Lin, I., & Wang, Z. (2015). Non-orthogonal multiple access for 5G: Solutions, challenges, opportunities, and future research trends. *IEEE Communications Magazine*, *53*(9), 74–81. DOI: 10.1109/MCOM.2015.7263349

Han, C., Harrold, T., Armour, S., Krikidis, I., Videv, S., Grant, P., & Haas, H. (2011). Green radio: Radio techniques to enable energy-efficient wireless networks. *IEEE Communications Magazine*, *49*(6), 46–54. DOI: 10.1109/MCOM.2011.5783986

Leung, V. C. M., Taleb, T., Chen, M., Magedanz, T., Wang, L.-C., & Tafazolli, R. (2014). Unveiling 5G wireless networks: Emerging research advances, prospects, and challenges [Guest Editorial]. *IEEE Network*, *28*(6), 3–5. DOI: 10.1109/MNET.2014.6963797

Parliament, U. K. (2025). (HC 201, Session 2019–21). The security of 5 G. House of Commons. https://committees.parliament.uk/work/134/the-security-of-5g/

Pirinen, P. (2014). A brief overview of 5G research activities. In *Proceedings of the 1st International Conference on 5G for Ubiquitous Connectivity* (pp. 17–22). https://doi.org/DOI: 10.4108/icst.5gu.2014.258061

Rusek, F., Persson, D., Lau, B. K., Larsson, E. G., Marzetta, T. L., Edfors, O., & Tufvesson, F. (2013). Scaling up MIMO: Opportunities and challenges with vast arrays. *IEEE Signal Processing Magazine*, *30*(1), 40–60. DOI: 10.1109/MSP.2011.2178495

Tse, D., & Viswanath, P. (2005). Fundamentals of wireless communication. Cambridge University Press & Assessment. https://www.cambridge.org/us/universitypress/subjects/engineering/wireless-communications/fundamentals-wireless-communication

Vaezi, M., Schober, R., Ding, Z., & Poor, H. V. (2019). Non-orthogonal multiple access: Common myths and critical questions. *IEEE Wireless Communications*, *26*(5), 174–180. DOI: 10.1109/mwc.2019.1800598

Wireless data from every light bulb | Harald Haas. (2011). [YouTube Video]. In *YouTube*. https://www.youtube.com/watch?v=NaoSp4NpkGg

Chapter 12
5G Technology:
Critical Research for Applications and Product Development

ABSTRACT

The rapid evolution of 5G technology has led to a surge in innovative applications, underscoring the need for targeted research to address critical challenges and unlock its full potential. This chapter provides a comprehensive framework for identifying technological areas that can drive innovation in 5G applications and product development. It aims to guide nations, organizations, academic institutions, and research entities in contributing to developing a robust and standardized 5G ecosystem. The chapter begins by examining key foundational technologies of 5G, including system ubiquity, interoperability, and IP-based network architectures. It highlights cognitive radio technology, integrated with artificial intelligence, as a pivotal component for enhancing terminal performance and optimizing system efficiency across diverse access systems. The discussion also addresses technological challenges, such as limited frequency spectrum resources and security vulnerabilities requiring focused research to ensure seamless implementation.

INTRODUCTION

Emerging 5G applications result in significant technological innovation being widely available. OECD analysis emphasizes that enhanced connectivity is pivotal for inclusive digital transformation and economic participation. *McKinsey identifies both the strategic potential and operational complexities of 5G deployment across industries,* (McKinsey & Company, 2020). For instance, Huawei outlines key 5G

DOI: 10.4018/979-8-3373-0497-7.ch012

applications with case-specific use cases in smart cities, transportation, and industrial automation, (Huawei, 2020).

To identify opportunities, we must find a specialized place to support the technical development of 5 G. This chapter discusses a framework that addresses the primary question: Where can technological regions help drive innovation? The solution shall involve collaboration among nations, firms, colleges, and a research institute to develop a genuine 5G standard. To begin, we reviewed the primary key technologies of fifth-generation mobile communication technology (5G).

Ubiquity and interoperability of the system are the primary specialized emphasis. A level I.P.-based network idea is evaluated. Cognitive stereo technology reaches its terminal, which contains artificial intelligence. The 5G technology aims to enhance system efficiency in various access systems terminology. Next, we identified technological difficulties, focusing on problems and security to address the limited frequency spectrum resources. Subsequently, we identified the innovation opportunity in a specialized field that had recently published a research article. We concluded that innovation potential lies in security, application issues, technological implementation, and network research. For instance, Nandini and Srivastava (2023) explore the major application domains and ongoing challenges in 5 G.

This chapter analyzes critical research on 5G applications from the Scopus database to inform the development of new industrial applications.

EVOLUTION OF THE NUMBER OF PUBLICATIONS ON THE TOPIC

The rapid expansion of 5G networks has sparked a surge in cybersecurity research, with the number of publications increasing from 69 in 2014 to over 2,600 by 2019. This reflects global concern over the vulnerabilities introduced by next-generation connectivity. Reputable journals, such as IEEE Communications Magazine and Sensors Switzerland, have become key platforms for disseminating these findings. Together, these efforts underline the urgency of securing 5G infrastructure amid growing digital threats.

Disciplines for Research in 5G Applications

The most crucial discipline for research in the area in terms of publications is the following (by the number of publications):

- Computer Science (5,388)
- Engineering (5,138)

- Physics and Astronomy (1,943)
- Materials Science (1,547)
- Mathematics (878)
- Chemistry (498)
- Energy (420)
- Decision Sciences (417)
- Social Sciences (285)
- Chemical Engineering (268)

Research Publications for New Applications

The rise of 5G has led to a surge in academic research on breakthrough technologies, including NOMA, millimeter-wave MIMO, and mobile edge computing. Millimeter wave MIMO systems in 5G require advanced signal processing to combat propagation and alignment challenges (Heath et al., 2016).

Influential studies by Wang et al. (2014), Dai et al. (2015), Mao et al. (2017), and others have shaped the theoretical foundations and practical applications of 5G. Highly cited articles in leading journals such as IEEE Communications Magazine highlight critical advances in architecture and signal processing. The METIS project and emerging work on edge and fog computing underscore the transformative role of 5 G. These contributions continue to define the evolution of next-generation mobile networks. Together, these contributions reflect a rich and evolving research landscape driving the global 5G transformation. The OECD Digital Economy Outlook presents comprehensive data on the impact of digital transformation across member countries. The ITU's report on 5G economic impact (ITU News, 2020) evaluates the financial benefits of 5G deployment across different countries. The OECD's analysis of SME digital transformation. The UNCTAD report underscores how equitable technological innovation is essential for harnessing the socioeconomic benefits of 5G globally (UNCTAD, 2021).

CRITICAL RESEARCH ACTIVITIES RELATED TO 5G APPLICATIONS

This section reviews the critical research activities related to 5G applications, which are essential to the development of 5G infrastructure. We examine the latest research and various projects, such as the European Union's 5 G-associated research projects on 5 G.

This critical 5G research sheds light on the building blocks and primary challenges of 5G technology, as well as how to address them. The studied references suggest that, in addition to capability-boosting technologies, 5G requirements provide ultra-reliable communications, low latency, and extensive device connectivity. The World Economic Forum highlights how 5G can revolutionize industrial operations by enabling real-time connectivity and automation (World Economic Forum, 2020).

Research on Architecture and Modeling

This sub-section focuses on the evolving landscape of research in 5G architecture and system modeling. As technology advances, it enables the simulation and analysis of complex network environments, enhancing the precision of design and validation processes. For instance, Ahmed et al. (2022) proposed a Stackelberg Game-Based approach for dynamic resource allocation in federated 5G edge networks. Architectural research now leverages digital tools to forecast system behavior and optimize performance. These virtual frameworks support innovation by testing new ideas before they are deployed in the real world. Ultimately, technology-driven modeling is essential for building scalable and efficient 5G infrastructures. Technology plays a vital role in the new era of architectural research. It enables the creation of a virtual world that triggers inferences and facilitates the drawing of conclusions from facts and findings.

UAVs integrated with 5G and beyond architectures open new frontiers for autonomous and remote operations (Banafaa et al., 2024). Banafaa et al. (2024) presented a comprehensive survey on the integration of UAVs in next-generation 5G and beyond wireless networks. Additionally, green communication strategies and cognitive radio aim to improve sustainability and spectrum efficiency in 5G networks.

NON-ORTHOGONAL MULTIPLE ACCESS FOR 5G

The growing demand for the Internet and the Internet of Things (IoT) presents challenging requirements for 5G wireless communications, including higher spectral efficiency and extensive connectivity. Unlike traditional orthogonal access technologies, NOMA can accommodate a large number of users through non-orthogonal resource allocation.

Non-orthogonal multiple access (NOMA) has received enormous attention for designing radio access techniques for fifth-generation (5G) wireless networks and beyond. Non-orthogonal multiple access (NOMA) is a technique that serves multiple users in the same resource block, such as a time slot, subcarrier, spreading code, or spatial domain.

In the story of wireless communications, from the first model (1G) to 4G, the various access patterns are the primary critical know-how that differentiates the different wireless methods. Frequency Division Multiple Access (FDMA) is typically used for 1 G networks. Time Division Multiple Access (TDMA) is generally used in 2G networks. Code Division Multiple Access (CDMA) is used in 3G networks. Orthogonal Frequency Division Multiple Access (OFDMA) is used for 4G, which is mainly an Orthogonal Multiple Access (OMA) scheme.

Orthogonal multiple access is necessary for easy single-user detection to separate the signals of various users. However, theoretically, OMA cannot always get the sum-rate capacity of multi-user wireless methods (Dai et al., 2015).

The 3GPP (2020) study provides a technical foundation for implementing non-orthogonal multiple access (NOMA) in 5G networks. It is not a government-sponsored standards organization like the ITU, but rather a forum for industry stakeholders to agree on standard specifications for products and services.

Enhanced Spectral Efficiency

The multi-user capability evaluation in NOMA (Non-Orthogonal Multiple Access) and OMA (Orthogonal Multiple Access) is compared in the pioneering work on these technologies, using two users in an additive white Gaussian noise (AWGN) channel model. The primary challenge to communication is the linear addition of wideband or white noise, characterized by a spectral density and a Gaussian amplitude distribution. The channel is an excellent example of AWGN (Cover & Thomas, 2005).

NOMA's non-orthogonal multiple access indicates the number of supported users or devices, which is not strictly limited by available resources and the scheduling graininess.

Thus, NOMA can accommodate more owners than OMA by utilizing non-orthogonal resource allocation; for example, MUSA (multi-user shared access) can continue to achieve reasonably good performance even when the overloading is 300%. (Cui et al., 2022).

Low Transmission Latency Plus Signaling Cost

For regular OMA with grant-based transmission, a person must initially send a scheduling petition to the starting station (B.S.). *Then, based on the request received, the B.S.* schedules uplink transmission and directs a grant with the downlink channel. This procedure results in significant latency and high signaling costs, which compromise the anticipated connectivity for 5 G.

The Software-Defined Multiple Access (SoDeMA) framework allows dynamic partitioning of time-frequency resources into tailored blocks—each with specific waveforms, power levels, pilot signals, and duplex modes—thus flexibly enabling the coexistence of OMA and NOMA in 5G networks to support diverse applications and services while addressing spectral efficiency, receiver complexity, and overloading control (Zhang et al., 2015a; Ding et al., 2017a).

GLOBAL TELECOMMUNICATION NETWORK (GTN) TESTING

The fifth-generation movable networks (5G) relied on new application designers for the Global Telecommunication Network (GTN). *The 5th generation reduces many barriers that prevent new uses for mobile networks by introducing higher data rates, lower latency, and significantly improved power efficienc,y* (Osseiran et al., 2014). 5G represents a technological breakthrough, not just the evolution of stereo technology.

The foundation of the 5G concept relies on a strong focus on the needs of various programs and technological scenarios. Therefore, test networks are essential for software development in 5G networks. The research presents a 5G examination system designed and deployed around Oulu, Finland, in conjunction with diverse partners actively involved in the development and specification of 5G technology (Ahokangas et al., 2019). The test system aims to serve many application developers by providing substantial test facilities in a carrier-grade, state-of-the-art network.

Applications Self-Interference Cancellation in 5G

Self-interference cancellation contradicts a continuing fundamental supposition in wireless network design; waves can only operate in half-duplex mode on the same channel. After enabling true in-band full duplex, which effectively doubles spectral efficiency, self-interference cancellation simplifies spectrum management. For example, Ahokangas et al. (2019) examined innovative business models for local 5G micro-operators within dynamic spectrum environments.

Not only does it make entire ecosystems like TD-LTE obsolete, but it also enables future networks to leverage the fragmented spectrum. The Mobile Economy 2023 report (GSMA, 2023) discusses the role of 5G in the global mobile ecosystem and its socioeconomic implications. The Future of Jobs Report (World Economic Forum, 2020) explores the impact of digital technologies, including 5G, on employment trends.

Unfortunately, this pressing global issue worsens in 5G networks (Sabharwal et al., 2014). *Blockchain integration into IoT and 5G ecosystems can address challenges of trust, security, and decentralization,* (Kshetri, 2017). Blockchain integration within 5G ecosystems enhances transparency and security in data transactions.

Self-interference cancellation enables the enhancement and evolution of 5G technologies toward denser, more diverse networks. It utilizes wireless communication systems, including increased link capacity, spectrum virtualization, any-division duplexing (ADD), novel relay solutions, and enhanced interference collaboration. Thus, self-interference cancellation has a tremendous impact on 5G networks and beyond, reflecting its fundamental nature (Sabharwal et al., 2014).

Channels Modeling and Measurement for the 5G

The fundamental challenge to calculating and modeling the propagation channel is primarily due to the wear of more significant frequencies and different bandwidths, together with the deployment of many larger antenna arrays. Ban, Li, Sim, and Wong (2016) analyzed the design of multi-antenna systems for future 4G/5G smartphones supporting multiple modes. Furthermore, expanding wireless website traffic requires different spectrums (e.g., higher frequencies in the mm-wave range). Cisco's report forecasts continued growth in IP traffic and highlights 5G as a key enabler of connected digital ecosystems (Cisco, 2020).

In addition, they are enabling technologies like large-scale MIMO and densification of network nodes, creating new demands for channel measurement and modeling of the spatial URL (Rappaport et al., 2013). Network densification is a critical trend in 5G evolution, enabling higher throughput and capacity (Bhushan et al., 2014).

Efficient as well as Practical Measurement

The measurement strategy has several components since the information is vital for extending and modifying free propagation models. It must record several frequency ranges, spatial consistency, 3D (spherical waves and elevation), and the brand-new communications model's tiny cell, device-to-device, and machine-to-machine communication (e.g., 60 GHz and above). *It must also record indoor and outdoor measurements and discuss feasible, practical use cases (such as crowded places and vehicle-to-vehicle/roadside communications).* It must also perform mm-wave measurements.

Spatial Mobility and Distributions

The current channel air is drop-based, e.g., the scattering surroundings are arbitrarily produced for every link. Nevertheless, the density of backlinks is likely to boost 5G methods. Regularly modeling these backlinks is essential, as it inherently supports the heterogeneous mobility performance of various network nodes, as well as research on Software-Defined Networking and Network Function Virtualization.

Software-Defined Networking (SDN) and Network Function Virtualization

Software-defined networking (SDN) and Network Function Virtualization (NFV) technologies are expected to play a crucial role in 5G networks. SDN and NFV are foundational for 5G architecture, offering flexible and scalable network services (Yong & Chen, 2015). *Virtualization and network software are essential ingredients for abstracting infrastructure energy and enabling the system's delivery as a service.* The tight coupling provides the flexibility needed to deliver network information on demand, creating a dynamic network service that works seamlessly to meet a wide range of use cases. In the preliminary network design, mobile primary networks could be used to design NFV and SDN.

This research presents the concept of a program-defined Radio Access Network (RAN) (Yong & Chen, 2015). *A RAN program is vital to help network slicing, RAN sharing, spectrum management, and other essential characteristics in 5G networks.* A software-defined network (SDN) enhances the efficiency and performance of enterprise networks.

SDN is considered a valuable skill for enhancing enterprise networks, the Internet, data centers, and metro networks. Software-Defined Networking (SDN) and Network Function Virtualization (NFV) are crucial in orchestrating mobile core networks. It is normal to utilize SDN in portable primary networks.

The key features of mobile core networks include packet forwarding and additional crucial tasks such as authentication, billing, and mobility control. The key question is precisely what the job of SDN encompasses within its scope. SDN controls system complexity; SDN concepts help tackle the various complexities of the next-generation heterogeneous RANs.

Summary

The following table provides an overview of key research areas shaping the architecture and modeling of 5G networks.

Table 1. Research on architecture and modeling

Topic	Description
Cellular Architecture Challenges	Limited RF spectrum and high energy consumption; indoor users suffer from signal loss due to building penetration, necessitating a redesign to separate indoor and outdoor scenarios.
Spatial Modulation	A low-complexity MIMO technique where data is encoded into the spatial position of the transmitting antenna.
Cognitive Radio Networks (CR)	Software-defined radios that dynamically access underutilized spectrum to enhance spectrum efficiency.
Mobile Femtocells	Small cells deployed in vehicles for improved connectivity on the move are functional in transportation.
Visible Light Communication (VLC)	Uses LEDs and photodiodes to provide both illumination and broadband data transmission.
Green Communications	Focus on reducing power consumption and CO2 emissions; indoor communication is emphasized for energy efficiency.
Non-Orthogonal Multiple Access (NOMA)	It supports multiple users in a single resource block, improves spectral efficiency, and supports more than OMA.
Comparison with OMA	OMA (e.g., FDMA, TDMA, CDMA, OFDMA) utilizes orthogonal allocation, whereas NOMA enables resource sharing with controlled interference.
Benefits of NOMA	Improved spectral efficiency, supports 300% overloading, and reduces latency and signaling costs with grant-free transmission.
SoDeMA (Software-Defined Multiple Access)	Framework to flexibly configure multiple access schemes per user/service; key for accommodating diverse 5G applications.
Future Coexistence of NOMA and OMA	Depending on user density and scenarios, both schemes may coexist for optimized performance.
Research Directions	Include NOMA codebook design, receiver complexity, theoretical modeling, and the development of grant-free NOMA systems.

RESEARCH ON RADIO RESOURCE MANAGEMENT (RRM)

This section highlights key research areas in Radio Resource Management (RRM) that are critical for optimizing the performance of 5G networks. As 5G introduces higher complexity with dense deployments, heterogeneous nodes, and new frequency bands, efficient management of radio resources becomes essential.

RRM encompasses a wide range of functionalities, including interference control, spectrum allocation, and energy efficiency.

Interference Management Within Heterogeneous Networks

Because multiple nodes are widely deployed within 5G networks, one strategy for enhancing resource utilization is to utilize readily available information in a spectrum-underlay fashion. Nevertheless, interference management is among the primary obstacles to the underlying communication of 5G networks.

One of the primary challenges for multi-tier heterogeneous networks is mitigating inter- and intra-tier, as well as inter-cell interference. The dense deployment and heterogeneity of wireless devices, as well as the resulting coverage and targeted traffic load imbalance from altering the transmit powers of various base stations (BSs), make them significantly more complicated than those in standard single-tier methods.

Full-Duplex Correspondence

In full-duplex (F.D.) communication, self-interference (S.I.) is defined as the primary limitation to achieving the theoretical gain of F.D. transmission (Sabharwal et al., 2014). *Despite the 7 S.I.* cancellation mechanisms, a specific amount stays in the product. Therefore, effective S.I. cancellation methods require eliminating recurring S.I. for a performance cap of F.D. transmission.

Furthermore, as pointed out, you can find new inter-cell and intra-cell interference sources in F.D. interaction because of simultaneous transmission/reception. Additionally, due to the excessively high transmit power of base stations (BSs), downlink-to-uplink interference becomes more intense than uplink-to-downlink interference (Simeone, Erkip, & Shamai, 2014).

Thus, to achieve maximum spectral efficiency, sophisticated channel estimation, interference management, and power allocation systems (utilizing advanced methods like transceiver beamforming and distributed space-time coding) necessitate the discussion of novel interference solutions. Also, the full-duplex (F.D.) transceivers must select the appropriate transmission modes (e.g., half-duplex or full-duplex) based on real-time system dynamics (Zhang et al., 2015a)

Energy-Harvesting Communication

EHNs (Energy Harvesting Networks) are defined as effective wireless charging and channel, energy, and time adaptation policies examined for RF energy harvesting. Standard multi-user scheduling schemes can result in power (or transmit) outages or effort overflows in RF EHNs, so effective multi-user booking schemes are necessary.

Additionally, effective CSI (channel state information) acquisition methods are necessary for HAPs (high-altitude platforms) to enhance transmit strength or the harvesting duration for directional power. A transfer was examined based on channel problems and issues, including distributed effects, interference management, and energy beamforming, taking into account the mobility of devices and energy resources.

Cloud-Ran

Deploying Cloud Radio Access Network (C-RAN) presents several challenges, including optimizing the use of processing resources, effectively utilizing the front haul backlinks, connecting Baseband Units (BBUs) with Remote Radio Heads (RRHs), and centralized control of the propagation signal[xxix]. The minimal front haul capability constrains the functionality of CRAN. Thus, front-haul-aware resource allocation schemes (including significant signal quantization and compression, as well as beamforming techniques) must maximize system throughput.

For optimal energy savings of the Cloud Radio Access Network (CRAN), RRHs (Remote Radio Heads) must function well, considering the establishment of BBUs in the BBU pool and network load (e.g., traffic needs from UEs/RRHs). Furthermore, to reduce scheduling complexity, the current source-sharing algorithms must be enhanced by considering the clustering of cells (e.g., RRHs) and baseband units (BBUs) in the cloud.

Wireless Network Virtualization

A fully functional WNV framework demands effective resource utilization, inter-slice isolation, and customizable slice resource allocation. The resource allocation algorithms must be created with the interests of heterogeneous and finite resources (e.g., spectrum, infrastructure, computing), the source needs of MVNOs (Mobile Virtual Network Operators), and fairness requirements.

Also, boot stamping issues and control signaling were resolved for spectrum virtualization. Problems related to source discovery, pricing-based allocation, isolation, and mobility management (e.g., seamless handover to various MVNOs/InP) opened new research opportunities.

Summary

Table 3 presents an overview of critical research directions in Radio Resource Management (RRM) for 5G networks, focusing on challenges and solutions in a highly dynamic and complex wireless environment. As 5G evolves to support massive connectivity, ultra-low latency, and enhanced reliability, efficient Radio Resource Management (RRM) becomes essential to sustain performance and scalability.

Table 2. Research on Radio Resource Management (RRM) in 5G

Research Area	Description
Interference Management in Heterogeneous Networks	The key challenge is due to dense, multi-tier deployment. Requires dynamic resource allocation, offloading decisions, CoMP, and device-based function selection.
Full-Duplex Communication (FD)	Self-interference (SI) is a significant barrier. SI cancellation, interference mitigation, and dynamic transmission mode switching (HD/FD) are needed.
Energy-Harvesting Communication	Requires effective scheduling, CSI acquisition, and adaptive energy policies for RF EHNs and mobile energy sources.
Cloud-RAN (C-RAN)	Challenges include limited front-haul capacity, coordination between the Baseband Unit (BBU) and Remote Radio Head (RRH), and clustering. Requires compression, beamforming, and optimized energy savings.
Wireless Network Virtualization (WNV)	Demands inter-slice isolation, fair resource allocation, spectrum and infrastructure sharing, and seamless mobility across MVNOs and InPs.
Channel and Strength Allocation	Essential for managing inter/intra-cell interference and dynamic usage of radio resources using methods like beamforming and coding.
Device Discovery & MAC Redesign	In high-density 5G environments with D2D and M2M, unit discovery and the MAC layer must be redesigned to handle large-scale access and interference.
Intra/Inter-Cell Interference	High power from base stations (BSs) leads to significant downlink-to-uplink interference; solutions involve space-time coding and adaptive power control.
Front-Haul-Aware Optimization	CRAN systems need resource allocation that respects front-haul limits while optimizing throughput and load balancing.
Mobility & Pricing-Based Allocation	WNV must address mobility handovers, resource discovery, and cost-based allocations while ensuring fair and efficient service delivery.

CLOUD TECHNOLOGIES FOR 5 G RADIO ACCESS NETWORKS

The evolution toward 5G mobile networks is characterized by an increasing number of wireless devices, enhanced device and service capabilities, and the requirement to access mobile services everywhere. Gartner's Top Strategic Technology Trends identifies 5G as a critical enabler of emerging technologies.

Two key enablers enable the realization of the 5G vision: very dense deployments and centralized processing.

Cognitive Radio Technology Research

As cloud technologies drive the centralization and flexibility of 5G radio access networks, cognitive radio systems emerge as a complementary innovation that enhances adaptability and intelligence at the edge. These systems enable dynamic spectrum access by allowing devices to sense their environment and autonomously adjust their communication parameters, making them essential for efficient and resilient 5G deployments. The 5G Business Readiness Report by Nokia analyzes the preparedness of global industries to adopt 5G innovations.

A cognitive radio system (CRS) is an air process that is aware of its geographical and operational environment, has developed policies, and maintains an internal state. Therefore, it can autonomously and dynamically adjust its functional protocols and parameters, and discover new ones based on its previous experience. Polese et al. (2017) examine the behavior of TCP and MP-TCP in 5G mmWave networks, highlighting protocol performance.

Conditions of Associated Regulation and Standardization

We summarize the conditions associated with regulation and standardization pursuits that are crucial for the success of any emerging technology. Finally, we remind you of a few essential investigation difficulties, particularly setup issues of cognitive stereo (C.R.). We concentrate specifically on R.F.'s front end, transceiver, digital-to-analog, and analog-to-digital interfaces, which are essential bottlenecks of CRS advancement.

Through this intelligent cycle of observation, decision-making, and reconfiguration, CRS systems enhance spectral efficiency, reduce interference, and optimize quality of service across heterogeneous networks. These systems are particularly effective in dynamic environments where traditional static allocation methods fall short. Moreover, CRS can support seamless interoperability between diverse Radio Access Technologies (RATs), ensuring reliable connectivity in multi-standard networks such as those incorporating 5G, LTE, and Wi-Fi. Selinis et al. (2018) explore the integration of LTE and Wi-Fi in the race toward full 5G adoption. Statistical projections indicate a continuous rise in global 5G subscriptions, reflecting accelerating adoption trends, (Petroc, 2025). For example, smart factories benefit from 5G integration by enabling intelligent decision-making and real-time automation (Lin, Lin, & Tung, 2016).

Researchers like Haykin (2005) and Akyildiz et al. (2006) have emphasized the transformative potential of cognitive radio in addressing spectrum scarcity and enabling autonomous network management. Ultimately, CRS exemplifies the convergence of artificial intelligence and telecommunications, laying the groundwork for more adaptive and resilient communication infrastructures. 5G supports the convergence of IoT, cloud computing, and big data through high-speed, low-latency connectivity (Lin, Lin, & Tung, 2016). Reducing latency in 5G networks is essential for mission-critical applications and real-time services.

Using a Software-Defined Stereo (SDR) Method

The reconfigurations are applied using a software-defined stereo (SDR) method. Furthermore, the CRS can see through its decisions to enhance its future choices. The outcomes of studying help each person acquire knowledge and make informed choices.

CRSs (cognitive radio systems) are categorized into heterogeneous and spectrum-sensing cognitive radio systems. The first category utilizes the system-centric strategy. Many operators work on several stereo entry networks (RANs) with different or the same radio access technologies (RATs)—the RAN repair frequency rings. The cognitive network optimizes stereo energy and increases the quality of service.

Spectrum-sensing cognitive Radio is the second type. RANs aided by different or the same RATs can discuss similar frequency bands with unoccupied subbands in an intelligent and synchronized manner. This kind of CRS is associated with most standardization tasks. For example, future smartphones will leverage multi-band 4G/5G antennas to support seamless transitions across network generations (Ban et al., 2016).

Summary

Cloud and cognitive radio technologies play a central role in enhancing the flexibility and intelligence of 5G Radio Access Networks (RANs). *These innovations enable dynamic spectrum access, efficient resource utilization, and adaptive system behavior*. Cloud-based solutions support centralized processing, while cognitive radios bring real-time environmental awareness and reconfiguration capabilities. Together, they address the challenges of scalability, latency, and energy efficiency.

The following table outlines the main components, functions, and implementation issues related to these technologies in 5G RANs.

Table 3. Cloud and cognitive radio technologies for 5G RANs

Topic	Description
Cloud Technologies for 5G RAN	5G relies on two enablers: dense deployments and centralized processing. Flexible functionality and centralized operations over heterogeneous backhaul networks increase strength and efficiency.
Cognitive Radio System (CRS) Overview	CRS adapts protocols based on geographic, operational, and policy awareness. Learns from past behavior and adapts dynamically using software-defined radio (SDR).
CRS Functional Capabilities	This includes awareness (via geolocation, CPC access, and spectrum sensing), decision-making, reconfiguration, and learning to improve spectrum efficiency.
CRS Components	Main components: intelligent management structures and reconfigurable radios. Adjusts frequency, power, modulation, and access technologies.
CRS Application Types	1) Heterogeneous CRS: multiple RANs using different RATs, optimize energy use and quality of service. 2) Spectrum-sensing CRS: share frequency bands using intelligent detection of available subbands.
Standardization and Regulation	Essential for the broad deployment of CRS, spectrum sensing types are the focus of most standardization efforts.
Challenges in CRS Implementation	Focus areas include RF front-end design, ADC/DAC interfaces, real-time decision algorithms, and integration into RAN infrastructure.
Learning and Reconfiguration	CRSs adapt protocols through continuous feedback and performance evaluation, using SDR frameworks to implement changes.
Goal of CRS	Enhance spectral efficiency, provide dynamic access, and support evolving user and network demands through intelligent decision-making.

RESEARCH ON 5G INTEGRATION AND REQUIREMENTS

The integration of 5G technology represents a pivotal advancement in modern telecommunications, addressing the growing demands for high-speed, low-latency, and scalable connectivity. Research in this field focuses on overcoming the technical, regulatory, and infrastructural challenges necessary for seamless implementation.

For example, Recent advances in optical wireless communications suggest strong potential for complementing 5G backhaul and high-speed access (Lagorio et al., 2021). Optical wireless communications face challenges in scalability and alignment but offer high-throughput solutions for 5G, (Ghassemlooy et al., 2015).

Key areas of investigation include network slicing, spectrum allocation, and the integration of 5G with IoT ecosystems and innovative technologies. This section examines the key requirements and innovations driving 5G adoption, providing insights into its transformative potential across various industries. The Cisco Annual Internet Report (Cisco, 2020) highlights projected global internet trends and provides valuable insight into 5G adoption. Researchers aim to unlock new economic and technological growth opportunities by addressing these aspects. Qualcomm's report on the 5G economy (Qualcomm, 2020) quantifies the expected contributions of 5G to global economic growth.

Moving from LTE to 5G for Connected Mobility

The fourth generation of communication technology has been commercially deployed for approximately five years. Although it is continually updated via new releases, LTE Advanced Pro Release thirteen is the most recent, and the fifth model has introduced improvements. The research examines how to present LTE system implementations compared to the original LTE demands. The objective is usually to determine key performance indicators with substandard implementations and, consequently, lend themselves to careful consideration when standardizing and designing the next model of wireless engineering.

Measurements

The measurements are collected using commercial radio system scanners and smartphones. Although LTE has very low air interface delays, the dimensions show that core system delays compromise the complete round-trip time layout requirement. Long-Term Evolution (LTE) is a common standard for wireless broadband communication for mobile devices, based on the GSM/EDGE and UMTS/HSPA technologies (Dahlman, Parkvall, & Sköld, 2014).

LTE's break-before-make handover implementation causes an information interruption of approximately 40 milliseconds at the median level during every handover. While this conforms to LTE needs, minimizing values is possible; however, break-before-make will not be adequate for connected mobility in cases such as streetcar safety, (Nguyen et al., 2020).

RESEARCH ON THE GENERIC ACCESS NETWORK (GAN)

The 2020 business launch states that 5G is expected to provide considerable benefits, including increased efficiency and capacity, reduced latency, and ubiquitous connectivity. For example, 5G technologies are instrumental in enabling smart manufacturing by supporting ultra-reliable low-latency communications (Cimini, Pirola, & Cavalieri, 2023).

The R.F. chain produces many system powers within telecom networks using a Generic Access Network (GAN). *The Rf chain is a protocol that extends mobile voice, data, and multimedia (IP Multimedia Subsystem/Session Initiation Protocol (IMS/SIP)) applications over IP networks.*

Unlicensed Mobile Access (UMA) is the term mobile carriers use to describe external IP access into their core networks. The latest generation system is called Wi-Fi Calling or VoWiFi by several handset manufacturers, including Apple and Samsung. It is a mirror by carriers like T-Mobile U.S. and Vodafone.

GAN's Outstanding Qualities

Gallium Nitride (GaN) stands out in radiofrequency (RF) engineering due to its superior energy density, high gain, and power-added efficiency (PAE), while also simplifying impedance matching in RF chains—enhancing system performance, particularly in base transceiver stations (BTS) (Kiesel, van Roessel, & Schmitt, 2020). *GaN's integration into 5G infrastructure supports high-performance and low-power applications, aligning with the ongoing evolution from 4G LTE toward full 5G deployment.* 5G infrastructure contributed significantly to real-time monitoring and public health responses during the COVID-19 pandemic (Li, Zhang, & Zhang, 2022).

Although 5G remains in the proof-of-concept phase, operators like Verizon have accelerated fixed wireless access trials since initial experiments began in 2013. Companies such as Qorvo leverage over 30 years of GaN expertise to support more than 20 global 5G trials, with a focus on industrial integration and defense applications. Moreover, advancements in deep mmWave, massive MIMO arrays, and beamforming continue in pre-commercial development, with chipset leaders like Intel and Qualcomm testing 28 GHz modems, including the X50, to meet future network demands. Intel's announcement of the first global 5G modem (Intel Newsroom, 2018) marks a significant milestone in the development of 5G hardware.

Applications for Generic Access Network (GaN)

Applications for Generic Access Network (GaN) are integral to advancing 5G technology, addressing the requirements of high frequency, efficiency, and scalability. GaN supports ultra-wideband linearization and high-power applications, making it ideal for massive MIMO and beamforming technologies, which enhance 5G network capacity and reliability. Its low voltage and small size enable efficient deployment in dense urban areas, including small cells and distributed antenna systems (DAS), essential for network densification. Bhushan et al. (2014) emphasized network densification as a central strategy in the evolution toward 5G wireless systems. Furthermore, GaN components excel in high-temperature environments while maintaining high reliability, facilitating their use in the automotive, industrial, and telecommunications sectors. These attributes make GaN a cornerstone for achieving the performance benchmarks required by 5G networks.

RESEARCH QUALITY OF SERVICE REQUIREMENTS

Several critical requirements that determine the quality of service (QoS) are being developed in 5G networks. The proposed quality of service requirements are based on functional requirements for 5G networks and traffic parameters for high-definition (HD) video and massive Machine-to-Machine (M2M) services, which were expected to face significant demand in 2020 (ITU News, 2020). The Cisco Annual Internet Report (Cisco, 2020) outlines forecasted global trends in mobile data traffic and 5G penetration.

Mobile Broadband in IoT, Big Data, Cloud, and SDN

Although the needs and standard specifications for 5G mobile broadband internet have yet to be finalized, the exploration and advancement of 5G technology have already begun. As a result, some 5G functions or subsystems are readily available.

This research offers a forward-looking perspective on the convergence of IoT, Cloud, big data, and SDN technologies, as well as the emergence of 5G mobile broadband networks. The investigation plans to demonstrate the specialized relationships between the solutions and the powerful applications and programs that emerge when two, three, or more of these technologies converge. Due to the rapid evolution of ICT (Information and Communication Technologies) and the continuous development of these five technologies, this research must be conducted annually to keep the associated information up to date with the main trends in ICT.

Commercial 5G is readily available, and IoT (Internet of Things) applications are being deployed everywhere, leveraging mobile broadband internet technologies. Furthermore, the fundamental Data produced by IoT applications has become the norm, and the Cloud is mainly used to calculate, retail, and virtualize network performance (NFV).

Necessary Network Infrastructure

These five ICT pillars—5G, Big Data, IoT, Cloud Computing, and SDN—are not only foundational to the digital transformation of industries but also mutually reinforcing. Mobile edge computing benefits from 5G's low latency and distributed processing capabilities. *Their integration enables end-to-end intelligent automation, network slicing, and dynamic resource allocation, especially in ultra-dense and mission-critical environments.* For example, SDN decouples the control and data planes, enabling programmable network behavior, while NFV virtualizes network functions to increase scalability and elasticity, (Mijumbi et al., 2016).

Cloud platforms, enhanced by AI and edge computing, facilitate real-time analytics of data streams generated by billions of IoT devices, supporting latency-sensitive applications such as autonomous vehicles, smart manufacturing, and telemedicine. Chiang and Zhang (2016) discussed emerging research opportunities at the intersection of fog computing and IoT within 5G contexts. Fog computing complements 5G by extending cloud functionalities closer to end users for lower latency, (Chiang & Zhang, 2016).

Big Data analytics provides insights into user behavior and network optimization, playing a vital role in predictive maintenance and energy efficiency, (Zhang et al., 2015a).

By 2020, this convergence has enabled what ITU News (2020) describes as "network intelligence," where AI-driven orchestration dynamically provisions services across hybrid infrastructures. The synergy among these technologies is not merely additive—it creates new capabilities and service models that were previously unattainable.

As we progress towards 6G, this ecosystem is expected to evolve further with the tighter integration of quantum computing, distributed ledger technologies, and next-generation AI. Deloitte's Global Telecommunications Outlook (Deloitte, 2020) outlines key trends and industry forecasts for the 5G ecosystem. In this context, the foundational interplay among 5G, SDN, Cloud, IoT, and Big Data continues to represent the core architectural paradigm for future digital societies.

The Roles of 5G, Cloud, Big Data, IoT, and Sdn

Although 5G standards are still evolving, research and partial implementations have been underway since 2020, with 5G and IoT technologies now widely deployed. These systems generate large volumes of data, which are analyzed via Big Data platforms and processed in the Cloud, supported by NFV for virtualization. To optimize costs and flexibility, Software-Defined Networking (SDN) is used to streamline infrastructure management. Ericsson's Mobility Report provides a forward-looking analysis of the global mobile network infrastructure, including 5G.

Combined, 5G, IoT, Cloud, Big Data, and SDN form an integrated ecosystem that enables real-time connectivity, scalability, and automation across various sectors. However, as Kiesel et al. (2020) point out, enterprises still seek reliable models to measure the profitability of 5G investments in production environments.

Table 4. Research on quality of service (quality of service) requirements in 5G networks

Topic	Description
Quality of Service (quality of service) Requirements	Based on 5G functional needs and traffic parameters (e.g., HD video, M2M). Managed via NFV with CQMF and CQCF components.
Network Function Virtualization (NFV)	Enables flexible and scalable radio access and core network deployment, thereby improving quality of service (quality of service) and performance.
Role of Virtualization & SDN	It supports efficient resource management and dynamic control across network layers, enabling diverse applications, from gaming to IoT.
5G & Mobile Broadband in IoT/Cloud/SDN Ecosystem	5G facilitates seamless IoT deployment, cloud computing supports data storage and processing, and SDN enables dynamic routing and cost optimization.
Big Data & Real-Time Processing	The Internet of Things (IoT) generates massive amounts of data, Big Data analytics extracts valuable insights, and Cloud Computing provides scalable infrastructure.
Converged ICT Technologies (2015–2020)	Emphasis on integrating 5G, Cloud, IoT, Big Data, and SDN to enable scalable, sustainable, and data-intensive communication networks.
IEEE Emerging Technology Trends	IEEE highlights 5G, Cloud, Big Data, IoT, and SDN as key drivers of telecom innovation.
Enterprise Adoption Challenges	Despite 5G's potential, enterprises face difficulty measuring ROI, prompting research into profitability frameworks (Kiesel et al., 2020).
ICT Synergy and Ecosystem	5G ensures high-speed connectivity; IoT enables device interaction; Big Data and Cloud manage and process data; SDN improves control.

TECHNICAL RELATIONSHIPS AMONG IOT, CLOUD, BIG DATA, & SDN

The research shows that IoT can generate big data using the four Vs: veracity, variety, velocity, and volume. Next, the cloud introduces big data storage and processing. SDN provides far more effective and adaptable networks for inter-cloud data transport. Advanced technologies, including machine learning analytics, are derived from Software-Defined Networking (SDN), Cloud Computing, and big data. The following figure illustrates the linkages between IoT, Cloud, Big Data, and SDN in 5G mobile broadband internet services (5G MBS).

The synergy between 5G, IoT, Big Data, Cloud Computing, and SDN enables modern technological ecosystems. 5G is the backbone for providing high-speed, low-latency connectivity, facilitating seamless data transfer among Internet of Things (IoT) devices. The Internet of Things (IoT) generates massive data streams, which Big Data processes by analyzing the volume, velocity, variety, and veracity of information to derive valuable insights.

Cloud Computing ensures efficient storage and computational power for this data, supporting scalable and flexible solutions. Software-defined networking (SDN) enhances the efficiency of network traffic management and resource allocation, ensuring optimal network performance. These technologies create a dynamic framework that enables cheaper, faster, and more insightful data transmission, driving innovation across various sectors.

5G technology facilitated pandemic prevention and control through its material characteristics, enabling both spatialization and de-spatialization affordances, (Li et al., 2022).

RESEARCH ON MACHINE-TO-MACHINE COMMUNICATIONS

M2M communications enable worldwide communication among intelligent devices, allowing for the monitoring of applications with minimal or no human intervention. The autonomous characteristics of M2M communications, which significantly reduce the cost of human resources, provide impetus for research in M2M communications in both industry and academic areas.

Machine-Type Communications Over 5G Systems

The most vital challenge and area for connectivity strategies is providing excellent wireless connectivity for machine-type communication over 5G systems. Why are machine-type communications over a 5 G system so critical and can become a complex issue?

Mobile interconnect networks can enhance production and quality in various areas, including automotive manufacturing machinery. They also inform product developers when and why certain products require maintenance, streamlining the products consumers purchase and ensuring all of them work efficiently.

An application uses wireless technology to monitor systems such as utility meters. It alerts the meter owner if certain elements have been tampered with, which is the most effective method for preventing fraud. The third application uses wireless networks to update digital billboards.

The industrial machine-to-machine market is rapidly evolving as enterprises increasingly recognize the value of connecting geographically disparate people, devices, sensors, and machines to corporate networks. Today, industries such as oil and gas, agriculture, the military, manufacturing, and public utilities utilize device-to-machine technologies for countless applications. Additionally, many companies use advanced and efficient data networking technologies to offer capabilities such as high-speed data transmission, mobile mesh networking, and 3G/4G cellular backhaul.

Telematics and in-vehicle entertainment are areas of focus for machine-to-machine developers.

Another use for talking products is perfect for manufacturing. This application may utilize remote control over heavy machinery in hazardous or remote locations, or monitor and control sensitive grids. We must offer reliable and fast connectivity, including 5G, and ensure the necessary performance of the apps.

Local Area Access in Industrial Machine-to-Machine Communications for 5G

Vast amounts of attached products, diverse program needs, deployment scenarios, and industrial designs will drive the development of the next-generation 5G networks. 5G networks are defined by massive MIMO, network slicing, and ultra-dense deployments enabling next-gen applications (Agiwal, Roy, & Saxena, 2016).

The Mobile Economy 2020 report (GSMA, 2020) provides insights into regional 5G rollout and mobile industry contributions to GDP.

As a result, 5G is crucial for the company to capitalize on its opportunities in 2020 and beyond. The International Telecommunication Union (ITU News, 2020) offers global statistics on digital development and access to 5G networks. A Stackelberg game model can optimize dynamic resource allocation in federated edge 5G networks (Ahmed et al., 2022).

Manufacturing machine-to-machine (M2M) communications will significantly increase the demand for connectivity. These specifications are particularly suited to the specific use of the situation and require personalized connectivity. This research creates feasible scenarios and value networks for 5G regional field manufacturing M2M marketing communications over the 2020-2030 period, spanning a ten-year timeframe.

Industrial M2M communications categorize subsequent needs and business opportunities by presenting statistics and forecasts utilizing various energy sources. The promising enablers of 5G technological innovation are examined through a feasible 5G specialized structure, delving into ongoing research. The research identified the techniques, situation preparation, and benefit system configurations before formulating the succeeding value and scenario networks. Selection interviews with business professionals are organized to address the current challenges of manufacturing online, future uncertainties, and market trends. The objective is to develop application forms that succeed in various scenarios and identify the most crucial functions and actors.

Understanding the Innovation in 5G for the Vehicle-to-Everything (V2x)

V2X, which stands for 'vehicle to everything,' is a term primarily used to describe a car's communication system, where data from sensors and other sources travels via high-bandwidth, low-latency, high-reliability links, paving the way for fully autonomous vehicles with the help of 5 G.

The term 5 G-V2X is often used to indicate the underlying technology compared to LTE-based V2X (LTE-V2X). In either case, C-V2X is the generic terminology that refers to the V2X technology using cellular technology, irrespective of the specific generation of technology.

According to Sparsh Gupta, the exponential growth of new developments in the communications domain (such as automobiles) is making them more connected, both with one another and with the planet. V2X, short for vehicle-to-everything, is the umbrella term for a car's communication process. Information from other sources travels and sensors through high-bandwidth, high-reliability links, and low latency, paving the way to completely independent vehicles using 5 G.

The 5G Automotive Association (5GAA) aims to enhance road safety, energy efficiency, and traffic management. Consequently, with the rise of 5G, businesses have accelerated their development in various aspects, and V2X is now a key area of focus.

A company's initial patent portfolio could guide informed business decisions in a specific area. Therefore, the research addresses several patents that are essential advancements in the 5G V2X region. Additionally, it aims to integrate various companies' research tasks with the current business plan for 5G V2X communication.

The declared essentiality of the 5G V2X/D2D SEPs cannot be established at this time, primarily because the technical specification underlying this URL is still under development. Nevertheless, these patents are likely to prove vital to 5G requirements—a case supported by numerous specialized reports by 3GPP, which highlight precisely how these patents utilize complex specifications in the future, (GreyB, 2025).

Most V2X/D2D patents focus on enhancing interconnectivity, reducing latency, and increasing communication speed. For instance, Ericsson's US10348444B2 minimizes latency through adaptive transmission formats dependent on vehicular velocity. This patent could even be confirmed as vital to 5G, as suggested by Technical Reports 21.866 and 38.812, (Serrano & Sorrentino, 2016).

A comparable situation is US9572193B2 by Nokia, which aims to improve Quality of Service in the D2D communication scenario. This feature may also be vital to the standard, as specified in TR 21.915. US9877337B2 and US10299235B2, as described by L.G., reduce overhead and effectively mitigate interference, respectively.

Critical Patents for 5G V2X Communication: While most patents focus on enhancing communications in a V2X/D2D setting, we have also encountered several other essential patents throughout our analysis. These patents, in particular, indicate the respective assignees' company objectives in the V2X domain.

Samsung's US10397297B2 claims a technique wherein an apparatus determines whether to use an adaptive VR streaming service based on maintaining a bandwidth state. The apparatus requests image information (or maybe video) based on precollected head tracking and bandwidth info. TR26.918 suggested another perspective, SEP. This patent is resolved because it enables different-quality content for every viewing angle, thereby conquering interruptions triggered by bandwidth shortages.

Samsung's declaration of the patent as an essential 5G standard is practical. Recently, it was announced that Samsung is operating the world's first remotely controlled 5G automobile using virtual reality (VR), which appears to complement Samsung's grand scheme for the 5G program.

Nokia's developments focus on creating the infrastructure to support 5G automobiles, a concept supported by US20180224846A1, which enhances the drivability of autonomous vehicles. This infrastructure involves detecting specific road traffic situations and transmitting the information to other vehicles to prevent crashes and control vehicle speeds by activating the cars in a higher alert mode. Additionally, US9955484B2 utilizes the mobile radio unit's location and position information to select the radio energy allocated to use products for interaction.

Innovations like these are consistent with Nokia's plans to offer answers "for enhanced traffic security and cooperative, connected as well as automated mobility."

Qualcomm's strategy differs from those of Nokia and Samsung. They are partnering with Chinese companies to deploy V2X fixes, which firms at the China Academy of Telecommunications Technology are currently developing. CATT's patent CN108347781B details how cars can communicate "automatic help information" with one another when a fleet of autonomous vehicles operates on the highway.

The following table provides a summary of ongoing research in Machine-to-Machine (M2M) and Vehicle-to-Everything (V2X) communications within 5G networks. These technologies play a critical role in enabling automation, intelligent connectivity, and real-time responsiveness across industries. M2M facilitates autonomous device interaction with minimal human input, while V2X enhances road safety and traffic efficiency through low-latency communication. With 5G's advanced capabilities, both M2M and V2X are becoming foundational components of industrial systems and connected mobility. This table outlines the key research areas, applications, and industry implications.

Table 5. Summary of 5G research in M2M and V2X communications

Research Area	Key Points	Applications and Implications
Machine-to-Machine (M2M) Communications	It facilitates autonomous device interaction with minimal human input, thereby enhancing industrial applications such as smart meters, remote machinery control, and manufacturing.	Used in utilities, agriculture, military, and manufacturing sectors to monitor, control, and automate processes efficiently.
Machine-Type Communications Over 5G	Ensures robust wireless connectivity for industrial machine communication. Supports use cases like smart metering and real-time manufacturing data.	Enhances production quality and informs predictive maintenance, a crucial aspect for connected industrial systems.
Local Area Access in Industrial M2M	Caters to diverse device requirements and environments. Supports personalized connectivity in industrial settings.	Enables regional manufacturing setups to implement reliable and flexible M2M networks.
Vehicle-to-Everything (V2X) Communications	Enables data exchange between vehicles and their surroundings using low-latency, high-bandwidth 5G links.	Supports autonomous driving, road safety enhancements, and traffic efficiency.
5G V2X Innovations and Patents	Focuses on reducing latency, increasing interconnectivity, and enhancing real-time decision-making. Key players include Ericsson, Nokia, Samsung, and Qualcomm.	Incorporated in self-driving cars, real-time VR-assisted vehicle control, and connected vehicle ecosystems.

RESEARCH ON MULTIPLE ANTENNAS FOR 5 G APPLICATIONS

This research has developed a hybrid antenna design for 4G/5G multiple-input multiple-output (MIMO) applications, comprising two modules: one for 4G and one for 5 G. The 4G antenna module is a two-antenna array (Ban et al., 2016).

According to these authors, the 4G antenna component is a two-antenna array that addresses the GSM850/900/1800/1900, UMTS2100, plus LTE2300/2500 running bands. On the other hand, the 5G antenna module is an eight-antenna array running in the 3.5 GHz band, effective at covering the C band (3400 3600 MHz), which may meet the need for a potential 5G program.

Moreover, in contrast to perfect, uncorrelated antennas in an eight-by-eight MIMO process, the 5G antenna module exhibits an excellent ergodic channel capability of ~40 b/s/Hz, only six b/s/Hz lower than the ideal case. In addition, this multi-mode hybrid antenna manufacturer usually has experimental outcomes like the S(scattering) parameter (used to characterize electrical networks using matched impedances), radiation pattern, antenna efficiency, and envelope correlation coefficient.

Eight-Port Orthogonal Antenna for 5G Applications

This research has developed an eight-port orthogonal antenna for Gg applications. A dual-polarized hybrid eight antenna array runs in the 2.6 GHz band (2550 2650 MHz) for 5G correspondence multi-input multi-output (MIMO) functioning present in the smartphone.

The recommended hybrid antenna array components, such as the smartphone's long edges, are symmetrically positioned. They manufacture four distinct antenna array types (C-shaped coupled-fed and L-shaped monopole slot) that display orthogonal polarization.

Thus, the coupling between the antenna array may be reduced, and the performance of the MIMO method may be enhanced. A prototype of the proposed eight-antenna array was constructed and evaluated. An effective impedance matching (ten dB return loss or better), desired cross-polarization discrimination (better than fifteen dB), and suitable isolation (better than 12.5 dB).

The envelope correlation coefficient and channel capability may also be calculated to assess the proposed antenna array's MIMO performance.

Development for 5G Base Station Antennas

This research primarily highlights the difficulties in improving base station antennas due to the challenges introduced by the fifth-generation (5G) wireless communication system (Zhang et al., 2015a). *Among the primary vital technologies, the Massive Multiple Input Multiple Output (M-MIMO) system can offer significant improvements in method capability and has been demonstrated both in theory and in functional networks, with the antenna aspect in the MIMO array review at the beginning*. The multi-system antenna for 5G, suitable for 4G and 3 G, requires the most crucial part; three different methods exist to produce a multibeam-specific antenna. In conclusion, the antenna assessment technique mentions brand-new content.

HIGH-FREQUENCY SUBSTRATE SUITABLE FOR 5G APPLICATIONS

This section examines the crucial role of high-frequency substrates in facilitating 5G applications. As 5G operates at much higher frequencies than previous generations, material selection becomes essential for maintaining signal integrity and minimizing loss. Advanced substrates are required to support millimeter-wave communication, ensuring reliable performance in compact, high-speed circuits. These substrates enable the development of efficient antennas and RF components vital for 5G networks. Understanding their properties is key to advancing next-generation wireless technologies.

Gallium-nitride (GaN) technology is crucial for the development of upcoming 5G networks. Gallium nitride (GaN) is a complex, mechanically stable, broadband-gap semiconductor. With higher strength, faster switching velocity, lower on-resistance, and higher thermal conductivity, power products based on gallium nitride (GaN) significantly outperform those based on silicon. In addition, gallium nitride crystals are grown on substrates, including sapphire, silicon carbide (SiC), and silicon (Si) [xxiv]. By raising a GaN epilayer and silicon, the current silicon manufacturing infrastructure eliminates the demand for costly, specialized manufacturing websites. Instead, it leverages being able to sell giant-diameter silicon wafers at a low cost.

GaN is a suitable substrate for advancing programs for all consumer electronics technologies. Therefore, the semiconductor market focuses on the potential for combining GaN and other substrates to enhance overall performance in various industries.

According to Siemens, gallium nitride (GaN) is significantly more power-efficient than silicon for 5G radio frequency (RF) applications. GaN is expected to replace silicon in 5G power amplifiers, especially in mm-wave 5G networks (Siemens, 2019). *Its ability to effectively manage better voltage in a significantly smaller area than similar sideways diffused MOSFETs (LDMOS) products makes it appealing.* Additionally, it can power up a much wider variety of mm-wave frequencies compared to regular silicon.

Most of the fundamental difference relates to the transistors' working voltage. GaN could be 50V or more significant, and 28nm CMOS is probably 1.8V, stated Keith Benson, Director of RF/Microwave Amplifier and Phased Array IC Solutions at Analog Devices. "In the end, power is voltage times electricity, along with a greater operating voltage, making heavy energy easier. Evaluating GaN compared to silicon is a complex task, as they differ significantly in their properties. GaN is still costly from a wafer expense, but the mask price is much less than cutting-edge CMOS" (Fogarty, 2019).

According to Siemens documents, the existence and advancement patterns of GaN differ across various markets, including wireless infrastructure, aerospace, defense, satellite communication, wired broadband internet, and other applications in the Industrial, Scientific, and Medical (ISM) band.

Yole proposes a comprehensive evaluation covering various emerging GaN players and over 600 GaN products developed and applied in programs such as radar, base transceiver stations, CATV, jammers, and VSAT.

This statement also includes a comprehensive evaluation of various scenarios associated with GaN-on-Si and GaN-on-SiC construction projects in 2020. Product sales likely did not rise significantly during the subsequent two years. Still, development continued, primarily driven by the improved adoption of GaN technologies within the wireless infrastructure and safeguard marketplaces.

The market strategies for GaN-on-SiC and GaN-on-Silicon technologies focus on optimizing performance and cost to meet specific application demands. GaN-on-SiC is primarily performance-driven, making it suitable for high-end applications such as wireless infrastructure and defense, where improved performance and reliability are crucial.

On the other hand, GaN-on-Silicon offers a cost-driven approach that is ideal for mass-market applications such as mobile handsets, RF energy, and CATV. It strikes a balance between performance and cost efficiency, enabling widespread adoption in cost-sensitive sectors. Together, these strategies illustrate the distinct roles of both technologies in addressing the diverse needs of the RF device market (Nasr, 2021).

CONCLUSION

Chapter 12 highlighted the crucial role of emerging applications in advancing the development and adoption of 5G technologies. With the proliferation of use cases across autonomous vehicles, industrial automation, healthcare systems, and smart city infrastructures, integrating 5G with other digital innovations is reshaping modern connectivity. 5G is projected to unlock new cross-sectoral value chains, fostering digital transformation across healthcare, manufacturing, and logistics (World Economic Forum, 2020).

This chapter particularly focused on technologies such as cognitive radio, Non-Orthogonal Multiple Access (NOMA), Software-Defined Networking (SDN), cloud computing, and Machine-to-Machine (M2M) communications. Their convergence enables ultra-reliable, low-latency communication, scalable device connectivity, and dynamic resource management. These innovations lay the groundwork for a seamless and intelligent digital ecosystem.

The innovation landscape rapidly evolves, from vehicular networks (V2X) and industrial IoT to energy-aware communication and global network testing systems. The table below summarizes the significant domains of emerging 5G applications, their strategic benefits, and corresponding challenges that demand ongoing research and development efforts.

Table 6. Strategic overview of emerging 5g applications

Application Domain	Key Technologies	Benefits	Challenges
Industrial IoT	Edge Computing, NOMA, MIMO	Real-time automation, fault prediction	Interference, high infrastructure cost
Autonomous Vehicles (V2X)	C-V2X, Beamforming	Low-latency navigation, safety systems	Latency management, spectrum fragmentation
Smart Grids & Energy	Cognitive Radio, AI	Energy efficiency, dynamic routing	Security risks, integration barriers
Healthcare & AR/VR	Full-Duplex, SDN	Remote diagnostics, immersive training	Bandwidth demand, device compatibility
Machine-to-Machine	M2M, IoT protocols	Reduced human intervention, scalability	Quality of service assurance, standardization issues
Network Management	SDN, NFV	Dynamic resource allocation, cost savings	Complex orchestration, security vulnerabilities

The advancement of 5G technology represents a pivotal moment in the evolution of telecommunications, offering unprecedented opportunities for innovation across applications and product development. This chapter has demonstrated the importance of targeted research in addressing critical challenges, including spectrum limitations, security vulnerabilities, and the seamless integration of diverse systems. By focusing on foundational technologies like cognitive radio, IP-based network architectures, and artificial intelligence, 5G has the potential to revolutionize industries and redefine global connectivity. Ruzomberka et al. (2023) highlight security vulnerabilities and resilience strategies in beyond 5 G networks.

One of the most significant takeaways is the importance of aligning technological innovation with real-world applications. Stakeholders can develop scalable and robust solutions that address the diverse needs of industries by identifying key areas such as security enhancements, network optimization, and energy-efficient implementation. The insights derived from recent research, including those found in the Scopus database, underscore the transformative potential of 5G in creating more intelligent, more efficient systems.

This chapter has also highlighted the role of collaboration among nations, organizations, academic institutions, and research entities in establishing a standardized 5G ecosystem. Such a unified approach will be instrumental in fostering innovation, driving industrial growth, and addressing the unique challenges posed by this next-generation technology.

In conclusion, the future of 5G lies in the convergence of advanced research and practical applications. By leveraging insights into security, interoperability, and cognitive technologies, stakeholders can harness 5 G's capabilities to build a more connected, innovative, and sustainable world. The path forward is clear: continued investment in research and development will unlock 5G's potential to transform industries and redefine global communication systems.

REFERENCES

Agiwal, M., Roy, A., & Saxena, N. (2016). Next generation 5G wireless networks: A comprehensive survey. *IEEE Communications Surveys and Tutorials*, *18*(3), 1617–1655. DOI: 10.1109/COMST.2016.2532458

Ahmed, J., Razzaque, M. A., Rahman, M. M., Alqahtani, S. A., & Hassan, M. M. (2022). A Stackelberg Game-Based Dynamic Resource Allocation in Edge Federated 5G Network. *IEEE Access: Practical Innovations, Open Solutions*, *10*, 10460–10471. DOI: 10.1109/access.2022.3144960

Ahokangas, P., Matinmikko-Blue, M., Yrjölä, S., Seppänen, V., Hämmäinen, H., Jurva, R., & Latva-aho, M. (2019). Business Models for Local 5G Micro-Operators. *IEEE International Symposium on Dynamic Spectrum Access Networks (DySPAN)*, 241–250. https://doi.org/DOI: 10.1109/TCCN.2019.2902547

Akyildiz, I. F., Lee, W.-Y., Vuran, M. C., & Mohanty, S. (2006). NeXt generation/dynamic spectrum access/cognitive radio wireless networks: A survey. *Computer Networks*, *50*(13), 2127–2159. DOI: 10.1016/j.comnet.2006.05.001

Ban, Y.-L., Li, C., Sim, C.-Y.-D., & Wong, K.-L. (2016). 4G/5G multiple antennas for future multi-mode smartphone applications. *IEEE Access: Practical Innovations, Open Solutions*, *4*, 2981–2988. DOI: 10.1109/ACCESS.2016.2582786

Banafaa, M., Ömer, P., Shayea, I., Alhammadi, A., Shamsan, Z., Razaz, M. A., Al-sagabi, M., & Al-Sowayan, S. (2024). A Comprehensive Survey on 5 G-and-Beyond Networks With UAVs. *IEEE Access: Practical Innovations, Open Solutions*, *12*, 7786–7826. DOI: 10.1109/access.2023.3349208

Bhushan, N., Li, J., Malladi, D., Gilmore, R., Brenner, D., Damnjanovic, A., Sukhavasi, R., Patel, C., & Geirhofer, S. (2014). Network densification: The dominant theme for wireless evolution into 5 G. *IEEE Communications Magazine*, *52*(2), 82–89. DOI: 10.1109/MCOM.2014.6736747

Chiang, M., & Zhang, T. (2016). Fog and IoT: An overview of research opportunities. *IEEE Internet of Things Journal*, *3*(6), 854–864. DOI: 10.1109/JIOT.2016.2584538

Cimini, C., Pirola, F., & Cavalieri, S. (2023). Challenges and expectations about 5G in the industry: Evidence from survey-based research. *International Journal of Manufacturing Technology and Management*, *37*(5/6), 492–511. DOI: 10.1504/ijmtm.2023.133688

Cisco. (2020). *Cisco Annual Internet Report (2018–2023)*. https://www.cisco.com/c/en/us/solutions/collateral/executive-perspectives/annual-internet-report/white-paper-c11-741490.pdf

Cover, T. M., & Thomas, J. A. (2005). *Elements of Information Theory*. John Wiley & Sons, Inc., DOI: 10.1002/047174882x

Cui, Y., Liu, P., Zhou, Y., & Duan, W. (2022). Energy-Efficient Resource Allocation for Downlink Non-Orthogonal Multiple Access Systems. *Applied Sciences (Basel, Switzerland)*, *12*(19), 9740. DOI: 10.3390/app12199740

Dahlman, E., Parkvall, S., & Sköld, J. (2014). *4G: LTE/LTE-Advanced for mobile broadband*. Elsevier.

Dai, L., Wang, B., Yuan, Y., Han, S., Chih-Lin, I., & Wang, Z. (2015). Non-orthogonal multiple access for 5G: Solutions, challenges, opportunities, and future research trends. *IEEE Communications Magazine*, *53*(9), 74–81. DOI: 10.1109/MCOM.2015.7263349

Ding, Z., Lei, X., Karagiannidis, G. K., Schober, R., Yuan, J., & Bhargava, V. K. (2017). A Survey on Non-Orthogonal Multiple Access for 5G Networks: Research Challenges and Future Trends. *IEEE Journal on Selected Areas in Communications*, *35*(10), 2181–2195. DOI: 10.1109/jsac.2017.2725519

Fogarty, K. (2019). GaN Versus Silicon For 5 G. *Semiconductor Engineering*. https://semiengineering.com/gan-versus-silicon-for-5g/

Ghassemlooy, Z., Arnon, S., Uysal, M., Xu, Z., & Cheng, J. (2015). Emerging optical wireless communications—Advances and challenges. *IEEE Journal on Selected Areas in Communications*, *33*(9), 1738–1749.

GPP. (2020). *Study on non-orthogonal multiple access (NOMA) for NR (TR 38.812)*. 3rd Generation Partnership Project. https://www.3gpp.org/DynaReport/38812.htm

Grey, B. (2025). Understanding the current innovation game in 5G V2X domain. *Greyb*. https://www.greyb.com/blog/5g-v2x-communication-research/

Haykin, S. S. (2005). *Modern wireless communications*. Internet Archive. https://archive.org/details/modernwirelessco0000hayk

Heath, R. W., Gonzalez-Prelcic, N., Rangan, S., Roh, W., & Sayeed, A. M. (2016). An overview of signal processing techniques for millimeter wave MIMO systems. *IEEE Journal of Selected Topics in Signal Processing*, *10*(3), 436–453. DOI: 10.1109/JSTSP.2016.2523924

Huawei (2020). *Bring 5.5G Reality 5g-Advanced. Huwawei.* https://www-file.huawei.com/-/media/corp2020/media-center/pdf/5g-advanced.pdf

Kiesel, R., van Roessel, J., & Schmitt, R. H. (2020). Quantification of economic potential of 5G for latency critical applications in production. *Procedia Manufacturing, 52*, 113–120. DOI: 10.1016/j.promfg.2020.11.021

Kshetri, N. (2017). Can Blockchain Strengthen the Internet of Things? *IT Professional, 19*, 68–72. DOI: 10.1109/mitp.2017.3051335

Lagorio, A., Cimini, C., Pirola, F., & Pinto, R. (2021). A Taxonomy of Technologies for Human-Centred Logistics 4.0. *Applied Sciences (Basel, Switzerland), 11*(20), 9661. DOI: 10.3390/app11209661

Li, G., Zhang, X., & Zhang, G. (2022). How the 5G Enabled the COVID-19 Pandemic Prevention and Control: Materiality, Affordance, and (De-)Spatialization. *International Journal of Environmental Research and Public Health, 19*(15), 8965.

Li, Y., & Chen, M. (2015). Software-Defined Network Function Virtualization: A Survey. *IEEE Access : Practical Innovations, Open Solutions, 3*, 2542–2553. DOI: 10.1109/access.2015.2499271

Lin, B.-S. P., Lin, F. J., & Tung, L.-P. (2016). The Roles of 5G Mobile Broadband in the Development of IoT, Big Data, Cloud and SDN. *Communications and Network, 08*(01), 9–21. DOI: 10.4236/cn.2016.81002

Mao, Y., You, C., Zhang, J., Huang, K., & Letaief, K. B. (2017). A survey on mobile edge computing: The communication perspective. *IEEE Communications Surveys and Tutorials, 19*(4), 2322–2358. DOI: 10.1109/COMST.2017.2745201

McKinsey & Company. (2020). *The 5G era New horizons for advanced electronics and industrial companies*. Mckinsey.com. https://www.mckinsey.com/~/media/mckinsey/industries/advanced%20electronics/our%20insights/the%205g%20era%20new%20horizons%20for%20advanced%20electronics%20and%20industrial%20companies/the-5g-era-new-horizons-for-advanced-electronics-and-industrial-companies.pdf

Mijumbi, R., Serrat, J., Gorricho, J.-L., Bouten, N., De Turck, F., & Boutaba, R. (2016). Network Function Virtualization: State-of-the-Art and Research Challenges. *IEEE Communications Surveys and Tutorials, 18*(1), 236–262. DOI: 10.1109/comst.2015.2477041

Nandini & Srivastava. D. (2023). 5G Technology—Its Applications and Challenges. *2023 7th International Conference on Computing, Communication, Control and Automation* (ICCUBEA, 1-4). https://doi.org/DOI: 10.1109/iccubea58933.2023.10392002

Nasr, I. (2021). RF GaN on Silicon: The Best of Two Worlds. *Microwave Journal.* https://www.microwavejournal.com/articles/37107-rf-gan-on-silicon-the-best-of-two-worlds

News, I. T. U. (2020). Setting the Scene for 5G: Opportunities and Challenges. *ITU Hub.* https://www.itu.int/hub/2020/03/setting-the-scene-for-5g-opportunities-challenges/

Nguyen, D. C., Pathirana, P. N., Ding, M., & Seneviratne, A. (2020). Blockchain for 5G and beyond networks: A state of the art survey. *Journal of Network and Computer Applications*, *166*, 102693. DOI: 10.1016/j.jnca.2020.102693

Osseiran, A., Boccardi, F., Braun, V., Kusume, K., Marsch, P., Maternia, M., Queseth, O., Schellmann, M., Schotten, H., Taoka, H., Tullberg, H., Uusitalo, M. A., Timus, B., & Fallgren, M. (2014). Scenarios for 5G mobile and wireless communications: The vision of the METIS project. *IEEE Communications Magazine*, *52*(5), 26–35. DOI: 10.1109/MCOM.2014.6815890

Petroc, T. (2025). *Forecast number of mobile 5G connections worldwide from 2021 to 2025 (in billions).* Statista. https://www.statista.com/statistics/691623/5g-mobile-connections-worldwide/

Polese, M., Mezzavilla, M., Rangan, S., & Zorzi, M. (2017). *Mobility Management for TCP in mmWave Networks*. https://doi.org/DOI: 10.1145/3130242.3130243

Qualcomm (2020). *Global update on spectrum for 4G & 5G*. Qualcomm Technologies, Inc. https://www.qualcomm.com/content/dam/qcomm-martech/dm-assets/documents/messaging_presentation_-_4g_5g_spectrums_-_december_2020.pdf

Rappaport, T. S., Sun, S., Mayzus, R., Zhao, H., Azar, Y., Wang, K., Wong, G. N., Schulz, J. K., Samimi, M., & Gutierrez, F. (2013). Millimeter wave mobile communications for 5G cellular: It will work! *IEEE Access : Practical Innovations, Open Solutions*, *1*, 335–349. DOI: 10.1109/access.2013.2260813

Ruzomberka, E., Love, D. J., Brinton, C. G., Gupta, A., Wang, C.-C., & Poor, H. V. (2023). Challenges and Opportunities for Beyond 5 G Wireless Security. *IEEE Security and Privacy*, *21*(5), 55–66. DOI: 10.1109/MSEC.2023.3251888

Sabharwal, A., Schniter, P., Guo, D., Bliss, D. W., Rangarajan, S., & Wichman, R. (2014). In-band full-duplex wireless: Challenges and opportunities. *IEEE Journal on Selected Areas in Communications*, *32*(9), 1637–1652. https://arxiv.org/pdf/1311.0456

Selinis, I., Katsaros, K., Allayioti, M., Vahid, S., & Tafazolli, R. (2018). The Race to 5G Era; LTE and Wi-Fi. *IEEE Access : Practical Innovations, Open Solutions*, *6*, 56598–56636. DOI: 10.1109/access.2018.2867729

Serrano, R. B., & Sorrentino, S. (2016). *Speed-dependent transmission format for vehicular transmission (U.S.* Patent No. 10,348,444 B2). U.S. Patent and Trademark Office. https://patents.google.com/patent/US10348444B2

Siemens (2019). *Industrial 5G For the industry of tomorrow*. siemens.com/industrial-5g. https://assets.new.siemens.com/siemens/assets/api/uuid:1afdad9c-539a-4cf9-9e3b-97be13a9e27d/presentation-5G-roundtable-sps19.pdf

Simeone, O., Erkip, E., & Shamai, S. (2014). Full-duplex cloud radio access networks: An information-theoretic viewpoint. *arXiv preprint arXiv:1405.2092*. https://arxiv.org/pdf/1405.2092

UNCTAD. (2021). *Technology and Innovation Report 2021: Catching Technological Waves—Innovation with Equity*. UN. https://digitallibrary.un.org/record/3926808

Wang, C.-X., Haider, F., Gao, X., You, X.-H., Yang, Y., Yuan, D., Aggoune, H., Haas, H., Fletcher, S., & Hepsaydir, E. (2014). Cellular architecture and key technologies for 5G wireless communication networks. *IEEE Communications Magazine*, *52*(2), 122–130. DOI: 10.1109/MCOM.2014.6736752

World Economic Forum. (2020). *The Impact of 5G: Creating New Value across Industries and Society*. World Economic Forum. https://www.weforum.org/reports/the-impact-of-5g-creating-new-value-across-industries-and-society

Zhang, J., Qiu, M., Tsai, C.-W., Hassan, M. M., & Alamri, A. (2015). Health-CPS: Healthcare cyber-physical system assisted by cloud and big data. *IEEE Systems Journal*, *11*(1), 88–95. DOI: 10.1109/jsyst.2015.2460747

ADDITIONAL READING

Ding, Z., Liu, Y., Choi, J., Sun, Q., Elkashlan, M., Chih-Lin, I., & Poor, H. V. (2017). Application of Non-Orthogonal Multiple Access in LTE and 5G Networks. *IEEE Communications Magazine*, *55*(2), 185–191. DOI: 10.1109/mcom.2017.1500657cm

Zhang, Z., Chai, X., Long, K., Vasilakos, A. V., & Hanzo, L. (2015). Full duplex techniques for 5G networks: Self-interference cancellation, protocol design, and relay selection. *IEEE Communications Magazine*, *53*(5), 128–137. DOI: 10.1109/MCOM.2015.7105651

Section 5
Competition

Chapter 13
5G Ecosystem and Supply Chain

ABSTRACT

The emergence of 5G technology is transforming global supply chain management through ultra-fast connectivity, low latency, and real-time automation. It enables enhanced visibility, predictive analytics, and seamless integration with IoT, AR, and cloud platforms. Businesses benefit from smart tracking, efficient logistics, and autonomous systems such as drones and AGVs. 5G also supports network slicing and edge computing, allowing tailored, high-performance services across complex operations. While deployment involves regulatory and cost challenges, the long-term advantages include improved resilience, cost efficiency, and collaborative agility. As 5G networks mature, their role in driving supply chain innovation becomes central to digital transformation.

INTRODUCTION

The advent of 5G technology is poised to revolutionize supply chains, offering transformative potential across industries. As businesses increasingly adopt 5G networks, driven by their ability to enable ultra-fast data speeds, reduced latency, and enhanced network slicing, the operational productivity of the supply chain process is expected to increase significantly. 5G technology is a transformative enabler for the digital supply chain, enhancing real-time monitoring and transparency.

Integrating 5G with IoT, augmented reality, and advanced analytics introduces unprecedented capabilities, such as real-time tracking, predictive analytics, and optimized logistics management. Industry leaders recognize the strategic advantage of establishing private 5G networks to ensure enhanced security, reliability, and efficiency. Private 5G networks offer businesses enhanced security, reduced latency, and increased operational reliability (EHTAdmin, 2020). While large multinational

DOI: 10.4018/979-8-3373-0497-7.ch013

organizations are at the forefront of these developments, the broader supply chain ecosystem is preparing for a paradigm shift in how goods are managed, monitored, and transported. Failing to consider the impact of 5G on supply chains could result in businesses falling behind their competitors in an increasingly data-driven and interconnected global economy.

Mangra et al. (2023) provide a comprehensive analysis of how 55 G-enabled technologies can transform the agricultural ecosystem, particularly in food supply chains, rural development, and climate resilience. Utilizing a transdisciplinary framework, the study addresses the integration of ecosystems, networks, and governance to create sustainable agricultural systems. The paper examines the entire food supply chain, encompassing natural resource management and production, as well as processing, distribution, and consumption. It further examines rural development through enhanced connectivity and cross-ecosystem collaboration, promoting local economic growth. Additionally, the authors discuss how 5G can support climate resilience by influencing land use and land cover (LULC) dynamics, facilitating precision agriculture and environmental monitoring (Mangra et al., 2023).

Lyu, Farooq, and Zhu (2025) investigate the cybersecurity vulnerabilities inherent in the 5G supply chain, emphasizing that 5G systems comprise interconnected software, hardware, and cloud components sourced from multiple vendors. This complex, multi-vendor ecosystem introduces significant risks, as mobile network operators have limited visibility and control over the cybersecurity practices of third-party suppliers. The authors highlight how recent cyberattacks have exploited weaknesses in supply chains, identifying this as a strategic threat to the confidentiality, integrity, and availability of 5G networks. They comprehensively analyze the 5G architecture and supply chain landscape, proposing a systematic risk assessment framework to manage supply chain risks. Their approach offers practical guidance for stakeholders to enhance the security and resilience of 5G infrastructures against evolving cyber threats (Lyu, Farooq, & Zhu, 2025).

Massive MIMO, a sophisticated antenna technology, is one of the crucial new solutions tin5 that enable effective spectrum utilization. Because radio waves need to be processed more quickly, Spectrum-sharing strategies, such as micro-licensing, ensure equitable access to 5G resources while fostering innovation.

THE 5G AND THE TRANSFORMATION OF THE SUPPLY CHAINS

The deployment of 5G technology marks a turning point in the evolution of global supply chains. With its ultra-fast connectivity, low latency, and capacity to support billions of connected devices, 5G enables real-time communication and data-driven decision-making across logistics networks. This technological leap facilitates greater automation, transparency, and responsiveness. As industries integrate 5G into their operations, supply chains become more agile, efficient, and customer-centric. McDonagh, Rodgers, and Petrie (2017) outlines how digital transformation strategies can generate significant returns when aligned with next-generation network technologies such as 5G. The following section explores how 5G is driving this transformation.

Two-thirds of businesses intend to deploy 5G by 2020, and functional productivity is observed when the critical elements are in place. Gröne and Ayyagari (2023) forecasts that by 2030, 5G could contribute up to $1.3 trillion to global GDP, particularly through gains in productivity and connectivity.

Apruzzese et al. (2023) examine the significant impact of 5G and complementary technologies, such as IoT, artificial intelligence, and blockchain, on transforming logistics and supply chain systems. Focusing on the transportation and logistics sector, the study demonstrates how 5G innovations can drive automation, facilitate seamless real-time data exchange, and enhance network connectivity. However, it also highlights significant challenges, such as the need for large-scale infrastructure development, heightened concerns over data security, and the substantial financial investment required for effective deployment.

The 5G Portable Broadband Internet Standard Format

When the 5G portable broadband internet standard format becomes available in urban areas worldwide, supply chain and strategy supervisors who lack a method may be at a disadvantage.

Industry experts anticipate the impact of 5G on business strategies and vehicle-to-vehicle communication, enabling the tracking of products as small as microchips. Pickers follow instructions from augmented reality programs that lead them to the correct bin within the factory. According to a Gartner survey, 66% of businesses intend to deploy 5G by 2020. All those businesses want 5G networks primarily employed for Internet of Things (IoT) marketing communications and video clips, with operational productivity as the primary key advantage.

Technological advancements in 5G, the fifth generation of cellular engineering, incorporate faster data speeds, reduced latency, and enhanced network slicing. Slicing enables the cellular signal to split among clients, devices, or systems.

Andrew Stevens, a senior investigation director with Gartner, told Supply Chain Dive that five could be the ubiquitous degree of a physical-to-electronic connectivity in monitoring terminology and product tracking.

According to Gartner, 5G is still in the early stages of growth in the short term, and most cellular network suppliers are not expected to complete end-to-end 5G infrastructure on their public networks until approximately 2025 to 2030. That means businesses are developing their 5G networks for individual facilities and specific uses to take advantage of the science ahead of their competitors, Sally Miller, CIO for DHL Supply Chain, told Supply Chain Dive.

Intel's 5G end-to-end strategy highlights a comprehensive approach to integrating cloud infrastructure, core networks, access networks, wireless technology, and smart devices. 5G's role as a backbone for IoT ecosystems is vital for developing smart warehouses and automated supply chains (Zhao, Askari, & Chen, 2021).

Integration of Intel's Architecture in the 5G Ecosystem

The 5G framework leverages Intel's architecture, including field-programmable gate arrays (FPGAs), software solutions, and embedded security technologies, to deliver advanced functionality and seamless connectivity. By aligning its infrastructure with the growing integration of artificial intelligence (AI), Intel aims to drive innovation across various verticals, including logistics, manufacturing, and retail. For instance, Bandyopadhyay et Al. (2023) analyzes the role of 5G in enabling the transition to smart manufacturing and Industry 4.0 ecosystems. This alignment supports digital transformation by promoting scalability, efficiency, and the convergence of AI and 5G. Huawei (2023) emphasizes the integration of cloud-native architectures as fundamental to maximizing 5G core network performance.

Figure 1. The 5G and the transformation of the supply chains

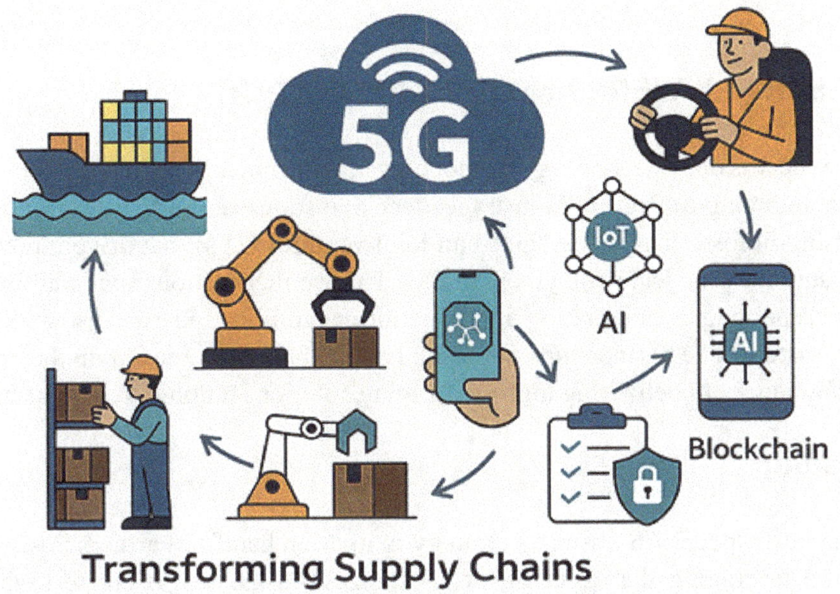

TOWARDS AN INTEGRATED 5 G ECOSYSTEM

Developing an integrated 5G ecosystem is a cornerstone of the global effort to enhance connectivity and enable advanced technologies. A thriving environment necessitates collaboration among industry leaders, business executives, and policymakers to meet the evolving needs of companies and consumers. Integrating 5G with AI-driven systems unlocks unprecedented opportunities for efficiency and scalability across industries.

This environment must support the seamless integration of 5G technologies, such as AI and the Internet of Things (IoT), as well as innovative infrastructure, which are increasingly vital for digital transformation across various industries. Integrating 5G with AI-driven systems unlocks unprecedented opportunities for efficiency and scalability across industries.

The improvement associated with a flourishing 5G environment effectively supporting companies' customers' requirements depends upon reasonable effort among leaders and business executives. To that conclusion, Samsung takes a leading role by accelerating its 5G development and fulfilling its part in bridging the current gap by introducing innovative 5 G ideas from the ground up to bring them to market.

Samsung is refining the edge of its 5G effort to align with the perspective of the 5G environment.

INDUSTRY PLAYERS AND COLLABORATION

The success of 5G technology lies not only in its technical advancements but also in the collaborative efforts of industry leaders. Samsung exemplifies this through its leadership in fostering partnerships and joint ventures (JVs) that drive innovation and ensure the practical deployment of 5 G. These collaborations span multiple industries, enabling the integration of production capabilities and expertise within the 5G ecosystem. The partnerships Samsung has established demonstrate the critical role of alliances in delivering high-performing solutions tailored to diverse needs.

Samsung

One key aspect of Samsung's strategy is its Merchandise Partnerships, which focus on leveraging the expertise of specialized businesses to create high-tech solutions. Through these joint ventures, Samsung accelerates the development of cutting-edge components and devices, ensuring a quicker time-to-market for innovative 5G technologies. By collaborating with businesses that possess deep domain expertise, Samsung enhances its capacity to develop world-class products and services in the 5G era.

Samsung's collaborative approach extends to its Movable Know-How Advancement Method, which partners with functional experts to incorporate the best technologies into its products and services. This approach strengthens Samsung's product portfolio and fosters the creation of comprehensive solutions that integrate seamlessly within the broader 5G ecosystem. By combining expertise in hardware, software, and network technologies, these partnerships ensure that Samsung remains at the forefront of 5G innovation.

The outputs of these collaborations extend beyond products to include programs and systems that enhance 5 G's utility and reach. Samsung optimizes the development process by forming strategic alliances with companies that specialize in specific areas, resulting in efficient and impactful solutions. This collaborative model is essential for achieving the goals set for 5G, including its deployment in the healthcare, manufacturing, and transportation industries, (Swan, 2021).

Samsung's commitment to motivating industry collaboration is pivotal to the profitable deployment and sustained growth of 5G technology. These alliances ensure that 5G solutions are practical, scalable, and adaptable to the complex needs of the modern digital economy. As the 5G ecosystem continues to evolve, Samsung's leadership in fostering industry partnerships serves as a benchmark for the collaborative efforts required to realize the full potential of transformative technology.

Samsung's leadership in 5 G encompasses building JVS and other industries, providing leaders with production levels within the 5G environment, which enables practical applications. Here are several of the alliances that Samsung has formed with partners, enabling the profitable deployment of 5G in existing areas. These JVS and the ensuing outputs in the item and program efforts are essential for the success of 5G moving forward.

Samsung's approach to merchandise partnerships highlights the importance of leveraging joint ventures (JVs) to accelerate innovation in the 5G ecosystem. By collaborating with businesses with specialized expertise in key domains, Samsung enables the development of cutting-edge components and solutions. These partnerships reduce the time-to-market for high-tech solutions, ensuring that advancements reach consumers and industries more quickly.

A central aspect of this strategy is Samsung's Movable Know-How Advancement Method, which brings together functional experts to integrate best-in-class technologies into its products and services. This collaborative model ensures that Samsung's products are innovative and optimized for practical use in the 5G environment. By combining the strengths of multiple partners, these partnerships deliver solutions that meet the demands of a rapidly evolving technological landscape, reinforcing Samsung's position as a leader in the industry.

Xilinx

Xilinx, Inc. (/ˈzaɪlɪŋks/ ZY-links) is an American technology company that develops highly flexible and adaptive processing platforms. The company invented the field-programmable gate array (FPGA), programmable system-on-chip (SoCs), and the adaptive compute acceleration platform (ACAP). The semiconductor company created the first fabless manufacturing model. These Xilinx products are utilized in various industries and technologies, including data centers, wired and wireless communications, AI/ML, automotive, industrial, consumer, aerospace, defense, broadcast, and professional AV.

Marvell

Marvell Technology Group is a Bermudan-domiciled semiconductor company founded in 1995. It has over 6,000 employees and 10,000 patents worldwide. Its annual revenue is $2.9 billion (FY19), and its headquarters are in Santa Clara, CA.

Given the importance of allowing each 4G LTE and 5G network to operate directly in the long term, the market requires processors customized to match the capabilities of these solutions.

Samsung and Marvell have developed joint initiatives for the development and launch of several generations of stereo and balanced airplane processors, which support 4G LTE and 5G NR solutions. These processors enable carriers to deploy multi-radio entry solutions to meet the ever-increasing information needs of consumers and emerging services.

A collaborative effort is essential. Next-generation networks will require an integrated and comprehensive strategy that aligns with the demands of 5G: near-zero latency, high data throughput, and enhanced density. However, they must maintain support for historical LTE and LTE Advanced capabilities. This post-resident functioning of LTE/LTE+ and 5G NR is a vital deployment necessity for ensuring cost-efficient deployment, which provides a seamless transition between 4G and 5G providers.

HPE (Hewlett-Packard Enterprise)

Samsung and HPE merge their strengths to develop best-in-class information management, edge-to-core infrastructure, and stereo networks on commercially accessible platforms. The partnership first addresses a standalone 5G primary software program and a virtual RAN (vRAN).

The first 5G networks around the planet launch inside a non-standalone (NSA) structure, in which the LTE enhanced packet core (EPC) manages each LTE and 5G RAN. The total energy of 5G is delivered to a next-generation center using a standalone (SA) strategy. Samsung has integrated several HPE virtualized network functions (VNFs) into its next-generation core, with industrial deployments set to begin in 2020.

The pattern of community virtualization is shifting away from the center within the RAN. Many top operators plan to use virtual trial RAN (vRAN) architectures that use traditional IT methods. The HPE EdgeLine EL8000 Converged Edge Product is a high-performance, standards-based wedge device for storage and computation at the system edge. Four HPEs and Samsung collaborated to deliver the EL8000, promoting virtualized Radio Access Network (vRAN) applications. Hope and

Samsung enable a quicker and smoother 5G transition through this partnership, leveraging the center's advantage.

Amdocs

Amdocs is a multinational corporation founded in Israel and headquartered in Chesterfield, Missouri. It has worldwide support and development centers, specializing in software and services for industries such as communications, media, financial services providers, and digital enterprises.

The five primary changes to the G system call for acceleration of the 5G program's shipping and delivery while leveraging the agility enabled by cloud networks. Samsung and Amdocs collaborate to integrate and onboard virtual network functions (VNFs), ensuring that community operators can utilize Samsung's 5G community solutions, which are designed on the Open Networking Automation Platform (ONAP).

AAmdocs' Expertise in deep ONAP, software-defined networks (SDN), and network function virtualization (NFV) enables a natural partnership with Samsung that advances Samsung's virtualization 5G community strategies. Beginning with Samsung's Virtualized Central Unit (VCU), the collaboration expands to incorporate parts of the resulting solutions that enable an elegant, end-to-end cloud-based ecosystem. The teams also explore driving solutions, enabling the rollout and integration of virtualized Radio Access Network (vRAN).

The collaborative initiatives of the remedies and all those enable marketing communications program suppliers (CSPs) to deploy innovative services across an agile and open environment.

Openet

Openet is a world-leading digital Business Support System (BSS) company that provides charging, policy, and data solutions, working with numerous mobile service providers.

The 5G community demands an efficient and reliable solution for managing real-time PC user policies, including policy enforcement. To that conclusion, Openet and Samsung collaborate to offer 5G community performances (NFs) within their primary application solutions. Six. The joint answer comprises Openet's cloud-native 5G software program wedge for real-time policy and charging systems. Its interface with Samsung's package core application items offers proper settings that optimize user encounters. Openet's experience with electronic business support systems (BSS) makes it uniquely suited to provide this essential component within the 5G community.

Zhilabs (A Samsung Company)

Zhilabs, recently acquired by Samsung Company, offers AI-powered analytics solutions that help service providers reduce network management complexity and operational costs.

It recognizes the premium price of an AI-based buyer analytics wedge; in late 2018, Samsung acquired Zhilabs, the designer of a real-time analytics wedge that delivers program suppliers the capability to fine-tune the buyer within 5 G. An AI-based community application analyzes clients' usage patterns. It screens the caliber of the service experience and the exponential information of 5G, which is essential to automating and intelligent community analytics programs.

Summary

The development and deployment of 5G technology rely on a complex network of stakeholders, each playing a key role in ensuring the successful implementation of the infrastructure and services.

Operators, such as Orange, AT&T, Verizon, MTN, and Airtel, are responsible for deploying and maintaining the core 5G infrastructure. They provide mobile and data services, serving as the backbone of the connectivity layer. These organizations invest heavily in spectrum acquisition, network expansion, and service innovation to ensure broad and reliable 5G coverage.

Service providers, including companies like Ericsson, Nokia, Huawei, AWS, and Microsoft Azure, enable the functional operation of networks by offering software platforms, cloud solutions, and integration services. Their expertise supports scalability, network orchestration, and seamless cloud-native transformations in 5G environments.

Equipment suppliers, such as Qualcomm, Intel, Samsung, ZTE, and Cisco, play a pivotal role in manufacturing the physical components of the 5G network. This includes advanced chipsets, antennas, modems, and base stations, which are essential for high-speed and low-latency communication.

ZTE Corporation (2023) details technological advancements in massive MIMO and beamforming to enhance 5G spectral efficiency and coverage.

End users are the ultimate beneficiaries of 5G capabilities. They range from industries like automotive (Tesla), healthcare (Philips), and smart manufacturing (Siemens) to everyday consumers. These stakeholders leverage 5G to enable smart mobility, real-time healthcare monitoring, industrial automation, immersive media, and more.

Understanding how these entities interact and support one another is key to grasping the full scope of the 5G value chain. This overview offers a simplified yet comprehensive perspective on the roles, responsibilities, and interdependencies that drive the 5G transformation forward.

The table below summarizes the main stakeholders in the 5G ecosystem. It highlights the different categories, provides key examples, and describes the primary roles and responsibilities of each group. This overview will help readers quickly grasp the structure and interconnections within the 5G value chain.

The development and deployment of 5G technology involve a diverse network of stakeholders, each playing a critical role in shaping the ecosystem. From infrastructure deployment to service delivery and hardware innovation, collaboration across industries is essential for realizing the full potential of 5 G. This table provides a structured overview of the main categories of participants in the 5G ecosystem, highlighting their specific functions and representative organizations. Understanding these roles helps clarify how 5G services are designed, delivered, and utilized. The table also illustrates the convergence between telecom, cloud, hardware, and end-user sectors in advancing 5G capabilities. It offers a structured summary of the primary actors involved in the 5G ecosystem. It categorizes the key players into four main groups: operators, service providers, equipment suppliers, and end users.

Table 1. Key stakeholders in the 5G ecosystem

Category	Examples	Roles & Responsibilities
Operators	Orange, AT&T, Verizon, MTN, Airtel	Deploy and manage 5G infrastructure; offer mobile/data services
Service Providers	Ericsson, Nokia, Huawei, AWS, Microsoft Azure	Provide network services, cloud integration, and platform tools
Equipment Suppliers	Qualcomm, Intel, Samsung, ZTE, Cisco	Develop and supply 5G chips, antennas, base stations, and routers
End Users	Automotive (Tesla), Healthcare (Philips), Industry 4.0 (Siemens), Consumers	Use 5G for smart mobility, telemedicine, automation, and streaming.

THE REGULATION OF 5G

The regulation of 5G plays a fundamental role in ensuring secure, efficient, and equitable deployment of next-generation wireless networks. As 5G technologies enable transformative applications across industries, regulatory bodies must address challenges related to spectrum allocation, privacy, cybersecurity, and international standards. Effective governance is crucial for striking a balance between innovation and public interest, as well as promoting fair competition. This section examines the primary regulatory frameworks influencing 5G deployment worldwide. It also highlights the importance of harmonization across countries to support interoperability and sustainable growth.

5G services place extra demands on network design and require increased concentration in some areas, resulting in more base stations, sites, and antennas.

Local and national authorities play a crucial role in creating the right conditions to secure necessary investments, making the 5G roll-out easier. Removing administrative hurdles related to permitting processes and regulations that send the right build and buying signals to market participants is a prerequisite. Collaboration between network operators, government, local authorities, municipalities, and the private sector is required to ensure the timely and efficient deployment of 5 G.

Qose and Zoltán (2024) explore the intersection of 5G technology, supply chain cybersecurity, and legal frameworks, emphasizing the growing importance of regulatory oversight in safeguarding national and economic security. The authors highlight that as 5G becomes central to digital transformation, legal measures—particularly within Europe—are being developed to assess supplier risks based on technical and geopolitical considerations. Despite these initiatives, many nations lack comprehensive legislation to effectively identify and manage high-risk vendors. The paper emphasizes that future legal requirements are expanding due to the increasing complexity of 5G networks and global political tensions, which may impact commercial freedoms and international trade, as risk assessments could lead to restrictions or bans on specific suppliers, (Qose & Zoltán, 2024).

In this section, we review the regulations in different regions around the globe.

The Regulations in the US

The United States has adopted a proactive approach to regulating 5G, aiming to maintain its global leadership in wireless innovation. Regulatory efforts are primarily led by the Federal Communications Commission (FCC), which oversees the allocation of spectrum, the deployment of infrastructure, and market competitiveness. These policies focus on accelerating the rollout of 5G, reducing regulatory barriers, and encouraging private sector investment. National security considerations also play a crucial role, particularly about foreign equipment providers. This section outlines the regulatory strategies shaping the 5G landscape in the U.S. and their implications for stakeholders.

The Federal Communications Commission (FCC) in the USA took significant steps in its order and report (FCC, 2016a) by developing new licensing, services, and regulations for specific 5G bands in the millimeter-wave spectrum, including methods for sharing between entrants and protecting incumbent spectrum users.

The Federal Communications Commission (FCC) established a three-tier sharing model for offering CBRS on a shared basis in the 3.5 GHz band. This design presents the next goal level of direct access licenses (PAL) in specific geographical areas known as census tracts for currently 3-season license periods, while ensuring that first-tier incumbent federal customers remain free from harmful interference.

The unit also introduces a final goal level for General Authorized Access (GAA) users, ensuring that better-priority-level incumbents and PAL license holders remain unaffected by potentially hazardous interference. GAA users can access areas of the band without incurring license costs, although they are not guaranteed to be interfered with or shielded from other drivers.

The mobile operator industry has become highly regulated, demonstrating that regulators significantly shape the ecosystem of mobile communication companies. The mobile industry has evolved from state-owned monopolies to a competitive setting, driven by the adoption of innovative regulatory systems that aim to promote advertising sharing and infrastructure competition.

The professional ecosystem primarily forms through regulatory mechanisms that allocate spectrum and rights to operators, and indirectly, other environmental stakeholders. At the same time, the environment is evolving significantly outside the mobile communication and connection domain, including single-handed developments that boost technologies and open the door to services based on electronic data produced, consumed, and discussed by businesses and individuals.

Components of Regulation of the Mobile Communication Ecosystem

It determines the essential components that regulate the mobile communication ecosystem, such as spectrum policy, competition, access to networks, content, infrastructure, and price regulation, facilitating control, harnessing benefits, and informing policy. From the operator's viewpoint, GSMA (2025) has pointed out the following regulatory elements: access regulation, consumer protection, economic regulation, competition enforcement, intellectual property, data and privacy safety, taxation, network security, resource management, and common accessibility and service.

The regulatory framework for 5G implementation plays a crucial role in ensuring equitable access, affordability, competition, and user privacy. Access Regulation is pivotal for enabling interconnection and interoperability, ensuring that stakeholders without direct infrastructure access can benefit from seamless connectivity, mainly through network slicing and small-cell deployments. Mukasa (2025) explores network slicing as a pivotal innovation for enabling differentiated services in enterprise 5G use cases. Pricing Regulation focuses on maintaining affordability by enforcing non-discriminatory pricing models between operators, ensuring high-quality services for end-users. Competition Regulation prevents market power abuse by promoting local operators catering to specific vertical sectors, fostering innovation and diversity in the 5G ecosystem. Privacy And Data Protection are increasingly critical as 5G accelerates data collection and usage, necessitating strict rules on data ownership and user confidentiality. Lastly, Authorization of Networks and Services, including spectrum licensing, provides a structured approach to granting access rights, supporting local licensing models that encourage diverse stakeholders to deploy small cell networks with guaranteed quality. These regulatory elements collectively support the sustainable and inclusive deployment of 5 G.

The regulatory landscape for 5G is evolving to accommodate new deployment models and localized network operations. Traditional spectrum licensing frameworks are being re-evaluated to better suit flexible and decentralized use. Micro licensing has emerged as a promising approach for managing local 5G networks in specific environments. This enables organizations to operate private networks with customized performance and governance. The table below summarizes the key regulatory elements shaping the 5G ecosystem (Matinmikko et al., 2018).

Table 2. Overview of regulatory element in 5G

Regulatory Element	Description
Spectrum Licensing	Includes traditional exclusive licensing and new micro licensing approaches.
Network Deployment	Supports local and private network installations for industrial and enterprise use.
Access Rights	Defines user access levels and operational permissions for localized spectrum use.
Governance Models	Introduces new structures for managing local 5G networks with tailored rules.
Policy Flexibility	Encourages adaptable regulatory frameworks for rapid 5G innovation.

Adapted from Matinmikko et al. (2018)

Auctions for Spectrum Prices

Kadena and Rajnai (2024) provide an overview of the transformative role of 5G technology in modern supply chain management, highlighting its ability to support real-time tracking and facilitate rapid decision-making across increasingly complex and globally distributed networks. The authors argue that the integration of 5G enhances communication between sensors, devices, and machines, enabling greater automation and visibility within the supply chain. The paper highlights key applications of 5G in supply chain operations, including warehousing, logistics, and stakeholder coordination, while addressing implementation challenges such as infrastructure readiness and interoperability. To overcome these obstacles, the authors advocate for a collaborative approach that involves multiple stakeholders to ensure the effective deployment and management of 5 G-enabled supply chains, (Kadena & Rajnai, 2024).

The predictions pertain to correct 5G networks and products, excluding those operators who will almost certainly market the latest 4G LTE (lLong Term Evolution technology, known as 4.5G, lsimilar to 5 G The variation does not matter to subscribers at first: Both 4.5G and 5G networks provide blistering speeds of a vast selection of megabits per second and gigabits per second. Nevertheless, there is a crucial distinction for the operator: price. Suppose operators can send a 500 Mbps wireless service more than 4.5 G. Why must they spend millions on capital expenditures (Capex) for tools and a stereo spectrum to allow 5G globally?

Solution to This Issue Concerning 5G's Higher Capacity

There is a generally recognized solution to 5G's demand for higher capacity. Compared to 4G, 5G is designed to deliver a hundredfold increase in traffic throughput and network efficiency through technologies such as massive MIMO, network slicing, and advanced spectrum utilization (Andrews et al., 2014; Boccardi et al., 2014). IDC (2023) estimates that global data traffic will double annually, driven in large part by 5G-connected devices and AI-enabled applications. For smaller players—third- or fourth-rank operators with surplus spectrum and fewer subscribers—it may be feasible to offer near-5G speeds on an optimized 4G network. However, for leading operators in countries like the United States, the Philippines, France, Ireland, Australia, and the United Kingdom, 5G adoption becomes imperative to maintain network performance.

Regarding capital expenditures (CAPEX), initial concerns over steep cost burdens are now being reassessed. Early predictions had forecasted a jump from 13% to 22% of revenue solely for partial deployment. However, as field trials matured in 2018, operators across North America, Europe, and Japan reported that 5 G-related CAPEX could remain modest, aided by prior investments in dense fiber infrastructure and 5 G-ready radio hardware, allowing software-based upgrades rather than wholesale replacements, (Mukasa, 2025).

Spectrum acquisition trends also reflect a shift from 3G-era spikes toward more sustainable pricing, akin to the 4G rollout, in contrast to the UK's 3G auctions which reached $3.50 per MHz-pop, 4G spectrum in the 800 MHz band sold for $0.60 per MHz-pop and 2.6 GHz bands for $0.07 (GSMA, 2025). Early 5G auctions in six countries showed values under $0.20 per MHz-pop, with two auctions pricing below $0.01. Still, exceptions exist: Italy's 2018 auction saw 700 MHz bands sell at $0.65 and mid-band 3.6–3.8 GHz frequencies fetch $0.42 per MHz-pop, well above regional expectations.

Ultimately, 5G is poised to become the foundational architecture for future digital connectivity. Although adoption may progress gradually over the next two years, its transformative potential in enhancing speed, capacity, and latency will eventually reshape global networks (Rapaport et al., 2013; Haider et al. 2011).

Regulatory Obstacles of Cognitive Radio and Market Mechanisms for Spectrum Management

The ITU WORKSHOP paper introduces the policy ramifications of technical advancements in the area, such as stereo technologies, and explains how these solutions can support high-quality industry competition.

It depends on the work currently being done in the Sport VIEWS (Spectrum Policies and Radio Technologies Viable in Emerging Wireless Societies) research project, funded by the European Commission (FP6).

According to the TU, the Workshops Survey examines emerging radio technologies, the economics of the spectrum, and the relevant characteristics of spectrum management. It also evaluates the problems facing them.

Pre-existing radio technologies, spectrum consumption, management, and present systems for spectrum allocation.

As indicated by the ITU WORKSHOP overview created in this examination, the primary focus is on innovation advancements in fixed, mobile (including roaming), and communication techniques.

Connections and dependencies with other spectrum consumption styles (radars and astronomy) and navigation systems were found at specific points. The impact of these various spectrum consumption types is noted in the engineering improvements in radio communications.

The focus is on frequencies under 6 GHz, encompassing the entire frequency range for mobile and terrestrial services. Numerous technical challenges remain, which are the main obstacles to the evolution of cognitive radio.

Table 3 provides a comprehensive overview of the critical components that influence the deployment and adoption of 5G technology. It examines the implications of 5G on supply chain management, capital expenditures (CapEx), spectrum pricing, and regulatory frameworks. Each topic highlights both the challenges, such as infrastructure readiness and regulatory complexity, and the emerging opportunities, like real-time visibility and shared infrastructure. The table provides valuable insights into how these factors collectively shape the 5G ecosystem. It also draws on current research and industry reports to support strategic planning and investment decisions.

Table 3. 5G technology overview: Supply chain, CapEx, spectrum, and regulatory challenges

Topic	Key Insights	Challenges	Opportunities	References
5G in Supply Chain Management	Enables real-time tracking, automation, and visibility	Infrastructure readiness, interoperability	Enhanced efficiency, stakeholder coordination	Kadena & Rajnai (2024)
CapEx for 5G	Initially expected to be high; actuals closer to 4G CapEx	Spectrum acquisition and infrastructure upgrades	Preloaded investments and shared infrastructure	Industry Analysis, 2018
Spectrum Pricing	Generally aligns with 4G pricing except in select markets	High costs in countries like Italy for mid-band spectrum	Affordable mmWave spectrum; auction competitiveness	ITU Reports, National Auctions
Regulatory and Cognitive Radio	Flexible spectrum use enhances innovation	Complex regulation, interference with other services	Dynamic access models and secondary markets	ITU WORKSHOP & SPORT VIEWS (FP6)

The primary regulatory issues related to a CR's awareness of its spectral environment are as follows:

Wideband Sensing

The radio must be able to assess actual spectrum use over a wide tuning range. The challenge with wideband spectrum sensing is combining a sizeable instantaneous bandwidth, sufficient measurement accuracy, and a vast amount of data pre-processing to identify opportunities. Businesses are increasingly leveraging 5G to enhance last-mile connectivity and improve delivery accuracy, (Ellwood, 2020).

Opportunity Identification

The radio must detect whether a frequency band is in use. Several signal detection methods are available for this purpose. They all suffer from the following trade-off in signal detection: higher precision in a wider monitored frequency band requires more processing time and a higher power supply.

Interference Prevention

If a frequency band is free and available to an opportunistic spectrum user, interference with a primary user is still possible. The CR might not detect the primary user's signal, but it does not know whether its transmission is detectable at its site. For this purpose, it would need the location information of all neighboring prior users, either provided by the primary network itself or through a costly network of sensors that monitor the spectrum for direct users and share this information with the CR.

Dynamic Coordination

Many methods to coordinate users of the same spectrum are currently under study. One method is described in section 3.5 of this report. • Spectrum policy compliance: Is transmission in a particular band permitted at a given location and time? As stated, a CR can verify this and adjust its transmission behavior accordingly. This feature requires the availability of some policy-based meta-language that translates policy rules into radio behavior controls.

Software Challenges

SDR challenges also exist for Cl. The decision-making and learning processes are still being studied (See Several DySPAN publications).

Figure 2. Main regulatory issues of the 5 G to a CR's awareness of its spectral environment

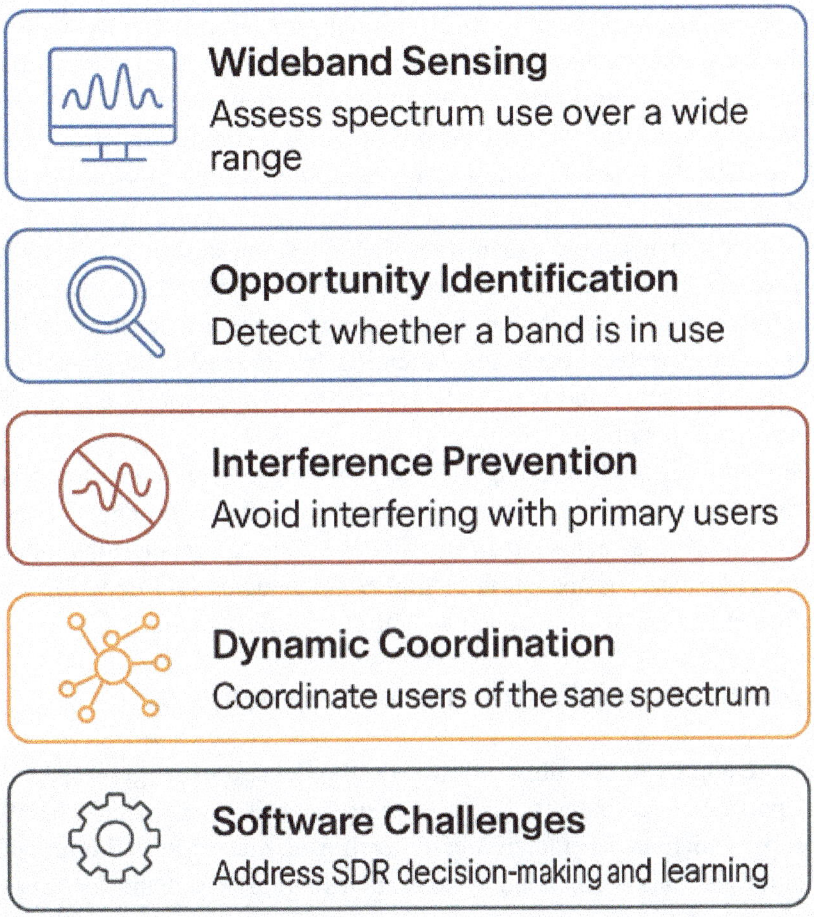

REGULATIONS FOR 5G: MICRO LICENSE FOR LOCALLY OPERATED NETWORKS

This section examines regulatory frameworks that facilitate localized 5G deployments through the issuance of micro-licenses. As 5G expands, national regulators are introducing new spectrum management approaches to enable private, locally operated networks, especially in industrial and enterprise settings. Micro licensing allows organizations to access specific spectrum bands in defined geographic areas without relying on national carriers. This fosters innovation, supports customized use cases, and encourages spectrum efficiency. The model is gaining momentum in sectors such as manufacturing, logistics, and smart campuses.

Micro licensing is a type of licensing for regional 5G community deployments, which involves authorization mechanisms to designate entry rights to the radio spectrum. This is a crucial style element shaping the future of the 5 G mobile communication industry. Micro-licensing provides a flexible spectrum allocation mechanism, enabling localized deployments among various stakeholders (Matinmikko et al., 2018).

These mechanisms must expand the sector beyond current MNOs while ensuring a reasonable competition atmosphere for each entrant and incumbent, and encouraging them to invest. An operator's importance in a spectrum band depends on their previous spectrum portfolio, indicating that increasing the chance to access a brand-new spectrum band is more beneficial to an operator with a significantly smaller spectrum portfolio

5G networks utilize a broad range of spectrum bands, including the low-frequency bands with current mobile allocations and the potential new frequency bands from 24 GHz to 86 GHz, as expected from WRC 19. Due to higher signal attenuation caused by radio wave propagation at higher frequencies, 5G networks can only provide limited coverage areas, such as within specific structures.

The Specificities of 5G Community Deployments

Standards approval structures, such as cross-border approvals, grant special access rights in portable interchanges. These structures consider the specific needs of 5G user groups, which are characteristically territorial, and address the requirements of particular verticals corresponding to regular mobile broadband services.

License Element, Description, and Role in Micro Licensing

Micro-licensing introduces a more flexible and localized approach to spectrum allocation and management, tailored to meet the diverse needs of various stakeholders. The Eligible Licensee criterion expands access beyond traditional mobile network operators (MNOs), enabling multiple organizations to apply for licenses. The Purpose Of Use emphasizes targeted mobile services with advanced interference management, a concept already proven in other radio services. Regarding License Duration, micro licenses are typically shorter (a few years) than traditional exclusive licenses, promoting adaptability and allowing for renewability.

The License Area is geographically constrained, significantly smaller than country-wide allocations, and often considers building-specific factors for indoor and outdoor usage. Technology Neutrality allows licensees to choose the technologies that comply with regulatory standards, offering greater flexibility in implementation. Transmission Parameters, such as block edge masks and transmit power limits, are set to vary based on deployment scenarios, including indoor small cells.

Selection mechanisms must be objective and transparent when awarding micro licenses, and competition must be promoted while spectrum allocations are managed effectively. Pricing encourages innovation and competition rather than maximizing regulatory revenue, with lower costs reflecting local spectrum availability. The Transferability of Rights outlines conditions for reselling or subleasing parts of the license, ensuring operational continuity for license holders.

Obligations specify deployment time frames, coverage targets, and infrastructure-sharing conditions, aligning with the localized nature of micro licenses. Lastly, Interference Coordination ensures protection from harmful interference while requiring cooperation among micro license holders and existing spectrum users to maintain predefined interference thresholds. These elements collectively support a dynamic, inclusive, and efficient framework for 5G network deployment.

The regulatory landscape for 5G is evolving to accommodate new deployment models and localized network operations. Traditional spectrum licensing frameworks are being re-evaluated to better suit flexible and decentralized use. Micro licensing has emerged as a promising approach for managing local 5G networks in specific environments. This enables organizations to operate private networks with customized performance and governance.

Table 4 presents a summary of the primary regulatory elements influencing the development of the 5G ecosystem. These elements are critical for enabling flexible, secure, and efficient network deployment across diverse contexts. The table outlines key aspects, including spectrum licensing models, access rights, and governance structures, that support locally operated networks. It also emphasizes the importance of regulatory adaptability to meet evolving technological and market demands. Together, these components form the foundation for effective 5G implementation.

Table 4. The key regulatory elements shaping the 5G ecosystem

Regulatory Element	**Updated Description**
Spectrum Licensing	Covers both exclusive licensing and micro licensing, enabling flexible access models.
Network Deployment	Facilitates the setup of private and locally operated networks tailored to specific needs and requirements.
Access Rights	Defines operational and user permissions for localized spectrum use.

continued on following page

Table 4. Continued

Regulatory Element	Updated Description
Governance Models	Establishes control frameworks for managing local networks with context-specific policies.
Policy Flexibility	Supports adaptable regulation to meet the innovation and deployment speed requirements of 5G environments.

COMPARISON OF LICENSING WITH INDIVIDUAL AND GENERAL AUTHORIZATION

This section provides a concise comparison between individual licensing and general authorization frameworks within the context of 5G deployment. As 5G networks require dynamic and flexible access to spectrum, understanding the regulatory implications of these two approaches becomes essential. Individual licensing offers more control and protection, whereas general authorization allows for broader and more rapid deployment. By evaluating both models, regulators and operators can better align their strategies with technical and market needs. The comparison below highlights the respective advantages and constraints of each approach.

The distinct approaches to spectrum allocation—individual Authorization, Micro Licensing, and General Authorization—highlight the unique benefits of micro licensing as a middle ground between exclusivity and openness. Individual Authorization involves issuing a limited number of licenses through auctions at high prices, accompanied by long-term obligations, thereby ensuring investment certainty and protection from interference. This approach, however, creates a static market with limited opportunities for new entrants.

In contrast, micro-licensing enables stakeholders, including new entrants, to access local rights while maintaining coordination among license holders. It offers a predefined level of interference protection, ensuring quality guarantees and fostering market inclusivity. By efficiently protecting incumbents, micro-licensing encourages spectrum availability and innovation while allowing flexible deployment strategies.

Conversely, General Authorization eliminates licensing fees and permits market entry for many users. This approach provides an open arena for experimentation and innovation but lacks guarantees for interference protection, making it less suitable for quality-sensitive applications.

By balancing accessibility, protection, and efficiency, micro-licensing bridges the gap between the exclusivity of individual authorization and the openness of general approval, catering to diverse needs in modern spectrum management. This approach is crucial for fostering innovation and inclusivity in the deployment of 5G networks.

MICRO LICENSING PROCESS

The Micro Licensing Process is a regulatory innovation designed to facilitate localized and flexible spectrum access for 5G networks. Unlike traditional national licenses, micro licenses enable small-scale operators, such as factories, hospitals, or campuses, to deploy private 5G networks tailored to specific needs. This process reduces bureaucratic barriers, shortens application timelines, and promotes efficient use of underutilized spectrum. It supports innovation, especially in Industry 4.0 contexts, by allowing dynamic and cost-effective deployment. The following section outlines the key steps and criteria involved in the micro licensing process.

The micro-licensing version involves a stepwise process to enhance any spectrum-discussing design. The micro licensing process for awarding and controlling micro licenses, which includes six measures, is outlined jointly, along with a few pointers to guide each phase. To begin, determine the ideal bands for micro-licensing. Micro licensing with neighborhood access rights is especially suitable for operations above the 6 GHz band, as propagation losses restrict network coverage and interference to local areas.

Operation in lower bands, primarily below 3 GHz, is significantly more complex, resulting in more substantial security margins in terms of separation distances between micro licensees, due to propagation characteristics.

Next, the band must be partitioned based on a specific band strategy to determine how the accessible bandwidth in each defined area is divided. A great deal is a specific frequency band covering a particular geographical area. When identifying the band program, consider making the ideal of the lot for exceptionally high detail, advanced technologies, harmonization, and other band programs, and ensuring that each licensee can deploy a purposeful, high-quality system with the band program.

Steps and Guidelines for Implementing Micro Licensing

The structured procedure for micro licensing, alongside practical guidelines, ensures effective implementation and adaptability. The Steps For Micro Licensing begin with identifying Suitable Spectrum Bands, focusing on frequencies supporting localized deployments. It is followed by determining a Specific Band Plan, which allocates available channels for each location, enabling efficient use of spectrum resources. Defining Micro License Parameters includes specifying conditions, such as interference protection and service quality guarantees, ensuring clarity for license holders.

Mechanisms for awarding licenses are carefully Designed to Handle Applications, allowing for fair and transparent allocation processes. Monitoring the use of issued licenses ensures compliance and addresses potential Conflicts That Arise During Operations. Finally, the procedure incorporates the Further Development of Micro Licensing, leveraging feedback and real-world experience to refine and adapt the framework.

The accompanying Guidelines emphasize actionable recommendations for each step. For instance, higher frequency bands above 6 GHz are recommended for their suitability in providing localized coverage. The guidelines advocate for flexible band channeling, robust interference protection mechanisms, and standardized service quality criteria. Auction mechanisms are proposed to handle varying numbers of applicants alongside provisions for reselling licenses to enhance flexibility.

Additionally, the guidelines highlight the importance of Rapid Response to Interference Issues and fostering coordination among license holders. Collecting stakeholder feedback is critical for continuously improving the micro-licensing process, ensuring it remains adaptable and effective in meeting the evolving needs of 5G deployments. This comprehensive approach promotes innovation, inclusivity, and efficient spectrum utilization.

The following table provides a structured overview of the micro licensing process, enabling localized 5G network deployment. It outlines each key step, from assessing local needs to monitoring regulatory compliance. This process is designed to ensure efficient spectrum use while minimizing interference with existing services. By providing clear guidelines, regulators aim to support agile and context-specific rollouts of 5 G. The table also highlights the collaborative role of stakeholders in the responsible management of spectrum resources.

Table 5. Procedure for micro licensing and guidelines

Step	Description
1. Identification of Local Needs	Determine specific requirements and scope for local 5G deployment.
2. Availability Check	Check the availability of spectrum in the desired area and frequency band.
3. Application Submission	Submit an application to the licensing authority with deployment details.
4. Assessment and Coordination	Evaluate the application, ensure no harmful interference, and coordinate with existing users.

continued on following page

Table 5. Continued

Step	Description
5. License Grant	Issue a local license with defined parameters such as duration, coverage, and technical constraints.
6. Compliance Monitoring	Ensure continued compliance with license terms and effective interference management.

Source: Adapted from Matinmikko M, Latvia-Aho M, Ahokangas, P Seppänen V., On regulations for 5G: Micro licensing for locally operated networks, Telecommunications Policy,2018, vol: 42 (8), pp: 622-635.

The Specific Cases of Governmental Strategies for the UK and India

This section examines several cases of governmental strategies implemented to support the deployment and regulation of 5G networks. Governments worldwide are taking proactive steps to ensure secure, efficient, and inclusive access to 5G technology. These strategies often involve spectrum allocation, investment incentives, and collaboration with industry stakeholders. By examining different national approaches, we gain insight into the diverse regulatory and economic frameworks shaping 5G ecosystems. The following cases highlight key policy choices and their implications for innovation and infrastructure development.

The section presents real-world cases that illustrate how governments address 5 G-related security and policy challenges. It highlights national strategies and actions taken to manage technological risks. We begin with the United Kingdom's official inquiry into the security of its 5G infrastructure.

The UK's Strategic Review of 5G Security

The United Kingdom has launched a comprehensive inquiry into the security implications of its 5G infrastructure, with particular scrutiny on the involvement of foreign vendors such as Huawei. This initiative reflects growing geopolitical and cybersecurity concerns as the UK navigates its digital future, striking a balance between national security and international partnerships.

UK Parliamentary Inquiry on Huawei and 5G Infrastructure

A subcommittee of the House of Commons Defence Committee is leading the inquiry, focusing on the potential risks associated with deploying telecommunications equipment supplied by foreign entities, particularly the Chinese firm Huawei.

The investigation aims to evaluate the long-term security implications of integrating Huawei's technology into the UK's 5G ecosystem.

Key Questions and National Security Concerns

Central to the inquiry is how the UK can reduce or eliminate its dependency on Huawei while identifying viable alternative suppliers. Additionally, the subcommittee is evaluating the potential impact of continued reliance on Huawei on the UK's strategic alliances, particularly its relationship with the United States.

Tobias Ellwood, chair of the Defence Committee, emphasized the transformative potential of 5G and the need to address national security concerns proactively. He stated that 5G will become integral to the country's communications infrastructure, necessitating a cautious approach to its implementation.

Huawei's Position and Government Assessments

Over the past eighteen months, the UK government and parliamentary bodies have examined cybersecurity risks associated with Huawei and concluded that a full ban on the company is not technically justified. Huawei, which has been active in the UK telecom sector for nearly two decades, reaffirmed its commitment to high cybersecurity standards and expressed its willingness to cooperate with UK authorities in the ongoing assessment.

This inquiry marks a critical step in shaping a secure and resilient 5G future for the United Kingdom.

India's Strategic Move Toward 5G Trials

India is preparing to take a decisive step in advancing its telecommunications landscape by initiating trials for fifth-generation (5G) cellular technology. The Ministry of Communications has set a short-term objective to launch these trials within 100 days, aiming for a mid-September timeline. While countries such as the United States, South Korea, and China have already begun commercial deployment, India is positioning itself for a national rollout by 2020. These developments signal the government's commitment to integrating next-generation connectivity into the country's digital ecosystem. The 5G initiative aligns with broader national goals of innovation, economic development, and global competitiveness.

Structured Summary: India's 5G Trial Preparations

On June 3, Union Communications Minister Ravi Shankar Prasad announced that India would hold a spectrum auction, including bands supporting 5G, within the calendar year. Although commercial 5G services have already launched in the U.S., South Korea, and China, India remains in the preparatory phase, focusing on comprehensive trials before commercial rollout.

Minister Prasad confirmed the government's intention to initiate 5G trials within 100 days of assuming office. These trials are a prerequisite for determining spectrum allocation and for validating India-specific use cases, especially in high-speed broadband and low-latency applications.

As part of its long-term strategy, the Indian government introduced a 224 crore initiative in 2018 to support affordable 5G research and innovation. Industry leaders, such as Ericsson, have already contributed to this goal by establishing a 5G testing facility at IIT Delhi, which will help build localized applications tailored to the country's socio-economic context.

With these steps, India aims not only to bridge the digital divide but also to shape global standards by leveraging its vast market potential and technological expertise.

Summary

As countries transition to 5G technology, national governments have adopted distinct strategies to address both deployment and cybersecurity challenges. These initiatives reflect not only technological priorities but also geopolitical considerations and economic ambitions. From formal inquiries into supply chain risks to government-funded trials and innovation programs, public authorities are playing a central role in shaping 5G ecosystems. The table below highlights key national approaches and the stakeholders involved in their implementation. It highlights how different governments are striking a balance between technological advancement, security, and strategic autonomy.

Table 6. Cases of governmental strategies for 5G security and deployment

Country	Strategy/Action	Focus Area(s)	Key Stakeholders	Remarks
United Kingdom	Launched an official inquiry into 5G infrastructure security	National security, supplier risk (esp. Huawei), U.S. relations	House of Commons Defence Committee, Huawei	Inquiry into Huawei's involvement in 5G: balancing tech leadership and geopolitical concerns
India	Government-initiated 5G spectrum auction and trials (announced June 2019)	Deployment roadmap, local innovation, timeline to 2020	Ministry of Communications, Ericsson, IIT Delhi	224 crore budget; focus on latency-sensitive and broadband applications

CONCLUSION

Chapter 13 provides a comprehensive overview of the 5G ecosystem and its impact on global supply chains, regulations, and industrial strategies. With technologies like Massive MIMO, cognitive radio, AI, IoT, and software-defined networking, the 5G infrastructure enables real-time tracking, enhanced automation, and efficient data communication. The chapter also emphasizes the importance of micro-licensing and localized spectrum management in enabling inclusive access to 5G networks, especially for private and industrial deployments. Strategic collaborations among technology leaders, such as Samsung, Intel, Marvell, and Amdocs, demonstrate how joint ventures and ecosystem alignment are pivotal to accelerating 5G adoption.

This table provides a structured overview of the main categories of stakeholders and illustrates their respective contributions. Understanding their roles is crucial for comprehending how 5G technologies are implemented and integrated into various industries. The synergy between these groups drives innovation and supports the evolution of next-generation digital infrastructures.

Table 7. Overview of 5G ecosystem stakeholders and their contributions

Stakeholder Category	Examples	Roles & Contributions
Operators	Orange, AT&T, Verizon, MTN	Deploy and manage 5G infrastructure; offer core services
Service Providers	Ericsson, Huawei, Microsoft Azure	Deliver platforms, network services, and orchestration
Equipment Suppliers	Qualcomm, Intel, Cisco, Samsung	Manufacture hardware: chipsets, antennas, base stations
End Users	Tesla, Siemens, Philips, Consumers	Utilize 5G for automation, IoT, mobility, and real-time applications

Future 5G networks are expected to disrupt the standard mobile communication marketplace by lowering the entry barrier for new entrants. They will reveal the necessary materials and technological developments to work on increased carrier frequencies and end-to-end community slices.

Alhuseini and Olama (2019) introduce a novel framework for 5G service value chains, focusing on ecosystem modeling, agile service delivery, and user-story automation. This framework supports the growing demand for decentralized, customizable 5G services tailored to enterprise and vertical market use cases. A key innovation is the programmable interaction between operators and customers, which facilitates the continuous refinement of services through customer-driven user stories. Laboratory testing validated the practical utility of this framework, specifically for private telephony and merchandise tracking applications (Alhuseini & Olama, 2019).

REFERENCES

Alhuseini, M. U., & Olama, M. M. (2019). 5G service value chain and network slicing framework using ecosystem modeling, agile delivery, and user-story automation. *IEEE Access : Practical Innovations, Open Solutions*, *7*, 110856–110873. DOI: 10.1109/access.2019.2933193

Andrews, J. G., Buzzi, S., Choi, W., Hanly, S. V., Lozano, A., Soong, A. C., & Zhang, J. C. (2014). What will 5G be? *IEEE Journal on Selected Areas in Communications*, *32*(6), 1065–1082. DOI: 10.1109/jsac.2014.2328098

Apruzzese, M., Bruni, M. E., Musso, S., & Perboli, G. (2023). 5G and companion technologies boost new logistics and supply chain business models. *Sustainability*, *15*(15), 11846. DOI: 10.3390/su151511846

Bandyopadhyay, S., Jain, P., Leing, J., & Prisacaru, S. (2023) *Navigating the three horizons of 5G business building*. McKinsey & Company. https://www.mckinsey.com/industries/technology-media-and-telecommunications/our-insights/navigating-the-three-horizons-of-5g-business-building

Boccardi, F., Heath, R. W., Lozano, A., Marzetta, T. L., & Popovski, P. (2014). Five disruptive technology directions for 5G. *IEEE Communications Magazine*, *52*(2), 74–80. DOI: 10.1109/mcom.2014.6736746

EHTAdmin. (2020). *Statement on Health Effects of 5G by Anthony B. Miller MD*. Environmental Health Trust. https://ehtrust.org/statement-on-health-effects-of-5g-by-anthony-b-miller/

Ellwood, D. W. (2020). The unreliable wisdom of hindsight. *The Political Quarterly*, *91*(1), 243–245. DOI: 10.1111/1467-923x.12801

Gröne, F., & Ayyagari, D. (2023). *Capturing more value from telecoms' connectivity services The state of 5G*. PwC. https://www.pwc.com/gx/en/industries/tmt/telecommunications/the-state-of-5G-capturing-more-value.html

GSMA. (2025). *The Mobile Economy 2025*. GSMA Intelligence. https://www.gsma.com/solutions-and-impact/connectivity-for-good/mobile-economy/wp-content/uploads/2025/02/030325-The-Mobile-Economy-2025.pdf

Haider, F., Wang, C.-X., Haas, H., Yuan, N. D., Wang, N. H., Gao, N. X., You, N. X.-H., & Erol Hepsaydir. (2011). Spectral efficiency analysis of mobile femtocell-based cellular systems. In 2011 *IEEE 13th International Conference on Communication Technology* (pp. 347–351). https://doi.org/DOI: 10.1109/ICCT.2011.6157895

Huawei (2023). *Huawei Launches Full-Series 5.5G Solutions to Bring 5.5G into Reality*. Huawei Technologies Co. https://www.huawei.com/en/news/2023/10/huawei5gsolution-mbbf2023

Kadena, E., & Rajnai, Z. (2024). 5G supply chain: An overview of applications and challenges. In *2024 IEEE 22nd World Symposium on Applied Machine Intelligence and Informatics (SAMI)* (pp. 000157–000162). IEEE. https://doi.org/DOI: 10.1109/SAMI60510.2024.10432858

Mangra, N., Behmann, F., Thakur, A., Popescu, A., Suciu, G., Jr., Giannattasio, G., Uppal, R., & Montlouis, W. (2023). 5 G-enabled agriculture ecosystem: Food supply chain, rural development, and climate resiliency. *IEEE*. https://ieeexplore.ieee.org/document/10028721

Matinmikko, M., Latva-aho, M., Ahokangas, P., & Seppänen, V. (2018). On regulations for 5G: Micro licensing for locally operated networks. *Telecommunications Policy*, *42*(8), 622–635. DOI: 10.1016/j.telpol.2017.09.004

Matinmikko, M., Latva-aho, M., Ahokangas, P., & Seppänen, V. (2018). On regulations for 5G: Micro licensing for locally operated networks. *Telecommunications Policy*, *42*(8), 622–635. DOI: 10.1016/j.telpol.2017.09.004

McDonagh, M., & Petrie, R. (2017). New Research from Accenture Strategy highlights Economic and Societal Impact of Investing in 5G Infrastructure. Accenture.

Mukasa, R. (2025). *5G deployment considerations*. Scribd. https://www.scribd.com/document/385172613/5g-Deployment-Considerations

Qose, S., & Zoltán, R. (2024). Supply chain in the context of 5G technology, security, and legal aspects. In *2024 IEEE 22nd World Symposium on Applied Machine Intelligence and Informatics (SAMI)* (pp. 000143–000148). https://doi.org/DOI: 10.1109/SAMI60510.2024.10432844

Rapaport, T. S., Sun, S., Mayzus, R., Zhao, H., Azar, Y., Wang, K., & Samimi, M. (2013). Millimeter wave wireless communications for 5G cellular: It will work! *IEEE Access: Practical Innovations, Open Solutions*, *1*, 335–349. DOI: 10.1109/access.2013.2260813

SWAN. (2021). R*adio Frequency Vulnerabilities White Paper*. Swan. https://cpb-eu-w2.wpmucdn.com/blogs.bristol.ac.uk/dist/6/635/files/2021/05/SWANWH2.pdf

Zhao, X., Askari, H., & Chen, J. (2021). Nanogenerators for smart cities in the era of 5G and Internet of Things. *Joule*, *5*(6), 1391–1431. DOI: 10.1016/j.joule.2021.03.013

ADDITIONAL READING

BEREC. (2025). BEREC Report on the 5G Ecosystem. BEREC. https://www.berec.europa.eu/en/document-categories/berec/reports/berec-report-on-the-5g-ecosystem

GSA. (2023). *Acquisition Guidance for Procuring 5G Technology*. www.gsa.gov. https://buy.gsa.gov/api/system/files/documents/GSA%20Acquisition%20Guidance%20for%20Procuring%205G%20Technology%20508_0.pdf

GSA. (2025). *5 G-Ecosystem February 2025*. GSA https://gsacom.com/paper/5g-ecosystem-february-2025/

Haneda, K., Tian, L., Asplund, H., Li, J., Wang, Y., Steer, D., Li, C., Balercia, T., Lee, S., Kim, Y., Ghosh, A., Thomas, T., Nakamura, T., Kakishima, Y., Imai, T., Papadopoulas, H., Rappaport, T., Maccartney, G., Samimi, M., & Sun, S. (2016). *Indoor 5G 3GPP-like Channel Models for Office and Shopping Mall Environments*. https://wides.usc.edu/Updated_pdf/haneda2016indoor.pdf

Heath, R. W., Gonzalez-Prelcic, N., Rangan, S., Roh, W., & Sayeed, A. M. (2016). An Overview of Signal Processing Techniques for Millimeter Wave MIMO Systems. *IEEE Journal of Selected Topics in Signal Processing*, *10*(3), 436–453. DOI: 10.1109/jstsp.2016.2523924

Iversen, M., Goldberg, T., & Pelson, J. Thiessen-MTRE, Andrew, Haan-Rivada, Brett (2025). *5G Supply Chain Assurance Panel*. NIST. https://csrc.nist.gov/presentations/2025/5g-supply-chain-assurance

Li, G., Zhang, X., & Zhang, G. (2022). How the 5G Enabled the COVID-19 Pandemic Prevention and Control: Materiality, Affordance, and (De-)Spatialization. *International Journal of Environmental Research and Public Health*, *19*(15), 8965. DOI: 10.3390/ijerph19158965

Luong, N. (2024). Forging the 5G future: Strategic imperatives for the US and its allies. Atlantic Council. https://www.atlanticcouncil.org/in-depth-research-reports/report/forging-the-5g-future-strategic-imperatives-for-the-us-and-its-allies/

Market Research Reports. (2015). *5G Wireless Ecosystem Forecast Report: 2015–2025*. Prweb.com. https://www.prweb.com/releases/5g_wireless_ecosystem_forecast_report_2015_2025/prweb12497540.htm

Medin, M., & Louie, G. (2019). *The 5G Ecosystem: Risks & Opportunities for DoD*. Defense Innovation Board. https://innovation.defense.gov/Portals/63/Templates/Updated%20Meeting%20Documents/5G%20UNCLASS%20PAPER_190404_FINAL.pdf

National Telecommunications and Information Administration (NTIA). (2025). *Public Wireless Supply Chain Innovation Fund*. NTIA. https://www.ntia.gov/funding-programs/public-wireless-supply-chain-innovation-fund

OECD. (n.d.). *Digital*. OECD. https://www.oecd.org/digital/5g-smart-infrastructure-policy.pdf

Rapaport, T. S., Sun, S., Mayzus, R., Zhao, H., Azar, Y., Wang, K., & Samimi, M. (2013). Millimeter wave wireless communications for 5G cellular: It will work! *IEEE Access : Practical Innovations, Open Solutions*, *1*, 335–349. DOI: 10.1109/access.2013.2260813

Rappaport, T. S., Xing, Y., MacCartney, G. R., Molisch, A. F., Mellios, E., & Zhang, J. (2017). Overview of millimeter wave communications for fifth-generation (5G) wireless networks—With a focus on propagation models. *IEEE Transactions on Antennas and Propagation*, *65*(12), 6213–6230. DOI: 10.1109/TAP.2017.2734243

Reportlinker. (2015, January 28). *The 5G Wireless Ecosystem: 2015 - 2025 - Technologies, Applications, Verticals, Strategies & Forecasts*. Prnewswire.com; Cision PR Newswire. https://www.prnewswire.com/news-releases/the-5g-wireless-ecosystem-2015---2025---technologies-applications-verticals-strategies--forecasts-300027177.html

Sutter, K. M. (2024). *Made in China 2025: Industrial Policies Issues for Congress*. Congress.gov. https://crsreports.congress.gov/product/pdf/IF/IF10964/6

U.S. Embassy San Jose. (2024). *Joint Statement on Enhancing Digital Economy Cooperation Between the United States of America and the Republic of Costa Rica*. U.S. Embassy in Costa Rica. https://cr.usembassy.gov/joint-statement-enhancing-digital-economy-cooperation-united-states-america-costa-rica/

Zhang, H., Liu, N., Chu, X., Long, K., Aghvami, H., & Leung, V. C. M. (2017). Network slicing based 5G and future mobile networks: Mobility, resource management, and challenges. *IEEE Communications Magazine*, *58*(11), 138–144. DOI: 10.1109/mcom.2017.1600940

Chapter 14
Strategies of Leading Companies in 5G

ABSTRACT

The strategies of leading companies in the 5G ecosystem reflect diverse and innovative approaches to capturing opportunities in this transformative technology. Spanning telecommunications giants, hardware manufacturers, and technology providers, these organizations invest in infrastructure, research, and partnerships to maintain a competitive edge. U.S.-based companies such as AT&T and Verizon focus on robust network deployment, while Chinese leaders like Huawei and ZTE spearhead advancements in 5G equipment. European players, including Nokia, Ericsson, and Orange, emphasize network solutions and service innovation. Meanwhile, Asian powerhouses like Samsung Electronics and NEC Corporation integrate 5G with cutting-edge technologies like IoT and AI, driving new capabilities and market applications.

INTRODUCTION

The strategies employed by leading companies in the 5G ecosystem showcase a diverse range of approaches to leveraging this transformative technology and driving innovation and industry leadership. These companies, spanning telecommunications giants, hardware manufacturers, and technology providers, are investing heavily in infrastructure, research, and partnerships to capture the vast opportunities presented by 5 G. The authors highlight the critical challenges developing countries face in shaping effective 5G strategies, stressing the importance of making balanced and

DOI: 10.4018/979-8-3373-0497-7.ch014

pragmatic policy decisions in light of technological, economic, and regulatory complexities (Forge & Vu, 2020).

The success of 5G depends on the collaboration of a diverse group of stakeholders, each playing a critical role in shaping the ecosystem. For example, Weisman-Pitts (2025) explores how 5G is reshaping key sectors such as finance, logistics, and industrial automation. From infrastructure deployment to end-user applications, these contributors ensure the smooth operation, innovation, and adoption of 5G technologies. The table below categorizes the key stakeholder groups, provides representative examples, and outlines their specific roles and contributions within the 5G value chain.

Table 1 provides an overview of the key stakeholders involved in the 5G ecosystem, highlighting their distinct roles and contributions in shaping the development, deployment, and regulation of next-generation wireless technologies.

Table 1. Overview of 5G ecosystem stakeholders and their contributions

Stakeholder Category	Examples	Roles & Contributions
Operators	Orange, AT&T, Verizon, MTN	Deploy and manage 5G infrastructure; offer core services
Service Providers	Ericsson, Huawei, Microsoft Azure	Deliver platforms, network services, and orchestration
Equipment Suppliers	Qualcomm, Intel, Cisco, Samsung	Manufacture hardware: chipsets, antennas, base stations
End Users	Tesla, Siemens, Philips, Consumers	Utilize 5G for automation, IoT, mobility, and real-time applications

From AT&T and Verizon in the United States, focusing on robust network deployment, to Huawei and ZTE in China spearheading advancements in 5G equipment, each organization brings unique strengths to the global market. European leaders, such as Nokia, Ericsson, and Orange, emphasize network solutions and service innovation. Asian powerhouses like Samsung Electronics and NEC Corporation are integrating 5G with cutting-edge technologies, including IoT and AI. Collectively, these strategies underscore a global race to define the 5G era, with companies striving to shape the future of connectivity and capture market dominance in this rapidly evolving landscape.

The authors highlight the critical challenges developing countries face in shaping effective 5G strategies, stressing the importance of making balanced and pragmatic policy decisions in light of technological, economic, and regulatory complexities (Forge & Vu, 2020). The authors highlight the critical challenges faced by developing countries in shaping effective 5G strategies, stressing the importance of making balanced and pragmatic policy decisions in light of technological, economic, and regulatory complexities (Forge & Vu, 2020), while broader economic analyses, such as that by IHS Markit for Qualcomm, project that 5G could enable up to $12.3 trillion in economic output by 2035 worldwide, underscoring the transformative potential of this technology, (Campbell, Morelli, & O'neil, 2017).

Douglas (2025) delivers a mid-cycle evaluation of 5G rollout performance, highlighting key technological benchmarks.

The 5G industry analysis is a comprehensive review of the twelve leading businesses, which are reputable providers of 5G solutions, including Samsung, AT&T, Verizon, Qualcomm, Orange, Ericsson, ZTE, Nokia, and Huawei. This section discusses the research tasks of these main businesses, as well as the services and products they have released and intend to release within the 5G URL.

Teece (2021) examines the strategic role of 5G patent portfolios in shaping technological leadership, emphasizing that the influence of a firm's intellectual property is determined more by the strategic relevance and technological significance of its patents than by sheer quantity. The study highlights how firms possessing critical 5G patents can shape industry standards, licensing dynamics, and policy decisions, securing competitive advantages in the global technology market. Furthermore, Teece discusses how national innovation policies should recognize the qualitative value of patent holdings, as they are central to promoting innovation and maintaining leadership in the evolving 5G landscape.

Lagorio et al. (2023) examine the role of 5G technology in revolutionizing logistics operations, focusing on its ability to improve real-time data flow, automation, and system connectivity. The study identifies specific logistics activities where 5G can offer significant enhancements, highlighting the critical barriers to its adoption, such as robust infrastructure, data protection concerns, and high implementation costs. The authors argue that while 5G has the potential to transform Logistics 4.0, overcoming these technological and financial challenges is essential for its successful deployment.

HUAWEI TECHNOLOGIES: LEADERSHIP AND INFLUENCE IN THE 5G ERA

Huawei Technologies Co., Ltd., headquartered in Shenzhen, China, stands as a dominant force in the global 5G infrastructure landscape. As a comprehensive technology and information solutions provider, the company operates in over 170 countries, with 36 joint innovation centers and 14 dedicated research and development (R&D) facilities. Huawei's infrastructure has played a central role in numerous 5G deployments across Asia-Pacific (APAC), Europe, and emerging markets, positioning the firm as a top supplier of network systems worldwide (Huawei Technologies, 2022).

Huawei's Core Strengths

One of Huawei's core strengths lies in its ability to deliver end-to-end 5G solutions, encompassing radio access networks, core networks, consumer devices, and intelligent software. This integrated model enhances operational efficiency for telecom operators and accelerates the global adoption of 5G technologies. The company has also actively contributed to international 5G standardization efforts, solidifying its presence in key industry consortia and shaping next-generation protocols, (ITU, 2022).

Investment in Innovation

A significant component of Huawei's strategy is its investment in innovation. The company holds one of the largest 5 G-related patent portfolios globally, including essential technologies such as network slicing, massive MIMO, and low-latency communication, (WIPO, 2023). These patents highlight Huawei's technical leadership and secure its role in setting the foundation for future wireless systems.

Significant Geopolitical Hurdles

Despite these advancements, Huawei has faced significant geopolitical hurdles. Concerns regarding cybersecurity and national sovereignty have led several Western nations, including the United States, the United Kingdom, and Australia, to restrict or ban Huawei equipment from their 5G rollouts. Medin and Louie (2019) analyzes the risks and opportunities of 5G for national security and critical infrastructure. The U.S. White House initiated a national strategy to coordinate federal efforts in advancing fifth-generation wireless technology, (McCabe, 2018). For its part, the UK Government prioritized investment in fiber and 5G as part of its Spring Budget, emphasizing the importance of digital infrastructure for economic growth, (Temperton, 2017).

These actions have presented challenges for Huawei's international partnerships, but have not deterred its commitment to technological progress (GSMA, 2025).

To maintain its competitive edge, Huawei continues to collaborate with global telecom providers, academic institutions, and industrial partners. Its focus includes enhancing community capacity by a factor of ten, improving coverage by ten decibels, and integrating artificial intelligence (AI) into network operations to streamline performance. This forward-looking strategy supports a broader transformation in enterprise communication and industrial digitalization.

Huawei's Comprehensive Approach

Overall, Huawei's comprehensive approach—combining R&D investment, technological innovation, global reach, and vertical integration—reinforces its central role in the 5G ecosystem. As the company navigates political uncertainties, its influence on the evolution of wireless infrastructure remains both substantial and enduring.

SAMSUNG ELECTRONICS (SOUTH KOREA): PIONEERING 5G DEVELOPMENT AND DEPLOYMENT

Samsung Electronics, headquartered in South Korea, has emerged as a global leader in the development and deployment of 5G technologies. The company initiated its 5G research projects as early as 2011, well before the worldwide commercialization of the technology. Over the past decade, Samsung has focused on key technical innovations, such as ultra-reliable low-latency communication (URLLC), which is vital for enabling mission-critical applications in sectors like autonomous vehicles and industrial automation.

Samsung's Strengths

Samsung's strengths lie in its ability to develop comprehensive 5G solutions that span the entire network architecture, from base stations to end-user devices. The company has played a pivotal role in advancing technologies such as network slicing, dynamic spectrum sharing, and millimeter-wave (mmWave) transmission. These innovations support high-throughput, low-latency, and scalable network environments, essential for modern digital ecosystems.

Contribution to Global 5G Standardization

Notably, Samsung has made significant contributions to global 5G standardization efforts through its active participation in organizations such as 3GPP. This involvement has ensured that its technologies are interoperable and aligned with international regulatory frameworks, promoting a cohesive and unified 5G ecosystem.

Samsung's Portfolio

Samsung's portfolio extends beyond infrastructure to include a wide range of 5 G-enabled consumer products, including smartphones, tablets, and fixed wireless access devices. The Galaxy series, for instance, integrates high-performance 5G capabilities, showcasing the company's ability to connect cutting-edge research and development with market-ready solutions. These devices not only meet consumer demand but also serve as testing grounds for new features and applications.

Several Emerging Use Cases

Furthermore, Samsung is driving adoption in emerging use cases such as smart cities, augmented reality (AR), and remote surgery by delivering the low latency and high reliability required for real-time operations. Its partnerships with telecom operators across the U.S., South Korea, India, and Europe have led to successful 5G trials and commercial rollouts, reinforcing its global impact. Chui, Collins, and Patel (2021) underscores the accelerating impact of IoT, enabled by 5G, on digital transformation across industries.

Sustained Investment in R&D

Samsung's sustained investment in R&D—amounting to billions annually—underscores its long-term vision for 5 G. The company is also exploring 6G technologies, reflecting a forward-looking approach to connectivity. By shaping industry standards, enabling advanced applications, and pushing the technological frontier, Samsung remains a key driver of the 5G revolution and a catalyst for global digital transformation.

ERICSSON (SWEDEN)

Ericsson has positioned itself as a global leader in 5G technology. It promises to become the sole vendor concentrating on every continent to establish 5G as a worldwide standard. By adopting this comprehensive approach, Ericsson aims to ensure that 5G technology is universally accessible and tailored to meet the diverse needs of various markets, paving the way for the next generation of wireless communication.

Development of Radio Prototypes

The development of radio prototypes marks a significant milestone in the company's 5G journey. Ericsson's 5G journey involves the development of its 5G radio Prototypes. These cutting-edge solutions allow operators to conduct real-world trials within their networks, providing valuable insights into the potential of 5G in various environments. These trials enable network operators to fine-tune their infrastructure and prepare for the seamless integration of 5G capabilities, ensuring optimal performance and user experience.

Ericsson's Extensive Expertise

Ericsson's extensive knowledge in the 5G domain has made it a strategic partner for companies worldwide. Its collaborations span multiple industries, fostering innovation and accelerating the deployment of 5G technology. These partnerships leverage Ericsson's advanced technologies, including dynamic spectrum sharing, ultra-reliable low-latency communication, and massive machine-type communication, to address the local market's requirements.

Focusing on the Global Standardization

By focusing on the worldwide standardization of 5G, Ericsson contributes significantly to harmonizing wireless technologies, ensuring interoperability across networks and devices. This approach drives efficiency and reduces costs for operators and end-users.

Furthermore, Ericsson's commitment to 5G extends beyond technology, with a focus on sustainability and scalability. Ericsson ensures that its 5G initiatives align with global innovation and environmental stewardship goals by developing energy-efficient solutions and enabling digital transformation across industries. Through its efforts, Ericsson continues to shape the future of wireless communication, delivering transformative technologies that empower businesses and individuals worldwide.

Speeds and a Superior User Experience

The Global mobile Suppliers Association (GSA, 2025) offers quantitative insights into global 5G deployment trends, spectrum usage, and vendor performance. 5G delivers unprecedented speeds and a superior user experience by utilizing existing mobile infrastructure, site deployments, and spectrum resources. In the meantime, mid-band and mmWave radios provide exceptional capacity, two-band utilization, and carrier aggregation. Effectively addressing their limited coverage, Ericsson

Spectrum Sharing offers a seamless solution for introducing 5G in low bands, eliminating the need for complex spectrum reconfiguration.

The efficient implementation of 5G does not require increased energy consumption. Ericsson achieves optimal performance through innovative deployment strategies, hardware modernization, AI-driven operations, and advanced software features. The dual-mode 5G Core enables unified management of 4G and 5G networks via cloud-native solutions, reducing complexity and minimizing operational costs. This architecture supports diverse revenue opportunities across industries and fosters innovation through an expansive partner ecosystem. Comprehensive coverage and the dual-mode 5G Cor rapidly transition to standalone 5 G. This approach ensures seamless connectivity between 4G and 5G networks, enhancing the user experience by providing six times faster access to higher data rates and enabling efficient network slicing.

At the Forefront of the Global 5G Market

Ericsson is at the forefront of the global 5G market, operating networks across four continents. Its 5G platform equips operators to accelerate the evolution from 4G to 5G, significantly enhancing network performance. By leveraging Ericsson's solutions, operators can streamline operations with AI-driven automation, optimize network complexity, and monetize innovative use cases in mobile broadband, fixed wireless access, and the Internet of Things (IoT). It drives growth and fosters innovation in the telecommunications industry (Ericsson, n.d). Sundarum (2025) provides a comprehensive overview of the global development of 5G, outlining the main growth trajectories and policy challenges.

AT&T (UNITED STATES): PIONEERING 5G DEPLOYMENT AND STRATEGY

AT&T Inc., one of the leading telecommunications operators in the United States, has positioned itself at the forefront of the national 5G rollout. The company launched its first gigabit-class 5G connections in late 2018 in select cities, including Waco (TX), Kalamazoo (MI), and South Bend (IN), marking a significant milestone in the development of next-generation mobile networks. Although the early service was primarily targeted at enterprise users, it laid the groundwork for broader consumer deployment across multiple bands and technologies.

AT&T 5G Speeds and Network Evolution

AT&T's 5G Plus service, based on millimeter-wave spectrum, has demonstrated early data speeds ranging from 200 to 400 Mbps, with peak testing rates surpassing 2 Gbps in markets such as Atlanta. These figures represented a notable improvement over 4G LTE speeds and were sufficient for ultra-fast downloads and low-latency applications. However, coverage limitations and signal availability posed challenges, prompting the company to invest in its nationwide 5G network using sub-6 GHz spectrum for broader accessibility, albeit at lower speeds.

AT&T 5G Devices and Consumer Access

Initially, AT&T's 5G access was limited to enterprise-focused devices such as the Netgear Nighthawk 5G Mobile Hotspot. The Galaxy S10 5G was the first smartphone offered by the carrier, initially available only to business clients. Consumer access expanded significantly in 2020 with the launch of 5G-compatible handsets, such as the Galaxy S20 Plus and S20 Ultra, which support both low-band 5G and mmWave-based 5 G. This diversification of device compatibility increased public adoption and improved service reach.

Strategic Outlook and Competitive Position

Despite its early entry into the 5G space, AT&T has encountered intense competition from rivals like T-Mobile, whose expansive low-band 5G network covers a larger population base. As of mid-2020, AT&T continued to scale its deployment, targeting national coverage while incorporating Dynamic Spectrum Sharing (DSS). DSS allows simultaneous support of LTE and 5G traffic within the same frequency bands, providing operational flexibility during the transition phase. While DSS does not increase overall network capacity, it enables a smoother migration toward a 5 G-dominant infrastructure.

Market Challenges and Future Directions

AT&T's challenge lies in balancing high-speed 5G Plus offerings in dense urban centers with reliable low-band 5G coverage across rural and suburban areas. The carrier's ongoing investments in spectrum acquisition, infrastructure upgrades, and software-defined network management are central to its strategy of delivering consistent and scalable 5G services. The Next Generation Mobile Networks Alliance (2020) outlines technical guidelines for building efficient and scalable 5G networks. Allied Market Research (2025) projects exponential growth in the wireless infrastructure market, largely driven by increased demand for 5G deployment. As the market matures, AT&T is expected to leverage its network assets, device ecosystem, and enterprise partnerships to solidify its position in the competitive 5G landscape.

BROADCOM CORPORATION: DRIVING 5G INNOVATION THROUGH SEMICONDUCTOR LEADERSHIP

Broadcom Inc. (NASDAQ: AVGO) is a global engineering company that designs, develops, and supplies semiconductor and infrastructure software solutions across key technology sectors. These include data centers, broadband, wireless, enterprise software, networking, and industrial markets. Broadcom's integrated ecosystem strategy supports the evolution of 5G with scalable, programmable, and high-performance chipsets. The company's flagship platforms, such as Monterey, Jericho 2, and Qumran 2a, are central to next-generation connectivity (Broadcom, 2023).

Advanced Switching Capabilities for 5G Networks

To meet 5G's stringent demands for ultra-low latency and high bandwidth, Broadcom has developed a comprehensive portfolio of switching solutions. These solutions include nanosecond-level synchronization, IEEE Time-Sensitive Networking (TSN) for traffic prioritization, and hierarchical network slicing. Their hardware enables end-to-end quality of service (QoS), as well as encapsulation for CPRI/LTE radios and IP/MPLS-based transport. These features support a scalable, jitter-controlled infrastructure with switching capabilities of up to 50 terabits per second (Tbps). Broadcom's merchant silicon model also enhances supply chain efficiency and economic scalability (Broadcom Inc., 2023).

Broadcom's 5 G-HD Wi-Fi Solutions for High-Density Environments

In response to growing demand in high-density areas, such as stadiums, campuses, and transportation hubs, Broadcom has launched its 5 G-HD Wi-Fi solution. The platform is built on the BCM49408 quad-core processor and BCM43465 Wave 2 11ac radio architecture. It supports dual-band (2.4GHz and 5GHz) and tri-band (2.4GHz, 5GHz, and 5GHz) configurations to enable maximum throughput and user capacity. This flexibility allows enterprises to deliver reliable, immersive connectivity in crowded venues.

Dual-Channel Architecture: Optimizing User Experience

The dual-channel architecture within Broadcom's 5 G-HD Wi-Fi ensures seamless service for both legacy and advanced devices. This setup mimics a two-lane highway—one for older, slower devices and one for newer, faster ones—thus preserving bandwidth and quality of service across user segments. Devices with 5 G-capable chipsets can fully utilize high-speed pathways, while older devices maintain stable access without compromising network performance for others.

Scaling With the Internet of Things (IoT)

With projections estimating over 30 billion connected devices by 2025 (Saadi & Mavrakis, 2020), Broadcom's 5 G-HD Wi-Fi and switching solutions are positioned to meet the exponential growth of the IoT. By enabling reliable and scalable connectivity in enterprise and consumer environments, Broadcom empowers sectors ranging from smart retail to automated logistics and industrial automation.

Strategic Outlook and Global Impact

Broadcom's emphasis on R&D, product integration, and high-performance networking positions the company as a crucial enabler in the 5G ecosystem. With a strong presence in global markets and ongoing collaborations with telecom providers and hardware manufacturers, Broadcom continues to drive innovation that supports the digital transformation of industries worldwide.

QUALCOMM: PIONEERING THE FUTURE OF 5G CHIPSETS

Qualcomm has positioned itself at the forefront of 5G innovation, focusing not on network deployment but on the core semiconductor technologies that enable mobile and connected devices to harness the capabilities of 5 G. While many competitors discuss potential applications, Qualcomm has already introduced practical solutions, including the X55 multi-mode 5G modem and an integrated 5G System-on-Chip (SoC), which are expected to accelerate the adoption of 5G in mid-range devices (Newsprints Team, 2024). PR Newsire (2019) forecasts that 5G baseband modem adoption will outpace previous generations due to faster integration in mobile devices.

Integrated 5G SoC: Expanding Accessibility

The X55 modem and Qualcomm's integrated 5G SoC address two critical barriers to widespread 5G adoption: cost and circuit board complexity. Most current 5G smartphones require a separate baseband chip and application processor, inflating production costs. Qualcomm's SoC combines both into one component, reducing space and power consumption, enabling cost-effective 5G integration into mid-tier phones. This shift is especially notable because, unlike the 4G transition, integrated 5G versions have emerged at a much faster pace.

R&D Investment and Leadership in Baseband Modems

Qualcomm continues to dominate the baseband modem market, holding nearly 49% revenue share. Its leadership stems from consistent, heavy investment in research and development—over 25% of its R&D budget in 2018 was allocated to pre-commercialization 5G projects. Qualcomm has historically capitalized on each generational transition in wireless networks, increasing its market share from 40% in 2010 to 66% in 2014 during the transition from 3G to 4 G. The company's licensing revenue also experienced double-digit growth over that period, further solidifying its role as a foundational player in mobile communications.

Strategic Diversification Beyond Smartphones

Despite a slowdown in smartphone growth and intensified competition, Qualcomm has expanded beyond mobile phones into adjacent sectors, including RF front-end components, the Internet of Things (IoT), networking, computing, and the automotive industry. These areas generated over $5 billion in revenue in fiscal year 2018 alone. This diversification is essential to stabilizing revenue streams and reducing dependency on mobile chip sales, which have faced pressure due to longer replacement cycles and saturated markets.

5G as a Growth Catalyst

Qualcomm's strategic bet on 5G could reverse its four-year revenue decline. According to Strategy Analytics, the adoption of 5G smartphone basebands is expected to outpace the 4G transition. By concentrating early efforts on smartphone integration before 5G Release 16 unlocks broader industrial applications, Qualcomm aims to regain dominance and revitalize its mobile chip division. The success of this approach may determine Qualcomm's leadership role not only in mobile but across the broader spectrum of connected technologies.

CISCO SYSTEMS

As one of the key players in global networking, Cisco Systems has strategically positioned itself to play a central role in the deployment and evolution of 5G infrastructure and services.

Company Background

Cisco Systems is a U.S.-based global technology leader renowned for its expertise in networking, cybersecurity, and cloud computing. With a long-standing reputation for innovation, Cisco has expanded its strategic vision to include the development and implementation of 5G infrastructure. Its "5G Right Now" initiative, launched at the Mobile World Congress in 2018, demonstrates Cisco's commitment to accelerating 5G adoption across service providers, enterprises, and the public sector.

Service Provider Enablement

Cisco supports service providers by offering solutions that enhance their ability to deliver 5G services efficiently and effectively. By enabling new revenue streams through advanced connectivity services, such as enhanced mobile broadband (eMBB) and ultra-reliable low-latency communication (URLLC), Cisco helps providers meet evolving customer demands while remaining competitive in the global 5G landscape.

Infrastructure Solutions

A core component of Cisco's 5G strategy is the development of scalable and secure network infrastructure. The company offers a comprehensive suite of high-performance networking equipment and cloud-native software, specifically designed for 5G environments. These tools ensure low latency, high bandwidth, and the flexibility required to support IoT, smart cities, and industrial applications.

Network Automation and Orchestration

Cisco places strong emphasis on automation to manage increasingly complex 5G networks. Through intent-based networking, software-defined networking (SDN), and network function virtualization (NFV), Cisco enables intelligent and adaptive networks. This automation is essential for reducing operational costs, optimizing traffic, and enabling real-time adjustments across distributed environments.

Outlook and Future Role in 5G

Cisco anticipates a significant expansion in connected devices, with forecasts projecting more than 30 billion devices by the end of the decade. To meet this demand, Cisco continues to invest in AI-driven analytics, edge computing, and orchestration platforms. These innovations position Cisco as a central player in the global 5G ecosystem, bridging the gap between technological capability and real-world implementation.

NEC CORPORATION (JAPAN)

Japan's NEC Corporation has emerged as a key player in the global 5G revolution, introducing innovative business concepts under the vision "5 G: A Future Beyond Imagination." NEC aims to create transformative market changes by integrating advanced communication technologies with cutting-edge digital solutions. According to SVP Toshimitsu Shimuzu, NEC's strategy focuses on developing collaborative business platforms and solutions that connect data from various industries and companies. This integration is powered by NEC's proprietary technologies, including artificial intelligence (AI), the Internet of Things (IoT), and other digital advancements, to deliver a seamless 5 G-enabled ecosystem.

In 2017, NEC Corporation, alongside other telecom manufacturers, played a pivotal role in advancing 5G by endorsing the initial 5 G New Radio (NR) Standard Format. This milestone marked the commencement of full-scale 5G NR development, positioning NEC as a leader in shaping the foundational framework for next-generation wireless communication.

NEC's commitment to 5G extends beyond standardization; the company actively works to produce intelligent, secure, and globally scalable solutions. Its focus includes improving interoperability between global mobile platforms and enhancing the security of 5G networks. This commitment underscores NEC's dedication to ensuring the success and reliability of 5G services across industries.

By integrating 5G with AI and IoT, NEC aims to drive innovation in smart cities, healthcare, and manufacturing. By combining its technological expertise with collaborative partnerships, NEC aims to enable seamless communication and connectivity, ultimately transforming the way industries operate and interact. NEC's vision for 5G pushes the boundaries of technological possibilities and sets a foundation for a brighter, more interconnected world.

NOKIA NETWORKS (FINLAND)

Nokia has also signed up for the rollout of 5G, as the business establishes and explores, along with partnering with various other entities, to make 5G reception as quickly as possible.

The business has an 8,000-hectare website to handle major 5G assessments. It collaborates with Deutsche Telekom and the Hamburg Port Authority on the 5G MoNArch project, whose primary goal is to gain real-world experience and knowledge from 5G networks. The industrial applications of traffic management could be enhanced using movable receptors and virtual reality (VR).

Nokia envisions 5G as a transformative enabler of business applications, digital value platforms, and advanced networking capabilities. 5G facilitates industrial automation and cognitive analytics for more intelligent operations by integrating predictive maintenance, workforce efficiency, and asset optimization. OC&C Strategy Consultants (2025) identifies critical strategic levers for capturing long-term value in the 5G ecosystem. Multi-cloud solutions, including local edge and distributed edge systems, complement high-performance networking to ensure seamless connectivity across the workforce, campuses, and automated systems. This ecosystem prioritizes dynamic security and robust connectivity, enabling various industries to achieve digital transformation. The holistic approach highlights the potential of 5G to drive innovation and operational efficiency.

The Nokia Bell Labs Future X industrial architecture outlines a comprehensive framework for leveraging 5G to enable advanced industrial capabilities. The architecture supports industrial automation, cognitive analytics, and digital operations by combining business applications such as predictive maintenance, workforce safety, and asset optimization with digital value platforms. Multi-cloud systems, including local and distributed edge solutions, ensure robust and scalable data processing and storage, while high-performance networking provides dedicated, seamless connectivity for diverse environments. This architecture emphasizes the integration of dynamic security measures to protect critical systems and data, offering a resilient foundation for digital transformation across industries. It highlights Nokia's vision for empowering industries to achieve effectiveness, innovation, and sustainability in the 5G era.

- A converged, service-aware, mission-critical, vast area network (WAN) comprises IP/MPLS, package optical, and package microwave platforms.
- An LTE wireless broadband internet system utilizes tiny cells and macrocells to extend capacity and coverage to offshore rigs, remote locations, processing facilities, and personnel in the area.

- Push-to-talk and push-to-video capabilities enable facilities to communicate with remote personnel in real-time.
- Compact, quickly deployable LTE strategies can provide connectivity for crisis and disaster recovery activities, establishing coverage in remote areas.
- Comprehensive protection management spans multi-technology and multi-vendor info and operational technology environment.
- Improve Efficiency By Having A Flexible, Service-Aware Network:

Move to digital oil and gasoline operations with Nokia's protected, dependable, high-bandwidth marketing and sales communications solutions.

Our converged networks cater to the distinct needs of petroleum and gasoline operators. They combine IP-based technologies and wireless broadband to supply the connectivity and bandwidth for substantial, detailed uses, including hands-free operation, remote monitoring, and real-time advantage control. These apps help efficiently and safely remove hydrocarbons from distant or remote areas. Reportlinker (2015) presents a forward-looking analysis of the 5G wireless ecosystem, including technologies, applications, and market forecasts.

TELECOM ITALIA

TIM confirms its leadership in the profound originality of 5 G. What about improving next-generation networks and providers by effectively finishing the very first link in Europe that is in a position to thoroughly beat the two Gbps velocity over a 5G living industrial community with twenty-six Gigahertz (GHz) millimeter wave (mmWave) wavelengths acquired through MiSE's 5G bid?

TIM achieved this essential milestone with Qualcomm and Ericsson Technologies, Inc., a subsidiary of Qualcomm Incorporated. This record also marks another turning point concerning millimeter-wave technology, which was previously demonstrated around Turin in 2017 through Italy's first 5G link. What about Rome, founded in 2018, with the first 5G video call in Europe? The pace turning point is achieved using an Askey CPE RTL0200 driven by a Qualcomm Snapdragon™ X50 5G Modem RF System.

This particular effect represents a tremendous turning point toward 5G evolution, kicking off the usage of 200 MegaHertz (MHz), acquired through MiSE's bid on the twenty-six GHz spectrum; aside from that additional 200 MHz, we have because of the bid's Club Use [1], for a complete quantity of 400 MHz. This technological advancement is expected to increase considerably by 2020. The 5G system can now utilize as much as 800 MHz and, later, as much as 1.000 MHz.

Due to their high-frequency bandwidth, 26 GHz mmWave frequencies can achieve Gbps speeds and provide a large amount of network bandwidth, which is of great value in advancing 5G Fixed Wireless Access (FWA) and 5G Indoor coverage.

The characteristics also enable new Industry 4.0 scenarios, according to providers, with outstanding high speed, minimal latency, and top reliability and security qualities.

TIM achieved a critical turning point with a regular 3GPP Non-Stand-Alone (NSA) solution, enabling equipment and community components from various suppliers to integrate and aggregate with TIM's 4G and 5G network, which was previously deployed in the 3.7 GHz spectrum.

Michele Gamberini, TIM's Chief Information and Technology Officer, commented: TIM continues to be an innovator within the millimeter-wave spectrum innovation along with this particular latest European 5G record, and the company confirms its technological leadership because of its effort toward development and groundbreaking community infrastructure, in a position to allow the whole country's digital transformation. This turning point paves the way for entirely new 5G remedies to grant fixed ultra-broadband to households, businesses, and public authorities that are not even close. This development focuses on coverage aimed at enhancing robotics and the hands-free operation of electronic devices within the intelligent production region. As a result, our customers are well-positioned to maximize the benefits of a broad range of integrated solutions, enabling them to participate in the digital society fully.

Emanuele Iannetti, Country Manager at Ericsson Italy, claims that 5G is the primary feature wedge of the next ten years. It will also enable Italy to become increasingly attractive and competitive on the global stage. We are delighted that TIM has selected Ericsson's 5G expertise to achieve this significant milestone, positioning our country at the forefront of the 5th-generation mobile network business landscape. Hence, Ericsson confirms its technological leadership and readiness to assume some current market demands.

ORANGE (FRANCE)

Orange, the leading French mobile network operator, is actively investing in the development of a connected world through three key areas of focus. First, it is enhancing mobile broadband to achieve speeds up to ten times faster than 4 G. Second, Orange is working on high-performance fixed internet access to complement fiber deployment in underserved regions. Third, it supports innovative applications that enable digital transformation across multiple industries.

Orange promotes its 5G network as a truly multi-service platform that can adapt to a wide range of devices. These include not only smartphones but also 360° immersive content, smart appliances, IoT devices, augmented reality tools, and autonomous vehicles. The company positions 5G as a transformative enabler for both consumers and enterprises, supporting real-time communication and industry innovation. Through its diverse 5G initiatives, Orange aims to reinforce its leadership in Europe's digital future.

VERIZON

Verizon, one of the largest telecommunications companies in the United States, has established itself as a pioneer in 5G deployment through a proactive and innovation-driven strategy. The company is heavily invested in building a robust and modern 5G infrastructure across major cities, aiming to deliver ultra-fast connectivity and low-latency services to its customers. Verizon was among the first to launch commercial 5G services in select markets, positioning itself as a leader in next-generation wireless technology.

Its strategic approach includes a combination of high-band millimeter wave and low-band spectrum to ensure both speed and broad coverage. Verizon also emphasizes the real-world applications of 5G, including smart cities, autonomous transportation, telemedicine, and augmented reality. The company believes that clear and forward-thinking policies are essential for 5G success, not just in the U.S., but globally. With this vision, Verizon continues to lead the way in 5G innovation and deployment within the telecommunications sector.

ZTE CORPORATION (CHINA)

The Chinese telecom ZTE's gear business also keeps pace with its 5G research and development (R&D) and testing efforts. Due to their capabilities and performance, various bottlenecks have been encountered by the 4G community within the Internet of Things. ZTE was the first provider to propose the Pre5G concept of its sequence remedy at that particular moment.

Nokia

The Future X structure leverages technologies such as the Industrial Internet of Things, edge computing, cloud computing, artificial intelligence, machine learning, virtual and augmented reality, and high-performance networks like 5G to drive significant output upgrades across various manufacturing sectors.

Our oil and gasoline structure enables you to produce a converged system that leverages the optimal blend of advanced solutions. It includes:

Enhance your operations by using broadband internet for marketing and communication networks. Legacy petroleum and gasoline networks are not designed to handle the highest volumes of real-time data produced by IoT and M2M devices with a high-quality program. Much apparatus inside the networks is approaching or even past end-of-life. Implementing electronic solutions in the gas and oil industries can be a time-consuming process.

Nokia helps you start your digital transformation with individual marketing communications infrastructures designed on LTE wireless broadband internet and IP/MPLS solutions. Collectively, the solutions offer the flexibility, scalability, resiliency, and predictability you need to enable secure marketing communications for new and existing mission-critical applications.

ORANGE S.A.

Orange introduced its first commercial 5G system in Romania in November 2019, initially in its three most significant urban areas. Orange estimates that most European countries will move away from a business community in 2020 as wavelengths are manufactured and offered by different regulatory authorities. They argue that this transformation is carried out alongside preexisting networks. Innovative, complex solutions created with 5G are expected to be released in 2023.

With a wiser, better-executing system and the fifth model of movable networks, our network is faster, more intelligent, and more effective. It achieves higher speed rates than 4G, with a homogeneous signal throughout the cell, and, starting from 2023, significantly smaller latency.

Because of these qualities, 5G enables seamless encounters and opens up new applications, such as leisure and education, with 360° immersive content that combines augmented and virtual reality.

As a substitute for repaired broadband internet, 5G's primary strength lies in the new frequency bands with optical fiber capability. Within just a few years, houses in certain areas will reap the benefits of high-speed broadband internet without the need for relocation, thanks to a fixed community. This also represents a genuine opportunity in areas where the infrastructure is underdeveloped. Wireless connectivity can benefit companies by connecting remote manufacturing sites, pop-up stores, mobile sites, smart community players, and urbanized infrastructure. The situation associated with a fixed community outage is also a viable alternative to fiber for critical applications, such as telemedicine.

The real-time industry is expected to utilize 5G, which is likely to make the most significant distinction and have a major impact on the B2B market. Science enables businesses to use the information when it is generated; therefore, they can continue to be much more agile, closer, and responsive to the consumer's must-haves.

As a multi-service system, 5G is expected to adapt to internet business procedures and devices by 2023, including augmented smartphones, industrial robots, refrigerators, sensors, cameras, and autonomous vehicles such as Google's cars.

THE DEVELOPMENT OF 5G WORLDWIDE

The rollout of 5G technology is progressing swiftly across the globe, becoming a foundational pillar of the digital economy. Its transformative potential spans multiple sectors, including industry, healthcare, mobility, and financial services. This evolution is not merely technical—it is socio-economic, marking a shift in how people, devices, and systems connect.

Economic Impact of 5G

Global investment in 5G infrastructure is projected to yield an estimated $13.2 trillion in economic output by 2035. Enhanced connectivity, smarter industry operations, and new digital business models will drive this vast contribution. In addition, 5G is expected to generate $2.1 trillion in GDP growth, particularly in emerging economies where the technology will bridge significant digital divides.

Employment and Industrial Transformation

The 5G revolution is expected to create 22.3 million new jobs globally, spanning high-tech, manufacturing, services, and creative industries. These jobs will emerge along the 5G value chain, including consumers, app developers, content creators, telecom operators, and original equipment manufacturers (OEMs). This evolution will empower a wide range of new services—from smart factories to autonomous transportation and telemedicine.

Strategic Role of Industry and Policy

The successful deployment of 5G requires strategic collaboration between governments, telecom operators, equipment manufacturers, and private enterprises. Leading the charge are organizations like the 3rd Generation Partnership Project (3GPP), which defines global specifications for mobile technologies, including 3G, 4G LTE, and now 5 G. Industry leaders such as Qualcomm are also pivotal, contributing foundational innovations that shape the 5G ecosystem.

A Catalyst for Global Connectivity

Beyond its technical promise, 5G represents a unifying platform that enhances global interconnectivity. With ultra-low latency, massive bandwidth, and real-time responsiveness, it will unlock the full potential of smart cities, connected healthcare, immersive content, and the Internet of Things. As 5G networks mature, they are expected to redefine how humans and machines interact, ushering in a new era of global digital transformation.

Summary

These players are not only advancing technological capabilities but also influencing policy, standardization, and market adoption on a global scale. Their investments in research and development, collaboration with governments, and alignment with emerging industry needs are crucial to the success of 5 G. As cases such as autonomous vehicles, smart factories, and telemedicine evolve, these companies will be central to enabling transformative digital experiences. This table helps contextualize their competitive positioning and long-term vision. Ultimately, their collective efforts are accelerating the transition to a more connected, efficient, and intelligent world.

Table 2. Overview of the business strategy of the 5 G industry players

Company	5G Highlights
Huawei (China)	Leader in 5G infrastructure, end-to-end solutions, vast patent portfolio, faces geopolitical challenges.
AT&T (USA)	Early 5G deployment, nationwide low-band and mmWave coverage, millimeter-wave speeds over 2 Gbps.
Broadcom (USA)	Provides a 5G switching portfolio, network slicing, high scalability, and enterprise Wi-Fi integration.

continued on following page

Table 2. Continued

Company	5G Highlights
Cisco Systems (USA)	Focus on 5G infrastructure, provider enablement, automation, Software-Defined Networking (SDN), and Network Function Virtualization (NFV) for orchestration and seamless integration.
Ericsson (Sweden)	Global 5G leader, radio prototypes, dynamic spectrum sharing, dual-mode 5G Core, energy-efficient.
Samsung (South Korea)	Early 5G R&D, millimeter-wave, network slicing, URLLC, devices, and infrastructure integration.
NEC Corporation (Japan)	AI, IoT integration, 5G NR standard contribution, secure, scalable solutions, and smart cities.
Nokia Networks (Finland)	Future X architecture, predictive maintenance, industrial automation, and cloud edge systems.
Orange S.A. (France)	European rollout, 5G as fiber substitute, 360° AR/VR, industrial applications.
Qualcomm Inc. (USA)	5G chipsets and System-on-Chip (SoC) solutions are R&D-intensive, with a focus on ecosystem development and key 5G patents, particularly for mid-range solutions.
Telecom Italia (Italy)	Europe's first 2 Gbps mmWave, Industry 4.0, collaboration with Ericsson and Qualcomm.
Verizon (USA)	National 5G deployment, mmWave leadership, substantial infrastructure investments.
ZTE Corporation (China)	Pre5G solutions, IoT bottlenecks addressed, R&D focused.

A REVIEW OF STRATEGIC INITIATIVES

The European Commission has taken strategic steps to structure the 5G market through its Horizon 2020 program. This initiative aimed to support research and innovation in 5G technologies across Europe. Key objectives included fostering scientific excellence, boosting industrial leadership, and addressing societal challenges. These goals guided significant collaborative research and development efforts. Horizon 2020 played a foundational role in shaping Europe's 5G ecosystem.

The European Commission has defined clear research priorities to guide new projects under its Horizon 2020 innovation framework, which ran from 2014 to 2020. These priorities aim to drive advancements in 5G technologies through targeted investments in collaborative research and development. Specific objectives include:

- Designing network architectures and protocols that achieve at least a tenfold increase in frequency reuse.

- Developing technologies to mitigate the "spectrum crunch" caused by exponential data traffic growth.
- Enabling efficient and flexible radio access infrastructure that supports seamless integration of fixed and mobile networks.
- Implementing real-time, adaptive radio resource allocation to ensure differentiated quality of service for diverse applications.
- Reducing overall system energy consumption while maintaining high performance.
- Achieving end-to-end latency as low as 1 millisecond to support ultra-responsive applications.

The most relevant strategic initiatives identified are the following.

METIS 2020

The METIS 2020 project (Mobile and Wireless Communications Enablers for the Twenty-Twenty Information Society) was a pivotal European Union initiative that laid the groundwork for the fifth-generation (5G) mobile and wireless communication systems. Launched in November 2012 and coordinated by Ericsson, METIS brought together a consortium of 29 partners, including major industry players, network operators, academic institutions, and research centers.

Key Objectives and Achievements:

- The METIS Foundation, aiming to develop a comprehensive 5G system concept, sought to meet the anticipated demands beyond 2020, addressing challenges such as increased data traffic, diverse service requirements, and the proliferation of connected devices, Huawei (2020).
    - Technical Targets: The project set ambitious goals, including achieving 1,000 times higher mobile data volume per area, 10 to 100 times higher typical user data rates, and 10 times longer battery life for low-power devices. Additionally, it targeted a fivefold reduction in end-to-end latency, aiming for 5 milliseconds for applications like road safety, (Monserrat et al., 2015).
    - Service Categories: METIS identified three primary 5G service categories: Extreme Mobile BroadBand (xMBB), Massive Machine-Type Communications (mMTC), and Ultra-reliable Machine-Type Communications (uMTC). These categories were designed to cater to a wide range of applications, from high-speed data services to reliable, low-latency communications for critical applications.

- Technological Innovations: The project introduced several key enablers, such as a lean system control plane for efficient signaling, dynamic radio access networks (RAN) for adaptable connectivity, localized content and traffic flow management, and a spectrum toolbox to enhance frequency agility and coexistence.
- Global Impact: METIS played a significant role in shaping the global 5G landscape by contributing to the development of channel models and system concepts that influenced standardization efforts. Its research outcomes have been instrumental in guiding subsequent projects and industry developments in 5G technology, (Parker, 2014).
- The Metis 5g Requirements:

The METIS 5G Vision and Its System-Level Contributions

The METIS project (Mobile and Wireless Communications Enablers for the Twenty-Twenty Information Society) is the largest FP7 initiative dedicated to laying the foundation for 5G systems and fostering consensus for standardization. METIS has proposed a comprehensive 5G concept built around three generic service types: extreme mobile broadband, massive machine-type communication (mMTC), and ultra-reliable machine-type communication (uMTC). These services are supported by four key technological enablers: a lean process management plane, an advanced radio access network, localized content and traffic flow optimization, and a dynamic spectrum toolbox.

Key Usage Scenarios and Challenges

METIS has defined five primary usage scenarios to highlight 5G challenges and capabilities:

1. High-Speed Mobility – Delivering ultra-fast data rates with low latency to ensure seamless connectivity for users in motion.
2. Dense Urban Environments – Supporting high user density in locations like stadiums, shopping malls, or concerts without degrading the quality of service.
3. High-Quality Experience on the Move – Ensuring reliable and responsive service for mobile users, particularly in vehicles.
4. Ultra-Reliable Low Latency Communication (URLLC) – Supporting mission-critical applications that require extremely low latency and high reliability.
5. Massive IoT Connectivity – Managing large volumes of connected devices, such as sensors and machines, in an efficient and scalable manner.

METIS Architectural Framework

METIS proposes a hybrid architectural framework that combines centralized and decentralized elements to support flexibility, scalability, and performance. This framework ensures efficient coordination between local and global resources, enhancing system reliability and adaptability.

Technology Components and Innovations

The METIS project has tested several key innovations in real-world and simulated environments. Notable technologies include:

- Interference Cancellation with FBMC/OQAM Waveforms – Improving signal clarity and bandwidth efficiency.
- SiGe BiCMOS Technology in RF Front Ends – Enhancing the performance of high-frequency analog components.
- DUPLO Project – Pioneering full-duplex transceivers operating on the same carrier frequency to double link capacity.
- PHYLAWS Project – Advancing physical-layer security for wireless communication through secrecy coding.
- CROWD Project – Developing energy-efficient, dense wireless networks with intelligent traffic and connectivity control.

These efforts collectively contribute to a robust and adaptable 5G ecosystem that can meet the demands of a future digital society.

PHY and MAC

Development of new PHY and MAC layer concepts better suited for heterogeneous transmissions in 5G: The 5GNOW (5th Generation Non-Orthogonal Waveforms for Asynchronous Signalling) is a project, is funded by the European Commission under FP7, aimed to develop new Physical (PHY) and Medium Access Control (MAC) layer concepts better suited for heterogeneous transmissions in 5G networks. This initiative challenged the traditional LTE and LTE-Advanced design principles of strict synchronism and orthogonality, recognizing that these constraints are not optimal for the diverse requirements of future wireless communications, including machine-type communications (MTC), fragmented spectrum, and heterogeneous network deployments.

Key Contributions:

- Non-Orthogonal Waveforms: 5GNOW explored alternative waveforms, including Generalized Frequency Division Multiplexing (GFDM), Filter Bank Multicarrier (FBMC), and Universal Filtered Multicarrier (UFMC), to support asynchronous and non-orthogonal transmissions, thereby enhancing flexibility and efficiency in diverse scenarios.
- Unified Frame Structure: The project proposed a unified frame structure capable of accommodating various traffic types, ranging from broadband human-centric services to sporadic MTC, within a single framework, facilitating efficient resource allocation and coexistence, (Wunder et al., n.d.).
- MAC Layer Innovations: 5GNOW developed a hybrid and hierarchical MAC layer design that supports synchronous and asynchronous traffic, enabling dynamic scheduling and improved support for heterogeneous services and devices, (Dryjanksi, 2019).
- The SOLDER project focuses on developing spectrum overlay techniques to efficiently aggregate heterogeneous frequency bands, enabling more dynamic and flexible spectrum usage. Meanwhile, the ICT I-JOINT initiative emphasizes the joint optimization of access and backhaul networks by integrating small cells, heterogeneous backhaul technologies, and centralized processing, improving overall network coordination and performance.
- The TROPIC project explores distributed computing and radio resource allocation within cooperative femtocell networks, enabling more efficient local processing and storage capabilities. Complementing this effort, MCN introduces cloud computing paradigms into mobile network deployment and operation, offering scalable infrastructure management through virtualized resources.
- COMBO addresses the convergence of fixed and mobile broadband infrastructures by proposing integrated access and aggregation models that ensure seamless user experience across heterogeneous networks. Additionally, MOTO explores direct terminal-to-terminal offloading techniques, which relieve core network congestion and support edge-based communication.
- Finally, PHYLAWS contributes to security by exploiting physical layer propagation characteristics to develop privacy-enhancing technologies. It offers a novel approach to securing wireless communication against eavesdropping and unauthorized access.

These projects reflect a holistic and multidisciplinary approach to tackling the complex demands of 5G and future mobile ecosystems.

The Dense Cooperative Wireless Cloud Network (DIVINE)

The Dense Cooperative Wireless Cloud Network (DIVINE) project employs a virtual relay-based self-contained wireless cloud with a straightforward and unambiguous interface to the terminal issue of wireless communications in heavily interfering ad hoc networks.

The Interworking and Joint Style of an Open Access and Backhaul Network Architecture for Small Cells based on Cloud Networks (iJOIN) initiative employs a centralized, open IT platform-based radio access network (RAN) concept with cloud infrastructure.

According to Pirinen (2014), early 5G research activities focused on converging various technologies, including spectrum expansion, network densification, and new air interface concepts, setting the foundation for enhanced mobile broadband and ultra-reliable low-latency communications.

DIVINE

Access and backhaul buildings are jointly created and optimized. The Distributed Resource, Storage, and Computing Allocation over Cooperative Femtocells (TROPIC) task seeks answers to the following questions: What communication and computation technologies are required for a converged, pervasive Femto network infrastructure and cloud computing? How much can the suggested Femto cloud methods effectively achieve in terms of spectral, energy, and service efficiency?

Heterogeneous Wireless Networks with Millimeter-Wave Small Cell Access and Backhauling (MiWaveS)

Beyond 2025, Heterogeneous Wireless Networks with Millimeter-Wave Small Cell Access and Backhauling (MiWaveS) will be an industry-driven, large-scale integration task. The project investigates and demonstrates crucial enabling technologies and functionalities that support the integration of millimeter-wave small cells in future heterogeneous networks. Its main interests are algorithms, networking functions, and incorporated antenna technologies and radio.

Innovative Dynamic spectrum 5G mobile networks Employing Licensed shared access(ADEL) task has focused on investigating the subsequent investigation issues inside the licensed shared access (LSA) paradigm: one) the robust and optimized allocation of spectral and energy materials at time scales from seconds to milliseconds, two) quality of service guarantees to the owners of all participating spectrum sharing networks, and three) Minimization of total energy expenditure in LSA networks.

Spectrum Overlay via aggregation of heterogeneous Dispersed bands (SOLDER) project's primary goal is to efficiently aggregate noncontinuous dispersed spectrum bands licensed to heterogeneous networks (Heterogeneous Radio) Access Technologies (h RATs) and thereby develop a brand-new spectrum overlay technology.

The TUCAN3G Project

The TUCAN3G project addresses the digital divide in rural and underserved regions of developing countries. It integrates 3G/4G femtocells with various low-cost backhaul options like WiLD, WiMAX, and VSAT. The solution is designed for outdoor deployment, making it robust for remote environments. Its focus is on affordability, scalability, and ease of implementation. The project demonstrates that reliable mobile connectivity can be both technically feasible and economically viable in challenging settings.

The TUCAN3G project proposes deploying 3 G and 4G femtocells alongside heterogeneous backhaul technologies, such as long-distance Wi-Fi (WiLD), Wi-MAX, and VSAT, in outdoor scenarios to provide economically sustainable and technologically feasible connectivity solutions for rural environments in developing countries (Martínez-Fernández et al., 2016).

The Cognitive Radio Standardization Initiative (CRS I)

The Cognitive Radio Standardization Initiative (CRSI) coordinates and supports both current and future projects. The cognitive radio and powerful spectrum access outcomes are exploited and prepared for standardization through a focused strategy.

Cognitive Radio for Satellite Communications (CoRaSat) researches powerful spectrum sharing by investigating, developing, and demonstrating relevant cognitive radio strategies for satellite networks. The intention is to demonstrate that adaptable spectrum use presents new business and benefits opportunities that outweigh its potential drawbacks.

Other Initiatives

The self-management goals of the Unified Heterogeneous Radio Access Networks (SEMAFOUR) project are to build multi-radio access technology (multi-RAT) and multilayer self-organizing networks (SON) and develop an integrated SON management system. The Massive MIMO for Efficient Transmission (MAMMOET) project aims to bring substantial MIMO from an extremely promising theoretical idea to a practically implementable level using revolutionary, inexpensive, and flexible hardware. High capability system Architecture with Remote radio heads

& Parasitic antenna arrays (HARP) task plans to recognize sent-out multi-antenna wireless access by combining stereo remote heads (RRHs) technology with electronically steerable passive array radiators (ESPARs) know-how.

The Massive MIMO for Efficient Transmission (MAMMOET) would enable radio-over-fiber connections to be propagated geographically over a vast region, provide dispersed access through radio-over-fiber, and offer a single, multi-antenna-like RF chain. Mobile Cloud Networking (MCN) enables the creation of cloud-based, flexible, and mobile networks on demand. Specific interest is in evolved packet methods as well as the lifecycle of their management. The Developing Mobile Online with Revolutionary Terminal-to-Terminal Offloading Technologies (MOTO) task aims to address 5G challenges by leveraging multiple offloading schemes, such as cellular-to-Wi-Fi, and opportunistically utilizing Device-to-Device (D2D) connections.

Development of New PHY and MAC Layer Concepts

The design of the new 5 G PHY (Physical) and MAC (Medium Access Control) layer concepts focuses on supporting heterogeneous transmissions. These concepts address the challenges of diverse data demands and variable transmission environments, ensuring efficiency and reliability in 5G networks, (5Gnow, 2022).

Solder

SOLDER aims to develop innovative spectrum overlay technology that enables the prominent aggregation of heterogeneous bands (HetBands). This approach enhances spectrum efficiency and supports the seamless operation of multiple frequency bands for next-generation networks.

I Join

The I JOIN project optimizes access and backhaul networks by jointly designing and integrating small cells, heterogeneous backhaul, and centralized processing. It ensures efficient resource utilization and robust connectivity in complex network architectures.

Tropic

TROPIC focuses on cloud computing and cooperative femtocell networks, addressing radio resource allocation and storage. By integrating these technologies, the project supports efficient network management and enhances overall system performance, (admin1, 2020). 17.14 Mcn

Mobile Cloud Networking (MCN) explores cloud computing applications for developing and operating future mobile networks. This project emphasizes scalable, flexible infrastructure for improved service delivery and network efficiency, (Mobile Could Networking, n.d.)

Combo

The COMBO project advances fixed and mobile converged broadband access and aggregation networks. This integrated approach optimizes resource usage and ensures seamless service delivery for fixed and mobile users.

Moto

MOTO focuses on mobile internet innovation through terminal-to-terminal offloading technologies. These technologies reduce network congestion and enhance the efficiency of data transmission across mobile devices, (FP7-Moto, 2025).

Phylaws

PHYLAWS explores physical concepts for wireless communications by leveraging propagation properties. The project aims to enhance security and reliability in wireless communication systems through advanced propagation modeling, (CoolUtils File Converters, 2025).

Summary

To support the evolution of 5G across Europe, the European Union has launched a series of strategic research and development initiatives. These projects aim to address the complex technological, infrastructural, and security challenges posed by next-generation wireless communication. Key areas of focus include spectrum efficiency, flexible network architecture, energy optimization, and secure data transmission. Each initiative contributes to building a coherent and interoperable 5G ecosystem aligned with EU digital transformation goals.

Programs like METIS 2020 lay the foundational framework for 5G scenarios, while others, such as 5GNOW and PHYLAWS, explore physical layer enhancements and wireless security. Projects like MCN and COMBO focus on integrating cloud technologies and converging fixed and mobile access systems. Collectively, these efforts enhance 5G readiness across both urban and rural environments, supporting a range of future applications, from autonomous systems to smart cities. The table below outlines the main initiatives and their specific contributions.

Table 3. A review of strategic 5G initiatives in Europe

Project Name	Purpose / Focus	Website
METIS 2020	Foundation for 5G communication systems, scenarios for various 5G applications.	www.metis2020.com
5GNOW	Development of PHY and MAC layers for heterogeneous transmissions.	www.5gnow.eu
SOLDER	Overlay technology for spectrum aggregation across HetBands.	www.ict-solder.eu
iJOIN	Joint access and backhaul optimization with centralized RAN.	www.ict-join.eu
TROPIC	Distributed computing and cooperative femtocells for RRA and storage.	www.ict-tropic.eu
MCN	Cloud Computing for Mobile Network Deployment and Operation.	www.mobile-cloud-networking.eu
COMBO	Integrated fixed and mobile broadband access/aggregation networks.	www.ict-combo.eu
MOTO	Terminal-to-terminal offloading to reduce congestion and improve data efficiency.	www.fp7-moto.eu
PHYLAWS	Enhancing Wireless Security Using Physical Layer Propagation Models.	www.phylaws-ict.org

CONCLUSION

Chapter 14 investigates how leading global companies are shaping the deployment and evolution of 5G technologies through strategic foresight and innovation. As 5G becomes the backbone of next-generation connectivity, industry leaders are developing customized strategies to harness its full potential. This chapter explores how telecom providers, equipment manufacturers, and semiconductor firms position themselves through research, global partnerships, and proprietary technologies. It also examines how companies respond to regulatory shifts, intellectual property challenges, and international market fragmentation while seeking to secure a competitive edge in the digital economy.

This chapter presents a detailed comparative analysis of the strategies employed by top-tier 5G players, including Huawei, Qualcomm, Samsung, Ericsson, and Nokia. These firms make unique contributions to the 5G value chain through innovations in chipsets, end-to-end infrastructure, standard-setting, and software-defined networking. These firms have reinforced their competitive advantage by investing in AI, mmWave technology, virtualization, and green energy solutions. Boston Consulting Group (2025) emphasizes the role of strategic management in leveraging 5G technologies for competitive advantage. Moreover, collaborations across the private and public sectors, as well as with global standardization bodies, enable them to accelerate 5G deployment and policy alignment worldwide.

However, challenges remain. Companies must navigate geopolitical tensions, address deployment costs, ensure interoperability, and integrate legacy systems. The following table summarizes these strategic initiatives:

In conclusion, the strategies adopted by leading companies in the 5G ecosystem demonstrate a global effort to enhance the transformative potential of this groundbreaking technology. Each organization contributes distinct strengths, from the innovative network solutions of European leaders like Nokia and Ericsson to the cutting-edge advancements in IoT and AI integration by Asian giants like Samsung and Huawei. U.S.-based firms, such as AT&T and Verizon, prioritize robust network deployment, while ZTE and Qualcomm focus on pushing the boundaries of 5 G-enabled technologies. These companies' investments in infrastructure, partnerships, and research and development highlight the collaborative and competitive nature of the 5G market.

As 5G becomes a critical driver of digital transformation, the efforts of these organizations underscore the importance of addressing challenges such as scalability, latency reduction, and security to unlock the full potential of this technology. Their innovations pave the way for enhanced connectivity and broader applications across industries, including smart cities, autonomous vehicles, and industrial automation.

The global race to define the 5G era will ultimately shape the future of connectivity, offering unprecedented opportunities for businesses and consumers.

REFERENCES

Admin1 (2020). *The best tips for choosing a broker.* Ict-Tropic.eu. https://www.ict-tropic.eu/

Allied Market Research. (2025). *Wireless Infrastructure Market Update 2025.* OpenPR. https://www.openpr.com/news/4020263/wireless-infrastructure-market-update-2025-projected

Boston Consulting Group. (2025). *Welcome to BCG.* BCG. https://www.bcg.com/

Broadcom Inc. (2023). *Enabling AI Infrastructure.* Broadcom. https://www.broadcom.com

Campbell, K., Morelli, B., & O'neil, B. (2017). *The 5G economy: How 5G technology will contribute to the global economy IHS Economics / IHS Technology.* https://cdn.ihs.com/www/pdf/IHS-Technology-5G-Economic-Impact-Study.pdf

Chui, M., Collins, M., & Patel, M. (2021). *The Internet of Things: Catching up to an accelerating opportunity.* McKinsey & Company. https://www.mckinsey.com/~/media/mckinsey/business%20functions/mckinsey%20digital/our%20insights/iot%20value%20set%20to%20accelerate%20through%202030%20where%20and%20how%20to%20capture%20it/the-internet-of-things-catching-up-to-an-accelerating-opportunity-final.pdf

CoolUtils File Converters. (2025). *Our File Converters.* Coolutils.com. https://www.coolutils.com/

Douglas, S. (2025). *5G Halftime Update: Spirent's 2025 5G Report.* Ospirent Communications. https://www.spirent.com/blogs/5g-halftime-update-spirents-2025-5g-report

Dryjanski, M. (2019). *Do we need next G's?* Grandmetric. https://www.grandmetric.com/wp-content/uploads/2019/10/UnifiedNetworks_World5GSummit.pdf?utm_source=chatgpt.com

Forge, S., & Vu, K. (2020). Forming a 5G strategy for developing countries: A note for policy makers. *Telecommunications Policy*, *44*(7), 101975. DOI: 10.1016/j.telpol.2020.101975

FP7-Moto. (2025). *FP7 Moto.* FP7-Moto.eu. https://fp7-moto.eu/

Gnow (2022*) 5gnow.eu – 5G mobile phones, tariffs and flat rates.* 5Gnow. https://www.5gnow.eu/

GSA. (2025). *5G Market Snapshot February 2025*. GSA. https://gsacom.com/paper/5g-market-snapshot-february-2025/

GSMA. (2025). *Spectrum Policy Trends 2025*. GSMA> https://www.gsma.com/connectivity-for-good/spectrum/wp-content/uploads/2025/02/GSMA-Spectrum-Policy-Trends-2025-1.pdf

Huawei (2020) *METIS: Striding towards 5G - Huawei Publications*. Huawei. https://www.huawei.com/en/huaweitech/publication/73/hw_403726?utm_source=chatgpt.com

ITU. (2022). *5G - Fifth generation of mobile technologies*. Www.itu.int. https://www.itu.int/en/mediacentre/backgrounders/Pages/5G-fifth-generation-of-mobile-technologies.aspx

Lagorio, A., Cimini, C., Pinto, R., & Cavalieri, S. (2023). 5G in logistics 4.0: Potential applications and challenges. *Procedia Computer Science*, *217*, 650–659. DOI: 10.1016/j.procs.2022.12.261

Martínez-Fernández, A., Vidal, J., Simó-Reigadas, J., Prieto-Egido, I., Agustín, A., Paco, J., & Rendón, Á. (2016). The TUCAN3G Project: Wireless technologies for isolated rural communities in developing countries based on 3G small cell deployments. *IEEE Communications Magazine*, *54*(7), 36–43. https://upcommons.upc.edu/server/api/core/bitstreams/95adf79b-a07d-4ab2-8aca-232231811e4a/content

McCabe, D. (2018, October 25). *White House directs task force to come up with 5G wireless strategy*. Axios. https://www.axios.com/2018/10/25/white-house-national-wireless-strategy-task-force-5g

Medin, M., & Louie, G. (2019). *The 5G Ecosystem: Risks & Opportunities for DoD. Defense Innovation Board.* https://innovation.defense.gov/Portals/63/Templates/Updated%20Meeting%20Documents/5G%20UNCLASS%20PAPER_190404_FINAL.pdf

Meesha Sundarum. (2025, April 16). *The State of 5G: Growth, Challenges, and Opportunities in 2025*. 5G Americas. https://www.5gamericas.org/the-state-of-5g-growth-challenges-and-opportunities-in-2025

Mobile Cloud Networking. (n.d.) *Mobile Cloud Networking in Online Casinos*. 777. https://mobile-cloud-networking.eu/

Monserrat, J. F., Mange, G., Braun, V., Tullberg, H., Zimmermann, G., & Bulakci, Ö. (2015). METIS research advances towards the 5G mobile and wireless system definition. *EURASIP Journal on Wireless Communications and Networking, 2015*(1). https://doi.org/DOI: 10.1186/s13638-015-0302-9

Newsprints Team. (2024). Qualcomm Product Strategy Guide. NextSprints.https://nextsprints.com/guide/qualcomm-product-strategy-guide

Newsire, P. R. (2019). *Global 5G Market Analysis and Outlook 2020-2026: Market Dynamics, Industry Leaders, Opportunities and Forecasts*. Business Insider. https://markets.businessinsider.com/news/stocks/global-5g-market-analysis-and-outlook-2020-2026-market-dynamics-industry-leaders-opportunities-and-forecasts-1028789220

OC&C Strategy Consultants. (2025). Reports. OC&C Strategy Consultants. https://www.occstrategy.com/en/reports/

Parker, T. (2014). *METIS task group lays out first 5G channel models*. Fierce Network. https://www.fierce-network.com/tech/metis-task-group-lays-out-first-5g-channel-models?utm_source=chatgpt.com

Pirinen, P. (2014). *A Brief Overview of 5G Research Activities*. IEEE Xplore., DOI: 10.4108/icst.5gu.2014.258061

Reportlinker. (2015, January 28). *The 5G Wireless Ecosystem: 2015 - 2025 - Technologies, Applications, Verticals, Strategies & Forecasts*. Prnewswire.com; Cision PR Newswire. https://www.prnewswire.com/news-releases/the-5g-wireless-ecosystem-2015---2025---technologies-applications-verticals-strategies--forecasts-300027177.html

Saadi, M., & Mavrakis, D. (2020). *5G AND AI THE FOUNDATIONS FOR THE NEXT SOCIETAL AND BUSINESS LEAP EXECUTIVE SUMMARY*. ABI Research. https://www.hsdf.org/wp-content/uploads/2021/04/5g-and-ai-report.pdf

Technologies, H. (2022). Annual Report 2022. Huawei. https://www.huawei.com/en/annual-report/2022/

Teece, D. J. (2021). Technological leadership and 5G patent portfolios: Guiding strategic policy and licensing decisions. *California Management Review*, *63*(3), 5–34. DOI: 10.1177/00081256211007584

Temperton, J. (2017, March 9). *Budget 2017: 5G and fibre broadband get funding boost in Spring Budget*. WIRED. https://www.wired.com/story/spring-budget-2017-5g-fibre-broadband

Weisman-Pitts, J. (2025). 5G Technology: A Comprehensive Report on Its Impact by 2025. *Global Banking and Finance Review*. https://www.globalbankingandfinance.com/5g-technology-a-comprehensive-report-on-its-impact-by-2025

WIPO. (2023). *World Intellectual Property Indicators 2023*. https://www.wipo.int/edocs/pubdocs/en/wipo-pub-941-2023-en-world-intellectual-property-indicators-2023.pdf

Wunder, G., Kasparick, M., Wild, T., Schaich, F., Chen, Y., Ten Brink, S., Gaspar, I., Michailow, N., Navarro, A., Fettweis, G., Cassiau, N., Kténas, D., Dryjanski, M., Pietrzyk, S., & Eged, B. (n.d.). *5GNOW: Application Challenges and Initial Waveform Results*. Retrieved July 23, 2025, from https://www.vodafone-chair.org/pbls/legacy/i-gaspar/5GNOW_Application_Challenges_and_Initial_Waveform_Results.pdf?utm_source=chatgpt.com

ADDITIONAL READINGS

5G. Initiative Team. (2020). NGMN 5G White Paper. NGMN Alliance. https://www.ngmn.org/wp-content/uploads/NGMN_5G_White_Paper_V1_0.pdf

Campbell, K., Morelli, B., & O'neil, B. (2017). *IHS Economics / IHS Technology The 5G economy: How 5G technology will contribute to the global economy*. https://www.qualcomm.com/content/dam/qcomm-martech/dm-assets/documents/5g_economy_ihs_economic_impact_analysis_january_2017.pdf

Consulting.us. (2019, March 26). *Global 5G rollout expected to reach one in seven by 2025*. Consulting.us. https://www.consulting.us/news/2006/global-5g-rollout-expected-to-reach-one-in-seven-by-2025

CORDIS. cordis.europa.eu. (2017). *PHYsical LAyer Wireless Security*. CORDIS | European Commission. https://cordis.europa.eu/project/id/317562

CORDIS. cordis.europa.eu. (2017). *Distributed computing, storage and radio resource allocation over cooperative femtocells*. CORDIS | European Commission. https://cordis.europa.eu/project/id/318784

European Commission. (n.d.). The EU Research and Innovation. European Commission. https://ec.europa.eu/research/horizon2020/

MCN. (n.d.). *Mobile Cloud Networking*. 777. https://www.mobile-cloud-networking.eu/site/

MOTO. (n.d.). *FP7-Moto*. FP7-Moto. https://www.fp7-moto.eu/

Chapter 15
5G Future

ABSTRACT

The competitive landscape of 5G technology is defined by its convergence with transformative innovations, including IoT, AI, and edge computing. These technologies rely on 5G's ultra-fast, low-latency, and high-capacity infrastructure to deliver scalable, real-time solutions. Industry leaders, including Oracle, Cisco, and Intel, drive this evolution through strategic R&D and partnerships, targeting key verticals such as healthcare, automotive, and manufacturing. Meanwhile, regulatory compliance and national policy frameworks significantly influence deployment speed and scope. Together, these dynamics underscore the importance of integrated strategies and cross-sector collaboration to maximize the economic and technological potential of 5G worldwide.

INTRODUCTION

The competitive 5G landscape is further shaped by the emergence of new technologies, such as AI, IoT, and edge computing, which rely heavily on robust 5G infrastructure. Companies increasingly focus on strategic diversification to cater to the growing demand for tailored 5G solutions across the healthcare, automotive, and manufacturing industries. Additionally, government policies and regulatory frameworks play a crucial role in shaping the pace and direction of 5G adoption globally. As the market matures, collaboration between technology providers and telecom operators is expected to accelerate innovation and drive global connectivity. These dynamics underline the critical importance of adaptability and strategic foresight in securing leadership in the 5G ecosystem. Patel, Chauhan, & Kapadiya (2012) foresee 5G as a transformative mobile technology, expected to integrate advanced

DOI: 10.4018/979-8-3373-0497-7.ch015

features such as multi-network connectivity and seamless handover between wireless technologies, with deployment anticipated around 2020.

The 5G Infrastructure advertised is divided and depends on new product launches, demonstrating the consequences of the most recent products. Thus, the leading business pioneers employ various methodologies, including high investments in R&D, new product launches, marketing activities, partnerships, joint ventures, mergers and acquisitions, alliances, collaborations, and other strategies to drive quality markets. Dominant players in the market are Oracle Corporation, Cisco Systems Inc., and Intel Corporation, among others.

TECHNOLOGY MARKET COMPETITIVE LANDSCAPE

Key players are adopting methodologies, such as investing in R&D (research and development), launching new products, and developing technology, to differentiate themselves as powerful rivals in the market. The global 5G Technology Market is a significant market with various worldwide and local players. Qualcomm, Intel, Ericsson, Samsung, and Qorvo are substantial players. Other key players in the market include AT&T, Huawei, Nokia, MediaTek, Cisco, Cavium, Analog Devices, Verizon Communications, SK Telecom, and T-Mobile.

In May 2019, Qualcomm and EE Brought 5G to the UK (United Kingdom) with the Launch of the First Commercial 5G Service. Premium 5G Devices Based on the Qualcomm Snapdragon 855 mobile platform, featuring the Snapdragon X50 5G Modem and RF (Radio Frequency) Front-End Solutions, Offer UK Consumers Multi-gigabit Peak Speeds and Extremely Low Dormancy.

In June 2019, Huawei acquired 46 5G contracts from 30 nations despite the US boycott. The nation's subtleties, which outline how to obtain the agreements, are not disclosed, as the official media has announced.

The competitive rivalry among all five 5G service business players is relatively high, owing to the presence of several leading players, such as AT&T, Verizon, and Nokia. These players' continually innovating offerings give them a competitive advantage. They are poised to make a significant impact in the marketplace through development, research, strategic partnerships, mergers, and acquisitions.

Mobile Network Operators (MNOs)

A few Mobile Network Operators (MNOs) have traditionally dominated the mobile connectivity industry, thanks to suitable infrastructure investments and long-range spectrum licenses. The mobile communication market recently changed. The MNO industry's dominance is wobbling with the arrival of over-the-top (OTT) solutions, which have become substitutes for MNOs' income-generating voice and text services, resulting in decreasing revenues for MNOs. The authors argue that future mobile networks must support IPv6 and flat IP architectures, ensuring higher efficiency and flexibility to meet the growing demand for robust and ubiquitous internet access (Patel, Chauhan, & Kapadiya, 2012).

These online services can run independently of the infrastructure, making the MNOs serve as little more than pipes to offer mobile broadband internet connectivity for the various OTT products that end-users experience. This particular substitutability of service shifts MNOs' revenue streams to internet giants. MNOs try to produce walled-garden services, emphasizing safety and security, which limits what makes end consumers the downfall. Salahdine, Han, and Zhang (2022) examine the security challenges introduced by enabling technologies in 5G, 6G, and beyond, highlighting that innovations like massive MIMO, network slicing, and mobile edge computing significantly increase the complexity of securing next-generation wireless networks.

Rivalry And Regulations

Regulators worldwide view 5G networks as a crucial enabler for new companies and are committed to promptly acquiring the new spectrum made available. Typically, compatibility, sharing, and other aspects of computer users are ensured between air services. Sophisticated spectrum-sharing strategies are likely to play a progressively crucial role in 5G, enabling MNOs to develop even more powerful sharing-based operations and opening the industry to new entrants.

According to its 5G Action Plan, the European Commission (EC) states that the possibility of spectrum sharing, which includes license-exempt use, should be maximized, as it typically supports development and industry entry in the transition to a European gigabit society. The EC encourages shared spectrum use, based on individual rights or general authorization of use, as an enabler for intensive and efficient range exploitation, particularly in the brand-new millimeter spectrum bands.

Huawei Leads the Global 5 G Market

The smartphone industry is entering an incredible era driven by 5G technologies. 2020 is expected to be the year of 5G Smartphones, with almost all Original Equipment Manufacturers (OEMs) likely to release at least 5 G Smartphones within the market.

In 2020, global 5G phone shipments were led by a few dominant manufacturers, with Huawei and Honor emerging as the top contributors due to their strong presence in key markets. Apple closely followed, leveraging its established ecosystem and brand loyalty to drive significant 5G adoption. Samsung maintained a competitive position, focusing on a diverse portfolio of devices catering to various price segments. Other major players, such as OPPO, Vivo, and Xiaomi, showcased strong growth, particularly in emerging markets driven by affordable and innovative 5G offerings. These trends reflect the dynamic competition in the 5G smartphone market and the increasing global demand for advanced connectivity features.

Huawei at its peak, directing all 5G smartphone product sales. Apple and Samsung, in second and third place after Huawei, will release their 5G smartphones in 2020, introducing specific 5G mid-range designs beyond their flagship lineup.

This analysis also states that 5G smartphones surpassed 270,000 devices worldwide in 2020. The study further claims that the penetration of entirely new 5G mobile phones increased by approximately 1% in 2019 to more than 15% in 2020.

- Huawei Is Leading the 5G Patent Race.

Huawei is leading the 5G patent race. One method of responding to this inquiry is to break down the number of licenses held or applied for by various organizations. With over 3,000 patent applications recorded, and more than 1,200 of these in reality, Chinese producer Huawei outperformed all competitors in a positioning study conducted by IPlytics and the Technical University of Berlin.

- Huawei And The Worldwide Race For 5G:

As the licensing procedure can take quite a while, the measurement also reflects which organizations have recently recorded licenses. The Finnish organization Nokia and the South Korean company Samsung have the most significant portion of their licenses previously granted, implying they have not applied for any new ones.

Producers from South Korea, a nation where 5G executes more quickly than anywhere else, wind up at the top of the list, as does Chinese supplier ZTE. U.S. organization Intel is up to date with the game, having previously acquired only a few licenses.

The 5G patent race is led by Huawei, which has filed the most patents globally, demonstrating its technological innovation dominance. Samsung and ZTE are closely followed, reflecting their significant investments in research and development that have enabled them to secure leadership in the 5G era. LG, Nokia, and Ericsson also hold notable positions, underscoring their substantial contributions to the development of 5G standards and infrastructure. Qualcomm and Intel, though further down the list, maintain a strategic presence with key patents, emphasizing their role in enabling 5G technologies. This competitive landscape underscores the importance of intellectual property in driving advancements and securing market leadership within the 5G ecosystem.

Samsung's Strategic Role in Shaping the 5G Era: A Blueprint for Emerging Leaders

As the global transition into the 5G era accelerates, Samsung has emerged as a critical force in shaping the telecommunications landscape. The company's alliances with industry giants such as HPE, Marvell, Openet, Intel, and Amdocs exemplify how strategic partnerships can fuel innovation and deployment. These collaborations have catalyzed the development of edge-to-core infrastructure, vRAN (virtualized Radio Access Networks), and next-generation RF chipsets—all foundational components of modern 5G systems.

This technological evolution underscores an essential lesson for the emerging generation of leaders: that sustainable progress demands a blend of technical excellence, collaborative acumen, and bold vision. One such example within our organization is our young and promising leader, who embodies these qualities with determination. Like Samsung, he recognizes that in an era defined by massive connectivity and automation, leadership is less about hierarchy and more about orchestrating an ecosystem.

- Samsung's edge lies in its multifaceted strategy:

Samsung's edge lies in its multifaceted strategy, which is developing lighter, more efficient base stations that reduce energy consumption by 25%; driving down the costs and complexity of deployment; and expanding the capabilities of Massive MIMO through collaborations with companies like Xilinx. The convergence of software and hardware, cloud infrastructure and edge computing, AI and IoT, all require strategic foresight—an attribute not just of large enterprises, but also of visionary individuals.

The young leader we refer to has shown remarkable alignment with this thinking. His initiatives to integrate next-gen platforms in leadership development echo Samsung's innovation philosophy. Just as Samsung merged its capabilities with Openet to enable real-time policy enforcement in 5G networks, our leader has worked to develop real-time learning systems for managerial performance. This approach transforms static training models into dynamic engines for growth, much like how network slicing revolutionizes user experiences in 5G environments.

- Samsung's role in the transformation of supply chains through 5G

Moreover, Samsung's role in transforming supply chains through 5G offers another layer of insight. As highlighted in their collaborations with logistics and automotive sectors, the intersection of connectivity and operational efficiency is crucial. Our young leader has also recognized this paradigm shift—designing programs that not only train but transform, embedding leadership directly into the fabric of organizational strategy.

Equally telling is Samsung's success in competitive landscapes. Amid regulatory uncertainties, patent wars, and global competition, Samsung has maintained resilience through adaptability. The rise of Open RAN, spectrum sharing, and decentralized architectures offers a glimpse into the decentralized yet interconnected nature of future leadership—a reality our leaders navigate with intuitive skill.

As 5G matures, especially in high-growth markets like China, the lesson becomes clear: success is not merely about scaling technology but about scaling trust, vision, and capabilities. The young leader referenced here is not just a spectator of this transformation; he is a participant, co-creating new frameworks for learning, growth, and strategic renewal.

In summary, Samsung's 5G journey exemplifies what is possible when innovation meets collaboration. It provides a blueprint—not just for enterprises, but for emerging leaders who must guide organizations through complexity and change. In this evolving narrative, our young and promising leader stands not at the edge, but at the center of what is to come next.

THE EVOLUTION OF THE MARKET SHARE OF DOMINANT PLAYERS

The 5G market has grown significantly since its early development stages, with a total valuation of USD 38 million in 2018 and a projected increase to USD 671.2 million in the United States by 2025. This impressive growth, marked by a compound annual growth rate (CAGR) of 41.9%, reflects the increasing demand for faster and more reliable connectivity.

Leading companies such as Huawei, Ericsson, Samsung, and Nokia are at the forefront of this transformation, investing heavily in R&D and expanding their infrastructure capabilities. Their global presence and strategic partnerships have enabled widespread 5G deployments across various sectors. These players dominate not only through innovation but also through their ability to adapt to regional regulatory and technological environments. The ongoing competition among them continues to fuel advancements and influence global market trends. As new use cases emerge in areas such as autonomous vehicles and the Internet of Things (IoT), market share distribution is likely to shift. Still, established leaders remain central to the evolution of 5 G.

The 5G market has experienced rapid growth, with a total valuation of USD 38 million in 2018. By 2025, it is projected to reach $ 671.2 million in the USA, reflecting a compound annual growth rate (CAGR) of 41.9% over the forecast period. This significant expansion underscores the growing adoption of 5G technologies across various industries. Dominant players, including Huawei, Ericsson, Samsung, and Nokia, are driving this growth through technological innovations and global deployments, thereby solidifying their market positions. The competition among these key players continues to shape the dynamics of the 5G landscape.

The statement also outlines the product sales and earnings produced by the worldwide 5G market. It is divided into numerous sections, including local, application, by type, country-level, and others. This segmentation allows a granular perspective on the sector, concentrating on federal government policies and altering its characteristics. Additionally, it assesses the research and development blueprints of businesses for improved product originality.

The three most prominent players, Nokia, Ericsson, and Huawei, account for approximately 30% of the industry's shares. By 2025, the 5G market is expected to be dominated by Nokia, Huawei, and Ericsson, collectively holding approximately 30% of the global market share. These leading companies leverage their technological expertise and innovation to drive advancements in 5G infrastructure and services. The remaining 70% of the market is distributed among other players, highlighting the competitive and fragmented nature of the industry. This segmentation underscores the vital role of strategic investments and partnerships in capitalizing on market opportunities. The expansion of mobile communications relies heavily on effective spectrum-sharing strategies, particularly in 5G networks. The forecast reflects the ongoing evolution and expansion of the 5G ecosystem across various sectors.

The Evolution of Market Dynamics in the Global 5G Equipment Industry

The global 5G equipment industry has continued to evolve rapidly since its commercial inception, shaped by technological breakthroughs, strategic partnerships, and geopolitical considerations. In 2025, the competitive landscape is characterized by a high level of regional specialization, intense R&D investments, and increasing demands for vendor diversification and supply chain resilience.

Samsung and several European vendors, such as Ericsson and Nokia, continue to lead in Time Division Duplex (TDD)-based infrastructure. However, Chinese manufacturers—including Huawei and ZTE—retain a technological edge, particularly in areas such as antenna engineering and massive MIMO (Multiple-Input Multiple-Output) systems. These advancements are critical for enabling the ultra-high-capacity and low-latency characteristics of 5G networks.

Notably, Naqvi et al. (2019) demonstrated a compact, high-performance antenna system capable of supporting both 4G and millimeter-wave 5G frequencies. Their system uses a Defected Ground Structure (DGS) and offers dual-band functionality at 3.8 GHz, 5.5 GHz, and 24.4–29.3 GHz, enabling efficient isolation and high gain. This innovation reflects the level of engineering sophistication emerging from Asian R&D hubs, particularly in China and South Korea.

Global Market Share and Consolidation Trends

According to historical data from the *Nikkei Asian Review*, Huawei led the global base station market in 2017, with a market share of 27.9%, followed by Ericsson (26.6%), Nokia (23.3%), and ZTE (13%). While these figures highlighted the dominance of Chinese and European firms at the onset of 5G, the situation in 2025 shows market fragmentation in some regions and consolidation in others (Okutsu, 2019).

In China, Huawei remains the top domestic supplier, though it has lost ground internationally due to sanctions and vendor bans in several Western countries. Nonetheless, its self-reliant chip development and integration of HarmonyOS with 5G base station platforms have helped retain strong leadership across the Global South.

Ericsson and Nokia, despite financial pressure and increasing R&D costs, continue to thrive through Open RAN partnerships and strong government support in Europe. Ericsson has expanded its influence in Latin America and Africa, leveraging strategic alliances with local operators and governments, promoting clean network policies.

Chinese Leadership in Advanced MIMO Systems

Chinese firms maintain a competitive lead in advanced MIMO technologies, which are essential for enhancing network capacity and user density in urban and industrial environments. At a global telecom conference in Milan in 2023, ZTE's product chief emphasized that Chinese vendors are several years ahead in real-world MIMO deployments, particularly in dynamic beamforming and AI-powered optimization (ZTE, 2023).

These capabilities enable superior spectrum efficiency and adaptability, two key enablers for next-generation use cases, such as autonomous vehicles and smart manufacturing. European players have invested in similar technologies, but deployment scale and production cost advantages still favor Chinese firms.

Japan's Shrinking Influence and Industry Consolidation

The Japanese telecom equipment sector has faced persistent challenges in competing with Chinese and European firms. Of Japan's three major mobile carriers, only NTT Docomo continues to prioritize domestic suppliers, such as NEC and Fujitsu, for its base stations. Nevertheless, even NTT Docomo maintains procurement relationships with Nokia, rooted in long-standing historical cooperation with European vendors.

In 2025, consolidation pressures are mounting in Japan. Without strong intellectual property leadership or a global market share, NEC and Fujitsu are increasingly viewed as potential acquisition targets or candidates for strategic mergers. This reflects a global pattern: as the 5G rollout matures and transitions toward 6G planning, smaller or regionally focused firms must scale or integrate to survive.

Conclusion

The market dynamics of 5G equipment manufacturing in 2025 highlight a shift toward global competition along strategic and technological lines. While Chinese firms continue to lead in MIMO and antenna innovation, European companies are regaining ground through collaborative open RAN strategies. Meanwhile, Japanese vendors risk obsolescence unless they can pivot through alliances or product differentiation.

This evolving landscape underscores the importance of R&D, vertical integration, and political alignment in shaping the future of global telecommunications infrastructure.

The Future of the Competitive Landscape of Telecom Operators

In 2025, the global telecom equipment market continues to be shaped by intense competition among key players, particularly in the 5G domain. Huawei remains the dominant force, holding approximately 31% of the global telecom infrastructure market. Despite geopolitical pressures and restrictions in several Western countries, Huawei's growth in emerging markets has solidified its leadership position (Maistre, 2025).

Nokia, while experiencing a 9% annual drop in overall sales, demonstrated resilience with a 10% increase in Q4 2024, primarily driven by rising demand in North America and India (Mukherjee, 2025). Ericsson similarly posted strong Q1 2025 results, benefiting from increased demand for its 5G equipment and software solutions despite competitive pressure and price sensitivity (Earnings Analyst, 2025).

Samsung, which once aimed to capture 20% of the 5G market by 2020, continues to struggle to gain a significant market share. The company still lags behind its rivals, although new opportunities in the U.S. and Europe could reignite its growth trajectory.

As the second wave of 5G infrastructure investment unfolds, industry analysts anticipate that operators will increasingly diversify their vendor base. However, legacy relationships and existing 4G infrastructure often influence continuity in vendor selection, giving Nokia, Ericsson, and Huawei an advantage in many regions.

The Future of 5G for R&D Spend

Back in 2018, Samsung pledged to invest $22 billion over three years in advancing 5G, the Internet of Things (IoT), and artificial intelligence (AI) (Reuters, 2018). *While a significant portion of the $16.51 billion allocated to research and development in 2019 likely went toward consumer electronics and semiconductors, this investment was nonetheless strategic.* According to Jan-Peter Kleinhans, a 5G researcher at the German think tank Stiftung Neue Verantwortung (SNV), these two sectors contributed to more than half of Samsung's $192 billion in revenue that year (Kleinhans, 2020).

At the time, clean network companies like Ericsson and Nokia were each spending approximately $4 billion annually on R&D. Kleinhans noted that for Samsung to make serious inroads into the 5G network infrastructure market, it would have to outspend these competitors significantly (Kleinhans, 2020).

Analyst Richard Gubbins suggested that Samsung's R&D spending could be primarily focused on expanding its portfolio of network products. Unlike Huawei, Ericsson, and Nokia, which offer end-to-end telecom infrastructure solutions, Samsung still relies on third-party suppliers for parts of the network, especially for backhaul systems. Backhaul plays a critical role in transporting data from base stations (which receive radio signals) to the broader core networks that connect end-user devices.

By 2025, Samsung is expected to have made substantial progress in closing this gap. It has increased its direct investment in telecom infrastructure R&D, focusing on building more complete, vertically integrated 5G network solutions. Samsung has also expanded partnerships to localize and secure its supply chain amid growing geopolitical pressure for clean networks. As global telecom markets continue to demand open and interoperable 5G systems, Samsung's strategic investment in R&D positions it as a serious competitor in both developed and emerging markets (Samsung, 2024).

Acquisitions

Samsung's long-term ambitions in the 5G infrastructure space may also involve strategic acquisitions to strengthen the software side of its network operations. In 2020, Jan-Peter Kleinhans suggested that the company could leverage its then-$78.53 billion in net cash reserves to acquire firms specializing in programmable networking components, particularly in support of Open Radio Access Network (Open RAN) development (Kleinhans, 2020).

Open RAN, a modular engineering standard designed to enable interoperability between network hardware and software, is expected to continue gaining traction in 2025. The model primarily benefits smaller players and startups, such as U.S.-based Altiostar Networks, Mavenir, and Parallel Wireless, which have received significant support from public-private funding initiatives. Notably, a $1 billion U.S. government-backed fund for Open RAN research and development was established (U.S. Department of Commerce, 2022). These initiatives have encouraged operators to adopt Open RAN systems within a seven-year window, enabling them to qualify for reimbursement when replacing high-risk vendors, such as Huawei.

Strategic acquisitions could position Samsung as a viable alternative to Huawei, noted Daryl Scholar, an analyst formerly with Ovum and now at MediaTech Insights. He pointed to Samsung's January 2020 acquisition of U.S.-based TeleWorld Solutions as a strong step toward expanding its 5G capabilities in the North American market (Scholar, 2020). However, he also warned that Samsung must be careful not to dilute or over-fragment its portfolio with unfocused acquisitions.

Despite skepticism from figures like former U.S. Attorney General William Barr, who once argued that building viable Open RAN platforms from scratch could take a decade (Barr, 2020), momentum has increased since then. While Barr had previously advocated for direct investment in European incumbents, such as Ericsson and Nokia, recent trends indicate that modular, open, and software-centric networks are increasingly favored by both operators and governments seeking to reduce vendor lock-in and enhance supply chain resilience.

In 2025, Samsung is actively exploring further acquisitions and strategic partnerships to accelerate its development of Open RAN. Industry observers expect a focus on edge computing, AI-based network orchestration, and virtualized radio access to dominate Samsung's M&A activities over the next few years.

The Future of the Competitive Landscape of Telecom Operators

By 2025, the global 5G telecom ecosystem is expected to have undergone significant evolution, driven by shifting market shares, geopolitical tensions, and increasing demand for open and flexible networks. Established players, including Huawei, Nokia, and Ericsson, continue to lead, while new challengers such as Mavenir, Altiostar, and Parallel Wireless gain momentum through Open RAN innovation. Huawei, despite facing ongoing U.S. sanctions, maintains a dominant market presence due to its vertically integrated business model and deep research and development (R&D) capabilities.

Nokia and Ericsson have leveraged bans on Chinese equipment in Europe and North America to expand their footprints. Nokia has focused on cloud-native solutions and edge computing, while Ericsson emphasizes network energy efficiency and private 5G networks for enterprises. Samsung, despite its ambitions, still struggles to break into the top tier, holding a modest market share. However, recent acquisitions and investments, including TeleWorld Solutions, aim to bolster its North American position.

Meanwhile, U.S.-based firms such as Mavenir and Altiostar are benefiting from national security-driven investments in Open RAN infrastructure, thereby reshaping the competitive playing field. Parallel Wireless, supported by U.S. funding and carrier partnerships, is also gaining recognition for its interoperable RAN solutions. These dynamics suggest that while incumbents retain their strength, innovation and the changing political context are opening the door for new actors.

Table 1 below provides an overview of the competitive positions and strategic moves of key telecom operators and companies shaping the 5G landscape in 2025.

Table 1. Competitive landscape of telecom operators

Operator / Company	Competitive Position and Strategic Moves
Samsung Electronics	Aimed to capture 20% of the 5G market by 2020; held only 3% in 2019. Faces intense competition from Huawei, Nokia, and Ericsson. Needs more R&D and acquisitions—recent acquisition: TeleWorld Solutions.
Huawei Technologies	Global leader with 28% market share in 2019. It maintains dominance through comprehensive hardware production and R&D strength despite U.S. restrictions.
Nokia Corp (Finland)	Strong in Europe, a key player in 4G and 5G rollouts. Benefits from Huawei bans in Western countries. Invests significantly in R&D.
Ericsson (Sweden)	One of the top global suppliers of 5G equipment. Focused on clean network infrastructure. Strong in both hardware and services.
Altiostar Networks (USA)	Specializes in openRAN. Aims to disrupt traditional telecom hardware with software-defined solutions. Receives support from U.S. openRAN initiatives.
Mavenir (USA)	Supports openRAN and software-driven mobile networks. Positioned to benefit from anti-Huawei policies.
Parallel Wireless (USA)	OpenRAN advocate. Develops interoperable RAN solutions. Poised for growth through U.S. funding and operator support.
TeleWorld Solutions (USA)	Acquired by Samsung in 2020 to strengthen 5G infrastructure service capabilities and expand U.S. presence.

THE FUTURE OF THE GLOBAL COMPETITIVE LANDSCAPE OF THE INDUSTRY

The global telecommunications industry in 2025 remains marked by fierce competition among key players as they strive to define the next era of connectivity through 5G and beyond. Industry leaders, including Huawei, Ericsson, Nokia, Samsung, and more recently, U.S. and Indian entrants, are leveraging advanced technologies, strategic acquisitions, and regional partnerships to assert their influence.

As 5G matures, its deployment is no longer simply about speed or latency improvements—it's about enabling the infrastructure for the Internet of Everything, edge computing, AI-powered orchestration, and industrial automation. Technologies such as massive MIMO, dynamic spectrum sharing, Open RAN, and network slicing are no longer experimental but are now foundational to operators' business models.

Fettweis (2016) anticipated that 5G would bring about not only ultra-fast data transmission but also fundamental shifts in how networks interact with real-world environments, enabling applications such as the Tactile Internet and massive IoT. As of 2025, this vision is partially realized, with private 5G networks being deployed in various industries, including manufacturing, logistics, mining, and smart cities.

A Fragmented but Rapidly Consolidating 5G Market

In 2020, the 5G services industry was highly fragmented, with regional telecoms making large-scale but isolated investments. This fragmentation presented opportunities for players to pursue strategic mergers and acquisitions. One early example was T-Mobile's $3.5 billion partnership with Ericsson in 2018, aimed at accelerating 5G deployment in the U.S. Similarly, BT's EE acquired 40 MHz of 3.4 GHz spectrum for $426.5 million to support future 5G services in the UK.

By 2025, this fragmentation has begun to give way to regional consolidation, as governments, especially in Europe and Asia, encourage interoperability through open standards and clean network policies. The rise of Open RAN has also enabled smaller vendors to enter the market without relying on vertically integrated giants, contributing to increased vendor diversity, particularly in Africa, Southeast Asia, and Latin America.

5G as a Disruptive Innovation

5G remains a textbook case of disruptive innovation, reshaping both hardware and software ecosystems across the industry. New form factors—such as foldable, rollable, and modular smartphones—are now mainstream in 2025, enabling users to integrate multiple functions (workstation, entertainment hub, AR/VR platform) into one device.

However, disruption is not limited to consumer design. The move from hardware-defined to software-defined networks has rendered legacy vendor strategies obsolete. Several companies failed to adapt, much like Motorola's dramatic fall during the 2G to 3 G transition or Japan's vendors missing the global 3G wave due to inward-focused innovation. Some current players risk obsolescence if they cling to outdated models.

Historically, each technological transition has reordered industry leadership:

- Motorola led the market during the analog and early 2G eras but lost nearly 80% of its global handset market share during the shift to 3G.
- Nokia, once dominant in 2G, stumbled in 3 G and collapsed in the 4G era.

- Samsung capitalized on the 3 G transition and solidified its position as a global leader by the time the 4G era arrived.
- Huawei emerged as a top global vendor in the 4G era, leveraging aggressive pricing and innovative fast followers.
- In the 5G era, Xiaomi and Reliance Jio disrupted cost structures with idevice, network, and service ecosystems.

By 2025, a competitive advantage will lie not only in technical superiority but also in platform thinking—the ability to orchestrate devices, cloud, data, and services into adaptive business models. Leaders in this space are no longer defined by national borders but by how well they enable digital transformation for entire economies.

Toward a Consolidation of the Industry With 5G (2025 Update)

In 2025, the 5G infrastructure, which was once highly fragmented, is rapidly consolidating as technological demands and capital intensity create significant barriers to entry. While the early years of 5G saw a proliferation of vendors, startups, and regional telecom providers, the industry is now coalescing around a group of major global players, including Samsung, Huawei, Ericsson, Nokia, Intel, Cisco, and NEC.

To survive in this capital-intensive environment, leading firms have pursued aggressive initiatives, including R&D investment, product innovation, strategic partnerships, joint ventures (JVs), mergers, and acquisitions. This consolidation is a response to the need for scalable, software-defined, and virtualized network architectures that can support edge computing, remote industrial control, and real-time data processing—core functionalities for the future of 5G and 6G, (Fettweis, 2016).

Examples of strategic moves include:

- Mavenir's 2020 partnership with Turkcell to deploy Open vRAN on the Turkcell Edge Cloud, enabling both standalone (SA) and non-standalone (NSA) 5G deployments, reflects a shift toward virtualized, vendor-neutral solutions (Mavenir, 2020).
- NEC's development of millimeter-wave (mmWave) MIMO antennas, announced in 2020, solved propagation and shielding challenges in indoor environments, enabling high-speed, high-capacity connections using the 28 GHz band, (Editorial Team, 2021).

These moves illustrate how vendors are not only expanding geographically but are also vertically integrating capabilities—from semiconductors to radio units to orchestration software—to deliver end-to-end 5G ecosystems.

Market Concentration and Strategic Dominance

Consolidation is a natural evolution in maturing technology sectors. In the case of 5G, this progression is accelerated by the scale and complexity of the required infrastructure. As of 2025, we observe a concentrated global market led by a few dominant firms that possess the capital, intellectual property, and cross-border partnerships necessary to meet the rising global demand.

This shift brings multiple advantages:

- Economies of scale reduce costs and improve deployment speed.
- Accelerated innovation in areas like network slicing, AI-driven orchestration, and mmWave spectrum utilization.
- Integrated service models combine devices, software, and infrastructure into cohesive, platform-based solutions.

However, this concentration also poses regulatory challenges. Authorities in the U.S., the EU, and parts of Asia now closely monitor M&A activity to prevent monopolistic behavior, promote innovation, and ensure security in critical communications infrastructure (European Commission, 2024).

The 5G market reflects trends seen in other sectors, such as home construction, point-of-sale software, and exhibition services, where fragmented industries eventually coalesce around dominant players who can deliver integrated and scalable services. For instance, in the telecom sector, services such as hardware rental, local content delivery, and portable data storage are now largely dependent on a handful of cross-national providers.

Strategic partnerships remain essential to maintaining dynamism. Instead of stifling competition, the leading players often work with regional ISPs, cloud providers, and AI startups to create agile, modular service offerings tailored to specific markets.

Ultimately, the consolidation of the 5G industry reflects not only economic logic but also the strategic imperative of global digital transformation. Firms that can scale infrastructure, build ecosystems, and align with national security and innovation policies are shaping the backbone of global connectivity. As demand continues to rise across sectors—from autonomous logistics to smart agriculture—these technologies play a pivotal role in driving the evolution from 5G to 6G and beyond.

Evolution of the Competitive Dynamic (2025 Update)

Unlike the 4G era, which was dominated by a handful of baseband chipset suppliers, the 5G market has introduced a broader, more diversified competitive landscape. From the early stages, companies such as Samsung LSI (Exynos series),

HiSilicon (Balong), Intel (XMM series), MediaTek (Dimensity), and Unisoc have entered the baseband market with aggressive R&D programs and differentiated technologies, (Strategy Analytics, 2019).

This expansion of players has accelerated innovation in network performance, spectrum efficiency, and device compatibility, while also driving down costs. Moreover, it has spurred greater consumer choice, especially in emerging markets. These advancements are accompanied by the necessity for new security frameworks, as the complexity of multi-vendor, cloud-based, and AI-driven networks increases the attack surface (Salahdine, Han, & Zhang, 2022). As a result, the 5G competitive dynamic is no longer solely about hardware dominance, but about system-level integration, security resilience, and global scalability.

The Critical Role of R&D for Competitiveness

By 2025, R&D investment will continue to be the cornerstone of competitive advantage in the 5G ecosystem. Leaders such as Huawei, Ericsson, Qualcomm, Samsung, and Nokia continue to invest billions of dollars annually in next-generation communication technologies, including ultra-reliable low-latency communication (URLLC), enhanced mobile broadband (eMBB), and massive machine-type communications (mMTC) (Kovtun et al., 2024).

These R&D efforts go beyond improving transmission speed—they focus on solving energy efficiency, signal interference, AI-driven automation, and cloud-native architectures. For example, Huawei has allocated over 20% of its annual revenue to research and development (R&D) since 2020, reinforcing its position as one of the most innovative players in the industry, (Huawei, 2025).

Firms that underinvest in R&D risk falling behind in a sector defined by rapid generational leaps and increasingly complex standards. R&D not only strengthens core product offerings but also enables first-mover advantages in patent ownership, standards influence, and global partnerships.

Economies of Scale and Scope as Strategic Imperatives

The ability to scale both horizontally (across geographies) and vertically (across the technology stack) is now a non-negotiable success factor in the 5G era. Economies of scale enable vendors to lower unit costs and serve growing global demand.

In contrast, economies of scope would allow firms to offer integrated solutions that combine basebands, antennas, software, and cloud services under a single platform.

As 5G becomes a foundation for smart cities, autonomous vehicles, industrial IoT, and immersive reality, cost-effective deployment is essential. The rivalry between Huawei, Ericsson, and Nokia—especially in emerging markets—is intensifying pricing pressure, driving widespread adoption (Sharma, 2023).

Smaller vendors are not excluded; many are carving out niche markets in areas like Open RAN software, edge AI, or cybersecurity. By focusing on specialized solutions and interoperability with larger systems, they contribute to a modular, collaborative 5G ecosystem.

A Race for Positioning in the 5G Era

By 2025, the race for 5G leadership has crystallized around a small group of technology powerhouses, with Qualcomm standing out for its foundational role in wireless innovation. Having already dominated in the 4G era through technologies like VoLTE, carrier aggregation, and LTE unlicensed, Qualcomm has leveraged its expertise in radio frequency (RF) front-end, chipsets, and millimeter-wave technologies to remain central in the 5G conversation.

At Mobile World Congress 2019, Qualcomm showcased its integration into flagship devices, such as the Samsung Galaxy S10 5G and the LG V50 ThinQ. By 2025, it is expected to have extended this presence into AI-driven 5G chipsets, mmWave small cells, and enterprise 5G modules, enabling cloud gaming, remote surgeries, and private 5G networks.

Qualcomm's success lies in its system-level solutions, which combine chipsets, RF components, and software to deliver turnkey 5G capabilities. Strategic collaborations with Microsoft, Google, Fujitsu, Vivo, and Nokia ensure Qualcomm remains indispensable to both consumer electronics and enterprise-grade applications.

As global connectivity shifts toward intelligent, adaptive, and autonomous networks, Qualcomm's continued investment in 6G research, AI at the edge, and terahertz communication suggests it will remain a cornerstone in the race for digital infrastructure leadership in the coming decade.

THE COMPETITIVE MOVES OF KEY COMPANIES IN THE 5G RACE

The 5G landscape has witnessed remarkable developments as companies vie to position themselves as leaders in this transformative technology. At the Mobile World Congress (MWC) 2019, numerous companies showcased their advancements, partnerships, and strategies aimed at capturing market share and driving innovation. Each player, from established industry giants to niche IoT baseband developers, has made significant moves to establish their foothold in the competitive 5G market. Strategy Analytics (2019) predicts that the 5G baseband market will experience substantial growth, driven by early positioning and strategic innovation from vendors.

Competitive Moves

MediaTek announced partnerships with Vivo and Oppo to bring 5G products powered by its Helio M70 system-on-chip (SoC). *The Helio M70 offers integrated 5G capabilities, catering to midrange devices and enabling MediaTek to strengthen its presence in the smartphone market.* The company plans to release its System-on-Chip (SoC) model, updated for the eyewear season, to expand its reach further and solidify its position as a key supplier for prominent smartphone brands.

Intel has also made significant strides with its multi-mode 5G platform, XMM8160. Positioned as a competitor to Qualcomm's X55 modem, Intel's platform is expected to support a range of diverse applications and ensure compatibility across various networks. Collaborating with network operators, Intel aims to qualify its 5G modem platform for widespread adoption. The company's 2020 roadmap reflects its commitment to delivering advanced solutions tailored to the evolving demands of 5G connectivity.

Samsung LSI secured significant design wins with the Galaxy Fold 5G and Galaxy X, utilizing 5 G with its Exynos 5100 5G modem. Although some regions will use Qualcomm's X50 modem in these devices, Samsung's dual supplier status highlights its ability to cater to varying regional requirements. The company is expected to further enhance its portfolio by integrating 5G capabilities into its System-on-Chip (SoC) editions, aligning with its long-term strategy to compete in both premium and mid-range segments.

HiSilicon, Huawei's semiconductor arm, showcased its Balong 5000 modem in the Huawei Mate X foldable device, highlighting and underscoring HiSilicon's expertise in modem technology, sustaining Huawei's competitive edge in the global 5G market. The Balong 5000 is a testament to Huawei's ability to deliver cutting-edge devices while advancing its integration capabilities in preparation for upcoming 5G SoC editions. Salahdine, Han, and Zhang (2023) argue that the evolution towards 5G, 6G, and beyond demands not only higher performance in terms of speed, latency, and connectivity, but also the integration of novel technologies such as terahertz communication, network slicing, and visible light communication to meet the growing expectations for ubiquitous and intelligent wireless services.

Niche players like Altair (part of Sony) and Sequans focus on cellular IoT baseband technologies, emphasizing 5 G's potential to revolutionize IoT applications. Sequans has partnered with TCL for 5G research and development, marking a strategic move to capitalize on the IoT opportunities presented by 5G over the next few years. The authors identify that while xG networks promise transformative connectivity, they also face significant challenges, including the need for new infrastructure, advanced virtualization, and the management of complex, high-frequency environments that underpin future intelligent communication ecosystems (Salahdine, Han, & Zhang, 2023).

Despite challenges like legal disputes with Apple and regulatory hurdles, Qualcomm continues to thrive, demonstrating resilience and innovation. The company's ability to deliver cutting-edge products, including the X50 and X55 modems, reinforces its leadership in the 5G baseband market. Strategy Analytics emphasizes that early efforts to establish market positions are critical for long-term success in the competitive 5G landscape.

Updated Moves

As 2025 unfolds, companies are doubling their strategies to capitalize on the 5G market. MediaTek's Helio M70 SoC is expected to gain traction as its early partnerships with Vivo and Oppo translate into tangible product launches. The mid-range segment, driven by MediaTek's offerings, continues to grow as more consumers adopt 5G devices.

Intel's XMM8160 platform is poised for broad deployment in collaboration with network operators, positioning it competitively against established players such as Qualcomm and Samsung. Samsung LSI, meanwhile, is likely to expand its presence by integrating 5G into its Exynos SoCs, further solidifying its dual-supplier strategy.

HiSilicon and Samsung LSI's focus on 5G SoCs highlights their ambitions to deliver integrated solutions that bridge the gap between modem and processor capabilities. These initiatives are shaping the competitive dynamics of the 5G market.

Niche players such as Sequans and Sony's Altair are poised to benefit from the rollout of 5G IoT applications. These applications will target specific industries and unlock new revenue streams. Their collaboration with partners like TCL underscores the importance of innovation and partnerships in niche markets.

The 5G baseband market is undergoing rapid transformation, with established players and emerging competitors vying to secure their positions. This competitive environment drives technological advancements, fosters innovation, and ensures that the benefits of 5G connectivity reach consumers and industries worldwide. As consolidation and competition continue, the 5G race remains a defining feature of the global technology landscape.

Table 2. Competitive moves of key companies in the 5G race

Company	Strategic Move in 5G Race
MediaTek	Partnered with Vivo and Oppo using Helio M70 SoC for midrange 5G devices; plans new SoC releases to expand market reach.
Intel	Developed multi-mode XMM8160 5G platform; collaborates with network operators to support diverse applications.
Samsung LSI	The secured design wins with Galaxy Fold 5G, utilizing the 5G Exynos 5100 modem; the aim is to integrate System-on-Chip (SoC) devices into a single SoC.
HiSilicon (Huawei)	Showcased Balong 5000 in Huawei Mate X; focuses on modem integration and cutting-edge 5G SoC editions.
Altair (Sony)	It focuses on IoT baseband technologies and targets 5G for niche applications in industrial sectors.
Sequans	It specializes in cellular IoT basebands and has partnered with TCL for 5G R&D in IoT-focused deployments.
Qualcomm	Continues leadership with X50 and X55 modems; resilient despite legal disputes and regulatory challenges.
Samsung (Devices)	Combines Qualcomm and Exynos modems in global devices; follows a dual-supplier strategy for flexibility.

THE EMERGENCE OF NEW BUSINESS MODELS WITH 5G

The rollout and maturation of 5G have triggered a fundamental transformation in global mobile communication, leading to the transformation of traditional business models and the emergence of new ecosystem-based strategies. These developments are not merely technological—they are economic and organizational shifts that redefine how value is created, delivered, and captured in a hyper-connected world.

According to Iivari et al. (2016), the adoption of ecosystemic business models is crucial for fostering collaboration and innovation in industrial contexts enabled by the Internet of Things and next-generation connectivity. 5G's capabilities—including ultra-low latency, massive machine-type communication, and enhanced mobile broadband—have accelerated this evolution by lowering market entry barriers, enabling real-time decision-making, and supporting decentralized architectures.

5G Supports Advanced Applications and Cross-Sector Integration

5G's unique attributes—high speed, low latency, and ultra-reliability—support advanced applications across diverse verticals, including autonomous vehicles, smart cities, eHealth, immersive media, and Industry 4.0. These applications are no longer experimental. By 2025, many will have transitioned to scaled commercial deployment, creating an integrated, inclusive, and interoperable digital ecosystem.

In 2024, the European Commission identified 5G as a primary enabler for the digital transformation of both society and the economy, citing its role in increasing the uptake of ultra-high-capacity networks. The true innovation lies in network slicing, which allows infrastructure providers to deliver tailored services to specific sectors on shared infrastructure, thereby balancing customization with efficiency.

New architectural designs, especially in the construction and urban development sectors, are being adapted to meet the unique visual and signal penetration demands of high-frequency 5G deployment (El-Shorbagy, 2021). This challenge has prompted collaboration between telecom companies and architects to harmonize functionality and aesthetics.

The Shift from Operator-Centric to Service-Driven Models

Historically, mobile ecosystems have been dominated by a few key players, including telecom operators and equipment manufacturers. In the 5G era, the ecosystem is expanding to include:

- Cloud providers (e.g., AWS, Azure)
- IoT platforms and device manufacturers
- Over-the-top (OTT) service providers (e.g., Netflix, Zoom)
- AI and data analytics firms
- Logiinnovative and innovative infrastructure companies

This broadening reflects a paradigm shift from infrastructure ownership to service orchestration and data monetization. OTT players, once peripheral, are now deeply embedded in infrastructure decisions and business models, often partnering with or bypassing operators to control the user experience directly.

Emergence of Cross-Sector and Hybrid Business Models

5G is dissolving traditional industry boundaries. Digital platforms now facilitate cross-sector value creation, with banks offering healthcare services, utilities managing smart homes, and retailers providing real-time delivery through autonomous systems. This shift is redefining organizational identities and strategic positioning.

By 2024–2025, business models will have matured around:

- Private 5G networks for enterprises
- Edge computing to support real-time processing close to the data source
- Managed IoT services offered by telecoms in partnership with cloud vendors
- Platform business models focused on data aggregation, AI, and service. brokerage

This new dynamic emphasizes scalability, interoperability, and speed of innovation. Stakeholders must now operate in fluid partnerships, co-developing use cases, sharing infrastructure, and distributing value across increasingly complex supply chains.

The Future: From Business Models to Business Ecosystems

The evolution of 5G business models signals a broader transition from single-firm strategies to dynamic ecosystems of partners, competitors, regulators, and users. This approach enables rapid innovation and resilience in the face of fast-changing technological and market conditions.

As of 2025, the winners in the 5G era are those organizations that:

- Embrace collaborative innovation
- Invest in software-defined arc. hitectures
- Prioritize customer-centric services over traditional CAPEX-heavy infrastructure
- Utilize data and AI as strategic assets to unlock new revenue streams

5G is no longer just a connectivity layer—it is the foundation for a new industrial age, driving everything from precision agriculture to real-time language translation. The key challenge ahead will be ensuring that this innovation remains inclusive, secure, and adaptable.

From Operator Dominance to Service Provider Outperformance

The traditional mobile communication ecosystem was once defined by clear and stable roles: content providers, internet service providers (ISPs), infrastructure vendors, device manufacturers, and integrated operators. These actors functioned in a vertically integrated manner, with mobile network operators (MNOs) controlling both service and infrastructure provisioning. Clean service providers and OTT platforms played a minor, often peripheral role in value creation.

However, the arrival of 5G has disrupted this equilibrium. With its ultra-low latency and massive connectivity features, 5G enables real-time interaction across complex verticals—from autonomous logistics to telemedicine—introducing new stakeholders into the ecosystem. Edge computing firms, AI-based analytics start-ups, cloud-native network function providers, and private 5G network operators are increasingly central to this dynamic landscape.

A pivotal transformation is the elevation of OTT platforms. Initially independent of telecom infrastructure, these platforms—such as streaming, messaging, and enterprise collaboration tools—now collaborate closely with telecom operators. Through partnerships and co-location agreements, they ensure higher service quality and reduced latency, especially for mission-critical applications.

Meanwhile, digital convergence is redefining industry boundaries. Traditional companies, such as banks, retailers, and energy providers, are integrating 5G capabilities to offer new services and enhance their offerings. Banks now operate telehealth platforms, and manufacturers run fully automated smart factories via private 5 G networks. These strategies reflect a broader ecosystemic evolution where services are co-created across sectors.

To thrive, all stakeholders—including traditional mobile network operators (MNOs)—must shift from static infrastructure models to adaptive, service-driven frameworks, prioritizing openness, modularity, and interconnectivity. The 5G era favors platform enablers over infrastructure controllers.

Figure 1. Trends of change in the 5G company business model

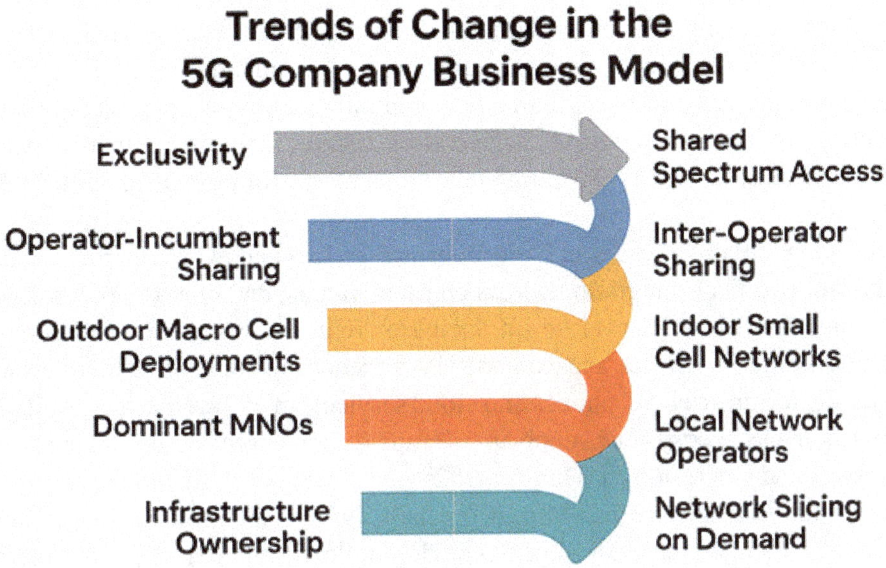

THE ASIAN RISING COMPETITION FROM CHINA AND SOUTH KOREA

Since the global rollout of 5G in 2019, China and South Korea have remained at the forefront of the 5G revolution. South Korea was the first country to achieve nationwide 5G deployment in early April 2019, marking a significant milestone in global telecommunications history. China quickly followed with massive infrastructure investments led by its three state-owned operators—China Mobile, China Unicom, and China Telecom, who together deployed over 2.5 million 5G base stations by 2024, (IMT, 2020).

Today, these two nations are not only leaders in infrastructure deployment but also standardization, research and development, hardware innovation, and global influence. Their accelerated progress is reshaping global competition and influencing the technology agendas of Europe, North America, and Southeast Asia.

Samsung, Vivo, Xiaomi, Huawei, and OPPO

In 2025, Asian mobile manufacturers—Samsung (South Korea), Huawei, Vivo, Xiaomi, and OPPO (China)—are expected to continue dominating the 5G smartphone market, both in terms of unit volume and technological innovation. Their devices now offer full integration with AI accelerators, edge computing compatibility, and satellite fallback connectivity, setting new benchmarks for next-generation mobile experiences.

Samsung remains a leader in both consumer devices and 5G infrastructure, benefiting from vertical integration across chipsets, radios, and cloud-native software platforms. Meanwhile, Huawei, despite facing international sanctions, has maintained strong domestic dominance and expanded its presence in Global South markets with affordable, feature-rich 5G phones and cloud services. In 2024, Huawei unveiled its 5G HarmonyOS 4 ecosystem, supporting seamless connectivity across smartphones, wearables, and smart homes (Huawei, 2024).

Vivo and OPPO have capitalized on their agility and growing international presence by focusing on mid-range 5G devices tailored for Africa, Latin America, and Southeast Asia. These companies are also pioneering camera-integrated AI and advanced cooling systems to support high-performance 5G gaming and streaming.

According to GSMA Intelligence (2025), more than 1.7 billion 5G smartphones are now in active use globally, representing over 60% of all smartphones in service. China alone accounts for more than 750 million 5G devices, with South Korea maintaining the world's highest 5G penetration rate (above 95%).

The competition from Asian firms has put pressure on Western companies, such as Apple, which continues to innovate primarily within premium segments, and smaller U.S. or European device makers that struggle to match the price-performance ratio of their Asian counterparts.

State-backed R&D investments, ecosystem support, and strategic vertical integration reinforce this regional lead. China's 5G push is now complemented by national initiatives in 6G research, positioning the country to maintain dominance in the next wave of mobile innovation.

China's Adoption of 5G in 2025

China's journey through successive generations of mobile technology reflects rapid technological adoption and market growth. The 3G era began in 2009, reaching a penetration rate of 35.5% after five years. However, the 3G market experienced disruption with the introduction of 4G in 2014, leading to an accelerated shift in user preferences. By 2017, 4G penetration in China had surged to 71.7%, with projections exceeding 85% by 2018, highlighting the country's swift transition to advanced communication technologies.

COLLABORATIVE EFFORTS IN THE 5G ERA

As 5G technology matures globally, cross-industry collaboration has become essential to accelerate deployment and unlock its full potential. In 2025, partnerships among telecom operators, technology providers, governments, and research institutions are driving innovation and addressing challenges related to infrastructure, security, and interoperability. These collaborative efforts support new use cases in smart cities, autonomous transport, and industrial automation. Joint ventures and public-private alliances are especially crucial in harmonizing global standards and fostering inclusive access to next-generation connectivity. The following section highlights key initiatives shaping this cooperative 5G ecosystem.

Interdependence Across Network Layers

5G networks are inherently complex, marked by high interdependencies between the radio access network (RAN), transport infrastructure, and core networks. To ensure optimal performance, chief technology officers (CTOs) must take a holistic view of system design and deployment, considering the entire 5G architecture as an integrated ecosystem. Unlike previous generations, 5G demands concurrent support for near-zero latency, ultra-dense connectivity, and massive data throughput, while maintaining backward compatibility with legacy LTE and LTE Advanced systems.

This co-residency of technologies creates both technical and financial challenges. While CTOs often approach deployment with a performance-centric lens, chief financial officers (CFOs) must analyze deployment through the prism of cost optimization, identifying key areas where operational and capital expenses can be aligned for maximum return on investment.

Strategic Investment and Deployment Trade-Offs

5G deployment must reconcile two cost dimensions: cost accelerators and cost optimizers. According to current industry analysis, Group A factors—such as active and passive infrastructure, RAN virtualization, and energy efficiency—offer the greatest cost deltas and thus the most significant savings opportunities. Meanwhile, Group B factors include technologies like multi-access edge computing (MEC), core network virtualization, and cloud-native orchestration, which may require higher upfront investments but serve as enablers for long-term cost containment and service innovation (GSMA Intelligence, 2024).

Automation and AI play a central role in this balancing act. Intelligent automation enables operators to reduce OPEX by simplifying network management, while enhancing service quality through self-optimization and real-time analytics. Such capabilities are key to achieving sustainable 5G profitability at scale.

Private Networks, Industrial Use Cases, and Public Sector Leadership

With full public 5G networks not expected until 2025–2030 in many countries (Gartner, 2023), a growing number of enterprises are deploying private 5G networks for tailored applications in logistics, manufacturing, mining, and healthcare. As Sally Miller, CIO of DHL Supply Chain, noted, early adoption of private networks allows firms to develop competitive advantages through low-latency, highly secure, and customizable connectivity.

China exemplifies this industrial transformation through state-led infrastructure investment and proactive innovation in cloud services, innovative city platforms, and edge AI. The integration of 5G into sectors such as public health, particularly during the COVID-19 pandemic, has demonstrated its effectiveness in supporting remote diagnostics and robotic medical procedures (Moglia et al., 2022). Such use cases underline the role of government-led 5G strategies in reinforcing national competitiveness.

Global Collaboration and Regulatory Influence

The future of 5G is shaped not only by technological innovation but also by cross-border collaboration, regulatory harmonization, and strategic public-private partnerships. Primary think tanks and government agencies have published white papers and frameworks emphasizing:

- The importance of spectrum policy alignment
- National 5G security standards
- Supply chain diversification
- Shared research and development frameworks

These efforts are crucial to countering geopolitical risk and fostering innovation ecosystems. Policymakers and researchers rely on these documents to navigate the regulatory, commercial, and societal implications of 5 G (Brookings Institution, 2024).

AI and Analytics as Catalysts for 5G Optimization

Tech giants like Intel are spearheading innovation at the intersection of 5G and AI. Intel's end-to-end architecture—from the cloud to the edge—emphasizes network slicing, AI-driven orchestration, and automation. Their hardware-software integration strategy combines FPGAs, security layers, and dynamic data analytics to optimize infrastructure performance and enable real-time adjustments.

Other firms are following suit. For example, Samsung's 2018 acquisition of Zhilabs, a provider of real-time analytics for user behavior and service quality, highlights the increasing importance of AI-powered insights in managing the exponential data growth characteristic of 5G networks. These intelligent platforms are key to delivering personalized services, proactive fault resolution, and operational transparency.

As Balika Sonthalia, a former VP at A.T. Kearney, emphasized, companies that ignore AI's role in 5G optimization risk being left behind in a market increasingly defined by data-centric agility and automation.

CONCLUSION

Chapter 15 examined the rapidly evolving landscape of 5G technology, highlighting its transformative impact across various industries. Applications ranging from healthcare and manufacturing to smart cities and edge computing are being increasingly integrated with emerging technologies such as AI and IoT, as 5 G is being widely adopted. As detailed throughout the chapter, the competitive positioning of global leaders such as Huawei, Intel, Cisco, Nokia, and Samsung has intensified. Through robust R&D, spectrum acquisition, and collaboration, these players are shaping the 5G era, aiming to meet the growing demands for ultra-fast, low-latency connectivity.

China's global 5G rollout leadership is also evident in its financial impact. Research forecasts that China's 5G industry revenue will reach 0.6 trillion Yuan in 2020, maintaining a growth rate of approximately 10% over the subsequent four years. This growth is supported by increasing users and continued investments in fundamental infrastructure. The 5G ecosystem is rapidly transforming global industries by enabling unprecedented speed, connectivity, and automation. As strategic investments and innovative use cases emerge, companies are reconfiguring their digital infrastructure to fully leverage the potential of 5 G. Key developments—from industrial applications to private networks—highlight the drive toward operational excellence and a competitive edge. Leaders like Intel and DHL exemplify how sector-specific strategies are accelerating the adoption of new technologies. This table outlines the pivotal insights and key events that have influenced the deployment and evolution of 5G worldwide.

Table 3. Strategic insights and events shaping the 5G ecosystem

Topic	Key Insights
Deployment Intentions	By 2020, 66% of businesses planned to deploy 5G, with primary uses including IoT communications and video, with a focus on operational productivity.
Industrial Use Cases	Applications include vehicle-to-vehicle communication, AR-based picking systems, and microchip-level tracking in supply chains.
5G Capabilities	Faster data speeds, ultra-low latency, and network slicing enable better device and system connectivity.
Private 5G Networks	Companies like DHL are exploring private 5G networks for enhanced security, quality, and control.
Intel's Strategy	Intel integrates 5G and AI across its platform—from cloud to devices—focusing on data centers, edge, and automation.
Intel CEO Vision	CEO Bob Swan emphasized the need to leverage Intel's full portfolio to drive AI and 5G innovation, targeting both return on investment (ROI) and industry-specific applications.
AI-Driven Optimization	Intel utilizes Xeon Scalable processors with AI enhancements to support predictive analytics and real-time diagnostics in industries such as healthcare and automotive.
Market Readiness	Most public 5G networks are expected to be fully deployed between 2025 and 2030; early movers will create a competitive advantage through facility-specific deployments.

REFERENCES

Analyst, E. (2025, April 16). *Ericsson's Q1 2025 Revenue Growth: Capitalizing on 5G Demand Amidst Competitive Challenges.* Ainvest. https://www.ainvest.com/news/ericsson-q1-2025-revenue-growth-capitalizing-5g-demand-competitive-challenges-2504

Editorial Team. (2021). *NEC Develops mmWave Distributed Antenna Radio Units to Improve Indoor 5G Performance.* everythingRF. https://www.everythingrf.com/News/details/11611-nec-develops-mmwave-distributed-antenna-radio-units-to-improve-indoor-5g-performance

El-Shorbagy, A. (2021). 5G Technology and the Future of Architecture. *Procedia Computer Science, 182*, 121–131. DOI: 10.1016/j.procs.2021.02.017

European Commission. (2024). *Competition Policy.* Competition-Policy.ec.europa.eu. https://competition-policy.ec.europa.eu/index_en

Fettweis, G. P. (2014). The tactile internet: Applications and challenges. *IEEE Vehicular Technology Magazine, 11*(3), 64–70. https://www.vodafone-chair.org/media/publications/legacy/gerhard-fettweis/The_Tactile_Internet_%E2%80%93_Applications_Challenges.pdf

Fettweis, G. P. (2016). 5G and the future of IoT. In *ESSCIRC Conference 2016: 42nd European Solid-State Circuits Conference* (pp. 21–24). Lausanne, Switzerland: IEEE. https://doi.org/DOI: 10.1109/ESSCIRC.2016.7598234

Huawei. (2025). *What's new.* Huawei. https://www.huawei.com/

Iivari, M. M., Ahokangas, P., Komi, M., Tihinen, M., & Valtanen, K. (2016). Toward Ecosystemic Business Models in the Context of the Industrial Internet. *Journal of Business Models, 4*(2), 42–59. DOI: 10.5278/ojs.jbm.v4i2.1624

IMT. (2020). *5G development in China.* IMT. https://www.itu.int/en/ITU-D/Regional-Presence/AsiaPacific/SiteAssets/Pages/Events/2024/5G%20Engagement%20and%20Experience%20Webinar%20Series/5G-Engagement-and-Experience-Webinar-Series/5G%20Development%20In%20China.pdf

Intelligence, G. S. M. A. (2025). *Global mobile trends: 5G penetration and usage statistics.* GSMA Intelligence. https://www.gsmaintelligence.com/research/global-mobile-trends-2025

Kleinhans, J.-P. (2020). *5G vs. National Security – A European Perspective*. Stiftung Neue Verantwortung. https://www.bundestag.de/resource/blob/908162/2-Stellungnahme-Jan-Peter-Kleinhans-Stiftung-Neue-Verantwortung.pdf

Kovtun, V., Krzysztof Grochla, Saad Aldosary, & Al-Maitah, M. (2024). Analysis of direct traffic at the transport protocol level in the WiMax-1/2 cluster oriented to offload the smart city's wireless ecosystem. *Royal Society Open Science, 11*(7). Advance online publication. DOI: 10.1098/rsos.240206

Maistre, R. L. (2025, March 20). *Don't expect a telco spending recovery in 2025 – Dell'Oro*. TelecomTV. https://www.telecomtv.com/content/access-evolution/don-t-expect-a-telco-spending-recovery-in-2025-dell-oro-52636/

Mavenir. (2020). *Mavenir and Turkcell partner on OpenRAN*. Mavenir. https://www.mavenir.com/press-releases/mavenir-and-turkcell-partner-on-openran/

Moglia, A., Georgiou, K., Marinov, B., Georgiou, E., Berchiolli, R. N., Satava, R. M., & Cuschieri, A. (2022). 5G in healthcare: From COVID-19 to future challenges. *IEEE Journal of Biomedical and Health Informatics, 26*(8), 4187–4196. DOI: 10.1109/JBHI.2022.3181205

Mukherjee, S. (2025, January 30). Nokia beats estimates as demand recovers, shares rise. *Reuters*. https://www.reuters.com/business/media-telecom/nokias-q4-profit-beats-estimates-2025-01-30/

Naqvi, S. I., Naqvi, A. H., Arshad, F., Riaz, M. A., Azam, M. A., Khan, M. S., Amin, Y., Loo, J., & Tenhunen, H. (2019). An integrated antenna system for 4G and millimeter-wave 5G future handheld devices. *IEEE Access : Practical Innovations, Open Solutions, 7*, 116555–116566. DOI: 10.1109/access.2019.2936513

Okutsu, A. (2019, June 10). *South Korea, US and Japan lead in 5G network deployment*. Nikkei Asia. https://asia.nikkei.com/Business/Business-trends/South-Korea-US-and-Japan-lead-in-5G-network-deployment

Patel, S., Chauhan, M., & Kapadiya, K. (2012). 5G: Future mobile technology - Vision 2020. *International Journal of Computer Applications, 54*(17), 6–10. DOI: 10.5120/8656-2264

Reuters. (2018). *Seeking new growth drivers, Samsung plans $22 billion spending on new tech*. Yahoo Finance. https://finance.yahoo.com/news/samsung-group-spend-22-billion-ai-5g-auto-050941607--finance.html

Salahdine, F., Han, T., & Zhang, N. (2022). Security in 5G and beyond recent advances and future challenges. *Security and Privacy, 6*(1). Advance online publication. DOI: 10.1002/spy2.271

Samsung (2024). *Investor Relations | Samsung*. Samsung Global. https://www.samsung.com/global/ir/

Sharma, D. C. (2023). *European Telecommunications Standards Institute (ETSI)*. sesei.eu. https://sesei.eu/wp-content/uploads/2023/05/All-about-ETSI_April-2023.pdf

Strategy Analytics. (2019). Predictive Analysis | Intelligent Analysis | Strategic Market Research | Strategy Analytics. https://www.strategyanalytics.com/

U.S. Department of Commerce. (2022). *U.S. Department of Commerce and Balkans Work Toward a Resilient and Secure 5G Ecosystem*. Commerce.gov. https://www.commerce.gov/news/blog/2022/07/us-department-commerce-and-balkans-work-toward-resilient-and-secure-5g-ecosystem

ZTE. (2023). *Focusing on GIS, ZTE Announced the 5 Gen of FWA Family at MWC 2023*. ZTE Devices. https://global.ztedevices.com/blogs/news/focusing-on-gis-zte-announced-the-5-gen-of-fwa-family-at-mwc-2023

ADDITIONAL READING

BEREC. (2025). *BEREC Report on the 5G Ecosystem*. BEREC. https://www.berec.europa.eu/en/document-categories/berec/reports/berec-report-on-the-5g-ecosystem

Brake, D. (2020). *A U.S. National Strategy for 5G and Future Wireless Innovation*. ITIF. https://itif.org/publications/2020/04/27/us-national-strategy-5g-and-future-wireless-innovation/

Campbell, K., Morelli, B., & O'neil, B. (2017). *IHS Economics / IHS Technology The 5G economy: How 5G technology will contribute to the global economy*. Qualcomm. https://www.qualcomm.com/content/dam/qcomm-martech/dm-assets/documents/5g_economy_ihs_economic_impact_analysis_january_2017.pdf

Center for a New American Security. (2019). *Securing Our 5G Future*. Cnas.org. https://www.cnas.org/publications/reports/securing-our-5g-future

CISA. (2020). *CISA 5G Strategy: Ensuring the Security and Resilience of 5G Infrastructure*. U.S. Department of Homeland Security. https://www.cisa.gov/sites/default/files/publications/cisa_5g_strategy_508.pdf

Cobb, M. (2024). *5G Security: Everything You Should Know For a Secure Network*. Networking. https://www.techtarget.com/searchnetworking/tip/5G-security-Everything-you-should-know-for-a-secure-network

Consulting.us. (2019). *Global 5G rollout expected to reach one in seven by 2025.* Consulting.us. https://www.consulting.us/news/2006/global-5g-rollout-expected-to-reach-one-in-seven-by-2025

DeMerse, M. (2025). *The Top 10 5G Events for Government in 2025.* Carahsoft. https://www.carahsoft.com/blog/carahsoft-the-top-10-5g-events-for-government-blog-2025

DeVine, M., & Gallagher, J. C. (2019). *Fifth-Generation (5G) Telecommunications Technologies: Issues for Congress.* Congress.gov. https://crsreports.congress.gov/product/pdf/r/r45485

Forge, S., & Vu, K. (2020). Forming a 5G strategy for developing countries: A note for policy makers. *Telecommunications Policy, 44*(7), 101975. DOI: 10.1016/j.telpol.2020.101975

GSA. (2023). *Acquisition Guidance for Procuring 5G Technology.* GSA. https://buy.gsa.gov/api/system/files/documents/GSA%20Acquisition%20Guidance%20for%20Procuring%205G%20Technology%20508_0.pdf

GSA. (2023). *GSA supports the National Strategy to Secure 5G with new acquisition guidance.* GSA. https://gsablogs.gsa.gov/technology/2023/03/28/gsa-supports-national-strategy-to-secure-5g-with-new-acquisition-guidance/

GSA. (2025). *5 G-Ecosystem February 2025.* GSA. https://gsacom.com/paper/5g-ecosystem-february-2025/

Martínez-Fernández, A., Vidal, J., Simó-Reigadas, J., Prieto-Egido, I., Agustín, A., Paco, J., & Rendón, Á. (2017). The TUCAN3G Project: Wireless technologies for isolated rural communities in developing countries based on 3G small cell deployments. *IEEE Communications Magazine, 54*(7), 36–43. https://upcommons.upc.edu/server/api/core/bitstreams/95adf79b-a07d-4ab2-8aca-232231811e4a/content

McKinsey & Company. (2021). *The Internet of Things: Catching up to an accelerating opportunity.* McKinsey & Company. https://www.mckinsey.com/~/media/mckinsey/business%20functions/mckinsey%20digital/our%20insights/iot%20value%20set%20to%20accelerate%20through%202030%20where%20and%20how%20to%20capture%20it/the-internet-of-things-catching-up-to-an-accelerating-opportunity-final.pdf

Ngor Luong. (2024). *Forging the 5G future: Strategic imperatives for the US and its allies.* Atlantic Council. https://www.atlanticcouncil.org/in-depth-research-reports/report/forging-the-5g-future-strategic-imperatives-for-the-us-and-its-allies/

PRWEB. (2015). *5G Wireless Ecosystem Forecast Report: 2015-2025*. RF Globalnet. https://www.prweb.com/releases/5g_wireless_ecosystem_forecast_report_2015_2025/prweb12497540.htm

Reportlinker. (2015, January 28). *The 5G Wireless Ecosystem: 2015 - 2025 - Technologies, Applications, Verticals, Strategies & Forecasts*. Prnewswire.com; Cision PR Newswire. https://www.prnewswire.com/news-releases/the-5g-wireless-ecosystem-2015---2025---technologies-applications-verticals-strategies--forecasts-300027177.html

Sundarum, M. (2025). *The State of 5G: Growth, Challenges, and Opportunities in 2025*. 5G Americas. https://www.5gamericas.org/the-state-of-5g-growth-challenges-and-opportunities-in-2025/

Teece, D. J. (2021). Technological leadership and 5G patent portfolios: Guiding strategic policy and licensing decisions. *California Management Review*, *63*(3), 5–34. https://hbsp.harvard.edu/product/CMR765-PDF-ENG

Weisman-Pitts, J. (2025). 5G Technology: A Comprehensive Report on Its Impact by 2025. *Global Banking and Finance Review*. https://www.globalbankingandfinance.com/5g-technology-a-comprehensive-report-on-its-impact-by-2025/

Conclusion

The comprehensive exploration of 5G technology presented in this book highlights its immense transformative potential, reshaping global telecommunications, fostering unprecedented innovation, and driving significant economic growth. As one of the most pivotal advancements in modern technology, 5G stands at the intersection of innovation and necessity, addressing the growing demand for connectivity, efficiency, and scalability across industries and societies.

From its foundational technological innovations to broad applications across diverse sectors, 5G redefines connectivity possibilities. Features like ultra-reliable low-latency communication (URLLC), massive machine-type communication (MTC), and enhanced mobile broadband (eMBB) have expanded the boundaries of what is achievable, enabling advancements in smart cities, autonomous vehicles, healthcare, industrial automation, and much more. 5G facilitates a new digital transformation era by revolutionizing industries' operations, creating opportunities for enhanced productivity, sustainability, and more intelligent decision-making.

The chapters in this book have explored the technical underpinnings of 5G and its deployment challenges, from spectrum allocation and regulatory frameworks to infrastructure development. These hurdles highlight the need for global collaboration among governments, businesses, and technology providers. Equitable access to 5G technology is crucial to bridging the digital divide, empowering underrepresented regions, and ensuring that the benefits of this revolutionary technology are shared universally.

Moreover, 5 G's applications extend beyond mere connectivity. They emphasize the fusion of advanced technologies such as artificial intelligence (AI), the Internet of Things (IoT), and edge computing, unlocking transformative capabilities in real-time data processing, predictive analytics, and resource optimization. Whether improving logistics with vehicle-to-vehicle (V2V) communication, optimizing energy distribution through smart grids, or enhancing customer experiences in financial services, 5G is at the heart of innovation and operational excellence.

The competitive dynamics within the 5G ecosystem also underscore its role in shaping global markets. Industry leaders like Huawei, Qualcomm, Samsung, Ericsson, and Nokia are driving progress through strategic investments, groundbreaking research, and collaborative partnerships. Their efforts highlight the competitive yet cooperative nature of the 5G landscape, where regulatory alignment and technological advancements must work in tandem to unlock the full potential of this transformative technology.

In conclusion, 5G is not merely a technological advancement but a fundamental enabler of global progress. Its capacity to connect people, industries, and systems on an unprecedented scale ensures that it remains a cornerstone of the digital age. 5G catalyzes a brighter, more connected, and sustainable future by fostering innovation, addressing critical challenges, and driving economic growth. As stakeholders across the globe continue to leverage the potential of 5G, its transformative impact will undoubtedly shape how we live, work, and connect for generations to come.

TECHNOLOGICAL EVOLUTION AND FOUNDATIONS

The evolution of 5G marks a pivotal moment in telecommunications, advancing connectivity far beyond its predecessors' capabilities. The foundational elements of 5G—ultra-reliable low-latency communication (URLLC), massive machine-type communication (MTC), and enhanced mobile broadband (eMBB)—collectively redefine the possibilities of wireless communication. These innovations enable transformative applications that span industries, supporting the rise of autonomous vehicles, smart cities, industrial IoT ecosystems, and remote healthcare solutions.

One of 5 G's defining features is its ability to offer unparalleled speed and reliability while maintaining scalability. URLLC ensures minimal latency, a critical requirement for applications like autonomous driving and industrial robotics, where real-time data processing is paramount. mMTC facilitates seamless communication among billions of IoT devices, enabling smart city infrastructures and advanced supply chain management. Meanwhile, eMBB supports high-bandwidth applications such as immersive augmented reality (AR) and virtual reality (VR) experiences, further pushing the boundaries of digital interaction.

The infrastructure supporting these capabilities plays a vital role in successfully deploying 5G. Technologies like small cells, edge computing, and advanced antenna systems ensure scalability and efficiency in high-demand environments. However, these advancements also underscore the need for efficient spectrum management. The allocation of low, mid, and high bands is crucial to balancing coverage and capacity, enabling 5G to meet diverse regional and industrial needs. Global collaboration among regulators, network operators, and technology providers is essential to optimize spectrum use and promote equitable access to this transformative technology.

Another cornerstone of 5G's evolution is its seamless integration with legacy technologies such as 3G and 4G, ensuring backward compatibility while setting the stage for a fully digital future. This compatibility facilitates a smoother transition for industries and consumers, reducing adoption barriers and maximizing the technology's reach. Additionally, the rise of 5G-compatible smartphones highlights the intersection of consumer needs and advanced network capabilities. Devices like the Samsung Galaxy S20 and Nokia 8.3 illustrate how manufacturers are innovating to harness 5G's potential, enabling faster speeds, enhanced performance, and richer user experiences.

These technological advancements underscore the collaborative effort required to bring 5G to life. From infrastructure deployment to regulatory frameworks, the evolution of 5G reflects the intersection of innovation, investment, and strategy. This foundation sets the stage for 5G's widespread adoption and positions it as a critical enabler of the digital age, paving the way for a more innovative, connected, and efficient global ecosystem.

Technological Perspectives and Requirements

The book's second section explores the technological advancements that form the backbone of 5G, emphasizing their role in meeting the ever-growing demands for connectivity across diverse sectors. Innovations like network slicing, beamforming, and spectrum virtualization are central to this exploration. Collectively, they enable scalable and high-performance networks capable of addressing complex industrial and societal needs.

Network slicing is revolutionary and allows multiple virtual networks to operate on the same physical infrastructure. This capability is transformative for sectors like healthcare, where critical applications such as remote surgery require ultra-reliable low-latency communication (URLLC). Simultaneously, other slices can support less demanding use cases like smart energy grids, ensuring optimal resource utilization and performance. Network slicing enables service providers to deliver tailored solutions, fostering greater efficiency and customization across industries.

Beamforming, another pivotal innovation, enhances the precision of signal delivery by directing radio waves toward specific devices. This advancement reduces interference, improves connection quality, and optimizes spectrum use, making it a cornerstone for applications in dense urban environments and industrial automation. The targeted nature of beamforming is especially beneficial for IoT ecosystems, where reliable communication between numerous devices is critical for seamless operation.

Spectrum virtualization plays a crucial role in addressing the challenges of spectrum allocation. By dynamically reallocating resources, this technology ensures that bandwidth is distributed efficiently across various use cases. This adaptability supports the diverse requirements of 5G, from low-band spectrum for wide-area coverage to high-band millimeter waves for high-capacity, low-latency applications. The ability to virtualize spectrum resources ensures that 5G networks remain flexible and future-proof, catering to evolving demands.

This section tracks the evolution from 1G to 5G and highlights how each generation has addressed new challenges and opportunities in mobile telecommunications. The leap from essential voice services to data-centric networks and ultra-connected ecosystems underscores 5 G's transformative nature. Its integration with advanced technologies like AI, cloud computing, and IoT accelerates global digital transformation, bridging communication gaps and fostering innovation.

The discussion also emphasizes the critical importance of strategic policies in driving 5G adoption and integration worldwide. Efficient spectrum management, equitable regulatory frameworks, and global collaboration among stakeholders are essential to overcoming deployment barriers. These policies ensure that 5G reaches its full potential, promotes equitable access, reduces the digital divide, and empowers underserved regions.

This section underscores 5 G's transformative power as a catalyst for innovation and connectivity by focusing on these technological perspectives and requirements. It demonstrates how integrating advanced technologies with strategic planning enables 5G to address the complex needs of industries, governments, and societies, paving the way for a more connected and dynamic global future.

The Applications

The third section delves into 5 G's transformative power, illustrating how its advanced capabilities are revolutionizing industries and reshaping the global economy. With gigabit-level data rates, ultra-reliable low-latency communication, and the capacity to connect billions of devices, 5G catalyzes innovation, driving operational efficiency, real-time decision-making, and enhanced customer experiences across various sectors.

Smart Cities represent one of the most compelling applications of 5G technology. By integrating IoT devices and advanced analytics, 5G enables real-time monitoring and management of urban infrastructure, including traffic systems, energy grids, and public safety networks. The high-speed and low-latency connectivity of 5G empowers city planners to deploy innovative solutions that reduce congestion, enhance sustainability, and improve the quality of life for residents. From intelligent lighting systems to waste management optimization, 5G is at the core of creating more livable and efficient urban environments.

In the **Automotive Industry**, 5G is pivotal in enabling autonomous vehicles. 5 G-powered vehicle-to-everything (V2X) communication allows cars to exchange real-time data with other vehicles, traffic infrastructure, and pedestrians. This capability enhances safety, reduces accidents, and optimizes traffic flow. The low-latency and high-reliability features of 5G are critical for autonomous driving, ensuring rapid response times and seamless interaction between vehicles and their surroundings.

5 G is also transforming Logistics And Supply Chain Management. Thanks to its capabilities, 5 G now allows for real-time tracking of shipments, predictive analytics, and advanced warehouse automation. These advancements enable businesses to optimize operations, reduce costs, and improve delivery times, offering a competitive edge in an increasingly dynamic global market.

In **Financial Services**, 5G supports adopting advanced technologies such as blockchain, AI, and real-time data analytics. With its high-speed connectivity, 5G enhances mobile banking experiences, supports instant transaction processing, and strengthens cybersecurity measures. Financial institutions leverage these capabilities to provide seamless and secure services, improving customer trust and satisfaction.

Despite its transformative potential, 5G deployment is not without challenges. High infrastructure costs and cybersecurity risks pose significant barriers. The dense interconnection of devices and networks increases vulnerabilities, requiring robust cybersecurity frameworks to mitigate risks. Additionally, the initial investment needed for 5G infrastructure can be prohibitive for some regions, potentially exacerbating digital divides.

Nevertheless, 5G's adaptability and scalability ensure widespread applicability from small-scale IoT implementations to large-scale industrial automation. Its integration with energy-efficient technologies further highlights its role in promoting sustainability and reducing environmental impact. By enabling more innovative industrial operations, 5G supports global efforts to achieve economic growth while addressing critical sustainability goals.

This section underscores how 5G's capabilities revolutionize industries, create more intelligent systems, and drive innovation. Addressing challenges and maximizing its transformative potential, 5G is a cornerstone for a more connected, efficient, and sustainable future.

Pioneering Research and Innovations

The fourth section of this book delves into the cutting-edge research driving the evolution of 5G, showcasing how technological advancements address the growing complexity of modern communication demands. Breakthroughs such as massive multiple-input multiple-output (MIMO), full-duplex communication, and millimeter-wave systems (mmWave) are redefining the capabilities of wireless networks, enabling unparalleled speed, efficiency, and scalability.

Massive MIMO, a cornerstone of 5G research, leverages large antenna arrays to enhance network capacity and efficiency. This technology allows multiple devices to communicate simultaneously, significantly improving spectral efficiency and ensuring robust connectivity in dense urban areas. Massive MIMO also plays a vital role in supporting data-intensive applications such as augmented reality (AR), virtual reality (VR), and industrial IoT systems by optimizing spatial multiplexing.

Full-duplex communication is another transformative innovation. It enables devices to send and receive data simultaneously on the same frequency band. This advancement doubles the efficiency of spectrum usage, reduces latency, and improves the performance of time-sensitive applications. Full-duplex technology is crucial for real-time remote healthcare and autonomous vehicle communication scenarios, where split-second data exchange can make a critical difference.

Millimeter-wave systems (Mmwave) expand the frequency spectrum available for 5G, enabling higher data transfer rates and supporting the increasing demand for bandwidth. While mmWave offers unprecedented capacity, its shorter range and susceptibility to obstacles require innovative solutions, such as small-cell deployments and advanced beamforming techniques. These systems are instrumental in powering high-speed connections for smart cities, next-generation entertainment platforms, and advanced industrial automation.

Emerging research also focuses on **Energy-Aware Communication** and resource optimization. Minimizing energy consumption is paramount, with 5G expected to connect billions of devices. Cloud-based radio access networks (C-RAN) enable centralized management, reducing operational costs and energy usage. **Spectrum Visualization** further enhances network efficiency by providing real-time insights into spectrum utilization, facilitating dynamic resource allocation to meet varying demands.

These advancements underscore the importance of interdisciplinary research and cross-sector collaboration in unlocking 5G's full potential. For instance, 5G enables seamless communication between vehicles, infrastructure, and pedestrians in the Automotive Sector, improving traffic safety and efficiency. In **Healthcare**, 5G powers telemedicine and remote surgeries, connecting patients and providers in real-time with minimal latency. Integrating 5G with AI and IoT in these fields highlights the transformative impact of collaborative innovation.

Furthermore, the ability to leverage 5G for advanced applications depends on robust research ecosystems that combine expertise from telecommunications, computer science, and engineering. By addressing challenges such as spectrum allocation, energy efficiency, and cybersecurity, researchers are laying the groundwork for 5G's long-term success.

This section illustrates how pioneering research meets the immediate needs of 5G deployment and shapes the technology's future trajectory. By fostering collaboration and innovation, these advancements ensure that 5G remains a dynamic force driving connectivity, productivity, and innovation across industries and societies worldwide.

Competitive Dynamics and Regulatory Frameworks

The fifth section explores the intricate interplay between competition and regulation within the rapidly evolving 5G ecosystem. As industry leaders like Huawei, Qualcomm, Ericsson, and Nokia spearhead technological advancements, their strategies underscore the critical role of innovation, partnerships, and diversification in capturing market leadership. The competitive landscape is marked by the emergence of private 5G networks, the integration of AI and IoT, and the pursuit of cutting-edge applications tailored to the unique demands of various industries.

Private 5 G networks transform traditional connectivity models by offering businesses enhanced control, reliability, and security. These networks allow enterprises to deploy tailored solutions, optimizing manufacturing, logistics, and healthcare operations. For instance, private networks enable real-time monitoring in industrial settings, improve supply chain transparency, and facilitate remote medical consultations with ultra-low latency. Companies like Ericsson and Nokia are at the forefront of developing private network solutions, highlighting their strategic focus on niche markets and enterprise-specific applications.

Integrating **AI and IoT** with 5G infrastructure further elevates its transformative potential. AI-driven analytics optimize network operations by predicting traffic patterns, enhancing resource allocation, and reducing energy consumption. IoT devices powered by 5G enable seamless communication across smart cities, autonomous vehicles, and connected healthcare ecosystems. This convergence of technologies expands 5 G's capabilities and accelerates the adoption of innovative solutions across industries.

Despite these advancements, **Regulatory Compliance** and **Spectrum Management** remain significant hurdles. Allocating spectrum resources is a contentious issue that requires careful coordination among governments, industry stakeholders, and regulatory bodies. Efficient spectrum use is critical for ensuring equitable access and avoiding interference, particularly in densely populated areas. Moreover, the absence of global standardization can impede the seamless deployment of 5G networks, emphasizing the need for collaborative policies that address regional disparities while fostering international alignment.

Collaborative Efforts between policymakers, technology providers, and telecom operators are crucial to overcoming these challenges. Governments play a pivotal role in facilitating spectrum auctions, setting regulatory standards, and incentivizing investments in infrastructure. Partnerships between public and private sectors can drive innovation while ensuring that 5G remains accessible to underserved regions, bridging digital divides and promoting inclusive growth.

The competitive strategies of leading companies also highlight the importance of **Mergers, Acquisitions, And Joint Ventures** in navigating the evolving 5G market. These initiatives enable firms to pool resources, share expertise, and expand their global footprint. For instance, collaborations between telecom operators and cloud service providers are accelerating the development of edge computing solutions, which enhance the performance of latency-sensitive applications.

Adaptability and foresight will be essential for maintaining leadership in the 5G ecosystem as the market matures. Companies that invest in research and development, embrace innovative partnerships, and align their strategies with regulatory frameworks will be best positioned to capitalize on emerging opportunities. At the same time, establishing robust regulatory policies will ensure that 5G's benefits are distributed equitably, driving sustainable growth across economies.

This section underscores the dynamic nature of the 5G ecosystem, where competition and regulation intersect to shape its trajectory. The 5G landscape is poised to deliver transformative connectivity solutions that empower industries, enhance global connectivity, and drive long-term economic progress by fostering collaboration, innovation, and compliance.

A FUTURE SHAPED BY 5G

As we reflect on this book's comprehensive exploration of 5G, we realize that this groundbreaking technology is more than just a leap in telecommunications. It represents a profound transformation across industries, societies, and economies. By bridging technological and societal gaps, 5G serves as the backbone of the digital age, empowering industries to innovate, addressing global challenges, and driving inclusive and sustainable growth.

The most remarkable attribute of 5G lies in its ability to foster connectivity at an unprecedented scale. By supporting ultra-reliable low-latency communication (URLLC), massive machine-type communication (MTC), and enhanced mobile broadband (eMBB), 5G enables the seamless integration of billions of devices, creating a more intelligent and interconnected world. This connectivity drives innovation across sectors, from autonomous vehicles and telemedicine to smart cities and energy-efficient industrial processes.

Moreover, 5G catalyzes digital transformation by unlocking new possibilities in artificial intelligence (AI), the Internet of Things (IoT), and advanced analytics. These synergies empower businesses to optimize operations, governments to enhance public services, and individuals to experience unparalleled convenience and efficiency. From predictive maintenance in manufacturing to real-time healthcare diagnostics, 5G lays the foundation for a brighter and more sustainable future.

Economic Development and Growth

The economic potential of 5G is equally transformative. Fostering innovation and operational efficiency, 5G drives productivity and opens avenues for entirely new business models. Private networks enable enterprises to tailor connectivity solutions to their unique needs, while logistics, energy, and financial services advancements create ripple effects throughout the global economy. It is estimated that the widespread adoption of 5G could add trillions of dollars to the worldwide GDP, underscoring its role as a key driver of economic growth.

At the same time, the equitable deployment of 5G is critical for bridging digital divides and ensuring its benefits reach underserved and developing regions. Strategic investments in infrastructure and spectrum management and collaborative efforts between governments and private stakeholders are essential for fostering inclusivity and creating opportunities for all.

Addressing Challenges and Building Resilience

While 5 G promises immense promise, challenges remain. Deploying 5G networks demands significant infrastructure investments, strategic spectrum allocation, and robust cybersecurity measures. Addressing these challenges requires global collaboration and forward-thinking policies prioritizing innovation and security. By establishing regulatory frameworks that encourage competition and standardization, stakeholders can ensure that 5G evolves in a secure, scalable, and sustainable manner.

A Vision for the Future

The transformative impact of 5G extends beyond technological innovation; it redefines how humanity interacts, collaborates, and progresses. It provides the foundation for tackling the world's most pressing challenges, from climate change and public health to economic inequality and global connectivity. By harnessing the potential of 5G, societies can create a future where technological progress aligns with environmental stewardship, economic inclusivity, and societal well-being.

This book concludes with a vision of 5G as a cornerstone of the digital era, paving the way for a brighter, more connected, and more sustainable world. By embracing 5 G's possibilities, industries, governments, and individuals can drive meaningful progress that benefits future generations. In doing so, 5G is not merely a technological milestone but a transformative force shaping the future of global connectivity and innovation.

Compilation of References

5G. co.uk. (2025). *The 5G Comparison Site*. 5g.co.uk. https://5g.co.uk/

Adewale, T., & Paul, J. (2024). *AI, 5G, and IoT: How These Technologies Are Creating the Perfect Storm for Smart Systems*. ResearchGate. https://www.researchgate.net/publication/385855348_AI_5G_and_IoT_How_These_Technologies_Are_Creating_the_Perfect_Storm_for_Smart_Systems

Admin. (2020). *The Evolution of Mobile Networks: From 1G to 5G – Centech 5G Hub*. Centech5ghub.com. https://centech5ghub.com/the-evolution-of-mobile-networks-from-1g-to-5g/

Admin1 (2020). *The best tips for choosing a broker*. Ict-Tropic.eu. https://www.ict-tropic.eu/

Agarwal, A., Agarwal, K., Agarwal, S., & Misra, G. (2019). 34-37 towards 5G Networks and Challenges. *American Journal of Electrical and Electronic Engineering, 7*(2), 34–37. https://doi.org/(NGMN Alliance, 2016)DOI: 10.12691/ajeee-7-2-2

Agiwal, M., Kwon, H., Park, S., & Jin, H. (2021). A Survey on 4G-5G Dual Connectivity: Road to 5G Implementation. *IEEE Access: Practical Innovations, Open Solutions, 9*, 16193–16210. DOI: 10.1109/ACCESS.2021.3052462

Agiwal, M., Roy, A., & Saxena, N. (2016). Next Generation 5G Wireless Networks: A Comprehensive Survey. *IEEE Communications Surveys and Tutorials, 18*, 1617–1655. DOI: 10.1109/COMST.2016.2532458

Agrawal, A., & Cleland-Huang, J. (2021). *RescueAR: Augmented Reality Supported Collaboration for UAV Driven Emergency Response Systems*. arXiv preprint arXiv:2110.00180.

Agrawal, V., Agrawal, S., Bomanwar, A., Dubey, T., & Jaiswal, A. (2023). Exploring the risks, benefits, advances, and challenges in internet integration in medicine with the advent of 5G technology: A comprehensive review. *Cureus*, *15*(11), e48767. DOI: 10.7759/cureus.48767

Ahmad, I., Shahabuddin, S., Kumar, T., Okwuibe, J., Gurtov, A., & Ylianttila, M. (2019). Security for 5G and Beyond. *IEEE Communications Surveys and Tutorials*, *21*(4), 1–1. DOI: 10.1109/comst.2019.2916180

Ahmed, S. F., Alam, M. S. B., Afrin, S., Rafa, S. J., Taher, S. B., Kabir, M., ... & Gandomi, A. H. (2024). Toward a secure 5G-enabled internet of things: A survey on requirements, privacy, security, challenges, and opportunities. IEEE Access, 12, 13125-13145.

Ahmed, J., Razzaque, M. A., Rahman, M. M., Alqahtani, S. A., & Hassan, M. M. (2022). A Stackelberg Game-Based Dynamic Resource Allocation in Edge Federated 5G Network. *IEEE Access: Practical Innovations, Open Solutions*, *10*, 10460–10471. DOI: 10.1109/access.2022.3144960

Ahokangas, P., Matinmikko-Blue, M., Yrjölä, S., Seppänen, V., Hämmäinen, H., Jurva, R., & Latva-aho, M. (2019). Business Models for Local 5G Micro-Operators. *IEEE International Symposium on Dynamic Spectrum Access Networks (DySPAN)*, 241–250. https://doi.org/DOI: 10.1109/TCCN.2019.2902547

Akyildiz, I. F., Lee, W.-Y., Vuran, M. C., & Mohanty, S. (2006). Next generation/dynamic spectrum access/cognitive radio wireless networks: A survey. *Computer Networks*, *50*(13), 2127–2159. DOI: 10.1016/j.comnet.2006.05.001

Al-Dujaili, M. J., & Al-dulaimi, M. A. (2022). Fifth-Generation Telecommunications Technologies: Features, Architecture, Challenges and Solutions. *Wireless Personal Communications*. DOI: 10.1007/s11277-022-09962-x

Alhuseini, M. U., & Olama, M. M. (2019). 5G service value chain and network slicing framework using ecosystem modeling, agile delivery, and user-story automation. *IEEE Access : Practical Innovations, Open Solutions*, *7*, 110856–110873. DOI: 10.1109/access.2019.2933193

Alkhatib, H., Faraboschi, P., Frachtenberg, E., Kasahara, H., Lange, D., Laplante, P., Merchant, A., Milojicic, D., Schwan, K., Alquraishi, M., Burgess, A., Forsyth, D., Iwata, H., Mcgeer, R., & Walz, J. (n.d.). *IEEE CS 2022 Report read on*. https://ieeecs-media.computer.org/assets/pdf/2022Report.pdf

Allied Market Research. (2025). *Wireless Infrastructure Market Update 2025.* OpenPR. https://www.openpr.com/news/4020263/wireless-infrastructure-market-update-2025-projected

Amali, C., & Ramachandran, B. (2018). Enabling Key Technologies and Emerging Research Challenges Ahead of 5G Networks: An Extensive Survey. *JOIV: International Journal on Informatics Visualization*, *2*(3), 133. DOI: 10.30630/joiv.2.3.128

Analyst, E. (2025, April 16). *Ericsson's Q1 2025 Revenue Growth: Capitalizing on 5G Demand Amidst Competitive Challenges.* Ainvest. https://www.ainvest.com/news/ericsson-q1-2025-revenue-growth-capitalizing-5g-demand-competitive-challenges-2504

Andrews, J. G., Buzzi, S., Choi, W., Hanly, S. V., Lozano, A., Soong, A. C. K., & Zhang, J. C. (2014). What will 5G be? *IEEE Journal on Selected Areas in Communications*, *32*(6), 1065–1082. DOI: 10.1109/JSAC.2014.2328098

Apruzzese, G., Colajanni, M., Ferretti, L., & Marchetti, M. (2023). 5G and Beyond: The Role of Emerging Technologies in Logistics and Supply Chain Management. *Sustainability*, *15*(15), 11846. DOI: 10.3390/su151511846

AT&T Business Editorial Team. (2018). *5 Ways 5G will Transform Healthcare | Improving patient experience with personalized, preventative care*. AT&T Business. https://www.business.att.com/learn/updates/how-5g-will-transform-the-healthcare-industry.html

Awwwards. (2019). *How 5G Will Change the Mobile App Development Landscape*. Awwwards.com. https://www.awwwards.com/5g-web-design-mobile.html

Azari, M. (2022). *Cellular, Wide-Area, and Non-Terrestrial IoT: A Survey on 5G Advances and the Road Toward 6G*. IEEE Communications Surveys & Tutorials; Institute of Electrical and Electronics Engineers (IEEE). https://www.academia.edu/100346983/Cellular_Wide_Area_and_Non_Terrestrial_IoT_A_Survey_on_5G_Advances_and_the_Road_To

Balador, A., Kouba, A., Cassioli, D., Foukalas, F., Severino, R., Stepanova, D., Agosta, G., Xie, J., Pomante, L., Mongelli, M., Pierini, P., Petersen, S., & Sukuvaara, T. (2018). Wireless Communication Technologies for Safe Cooperative Cyber Physical Systems. *Sensors (Basel)*, *18*(11), 4075. DOI: 10.3390/s18114075

Banafaa, M., Ömer, P., Shayea, I., Alhammadi, A., Shamsan, Z., Razaz, M. A., Alsagabi, M., & Al-Sowayan, S. (2024). A Comprehensive Survey on 5 G-and-Beyond Networks With UAVs. *IEEE Access: Practical Innovations, Open Solutions*, *12*, 7786–7826. DOI: 10.1109/access.2023.3349208

Banafaa, M., Ömer, P., Shayea, I., Alhammadi, A., Shamsan, Z., Razaz, M. A., Majid, A., & Al-Sowayan, S. (2024). A comprehensive survey on 5 G-and-beyond networks with UAVs: Applications, emerging technologies, regulatory aspects, research trends, and challenges. *IEEE Access : Practical Innovations, Open Solutions*, *12*, 7786–7826. DOI: 10.1109/ACCESS.2023.3349208

Bandyopadhyay, S., Jain, P., Leing, J., & Prisacaru, S. (2023) *Navigating the three horizons of 5G business building*. McKinsey & Company. https://www.mckinsey.com/industries/technology-media-and-telecommunications/our-insights/navigating-the-three-horizons-of-5g-business-building

Ban, Y.-L., Li, C., Sim, C.-Y.-D., & Wong, K.-L. (2016). 4G/5G multiple antennas for future multi-mode smartphone applications. *IEEE Access: Practical Innovations, Open Solutions*, *4*, 2981–2988. DOI: 10.1109/ACCESS.2016.2582786

Bhushan, N., Li, J., Malladi, D., Gilmore, R., Brenner, D., Damnjanovic, A., Sukhavasi, R., Patel, C., & Geirhofer, S. (2014). Network densification: The dominant theme for wireless evolution into 5 G. *IEEE Communications Magazine*, *52*(2), 82–89. DOI: 10.1109/MCOM.2014.6736747

Bishen, A. (2019). *5G and the Cloud.* 5G Americas. https://www.5gamericas.org/5g-and-the-cloud

Bishen, A. (2019). *5G and the Cloud.* 5G Americas. https://www.5gamericas.org/5g-and-the-cloud/

Boccardi, F., Heath, R. W., Lozano, A., Marzetta, T. L., & Popovski, P. (2014). Five disruptive technology directions for 5G. *IEEE Communications Magazine*, *52*(2), 74–80. DOI: 10.1109/mcom.2014.6736746

Boston Consulting Group. (2025). *Welcome to BCG.* BCG. https://www.bcg.com/

Broadcom Inc. (2023). *Enabling AI Infrastructure*. Broadcom. https://www.broadcom.com

Campbell, K., Morelli, B., & O'neil, B. (2017). *The 5G economy: How 5G technology will contribute to the global economy IHS Economics / IHS Technology*. https://cdn.ihs.com/www/pdf/IHS-Technology-5G-Economic-Impact-Study.pdf

Carrillo, D., Kalalas, C., Raussi, P., Michalopoulos, D. S., Rodriguez, D. Z., Kokkoniemi-Tarkkanen, H., Ahola, K., Nardelli, P. H. J., Fraidenraich, G., & Popovski, P. (2023). Boosting 5G on Smart Grid Communication: A Smart RAN Slicing Approach. *IEEE Wireless Communications*, *30*(5), 170–178. DOI: 10.1109/MWC.004.2200079

Cawfield, D. (2016) *Manager Insight: Patience pays off in Manulife balancing act.* Morningstar Canada. https://www.investmentexecutive.com/news/products/manager-insight-patience-pays-off-in-manulife-balancing-act/

Chataut, R., & Akl, R. (2020). *Massive MIMO Systems for 5G.* https://doi.org/DOI: 10.32545/encyclopedia202006.0016.v2

Chen, Y., Zhang, H., Chen, J., Guo, Y., Jiang, P., Gao, F., Bao, H., & Huang, X. (2022). Thermally Conductive but Electrically Insulating Polybenzazole Nanofiber/Boron Nitride Nanosheets Nanocomposite Paper for Heat Dissipation of 5G Base Stations and Transformers. *ACS Nano, 16*(9), 14323–14333. DOI: 10.1021/acsnano.2c04534

Chhabra, A. & Rogers Communications Inc. (2023). *Convergence With 5G Applications and Use Case Monetization (2023).* SCTE TechExpo. https://www.nctatechnicalpapers.com/Paper/2023/3629_Chhabra_5105_paper

Chiang, M., & Zhang, T. (2016). Fog and IoT: An overview of research opportunities. *IEEE Internet of Things Journal, 3*(6), 854–864. DOI: 10.1109/JIOT.2016.2584538

Chin, W. H., Fan, Z., & Haines, R. (2014). Emerging technologies and research challenges for 5G wireless networks. *IEEE Wireless Communications, 21*(2), 106–112. DOI: 10.1109/MWC.2014.6812298

Chowdhury, M. Z., Fujii, T., Muntean, G.-M., Choi, J.-W., & Araniti, G. (2018). Emerging Small Cell Wireless Technologies for 5G: Architectures and Applications. *Wireless Communications and Mobile Computing, 1–2.* DOI: 10.1155/2018/8969264

Chui, M., Collins, M., & Patel, M. (2021). *The Internet of Things: Catching up to an accelerating opportunity.* McKinsey & Company. https://www.mckinsey.com/~/media/mckinsey/business%20functions/mckinsey%20digital/our%20insights/iot%20value%20set%20to%20accelerate%20through%202030%20where%20and%20how%20to%20capture%20it/the-internet-of-things-catching-up-to-an-accelerating-opportunity-final.pdf

Cimini, C., Pirola, F., & Cavalieri, S. (2023). Challenges and expectations about 5G in the industry: Evidence from survey-based research. *International Journal of Manufacturing Technology and Management, 37*(5/6), 492–511. DOI: 10.1504/ijmtm.2023.133688

Cisco. (2020). *Cisco Annual Internet Report (2018–2023).* https://www.cisco.com/c/en/us/solutions/collateral/executive-perspectives/annual-internet-report/white-paper-c11-741490.pdf

Cisco. (2020). *Cisco Annual Internet Report (2018–2023) White Paper*. Cisco. https://www.cisco.com/c/en/us/solutions/collateral/executive-perspectives/annual-internet-report/white-paper-c11-741490.html

Colin. (2022, April 12). *No Moment's too big for all-new Lexus LX 600 SUV*. Reel 360 News. https://reel360.com/article/no-moments-too-big-for-all-new-lexus-lx-600-suv/

CoolUtils File Converters. (2025). *Our File Converters*. Coolutils.com. https://www.coolutils.com/

Cover, T. M., & Thomas, J. A. (2005). *Elements of Information Theory*. John Wiley & Sons, Inc., DOI: 10.1002/047174882x

Cui, Y., Liu, P., Zhou, Y., & Duan, W. (2022). Energy-Efficient Resource Allocation for Downlink Non-Orthogonal Multiple Access Systems. *Applied Sciences (Basel, Switzerland)*, *12*(19), 9740. DOI: 10.3390/app12199740

Dahlman, E., Parkvall, S., & Sköld, J. (2014). *4G: LTE/LTE-Advanced for mobile broadband*. Elsevier.

Dai, L., Wang, B., Yuan, Y., Han, S., Chih-Lin, I., & Wang, Z. (2015). Non-orthogonal multiple access for 5G: Solutions, challenges, opportunities, and future research trends. *IEEE Communications Magazine*, *53*(9), 74–81. DOI: 10.1109/MCOM.2015.7263349

De Looper, C. (2019). *What Is 5G? Digital Trends*. Yale University. https://archive-yaleglobal.yale.edu/content/what-5g-digital-trends

de Pablos, O., & Patricia, Z. Xi (2023). 5G, Artificial Intelligence, and Next Generation Internet of Things: Digital Innovation for Green and Sustainable Economies. IGI Global.

Deloitte. (2025). *5G: The chance to lead for a decade*. 5G Americas. https://www.5gamericas.org/5g-the-chance-to-lead-for-a-decade/

Deutsche Telekom. (2019). *Highlights - Deutsche Telekom AG Interim Report 2019*. Deutsche Telekom. https://report.telekom.com/interim-report-q2-2019/at-a-glance/highlights.html

DIGI. (20250. *Go-To 5G Connectivity*. Digi International Inc. https://www.digi.com/solutions/by-technology/5g

Ding, A. Y., & Janssen, M. (2018). *Opportunities for Applications Using 5G Networks: Requirements, Challenges, and Outlook.* In *ICTRS '18: Proceedings of the Seventh International Conference on Telecommunications and Remote Sensing* (pp. 27–34). https://doi.org/DOI: 10.1145/3278161.3278166

Ding, Z., Lei, X., Karagiannidis, G. K., Schober, R., Yuan, J., & Bhargava, V. K. (2017). A Survey on Non-Orthogonal Multiple Access for 5G Networks: Research Challenges and Future Trends. *IEEE Journal on Selected Areas in Communications, 35*(10), 2181–2195. DOI: 10.1109/jsac.2017.2725519

Dixit, E., & Rani, S. (2025). *Integrated Trends, Opportunities, and Challenges of 5G and Internet of Things.* In *Current and Future Cellular Systems*, IEEE, 139–152. https://doi.org/DOI: 10.1002/9781394256075.ch7

Douglas, S. (2025). *5G Halftime Update: Spirent's 2025 5G Report.* Ospirent Communications. https://www.spirent.com/blogs/5g-halftime-update-spirents-2025-5g-report

Dryjanski, M. (2019). *Do we need next G's?* Grandmetric. https://www.grandmetric.com/wp-content/uploads/2019/10/UnifiedNetworks_World5GSummit.pdf?utm_source=chatgpt.com

Dutta, A., & Hammad, E. (2020). *5G security challenges and opportunities: A system approach.* In 2020 IEEE 3rd 5G World Forum (5GWF) (pp. 109–114). IEEE. https://doi.org/DOI: 10.1109/5GWF49715.2020.9221122

Editorial Team. (2021). *NEC Develops mmWave Distributed Antenna Radio Units to Improve Indoor 5G Performance.* everythingRF. https://www.everythingrf.com/News/details/11611-nec-develops-mmwave-distributed-antenna-radio-units-to-improve-indoor-5g-performance

Egea-Roca, D., Arizabaleta-Diez, M., Pany, T., Antreich, F., Lopez-Salcedo, J. A., Paonni, M., & Seco-Granados, G. (2022). GNSS User Technology: State-of-the-Art and Future Trends. *IEEE Access : Practical Innovations, Open Solutions, 10*, 39939–39968. DOI: 10.1109/access.2022.3165594

EHTAdmin. (2020). *Statement on Health Effects of 5G by Anthony B. Miller MD.* Environmental Health Trust. https://ehtrust.org/statement-on-health-effects-of-5g-by-anthony-b-miller/

Ellwood, D. W. (2020). The unreliable wisdom of hindsight. *The Political Quarterly, 91*(1), 243–245. DOI: 10.1111/1467-923x.12801

El-Shorbagy, A. (2021). 5G Technology and the Future of Architecture. *Procedia Computer Science, 182*, 121–131. DOI: 10.1016/j.procs.2021.02.017

Ericsson (2023). *Ericsson Mobility Report.* Ericsson.com. https://d110erj175o600.cloudfront.net/wp-content/uploads/2023/06/21123003/Ericsson-Mobility-Report-June-2023.pdf

Ericsson. (2019). *5G for business: a 2030 market compass. Setting a direction for 5G-powered B2B opportunities.* Ericsson.com. https://d110erj175o600.cloudfront.net/wp-content/uploads/2019/10/the-5g-for-business-a-2030-compass-report-2019.pdf

Ericsson. (2020). *3GPP releases 16 & 17 overview – 5G NR evolution.* Eriicsson.com. Retrieved from https://www.ericsson.com/en/reports-and-papers/ericsson-technology-review/articles/5g-nr-evolution

Ericsson. (2024a). *5G device outlook.* Ericsson.com. https://www.ericsson.com/en/reports-and-papers/mobility-report/dataforecasts/5g-device-outlook

Ericsson. (2024b). *5G network coverage outlook.* Ericsson.com. https://www.ericsson.com/en/reports-and-papers/mobility-report/dataforecasts/network-coverage

Ericsson. (2024c). *Exploring how traffic patterns drive network evolution.* Ericsson.com. https://www.ericsson.com/en/reports-and-papers/mobility-report/articles/traffic-patterns-drive-network-evolution

Ericsson. (2024d). *Mobile data traffic forecast.* Ericsson.com. https://www.ericsson.com/en/reports-and-papers/mobility-report/dataforecasts/mobile-traffic-forecast

Ericsson. (2024e). *Mobile subscriptions outlook.* Ericsson.com. https://www.ericsson.com/en/reports-and-papers/mobility-report/dataforecasts/mobile-subscriptions-outlook

Ericsson. (2025). *Ericsson Fixed Wireless Access Handbook 2025. GSA.* https://gsacom.com/paper/ericsson-fixed-wireless-access-handbook-2025/

Ericsson. (2025). *Ericsson Mobility Report | Read the latest edition. Ericsson.com.* https://www.ericsson.com/en/mobility-report/reports/june-2020/fwa-outlook

ETSI. (2017). *5G; Study on Scenarios and Requirements for Next Generation Access Technologies (3GPP TR 38.913 version 14.2.0 Release 14).* 3GPP. https://www.etsi.org/deliver/etsi_tr/138900_138999/138913/14.02.00_60/tr_138913v140200p.pdf

European Commission. (2024). *Competition Policy.* Competition-Policy.ec.europa.eu. https://competition-policy.ec.europa.eu/index_en

Evans, A. (2017). *Intel accelerates the future with the world's first global 5G modem.* Microwave Journal. https://www.microwavejournal.com/articles/27684-intel-accelerates-the-future-with-worlds-first-global-5g-modem

Evercom (2025). *5G LTE Antenna.* Evercomtech.com. https://www.evercomtech.com/5g-lte-antenna

Everipedia. (2018). *5G.* Everipedia. https://everipedia.io/wiki/5g/

Explained, E. M. F. (2019) *5G and EMF Explained.* AMTA. https://emfexplained.info/site/misc/emf/downloads/5G&EMF%20Explained_AMTA_23Aug_2019_20.pdf

Fettweis, G. P. (2014). The tactile internet: Applications and challenges. *IEEE Vehicular Technology Magazine, 11*(3), 64–70. https://www.vodafone-chair.org/media/publications/legacy/gerhard-fettweis/The_Tactile_Internet_%E2%80%93_Applications_Challenges.pdf

Fettweis, G. P. (2016). 5G and the future of IoT. In *ESSCIRC Conference 2016: 42nd European Solid-State Circuits Conference* (pp. 21–24). Lausanne, Switzerland: IEEE. https://doi.org/DOI: 10.1109/ESSCIRC.2016.7598234

Fieldfisher (2020). *Impact of 5G on AI and industrial IoT applications.* Fieldfisher.com. https://res.cloudinary.com/fieldfisher/image/upload/v1610106432/PDF-Files/The_impact_of_5G_on_AI_December_2020_glpw1r.pdf

Finley, K. (2018). *What is 5G? The Complete Guide to When, Why, and How.* Wired; WIRED. https://www.wired.com/story/wired-guide-5g/

Fischler, J. (2020, January 6). *Where is 5G available in the US? The biggest providers, including Verizon and T-Mobile, all offer 5G.* (Updated for 2025). Lifewire. https://www.lifewire.com/5g-availability-us-4155914

FlytBase. (2025). *Drone for Disaster Management – How Drones are Used for Emergency Response.* Www.flytbase.com. https://www.flytbase.com/blog/drone-disaster-relief

Fogarty, K. (2019). GaN Versus Silicon For 5 G. *Semiconductor Engineering.* https://semiengineering.com/gan-versus-silicon-for-5g/

Forge, S., & Vu, K. (2020). Forming a 5G strategy for developing countries: A note for policy makers. *Telecommunications Policy, 44*(7), 101975. DOI: 10.1016/j.telpol.2020.101975

Fowdur, T. P., Indoonundon, M., & Milovanovic, D. A. (2024). *5G NR modelling in MATLAB: Network architecture, protocols, and physical layer.* MathWorks. https://www.mathworks.com/academia/books/5g-nr-modelling-in-matlab-fowdur.html

FP7-Moto. (2025). *FP7 Moto.* FP7-Moto.eu. https://fp7-moto.eu/

Frieden, R. (2019). The evolving 5G case study in spectrum management and industrial policy. *Telecommunications Policy, 43*(9), 101826. DOI: 10.1016/j.telpol.2019.04.001

Ghassemlooy, Z., Arnon, S., Uysal, M., Xu, Z., & Cheng, J. (2015). Emerging optical wireless communications—Advances and challenges. *IEEE Journal on Selected Areas in Communications, 33*(9), 1738–1749.

Global Mobile Suppliers Association (GSA). *(2015). The Road to 5G: Drivers, Applications, Requirements and Technical Development.* https://gsacom.com/content/uploads/2015/11/151118-GSA-the-Road-to-5G.pdf

Gnow (2022*) 5gnow.eu – 5G mobile phones, tariffs and flat rates.* 5Gnow. https://www.5gnow.eu/

Goasduff, L. (2021). *Gartner Forecasts Worldwide 5G Network Infrastructure Revenue to Grow 39% in 2021.* Gartner. https://www.gartner.com/en/newsroom/press-releases/2021-08-04-gartner-forecasts-worldwide-5g-network-infrastrucutre-revenue-to-grow-39pc-in-2021

GPP – The Mobile Broadband Standard Partnership Project. (n.d.). 3GPP. https://www.3gpp.org

GPP. (2020). *About 3GPP*. 3gpp.org. https://www.3gpp.org/about-3gpp

GPP. (2020). *Study on non-orthogonal multiple access (NOMA) for NR (TR 38.812).* 3rd Generation Partnership Project. https://www.3gpp.org/DynaReport/38812.htm

Greene, B. (2018, December 9). *5G Latency - Reality Checks.* SENKI. https://www.senki.org/5g-latency-reality-checks/

Grey, B. (2025). Understanding the current innovation game in 5G V2X domain. *Greyb.* https://www.greyb.com/blog/5g-v2x-communication-research/

Gröne, F., & Ayyagari, D. (2023). *Capturing more value from telecoms' connectivity services The state of 5G.* PwC. https://www.pwc.com/gx/en/industries/tmt/telecommunications/the-state-of-5G-capturing-more-value.html

GSA. (2025). *5G Market Snapshot February 2025.* GSA. https://gsacom.com/paper/5g-market-snapshot-february-2025/

GSMA. (2020). *5G Spectrum Requirements and Deployment Strategies.* GSMA Intelligence. https://www.gsma.com

GSMA. (2025). *Spectrum Policy Trends 2025*. GSMA> https://www.gsma.com/connectivity-for-good/spectrum/wp-content/uploads/2025/02/GSMA-Spectrum-Policy-Trends-2025-1.pdf

GSMA. (2025). *The Mobile Economy 2025*. GSMA Intelligence. https://www.gsma.com/solutions-and-impact/connectivity-for-good/mobile-economy/wp-content/uploads/2025/02/030325-The-Mobile-Economy-2025.pdf

Gupta, A., & Jha, R. K. (2015). A Survey of 5G Network: Architecture and Emerging Technologies. *IEEE Access : Practical Innovations, Open Solutions*, *3*(1), 1206–1232. DOI: 10.1109/access.2015.2461602

Gupta, A., & Jha, R. K. (2015). A survey of 5G network: Architecture and emerging technologies. *IEEE Access : Practical Innovations, Open Solutions*, *3*, 1206–1232. DOI: 10.1109/ACCESS.2015.2461602

Haider, F., Wang, C.-X., Haas, H., Yuan, N. D., Wang, N. H., Gao, N. X., You, N. X.-H., & Erol Hepsaydir. (2011). Spectral efficiency analysis of mobile femtocell-based cellular systems. In 2011 *IEEE 13th International Conference on Communication Technology* (pp. 347–351). https://doi.org/DOI: 10.1109/ICCT.2011.6157895

Hajlaoui, E., Zaier, A., Khlifi, A., Ghodhbane, J., Hamed, M. B., & Sbita, L. (2020, September 1). *4G and 5G technologies: A Comparative Study*. IEEE Xplore. https://doi.org/DOI: 10.1109/ATSIP49331.2020.9231605

HandWiki. (2025). 5G. Handwiki.org. https://handwiki.org/wiki/5G

Haneda, K., Tian, L., Asplund, H., Li, J., Wang, Y., Steer, D., Li, C., Balercia, T., Lee, S., Kim, Y., Ghosh, A., Thomas, T., Nakamura, T., Kakishima, Y., Imai, T., Papadopoulas, H., Rappaport, T., Maccartney, G., Samimi, M., & Sun, S. (2016). *Indoor 5G 3GPP-like Channel Models for Office and Shopping Mall Environments*. https://wides.usc.edu/Updated_pdf/haneda2016indoor.pdf

Han, S., I., C. L., Xu, Z., & Rowell, C. (2015). Large-scale antenna systems with hybrid analog and digital beamforming for millimeter-wave 5 G. *IEEE Communications Magazine*, *53*(1), 186–194. DOI: 10.1109/mcom.2015.7010533

Haykin, S. S. (2005). *Modern wireless communications*. Internet Archive. https://archive.org/details/modernwirelessco0000hayk

Heath, R. W., Gonzalez-Prelcic, N., Rangan, S., Roh, W., & Sayeed, A. M. (2016). An Overview of Signal Processing Techniques for Millimeter Wave MIMO Systems. *IEEE Journal of Selected Topics in Signal Processing*, *10*(3), 436–453. DOI: 10.1109/jstsp.2016.2523924

Heath, R. W., Gonzalez-Prelcic, N., Rangan, S., Roh, W., & Sayeed, A. M. (2016). An overview of signal processing techniques for millimeter wave MIMO systems. *IEEE Journal of Selected Topics in Signal Processing*, *10*(3), 436–453. DOI: 10.1109/JSTSP.2016.2523924

Henley Business School. (2021). *Henley research to explore factors impacting Huawei 5G rollout*. Henley Business School. https://www.henley.ac.uk/news/2021/henley-research-to-explore-factors-impacting-huawei-5g-rollout

Holtrup, G., Lacube, W., David, D. P., Mermoud, A., Bovet, G., & Lenders, V. (2021, August 20). *5G System Security Analysis*. ArXiv.org. https://doi.org//arXiv.2108.08700DOI: 10.48550

Horwitz, J. (2019). *The definitive guide to 5G low, mid, and high band speeds*. VentureBeat. https://venturebeat.com/mobile/the-definitive-guide-to-5g-low-mid-and-high-band-speeds/

Hossain, E., & Hasan, M. (2015). 5G cellular: Key enabling technologies and research challenges. *IEEE Instrumentation & Measurement Magazine*, *18*(3), 11–21. DOI: 10.1109/MIM.2015.7108393

Huawei (2020) *METIS: Striding towards 5G - Huawei Publications*. Huawei. https://www.huawei.com/en/huaweitech/publication/73/hw_403726?utm_source=chatgpt.com

Huawei (2020). *Bring 5.5G Reality 5g-Advanced. Huwawei.* https://www-file.huawei.com/-/media/corp2020/media-center/pdf/5g-advanced.pdf

Huawei (2023). *Huawei Launches Full-Series 5.5G Solutions to Bring 5.5G into Reality*. Huawei Technologies Co. https://www.huawei.com/en/news/2023/10/huawei5gsolution-mbbf2023

Huawei. (2025). *What's new*. Huawei. https://www.huawei.com/

Huawei. (2025). *White Paper: 5G Network Architecture - A High-Level Perspective.* HuaweiTech. https://www.huawei.com/en/huaweitech/industry-insights/outlook/mobile-broadband/insights-reports/5g-network-architecture

Iivari, M. M., Ahokangas, P., Komi, M., Tihinen, M., & Valtanen, K. (2016). Toward Ecosystemic Business Models in the Context of the Industrial Internet. *Journal of Business Models*, *4*(2), 42–59. DOI: 10.5278/ojs.jbm.v4i2.1624

IMT. (2020). *5G development in China.* IMT. https://www.itu.int/en/ITU-D/Regional-Presence/AsiaPacific/SiteAssets/Pages/Events/2024/5G%20Engagement%20and%20Experience%20Webinar%20Series/5G-Engagement-and-Experience-Webinar-Series/5G%20Development%20In%20China.pdf

innovationoecd. (2017). *OECD Digital Economy Outlook 2017: Presentation at Global Parliamentary Network, 11 October 2017.* SlideShare; Slideshare. https://www.slideshare.net/slideshow/oecd-digital-economy-outlook-2017-presentation-at-global-parliamentary-network-11-october-2017/80686796

Intelligence, G. S. M. A. (2020). *The Mobile Economy 2020.* GSMA Intelligence. https://www.gsmaintelligence.com/research/the-mobile-economy-2020

Intelligence, G. S. M. A. (2023). *The Mobile Economy 2023.* GSMA Intelligence. https://www.gsma.com/solutions-and-impact/connectivity-for-good/mobile-economy/wp-content/uploads/2023/03/270223-The-Mobile-Economy-2023.pdf

Intelligence, G. S. M. A. (2025). *Global mobile trends: 5G penetration and usage statistics.* GSMA Intelligence. https://www.gsmaintelligence.com/research/global-mobile-trends-2025

International Telecommunication Union Radiocommunication Sector (ITU-R). *(2017). Minimum requirements related to technical performance for IMT-2020 radio interface(s)* (Report ITU-R M.2410-0). Itu.net. https://www.itu.int/pub/R-REP-M.2410-2017

IoT Analytics. (2018). *The Rise of Connected Devices and the Role of 5G Networks.* Retrieved from https://iot-analytics.com

ITU. (2018). *Setting the scene for 5G: Opportunities & challenges.* ITU. https://www.itu.int/en/ITU-D/Documents/ITU_5G_REPORT-2018.pdf

ITU. (2020). *Measuring digital development: Facts and figures 2020.* International Telecommunication Union. https://mpt.gov.by/en/node/6911/printable/pdf#:~:text=ITU%E2%80%99s%20Measuring%20digital%20development%3A%20Facts%20and%20figures%202020,Latest%20figures%20show%20a%20wide%20urban-rural%20connectivity%20gap

ITU. (2021). *Measuring digital development: Facts and figures 2021.* International Telecommunication Union. https://mpt.gov.by/en/node/6911/printable/pdf#:~:text=ITU%E2%80%99s%20Measuring%20digital%20development%3A%20Facts%20and%20figures%202020,Latest%20figures%20show%20a%20wide%20urban-rural%20connectivity%20gap

ITU. (2022). *5G - Fifth generation of mobile technologies*. Www.itu.int. https://www.itu.int/en/mediacentre/backgrounders/Pages/5G-fifth-generation-of-mobile-technologies.aspx

ITU. (2022). *5G - Fifth generation of mobile technologies*. Www.itu.int. https://www.itu.int/en/mediacentre/backgrounders/Pages/5G-fifth-generation-of-mobile-technologies.aspx

ITU-R. (2015). *IMT Vision – Framework and overall objectives of the future development of IMT for 2020 and beyond*. International Telecommunication Union. https://www.itu.int/en/publications/Documents

Jaber, M., Imran, M. A., Tafazolli, R., & Tukmanov, A. (2016). 5G backhaul challenges and emerging research directions: A survey. *IEEE Access : Practical Innovations, Open Solutions*, *4*, 1743–1766. DOI: 10.1109/ACCESS.2016.2556011

Jhon, R. (2025). *What are the frequency bands for FR1 and FR2?* Teletopix.org. https://teletopix.org/what-are-the-frequency-bands-for-fr1-and-fr2/#gsc.tab=0

Jovović, I., Forenbacher, I., & Periša, M. (n.d.). *Massive Machine-Type Communications: An Overview and Perspectives Towards 5G*. https://www.fpz.unizg.hr/ikp/upload/RCITD_2015_Massive%20Machine%20Type%20Communications%20An%20Overview%20and%20Perspectives%20Towards%205G.pdf

Kadena, E., & Rajnai, Z. (2024). 5G supply chain: An overview of applications and challenges. In *2024 IEEE 22nd World Symposium on Applied Machine Intelligence and Informatics (SAMI)* (pp. 000157–000162). IEEE. https://doi.org/DOI: 10.1109/SAMI60510.2024.10432858

Kantenna Technology Limited. (2025). *HOME*. Kantennatech.com. https://www.kantennatech.com/

Kerner, S. M., & Ferguson, K. (2024, December 17). *5 predictions about 5G adoption in 2023 and beyond | TechTarget*. WhatIs.com. https://www.techtarget.com/whatis/feature/5-Predictions-about-5G-Adoption-in-2021-and-Beyond

Keshta, I. (2025). A cloud-assisted key agreement protocol for the E-healthcare system. *PLoS One*, *20*(6), e0322313. DOI: 10.1371/journal.pone.0322313

Kholidy, H. A. (2021). Multi-Layer Attack Graph Analysis in the 5G Edge Network Using a Dynamic Hexagonal Fuzzy Method. *Sensors (Basel)*, *22*(1), 9. DOI: 10.3390/s22010009

Kiesel, R., van Roessel, J., & Schmitt, R. H. (2020). Quantification of economic potential of 5G for latency critical applications in production. *Procedia Manufacturing, 52*, 113–120. DOI: 10.1016/j.promfg.2020.11.021

Kim, E.-S., Sung, M., Lee, J. H., Lee, J. K., Cho, S.-H., & Kim, J. (2020). Coverage Extension of Indoor 5G Network Using RoF-Based Distributed Antenna System. *IEEE Access : Practical Innovations, Open Solutions, 8*, 194992–194999. DOI: 10.1109/access.2020.3033592

Kleinhans, J.-P. (2020). *5G vs. National Security – A European Perspective*. Stiftung Neue Verantwortung. https://www.bundestag.de/resource/blob/908162/2-Stellungnahme-Jan-Peter-Kleinhans-Stiftung-Neue-Verantwortung.pdf

Kovtun, V., Krzysztof Grochla, Saad Aldosary, & Al-Maitah, M. (2024). Analysis of direct traffic at the transport protocol level in the WiMax-1/2 cluster oriented to offload the smart city's wireless ecosystem. *Royal Society Open Science, 11*(7). Advance online publication. DOI: 10.1098/rsos.240206

Kshetri, N. (2014). The emerging role of Big Data in key development issues: Opportunities, challenges, and concerns. *Big Data & Society, 1*(2). DOI: 10.1177/2053951714564227

Kshetri, N. (2017). Can Blockchain Strengthen the Internet of Things? *IT Professional, 19*, 68–72. DOI: 10.1109/mitp.2017.3051335

Lagorio, A., Cimini, C., Pinto, R., & Cavalieri, S. (2023). 5G in Logistics 4.0: Potential applications and challenges. *Procedia Computer Science, 217*, 650–659. DOI: 10.1016/j.procs.2022.12.261

Lagorio, A., Cimini, C., Pirola, F., & Pinto, R. (2021). A Taxonomy of Technologies for Human-Centred Logistics 4.0. *Applied Sciences (Basel, Switzerland), 11*(20), 9661. DOI: 10.3390/app11209661

Larsson, E. G., Edfors, O., Tufvesson, F., & Marzetta, T. L. (2014). Massive MIMO for next-generation wireless systems. *IEEE Communications Magazine, 52*(2), 186–195. https://ieeexplore.ieee.org/document/6736761. DOI: 10.1109/MCOM.2014.6736761

Lee, T. (2024). *Advanced Packaging for 5G in RF and Analog Mixed Signal*. IEEE. https://eps.ieee.org/images/files/HIR_Presentations/HIR2024_5G_Lee_Final.pdf

Lema, M. A., Laya, A., Mahmoodi, T., Cuevas, M., Sachs, J., Markendahl, J., & Dohler, M. (2017). Business Case and Technology Analysis for 5G Low Latency Applications. *IEEE Access : Practical Innovations, Open Solutions*, 1–1. DOI: 10.1109/access.2017.2685687

Le, N. T., Hossain, M. A., Islam, A., Kim, D., Choi, Y.-J., & Jang, Y. M. (2016). Survey of Promising Technologies for 5G Networks. *Mobile Information Systems*, *2016*, 1–25. DOI: 10.1155/2016/2676589

Li, C., Tang, Y., & Luo, F. (2022). Research on Internet Regional Security Based on 5G Wireless Communication Technology. *2022 International Conference on Smart City and Green Energy (ICSCGE)*. https://doi.org/DOI: 10.1109/ICSCGE58152.2022.00008

Li, D. (2019). 5G and Intelligence Medicine—How the Next Generation of Wireless Technology Will Reconstruct Healthcare? *Precision Clinical Medicine*, *2*(4). Advance online publication. DOI: 10.1093/pcmedi/pbz020

Li, G., Zhang, X., & Zhang, G. (2022). How the 5G Enabled the COVID-19 Pandemic Prevention and Control: Materiality, Affordance, and (De-)Spatialization. *International Journal of Environmental Research and Public Health*, *19*(15), 8965.

Lin, B.-S. P., Lin, F. J., & Tung, L.-P. (2016). The Roles of 5G Mobile Broadband in the Development of IoT, Big Data, Cloud and SDN. *Communications and Network*, *08*(01), 9–21. DOI: 10.4236/cn.2016.81002

Li, Y., & Chen, M. (2015). Software-Defined Network Function Virtualization: A Survey. *IEEE Access : Practical Innovations, Open Solutions*, *3*, 2542–2553. DOI: 10.1109/access.2015.2499271

Maistre, R. L. (2025, March 20). *Don't expect a telco spending recovery in 2025 – Dell'Oro*. TelecomTV. https://www.telecomtv.com/content/access-evolution/don-t-expect-a-telco-spending-recovery-in-2025-dell-oro-52636/

Mangra, N., Behmann, F., Thakur, A., Popescu, A., Suciu, G., Jr., Giannattasio, G., Uppal, R., & Montlouis, W. (2023). 5 G-enabled agriculture ecosystem: Food supply chain, rural development, and climate resiliency. *IEEE*. https://ieeexplore.ieee.org/document/10028721

Manivasagam, G., & Shukla, M. (2023). Massive MIMO in 5G Networks: Trends and Challenges. *2023 International Conference on Power Energy, Environment & Intelligent Control (PEEIC)*, 1456–1460. DOI: 10.1109/peeic59336.2023.10450543

Mao, Y., You, C., Zhang, J., Huang, K., & Letaief, K. B. (2017). A survey on mobile edge computing: The communication perspective. *IEEE Communications Surveys and Tutorials*, *19*(4), 2322–2358. DOI: 10.1109/COMST.2017.2745201

Margaris, P. (2019, January 8). *Differentiating With 5G Security: How Mobile Service Providers Can Become Secure Business Enablers*. Palo Alto Networks Blog. https://www.paloaltonetworks.com/blog/2019/01/differentiating-5g-security-mobile-service-providers-can-become-secure-business-enablers/

Marriwala, N. Tripathi, C. C., Kumar, D., & Jain, S. (2020). Mobile Radio Communications and 5G Networks. In *Lecture notes in networks and systems*. Springer International Publishing. https://doi.org/DOI: 10.1007/978-981-15-7130-5

Martínez-Fernández, A., Vidal, J., Simó-Reigadas, J., Prieto-Egido, I., Agustín, A., Paco, J., & Rendón, Á. (2016). The TUCAN3G Project: Wireless technologies for isolated rural communities in developing countries based on 3G small cell deployments. *IEEE Communications Magazine*, *54*(7), 36–43. https://upcommons.upc.edu/server/api/core/bitstreams/95adf79b-a07d-4ab2-8aca-232231811e4a/content

Matinmikko, M., Latva-aho, M., Ahokangas, P., & Seppänen, V. (2018). On regulations for 5G: Micro licensing for locally operated networks. *Telecommunications Policy*, *42*(8), 622–635. DOI: 10.1016/j.telpol.2017.09.004

Matterport (2025). Explore the business benefits of 3D tours. Matterport.

Mavenir. (2020). *Mavenir and Turkcell partner on OpenRAN*. Mavenir. https://www.mavenir.com/press-releases/mavenir-and-turkcell-partner-on-openran/

McCabe, D. (2018, October 25). *White House directs task force to come up with 5G wireless strategy*. Axios. https://www.axios.com/2018/10/25/white-house-national-wireless-strategy-task-force-5g

McClellan, M., Cervelló-Pastor, C., & Sallent, S. (2020). Deep Learning at the Mobile Edge: Opportunities for 5G Networks. *Applied Sciences (Basel, Switzerland)*, *10*(14), 4735. DOI: 10.3390/app10144735

McDonagh, M., & Petrie, R. (2017). New Research from Accenture Strategy highlights Economic and Societal Impact of Investing in 5G Infrastructure. Accenture.

McKinsey & Company. (2020). *The 5G era New horizons for advanced electronics and industrial companies*. Mckinsey.com. https://www.mckinsey.com/~/media/mckinsey/industries/advanced%20electronics/our%20insights/the%205g%20era%20new%20horizons%20for%20advanced%20electronics%20and%20industrial%20companies/the-5g-era-new-horizons-for-advanced-electronics-and-industrial-companies.pdf

Medin, M., & Louie, G. (2019). *The 5G Ecosystem: Risks & Opportunities for DoD. Defense Innovation Board.* https://innovation.defense.gov/Portals/63/Templates/Updated%20Meeting%20Documents/5G%20UNCLASS%20PAPER_190404_FINAL.pdf

Meesha Sundarum. (2025, April 16). *The State of 5G: Growth, Challenges, and Opportunities in 2025.* 5G Americas. https://www.5gamericas.org/the-state-of-5g-growth-challenges-and-opportunities-in-2025

Meza, R. R. P. (2007). *Interference and power control in wireless ad hoc networks.* University of New Mexico UNM Digital Repository. https://digitalrepository.unm.edu/cgi/viewcontent.cgi?article=1202&context=ece_etds

Mijumbi, R., Serrat, J., Gorricho, J.-L., Bouten, N., De Turck, F., & Boutaba, R. (2016). Network Function Virtualization: State-of-the-Art and Research Challenges. *IEEE Communications Surveys and Tutorials, 18*(1), 236–262. DOI: 10.1109/comst.2015.2477041

Miller, M. J. (2020, October 22). *Gartner's Top Strategic Technology Trends for 2021.* PCMAG; PCMag. https://www.pcmag.com/news/gartners-top-strategic-technology-trends-for-2021

Mobile Cloud Networking. (n.d.) *Mobile Cloud Networking in Online Casinos.* 777. https://mobile-cloud-networking.eu/

Moglia, A., Georgiou, K., Marinov, B., Georgiou, E., Berchiolli, R. N., Satava, R. M., & Cuschieri, A. (2022). 5G in healthcare: From COVID-19 to future challenges. *IEEE Journal of Biomedical and Health Informatics, 26*(8), 4187–4196. DOI: 10.1109/JBHI.2022.3181205

Monserrat, J. F., Mange, G., Braun, V., Tullberg, H., Zimmermann, G., & Bulakci, Ö. (2015). METIS research advances towards the 5G mobile and wireless system definition. *EURASIP Journal on Wireless Communications and Networking, 2015*(1). https://doi.org/DOI: 10.1186/s13638-015-0302-9

Mujahid. (2024, August 19). *The Future of 5G and Beyond: Transforming Connectivity.* Zoeteck.com. https://zoeteck.com/the-future-of-5g-and-beyond-transforming-connectivity/

Mukasa, R. (2025). *5G deployment considerations.* Scribd. https://www.scribd.com/document/385172613/5g-Deployment-Considerations

Mukherjee, S. (2025, January 30). Nokia beats estimates as demand recovers, shares rise. *Reuters.* https://www.reuters.com/business/media-telecom/nokias-q4-profit-beats-estimates-2025-01-30/

Nandini & Srivastava. D. (2023). 5G Technology—Its Applications and Challenges. *2023 7th International Conference on Computing, Communication, Control and Automation* (ICCUBEA, 1-4). https://doi.org/DOI: 10.1109/iccubea58933.2023.10392002

Naqvi, S. I., Naqvi, A. H., Arshad, F., Riaz, M. A., Azam, M. A., Khan, M. S., Amin, Y., Loo, J., & Tenhunen, H. (2019). An integrated antenna system for 4G and millimeter-wave 5G future handheld devices. *IEEE Access : Practical Innovations, Open Solutions*, *7*, 116555–116566. DOI: 10.1109/access.2019.2936513

Narayanan, S., Liberg, O., Höglund, A., Tsolkas, D., Alonso, L., Passas, N., & Merakos, L. (2024). Relaying Solutions For 5G-IoT Applications: A 3GPP Perspective. *IEEE Communications Standards Magazine*, *8*(2), 28–35. DOI: 10.1109/mcomstd.0005.2300019

Nasr, I. (2021). RF GaN on Silicon: The Best of Two Worlds. *Microwave Journal*. https://www.microwavejournal.com/articles/37107-rf-gan-on-silicon-the-best-of-two-worlds

News, I. T. U. (2020). Setting the Scene for 5G: Opportunities and Challenges. *ITU Hub*. https://www.itu.int/hub/2020/03/setting-the-scene-for-5g-opportunities-challenges/

Newsire, P. R. (2019). *Global 5G Market Analysis and Outlook 2020-2026: Market Dynamics, Industry Leaders, Opportunities and Forecasts*. Business Insider. https://markets.businessinsider.com/news/stocks/global-5g-market-analysis-and-outlook-2020-2026-market-dynamics-industry-leaders-opportunities-and-forecasts-1028789220

Newsprints Team. (2024). Qualcomm Product Strategy Guide. NextSprints.https://nextsprints.com/guide/qualcomm-product-strategy-guide

NGMN. (2015). *NGMN 5G White Paper*. NGMN. https://www.ngmn.org/work-programme/5g-white-paper.html

Nguyen, D. C., Pathirana, P. N., Ding, M., & Seneviratne, A. (2020). Blockchain for 5G and beyond networks: A state of the art survey. *Journal of Network and Computer Applications*, *166*, 102693. DOI: 10.1016/j.jnca.2020.102693

Nokia Corporation. (2025). *Nokia 8.3 5G specifications. HMD - Human Mobile Devices*. https://www.hmd.com/en_us/nokia-8-3-5g/specs

Nokia. (2019). *SETAR and Nokia bring 5G to Aruba in end-to-end deal*. Noki. https://www.nokia.com/newsroom/setar-and-nokia-bring-5g-to-aruba-in-end-to-end-deal/

Nokia. (2020). *Nokia 5G Readiness Report: Forecast $8 Trillion Impact on Global GDP by 2030.* Benton Foundation. https://www.benton.org/headlines/nokia-5g-readiness-report-forecast-8-trillion-impact-global-gdp-2030

Nwajana, A. O., & Ihianle, I. K. (2021). *Handbook of Research on 5G Networks and Advancements in Computing*. IGI Global.

O'Connell, E., Moore, D., & Newe, T. (2020). Challenges Associated with Implementing 5G in Manufacturing. *Telecom, 1*(1), 48–67. MDPI. https://doi.org/DOI: 10.3390/telecom1010005

OC&C Strategy Consultants. (2025). Reports. OC&C Strategy Consultants. https://www.occstrategy.com/en/reports/

Odida, M. (2024). The Evolution of Mobile Communication: A Comprehensive Survey on 5G Technology. *Journal of Sensor Networks and Data Communications, 4*(1), 01-11. https://doi.org/DOI: 10.33140/jsndc.04.01.06

OECD. (2021a). *The digital transformation of SMEs*. https://www.oecd.org/ OECD. https://www.oecd.org/en/publications/the-digital-transformation-of-smes_bdb9256a-en.html

Okutsu, A. (2019, June 10). *South Korea, US and Japan lead in 5G network deployment.* Nikkei Asia. https://asia.nikkei.com/Business/Business-trends/South-Korea-US-and-Japan-lead-in-5G-network-deployment

Osseiran, A., Boccardi, F., Braun, V., Kusume, K., Marsch, P., Maternia, M., Queseth, O., Schellmann, M., Schotten, H., Taoka, H., Tullberg, H., Uusitalo, M. A., Timus, B., & Fallgren, M. (2014). Scenarios for 5G mobile and wireless communications: The vision of the METIS project. *IEEE Communications Magazine, 52*(5), 26–35. DOI: 10.1109/mcom.2014.6815890

Paglierani, P., Neokosmidis, I., Rokkas, T., Meani, C., Nasr, K. M., Moessner, K., & Sayyad Khodashenas, P. (2019). Techno-economic analysis of 5G immersive media services in cloud-enabled small cell networks: The neutral host business model. *Transactions on Emerging Telecommunications Technologies, 31*(2), e3746. DOI: 10.1002/ett.3746

Panetta, K. (2019, October 21). *Gartner Top 10 Strategic Technology Trends for 2020.* Gartner. https://www.gartner.com/smarterwithgartner/gartner-top-10-strategic-technology-trends-for-2020

Park, Seongmin, Chopi, Bomin, Yim, Kangbin (2020). Vestiges of Past Generation: Threats to 5G Core Network. *Advances in Intelligent Systems and Computing*, 468–480. https://doi.org/DOI: 10.1007/978-3-030-50399-4_45

Parker, T. (2014). *METIS task group lays out first 5G channel models.* Fierce Network. https://www.fierce-network.com/tech/metis-task-group-lays-out-first-5g-channel-models?utm_source=chatgpt.com

Patel, S., Chauhan, M., & Kapadiya, K. (2012). 5G: Future mobile technology - Vision 2020. *International Journal of Computer Applications*, *54*(17), 6–10. DOI: 10.5120/8656-2264

Patwary, M. N., Nawaz, S. J., Rahman, M. A., Sharma, S. K., Rashid, M. M., & Barnes, S. J. (2020). The potential short- and long-term disruptions and transformative impacts of 5G and beyond wireless networks: Lessons learnt from developing a 5G testbed environment. *IEEE Access : Practical Innovations, Open Solutions*, *8*, 11352–11379. DOI: 10.1109/ACCESS.2020.2964673

Petroc, T. (2025). *Forecast number of mobile 5G connections worldwide from 2021 to 2025 (in billions).* Statista. https://www.statista.com/statistics/691623/5g-mobile-connections-worldwide/

Pirinen, P. (2014, November 1). A brief overview of 5G research activities. *IEEE Xplore*. https://doi.org/DOI: 10.4108/icst.5gu.2014.258061

Polese, M., Mezzavilla, M., Rangan, S., & Zorzi, M. (2017). *Mobility Management for TCP in mmWave Networks*. https://doi.org/DOI: 10.1145/3130242.3130243

Polese, M., Jana, R., & Zorzi, M. (2017). TCP and MP-TCP in 5G mmWave Networks. *IEEE Internet Computing*, *21*(5), 12–19. DOI: 10.1109/MIC.2017.3481348

Popovski, P., Trillingsgaard, K. F., Simeone, O., & Durisi, G. (2018). 5G Wireless Network Slicing for eMBB, URLLC, and mMTC: A Communication-Theoretic View. *IEEE Access: Practical Innovations, Open Solutions*, *6*, 55765–55779. DOI: 10.1109/access.2018.2872781

Qose, S., & Zoltán, R. (2024). Supply chain in the context of 5G technology, security, and legal aspects. In *2024 IEEE 22nd World Symposium on Applied Machine Intelligence and Informatics (SAMI)* (pp. 000143–000148). https://doi.org/DOI: 10.1109/SAMI60510.2024.10432844

Qualcomm (2020). *Global update on spectrum for 4G & 5G.* Qualcomm Technologies, Inc. https://www.qualcomm.com/content/dam/qcomm-martech/dm-assets/documents/messaging_presentation_-_4g_5g_spectrums_-_december_2020.pdf

Qualcomm. (2017). *Wireless Technology & Innovation | Mobile Technology | Qualcomm.* Qualcomm. https://www.qualcomm.com/

RantCell. (2019). Rantcell.com. https://rantcell.com/do-autonomous-vehicles-need-5g.html

Rapaport, T. S., Sun, S., Mayzus, R., Zhao, H., Azar, Y., Wang, K., & Samimi, M. (2013). Millimeter wave wireless communications for 5G cellular: It will work! *IEEE Access : Practical Innovations, Open Solutions*, *1*, 335–349. DOI: 10.1109/access.2013.2260813

Reportlinker. (2015, January 28). *The 5G Wireless Ecosystem: 2015 - 2025 - Technologies, Applications, Verticals, Strategies & Forecasts*. Prnewswire.com; Cision PR Newswire. https://www.prnewswire.com/news-releases/the-5g-wireless-ecosystem-2015---2025---technologies-applications-verticals-strategies--forecasts-300027177.html

Research, A. B. I. (2025). *Top industries/applications driving 5G market growth from 2025 to 2030*. Abiresearch.com. https://www.abiresearch.com/blog/5g-market-top-industries-applications-2025-to-2030

Reuters. (2018). *Seeking new growth drivers, Samsung plans $22 billion spending on new tech*. Yahoo Finance. https://finance.yahoo.com/news/samsung-group-spend-22-billion-ai-5g-auto-050941607--finance.html

Robert Bosch Gmb, H. (2012). *CAN with Flexible Data-Rate Specification Version 1.0*. Robert Bosch GmbH. https://www.bosch-semiconductors.com/media/ip_modules/pdf_2/can/can_fd_spec.pdf

Ruzomberka, E., Love, D. J., Brinton, C. G., Gupta, A., Wang, C.-C., & Poor, H. V. (2023). Challenges and Opportunities for Beyond 5 G Wireless Security. *IEEE Security and Privacy*, *21*(5), 55–66. DOI: 10.1109/MSEC.2023.3251888

Rysavy, P. (2018). *Industry Voices—Rysavy: Midband spectrum for 5G is needed now*. Fierce Network. https://www.fierce-network.com/5g/industry-voices-rysavy-mid-band-spectrum-for-5g-needed-now

Saadi, M., & Mavrakis, D. (2020). *5G AND AI THE FOUNDATIONS FOR THE NEXT SOCIETAL AND BUSINESS LEAP EXECUTIVE SUMMARY*. ABI Research. https://www.hsdf.org/wp-content/uploads/2021/04/5g-and-ai-report.pdf

Sabharwal, A., Schniter, P., Guo, D., Bliss, D., Rangarajan, S., & Wichman, R. (2014). In-band full-duplex wireless: Challenges and opportunities. *IEEE Journal on Selected Areas in Communications*, *32*(9), 1637–1652.

Sadek, A. (2024). *AI and ERP are powerful together in the medical manufacturing industry*. Expert ERP Solutions - MEBSCO - [ERP Solutions] Your Trusted Digital Transformation Partner. https://mebsco.com/2024/11/07/ai-and-erp/

Sag, A. Townsend, W., & Moorhead, P. (2025). *The G2 on 5G Podcast - Episode 219: New Intel CEO, 5G Advanced, Converged Connectivity*. Moor Insights & Strategy. https://moorinsightsstrategy.com/g2-on-5g/the-g2-on-5g-episode-219-new-intel-ceo-5g-advanced-converged-connectivity/

Salahdine, F., Han, T., & Zhang, N. (2022). Security in 5G and beyond recent advances and future challenges. *Security and Privacy*, *6*(1). Advance online publication. DOI: 10.1002/spy2.271

Salih, A. A., Zeebaree, S. R. M., Abdulraheem, A. S., Zebari, R. R., Sadeeq, M. A. M., & Ahmed, O. M. (2020). *Evolution of mobile wireless communication to the 5G revolution*. Technology Reports of Kansai University 62(5):2139-2151 ResearchGate.

Samsung (2024). *Investor Relations | Samsung*. Samsung Global. https://www.samsung.com/global/ir/

Satariano, A., & Bayer, E. (2019). MWC 2019: Huawei Politics, Robots and Speedy 5G. *The New York Times*. https://www.nytimes.com/2019/02/27/technology/mwc-2019-5g-huawei.html

Sauter, M. (2021). *From GSM to LTE-Advanced Pro and 5G*. DOI: 10.1002/9781119714712

Sauter, M., & Wiley, J. (2021). *From GSM to LTE-Advanced Pro and 5G : an introduction to mobile networks and mobile broadband*. Wiley.

Schroth, L. (2020). *5G and Drones: What is 5G?* Drone Industry Insights. https://droneii.com/5g-and-drones-part-1-what-is-5g

Seils, D. M., Christner, J., & Jin, R. (2010). *Deploying Cisco Wide Area Application Services* (2nd ed.). Cisco Press., https://www.ciscopress.com/store/deploying-cisco-wide-area-application-services-9781587059124

Selinis, I., Katsaros, K., Allayioti, M., Vahid, S., & Tafazolli, R. (2018). The Race to 5G Era: LTE and Wi-Fi. *IEEE Access : Practical Innovations, Open Solutions*, *6*, 56598–56636. DOI: 10.1109/ACCESS.2018.2867729

Selinis, I., Katsaros, K., Allayioti, M., Vahid, S., & Tafazolli, R. (2018). The Race to 5G Era; LTE and Wi-Fi. *IEEE Access : Practical Innovations, Open Solutions*, *6*, 56598–56636. DOI: 10.1109/access.2018.2867729

Serrano, R. B., & Sorrentino, S. (2016). *Speed-dependent transmission format for vehicular transmission (U.S.* Patent No. 10,348,444 B2). U.S. Patent and Trademark Office. https://patents.google.com/patent/US10348444B2

Sharma, D. C. (2023). *European Telecommunications Standards Institute (ETSI)*. sesei.eu. https://sesei.eu/wp-content/uploads/2023/05/All-about-ETSI_April-2023.pdf

Sharma, P. (2013). Evolution of Mobile Wireless Communication Networks-1G to 5G as well as Future Prospective of Next Generation Communication Network. *IJCSMC, 2*(8), 47–53.

Sharma, P., Jain, S., Gupta, S., & Chamola, V. (2021). Role of machine learning and deep learning in securing 5G-driven industrial IoT applications. *Ad Hoc Networks, 123*, 102685. DOI: 10.1016/j.adhoc.2021.102685

Shayea, I., Ergen, M., Hadri Azmi, M., Aldirmaz Çolak, S., Nordin, R., & Daradkeh, Y. I. (2020). Key challenges, drivers, and solutions for mobility management in 5G networks: A survey. *IEEE Access : Practical Innovations, Open Solutions, 8*, 172534–172552. DOI: 10.1109/ACCESS.2020.3023802

Shehab, M. J., Kassem, I., Kutty, A. A., Kucukvar, M., Onat, N., & Khattab, T. (2022). 5G networks towards smart and sustainable cities: A review of recent developments, applications and future perspectives. *IEEE Access : Practical Innovations, Open Solutions, 10*, 2987–3006. DOI: 10.1109/ACCESS.2021.3139436

Shrivastava, S., Rana, A., & Taneja, A. (2025). *Driving Next Generation IoT with 5G and Beyond*. 33–64. https://doi.org/DOI: 10.1002/9781394256075.ch3

Siemens (2019). *Industrial 5G For the industry of tomorrow*. siemens.com/industrial-5g. https://assets.new.siemens.com/siemens/assets/api/uuid:1afdad9c-539a-4cf9-9e3b-97be13a9e27d/presentation-5G-roundtable-sps19.pdf

Simeone, O., Erkip, E., & Shamai, S. (2014). Full-duplex cloud radio access networks: An information-theoretic viewpoint. *arXiv preprint arXiv:1405.2092*. https://arxiv.org/pdf/1405.2092

Simnovus (2024). *Sustainability in 5G RAN: A Key to Green Connectivity*. Simnovus. https://simnovus.com/sustainability-in-5g-ran-a-key-to-green-connectivity/

Singh, K., & Shivinder Devra. (2017, October 12). *Evolution of Cellular Network: From 1G to 5G*. Academia.edu. https://www.academia.edu/34839361/Evolution_of_Cellular_Network_From_1G_to_5G

Singh, V. P., Singh, M. P., Hegde, S., & Gupta, M. (2024). Security in 5G Network Slices: Concerns and Opportunities. *IEEE Access : Practical Innovations, Open Solutions, 12*, 52727–52743. DOI: 10.1109/ACCESS.2024.3386632

Sowa, L. (2020, September 15). *Advocates for New Wireless Technologies Claim Sooner Rollout, Explicate Exciting 5G Attributes – Broadband Breakfast.* Broadband Breakfast.

Strategy Analytics. (2019). Predictive Analysis I Intelligent Analysis I Strategic Market Research I Strategy Analytics. https://www.strategyanalytics.com/

Strategy, A. (2021). *The Impact of 5G on the U.S. & European Economies*. Qualcomm. https://www.qualcomm.com/content/dam/qcomm-martech/dm-assets/documents/accenture_5g_economic_impact_us_eu_overview_slides.pdf

Sun, H., Hu, R. Q., & Qian, Y. (2023). Open Issues and Future Directions for 5G and Beyond Wireless Networks. In 5G and Beyond Wireless Communication Networks, *IEEE*, 155–156. https://doi.org/DOI: 10.1002/9781119089469.ch12

Sun, Y., Chen, J., Wang, Z., Peng, M., & Mao, S. (2022). Enabling Mobile Virtual Reality with Open 5G, Fog Computing, and Reinforcement Learning. *IEEE Network*, *36*(6), 142–149. DOI: 10.1109/MNET.010.2100481

Svensson, T., Li, J., Wanlu, S., Lakshamana, T. R., Ström, E., & Yutao, S. (2020) *Information Society (METIS)*. (2020). Research.chalmers.se. https://research.chalmers.se/en/project/5261

SWAN. (2021). R*adio Frequency Vulnerabilities White Paper*. Swan. https://cpb-eu-w2.wpmucdn.com/blogs.bristol.ac.uk/dist/6/635/files/2021/05/SWANWH2.pdf

Tahir, M., Habaebi, M. H., Dabbagh, M., Mughees, A., Ahad, A., & Ahmed, K. I. (2020). A review of the application of Blockchain in 5G and beyond networks: Taxonomy, field trials, challenges and opportunities. *IEEE Access : Practical Innovations, Open Solutions*, *8*, 115876–115904. DOI: 10.1109/ACCESS.2020.3003020

Tahir, M., Habaebi, M. H., Dabbagh, M., Mughees, A., Ahad, A., & Ahmed, K. I. (2020). A Review on Application of Blockchain in 5G and Beyond Networks: Taxonomy, Field-Trials, Challenges and Opportunities. *IEEE Access : Practical Innovations, Open Solutions*, *8*, 115876–115904. DOI: 10.1109/access.2020.3003020

Taleb, T., Samdanis, K., Mada, B., Flinck, H., Dutta, S., & Sabella, D. (2017). On Multi-Access Edge Computing: A Survey of the Emerging 5G Network Edge Cloud Architecture and Orchestration. *IEEE Communications Surveys and Tutorials*, *19*(3), 1657–1681. DOI: 10.1109/COMST.2017.2705720

Tao, X., & Chen, L. (2022). *5G NR indoor localization with smartphones*. 25–25. https://doi.org/DOI: 10.1117/12.2640925

Taylor, P. (2025). *Forecast number of mobile 5G subscriptions worldwide from 2023 to 2028 (in billions)*. Statista. https://www.statista.com/statistics/760275/5g-mobile-subscriptions-worldwide/

Team Roar. (2017). *All You Need To Know About 5G*. Roar Media. https://archive.roar.media/english/tech/sponsored/all-you-need-to-know-about-5g

Technologies, H. (2022). Annual Report 2022. Huawei. https://www.huawei.com/en/annual-report/2022/

Teece, D. J. (2021). Technological leadership and 5G patent portfolios: Guiding strategic policy and licensing decisions. *California Management Review*, *63*(3), 5–34. DOI: 10.1177/00081256211007584

Teixeira, L., Almeida, P., Rodrigues, M., Antunes, D., Gomes, D., & Aguiar, R. L. (2024). *Shallow vs*. Deep Learning: Prioritizing Efficiency in Next Generation Networks. 2024 11th International Conference on Future Internet of Things and Cloud (FiCloud), Vienna, Austria, 308–315. https://doi.org/DOI: 10.1109/FiCloud62933.2024.00055

TELCOMA. (2025). *TELCOMA 5G Certifications and 5G Training Courses for Professionals including 6G, 4G-LTE, IoT, Cloud*. Telcoma. Telcomaglobal.com. https://www.telcomaglobal.com/p/5g-core-network-5g

Temperton, J. (2017, March 9). *Budget 2017: 5G and fibre broadband get funding boost in Spring Budget*. WIRED. https://www.wired.com/story/spring-budget-2017-5g-fibre-broadband

Thompson, J., Ge, X., Wu, H.-C., Irmer, R., Jiang, H., Fettweis, G., & Alamouti, S. (2014). 5G wireless communication systems: Prospects and challenges [Guest Editorial]. *IEEE Communications Magazine*, *52*(2), 62–64. DOI: 10.1109/mcom.2014.6736744

Tian, M.-W., Wang, L., Yan, S.-R., Tian, X.-X., Liu, Z.-Q., & Rodrigues, J. J. P. C. (2019). Research on financial technology innovation and application based on 5G network. *IEEE Access : Practical Innovations, Open Solutions*, *7*, 138614–138623. DOI: 10.1109/ACCESS.2019.2936860

U.S. Department of Commerce. (2022). *U.S. Department of Commerce and Balkans Work Toward a Resilient and Secure 5G Ecosystem*. Commerce.gov. https://www.commerce.gov/news/blog/2022/07/us-department-commerce-and-balkans-work-toward-resilient-and-secure-5g-ecosystem

UNCTAD. (2021). *Technology and Innovation Report 2021: Catching Technological Waves—Innovation with Equity*. UN. https://digitallibrary.un.org/record/3926808

Usmonov, B., Rustamov, U., & Shukurillaevich, U. (2019). *5G Technology Evolution*. 978-1-7281-2564-0/19/$31.00 ©2019 IEEE. https://www.academia.edu/43046535/5G_Technology_Evolution

Varga, P., Peto, J., Franko, A., Balla, D., Haja, D., Janky, F., Soos, G., Ficzere, D., Maliosz, M., & Toka, L. (2020). 5G support for Industrial IoT Applications— Challenges, Solutions, and Research gaps. *Sensors (Basel)*, *20*(3), 828. DOI: 10.3390/s20030828

Verizon. (2020). *What is the Latency of 5G?* Verizon. https://www.verizon.com/about/our-company/5g/5g-latency?msockid=3c2aa922788360e430f3bcaa79416152

Wang, C.-X., Haider, F., Gao, X., You, X.-H., Yang, Y., Yuan, D., Aggoune, H., Haas, H., Fletcher, S., & Hepsaydir, E. (2014). Cellular architecture and key technologies for 5G wireless communication networks. *IEEE Communications Magazine*, *52*(2), 122–130. DOI: 10.1109/MCOM.2014.6736752

Wazid, M., Das, A. K., Shetty, S., Gope, P., & Rodrigues, J. J. P. C. (2021). Security in 5 G-enabled Internet of Things communication: Issues, challenges, and future research roadmap. *IEEE Access : Practical Innovations, Open Solutions*, *9*, 4466–4489. DOI: 10.1109/ACCESS.2020.3047895

Weisman-Pitts, J. (2025). 5G Technology: A Comprehensive Report on Its Impact by 2025. *Global Banking and Finance Review*. https://www.globalbankingandfinance.com/5g-technology-a-comprehensive-report-on-its-impact-by-2025

Wikipedia. (n.d.). *Multi-user MIMO*. Wikipedia. https://en.wikipedia.org/wiki/Multi-user_MIMO

WIPO. (2023). *World Intellectual Property Indicators 2023*. https://www.wipo.int/edocs/pubdocs/en/wipo-pub-941-2023-en-world-intellectual-property-indicators-2023.pdf

Won, H., Hong, Y.-K., Choi, M., Garcia, H., Shin, D., Yoon, Y.-S., Lee, K., Xin, H., & Yeo, C.-D. (2022). Microwave absorption performance of M-type hexagonal ferrite and MXene composite in Ka and V bands (5G mmWave frequency bands). *Journal of Magnetism and Magnetic Materials*, *560*, 169523. DOI: 10.1016/j.jmmm.2022.169523

World Economic Forum. (2020). *The Impact of 5G: Creating New Value across Industries and Society*. World Economic Forum. https://www.weforum.org/reports/the-impact-of-5g-creating-new-value-across-industries-and-society

Wunder, G., Kasparick, M., Wild, T., Schaich, F., Chen, Y., Ten Brink, S., Gaspar, I., Michailow, N., Navarro, A., Fettweis, G., Cassiau, N., Kténas, D., Dryjanski, M., Pietrzyk, S., & Eged, B. (n.d.). *5GNOW: Application Challenges and Initial Waveform Results*. Retrieved July 23, 2025, from https://www.vodafone-chair.org/pbls/legacy/i-gaspar/5GNOW_Application_Challenges_and_Initial_Waveform_Results.pdf?utm_source=chatgpt.com

Yang, H., Chen, L., Liu, H., & Zhu, G. (2024). Dynamic Feature-fused Localization with Smartphones Exploiting 5G NR SSB and Wi-Fi for Indoor Environments. *IEEE Transactions on Instrumentation and Measurement*, *73*, 1–14. DOI: 10.1109/tim.2024.3352695

Zanella, A., Bui, N., Castellani, A., Vangelista, L., & Zorzi, M. (2018). Internet of Things for Smart Cities. *IEEE Internet of Things Journal*, *1*(1), 22–32. https://ieeexplore.ieee.org/document/6740844

Zhang, Z., & Patel, K. (2023, July 9). *ANUP Implementation in 5G with BGP Signaling*. Ietf.org. https://www.ietf.org/archive/id/draft-zzhang-dmm-anup5g-signaling-00.html

Zhang, J., Qiu, M., Tsai, C.-W., Hassan, M. M., & Alamri, A. (2015). Health-CPS: Healthcare cyber-physical system assisted by cloud and big data. *IEEE Systems Journal*, *11*(1), 88–95. DOI: 10.1109/jsyst.2015.2460747

Zhang, R., Zhou, W., & Hu, H. (2021). Towards 5G Security Analysis against Null Security Algorithms Used in Normal Communication. *Security and Communication Networks*, *2021*, 1–15. DOI: 10.1155/2021/4498324

Zhang, Z., & Wang, Q. (2021). Application Status and Prospects of 5G Technology in Distribution Automation Systems. *Wireless Communications and Mobile Computing*, *2021*, 1–9. DOI: 10.1155/2021/5553159

Zhao, X., Askari, H., & Chen, J. (2021). Nanogenerators for smart cities in the era of 5G and Internet of Things. *Joule*, *5*(6), 1391–1431. DOI: 10.1016/j.joule.2021.03.013

ZTE. (2023). *Focusing on GIS, ZTE Announced the 5 Gen of FWA Family at MWC 2023*. ZTE Devices. https://global.ztedevices.com/blogs/news/focusing-on-gis-zte-announced-the-5-gen-of-fwa-family-at-mwc-2023

About the Author

Vincent Sabourin is a full professor of strategic management at the School of Management (ESG) of the University of Quebec in Montreal (UQAM). He has been teaching at multiple executive levels, including the MBA for Science and Engineering. He served as director of the Department of Strategy, Social, and Environmental Responsibility and was a commissioner on the University's Academic Affairs Commission. Dr. Sabourin's expertise in strategic management is widely recognized. He serves as Chairman of the Consortium for Innovation, a nonprofit research and educational network devoted to innovation and management. Governments and businesses regularly consult him for his strategic insight. He led the Promising Plastics project for the Canadian Federation of Plastic Producers and has delivered keynote addresses at numerous conferences on advanced plastics, innovative technologies, and the Internet of Things. His reputation as a strategy expert was confirmed when he was selected as one of 25 Canadian experts in strategic planning to participate in a doctoral dissertation at HEC Montréal. He has been invited as a distinguished speaker at global forums, including the International Federation of Information Technology for Travel and Tourism (IFITT) and the Academy of Management of China. Professor Sabourin's academic influence extends well beyond the classroom. His research, published in books, chapters, peer-reviewed journals, and conference proceedings, has made significant contributions to the fields of strategy and innovation management. His most recent books, including Strategic Management for the Hospitality and Tourism Industry: Developing a Competitive Advantage (CRC Press, 2022), IoT Benefits and Growth Opportunities for the Telecom Industry (CRC Press, 2022), and The Future of the Medical Plastics Market: Opportunities for Growth (Routledge, 2021), are not just theoretical, but offer practical insights. He is also the editor of Strategic Monitoring and Innovation Management Practices (PUQ, 2022), which includes a chapter on 5G technology. Significant achievements mark Professor Sabourin's academic journey. He holds a diploma in industrial psychology from McGill University, a master's in administration from HEC Montreal, and a Ph.D. in strategic management from McGill University, where he studied under the esteemed Professor Henry Mintzberg. He also published several book chapters, including Blockchain and the management transformation of business models in the financial services industry (2020), a chapter commissioned by the publisher. He authored "Identification and Categorization of Disruptive Innovations According to the Strategic Scope of the Firm" in the Handbook of Research on Disruptive Technology and Critical Explorations for Innovation Management, published by IGI Global. He also contributed the chapter "Cleantech and Water Treatment" as a case of disruptive innovation in the same volume published by IGI Global.

Index

Symbols

5G Deployment 54, 64, 66, 72, 77, 78, 80, 83, 87, 88, 89, 90, 91, 92, 94, 96, 100, 102, 105, 111, 112, 113, 128, 129, 132, 161, 177, 185, 198, 203, 217, 262, 263, 280, 302, 327, 330, 347, 349, 363, 395, 406, 408, 415, 425, 426, 428, 436, 439, 440, 450, 470, 478, 481, 483

5G Research 86, 87, 98, 116, 345, 350, 371, 411, 423, 436, 445, 454, 476

5G Rollout 32, 91, 97, 98, 99, 103, 105, 106, 108, 132, 198, 207, 266, 315, 368, 421, 426, 455, 465, 485, 490

5G Spectrum 32, 52, 91, 140, 196, 197, 198, 223, 224, 225, 412

5G Strategy 95, 431, 452, 489, 490

5G Technology 1, 3, 8, 9, 10, 13, 15, 18, 19, 20, 28, 31, 35, 37, 38, 41, 43, 49, 50, 58, 59, 61, 73, 74, 75, 77, 78, 79, 82, 87, 88, 89, 90, 91, 95, 104, 112, 117, 121, 123, 124, 136, 140, 153, 157, 158, 159, 161, 162, 163, 165, 169, 191, 192, 210, 231, 232, 233, 237, 238, 240, 241, 242, 246, 250, 251, 253, 254, 256, 257, 258, 260, 262, 265, 266, 269, 270, 271, 272, 273, 274, 278, 281, 283, 287, 290, 292, 295, 297, 298, 301, 302, 303, 304, 306, 310, 311, 313, 318, 319, 330, 347, 348, 350, 352, 361, 364, 367, 375, 379, 385, 386, 390, 394, 395, 396, 398, 400, 401, 409, 411, 415, 416, 421, 424, 425, 438, 442, 452, 454, 455, 457, 458, 483, 485, 487, 489, 490, 491

A

Autonomous Vehicles 1, 3, 8, 15, 16, 21, 26, 29, 47, 63, 73, 80, 85, 122, 127, 138, 139, 148, 154, 155, 161, 162, 163, 168, 170, 178, 181, 182, 183, 190, 195, 197, 204, 206, 209, 212, 213, 214, 225, 235, 247, 250, 258, 260, 262, 263, 290, 293, 294, 318, 333, 334, 339, 365, 369, 370, 371, 374, 375, 423, 436, 438, 439, 450, 462, 464, 474, 478

C

Chipsets 37, 38, 39, 50, 58, 59, 79, 394, 413, 420, 428, 429, 440, 450, 461, 474, 482

Competitive Landscape 91, 457, 458, 460, 463, 465, 468, 469, 472

D

Digital Transformation 1, 29, 45, 59, 86, 87, 115, 142, 162, 163, 169, 175, 177, 190, 210, 226, 232, 240, 247, 258, 261, 268, 281, 292, 294, 304, 305, 318, 347, 349, 365, 375, 385, 387, 388, 389, 396, 424, 425, 429, 433, 435, 436, 437, 439, 449, 450, 471, 472, 478

E

Edge Computing 3, 4, 9, 15, 19, 21, 26, 34, 41, 56, 58, 63, 65, 79, 80, 85, 86, 123, 148, 149, 150, 154, 155, 157, 158, 178, 181, 182, 212, 219, 223, 241, 259, 260, 261, 264, 265, 268, 269, 291, 318, 328, 338, 349, 365, 375, 379, 385, 432, 437, 457, 459, 461, 467, 468, 469, 471, 479, 480, 482, 483, 485

eMBB 1, 10, 12, 16, 17, 18, 19, 37, 41, 46, 47, 56, 57, 68, 121, 122, 132, 136, 137, 138, 139, 148, 153, 154, 155, 157, 170, 182, 211, 232, 263, 269, 288, 431, 473

Energy Efficiency 29, 137, 145, 154, 159, 166, 177, 188, 204, 209, 210, 214, 215, 216, 217, 218, 221, 223, 248, 281, 288, 289, 325, 326, 327, 332, 333, 334, 342, 355, 356, 361, 365,

369, 375, 468, 473, 483

F

Full-Duplex Communication 209, 215, 216, 217, 221, 223, 224, 325, 326, 335, 340, 341, 358

G

Global Connectivity 49, 50, 80, 85, 113, 161, 162, 204, 319, 375, 439, 457, 472, 474

I

Industrial Automation 4, 10, 16, 29, 35, 47, 52, 57, 80, 85, 86, 112, 122, 138, 141, 154, 155, 169, 171, 196, 202, 203, 205, 206, 207, 213, 224, 231, 232, 251, 257, 305, 334, 348, 374, 394, 420, 423, 429, 433, 440, 451, 469, 483
Industry Collaboration 136, 390, 483
Industry Leadership 419, 470
Infrastructure Readiness 86, 111, 112, 221, 398, 400, 401
IoT Applications 114, 140, 155, 157, 160, 232, 234, 258, 268, 281, 284, 285, 321, 365, 476
IoT Ecosystems 57, 124, 142, 154, 163, 190, 204, 213, 235, 264, 280, 362, 388
IoT Integration 58, 79, 440
ITU-R 2, 16, 32, 46, 56, 152, 161, 195, 196, 198, 199, 200, 209, 210, 223, 224, 226

L

Logistics 13, 20, 22, 27, 28, 32, 38, 48, 56, 58, 60, 79, 112, 121, 122, 137, 139, 150, 154, 155, 159, 162, 175, 190, 191, 196, 203, 204, 226, 233, 242, 251, 254, 260, 264, 276, 278, 282, 285, 287, 288, 289, 290, 291, 292, 293, 294, 295, 296, 303, 304, 305, 318, 319, 320, 321, 375, 379, 385, 387, 388, 398, 404, 414, 420, 421, 429, 453, 461, 470, 472, 480, 484
Low-Latency Communication 1, 58, 68, 79, 80, 86, 121, 122, 141, 154, 155, 168, 170, 190, 202, 218, 232, 238, 245, 247, 280, 288, 294, 328, 371, 375, 394, 422, 423, 425, 431, 473

M

Massive MIMO 22, 23, 32, 41, 46, 62, 64, 66, 86, 91, 135, 141, 143, 156, 209, 216, 217, 221, 223, 224, 285, 325, 326, 332, 333, 336, 337, 340, 342, 363, 364, 368, 386, 394, 399, 412, 422, 446, 447, 459, 461, 463, 470
Millimeter Wave 17, 18, 40, 46, 52, 125, 199, 337, 340, 344, 349, 378, 380, 415, 416, 417, 434, 436
MMTC 10, 16, 19, 20, 21, 27, 41, 46, 47, 56, 57, 68, 80, 121, 122, 136, 137, 138, 139, 143, 148, 153, 154, 155, 157, 211, 232, 238, 288, 334, 441, 442, 473
Mobile Connectivity 18, 37, 39, 141, 292, 315, 446, 458

N

Network Architecture 4, 31, 40, 41, 63, 65, 80, 123, 124, 127, 139, 156, 164, 185, 196, 217, 221, 297, 326, 423, 445, 448
Network Optimization 269, 333, 365, 376
Network Slicing 1, 4, 20, 27, 41, 63, 64, 65, 66, 80, 83, 88, 89, 123, 127, 128, 135, 141, 148, 154, 157, 158, 161, 162, 163, 164, 167, 168, 169, 170, 171, 179, 182, 184, 185, 187, 190, 211, 218, 223, 224, 241, 248, 249, 257, 263, 266, 267, 268, 269, 280, 288, 302, 315, 328, 332, 334, 336, 339, 354, 362, 365, 368, 385, 387, 397, 399, 414, 417, 422, 423, 426, 428, 439, 440, 459, 461, 470, 472, 475, 478, 485, 486
Nokia 8.3 38, 39, 57, 58, 60, 79

P

Predictive Analytics 238, 287, 319, 385, 486
Private 5G Networks 179, 183, 246, 385, 407, 468, 470, 474, 479, 484, 486

R

Real-Time Tracking 253, 287, 293, 318, 385, 398, 401, 412
Regional Disparities 112
Regulatory Frameworks 29, 66, 73, 74, 78, 86, 111, 180, 181, 258, 395, 398, 400, 403, 423, 457

S

Samsung Galaxy S20 37, 38, 39, 57, 58, 79
Smart Cities 1, 3, 10, 11, 15, 16, 20, 29, 33, 45, 57, 63, 80, 86, 87, 137, 139, 141, 154, 162, 163, 164, 168, 170, 177, 190, 204, 207, 213, 214, 216, 217, 225, 231, 232, 245, 248, 249, 250, 254, 257, 258, 260, 261, 264, 269, 274, 284, 289, 348, 415, 424, 431, 432, 436, 439, 440, 449, 450, 470, 474, 478, 483, 485
Smartphones 14, 37, 38, 39, 40, 41, 42, 43, 57, 58, 59, 61, 79, 85, 87, 102, 132, 165, 174, 176, 208, 236, 261, 275, 297, 312, 353, 360, 362, 424, 430, 436, 438, 459, 460, 470, 482
Spectrum Allocation 6, 49, 50, 65, 73, 113, 161, 195, 196, 197, 203, 212, 213, 218, 221, 223, 224, 356, 362, 395, 400, 404, 406, 409, 411
Spectrum Virtualization 214, 325, 327, 336, 353, 357
Sub-6 GHz 9, 17, 18, 40, 42, 58, 68, 70, 79, 135, 166, 196, 197, 198, 199, 200, 201, 224, 332, 427
Supply Chain 13, 38, 56, 203, 226, 233, 251, 252, 282, 287, 288, 289, 290, 291, 293, 294, 295, 296, 320, 385, 386, 387, 388, 396, 398, 400, 401, 411, 414, 415, 416, 417, 428, 463, 466, 467, 484
Sustainability 116, 169, 204, 205, 208, 226, 233, 237, 239, 288, 318, 319, 320, 350, 414, 425, 433

T

Telecommunications 1, 2, 6, 26, 28, 29, 34, 37, 50, 59, 60, 63, 66, 68, 72, 73, 75, 76, 77, 78, 81, 82, 83, 86, 90, 96, 99, 102, 103, 124, 142, 161, 162, 163, 164, 172, 173, 174, 175, 176, 189, 190, 193, 199, 204, 207, 226, 248, 249, 254, 257, 258, 259, 262, 263, 267, 268, 269, 283, 289, 298, 316, 338, 344, 360, 361, 364, 365, 371, 375, 409, 410, 414, 415, 417, 419, 426, 436, 452, 461, 465, 469, 481, 489, 490

U

Urban Connectivity 49, 71, 196
URLLC 1, 10, 16, 21, 27, 41, 46, 47, 56, 57, 65, 68, 80, 121, 122, 125, 134, 136, 138, 139, 142, 148, 152, 153, 154, 155, 157, 170, 180, 181, 182, 183, 184, 211, 218, 232, 238, 247, 264, 269, 288, 294, 423, 431, 440, 442, 473

www.ingramcontent.com/pod-product-compliance
Lightning Source LLC
LaVergne TN
LVHW061530290926
849162LV00004B/733